Sediment Toxicity Assessment

Sediment Toxicity Assessment

Edited by

G. Allen Burton, Jr.

LEWIS PUBLISHERS
Boca Raton Ann Arbor London Tokyo

Library of Congress Cataloging-in-Publication Data

Burton, G. Allen
Sediment toxicity assessment/edited by G. Allen Burton
p. cm.
Includes bibliographical references and index.
ISBN 0-87371-450-4
1. Sediments (Geology)--Evaluation.. 2. Pollution--Environmental aspects--Evaluation. 3. Toxicity testing. 4.Toxicology. I. Burton, G. Allen.
QE471.2.S39 1992
628.1'61--dc20 91-44273
CIP

Direct all inquiries to CRC Press, Inc. 2000 Corporate Blvd., N. W., Boca Raton, Florida 33431

LEWIS PUBLISHERS, INC.
121 South Main Street, Chelsea, MI 48118

PRINTED IN THE UNITED STATES OF AMERICA
1 2 3 4 5 6 7 8 9 0

Printed on acid-free paper

Preface

In the past few years there has been a dramatic increase in research and regulatory activity dealing with contaminated sediments of streams, lakes, and nearshore, coastal environments. This increase has likely resulted from the growing recognition that sediments in many areas have elevated levels of toxicants which are possibly impacting the local ecosystem. Unfortunately, our methods for determining whether or not the sediments are causing a detrimental impact have, until recently, been rather crude, indirect, or circumstantial. It is apparent that methods exist now which effectively assess the quality of sediment and its interaction with the ecosystem. Reliable sediment quality assessments can only be accomplished, however, by using a number of methods which integrate physical habitat, chemical, and biological components of the ecosystem. The ecosystem significance of sediment contamination cannot be accurately measured or predicted, at this point in time, by only using ecosystem components. The following chapters focus on assessing the toxicity of sediments, therefore, emphasizing biological components and their interaction with physical and chemical factors,

It is apparent that the science of sediment toxicology is progressing rapidly. Hopefully, this endeavor will aid the process!

G. Allen Burton, Jr., Ph.D.
Associate Professor
Department of Biological Sciences
Wright State University
Dayton, OH 45435

Preface

Acknowledgments

The preparation of this book was made possible by the generous support of the institutions which the authors represent and their respective grant support. The quality of the contributions was assured by the extensive and timely reviews of the following individuals:

W. H. Benson, Department of Pharmacology, University of Mississippi, University, Mississippi

R. M. Burgess, Science Applications International Corporation, Narragansett, Rhode Island

R. S. Carr, National Fisheries Contaminant Research Field Station, U. S. Fish and Wildlife Service, Corpus Christi, Texas

P. M. Chapman, EVS Consultants, North Vancouver, British Columbia, Canada

W. H. Clements, Department of Fisheries and Wildlife Biology, Colorado State University, Fort Collins, Colorado

J. J. Coyle, National Fisheries Contaminant Research Center, U. S. Fish and Wildlife Service, Columbia, Missouri

P. Crocker, Water Quality Division, U. S. Environmental Protection Agency, Dallas, Texas

D. A. Dawson, Department of Animal Science, University of Tennessee, Knoxville, Tennessee

T. DeWitt, Hatfield Marine Science Center, U. S. Environmental Protection Agency, Newport, Oregon

R. J. Diaz, Virginia Institute of Marine Science, The College of William and Mary, Gloucester Point, Virginia

K. L. Dickson, Institute of Applied Sciences, University of North Texas, Denton, Texas

E. J. Englund, Environmental Monitoring and Support Laboratory, U. S. Environmental Protection Agency, Las Vegas, Nevada

W. R. Gala, Environmental Health Center, Chevron USA, Richmond, California

T. C. Ginn, PTI Environmental Services, Bellevue, Washington

D. Gunnison, Waterways Experiment Station, U. S. Army Corps of Engineers, Vicksburg, Mississippi

J. F. Hall, Port Arthur Research Laboratories, Texaco, Port Arthur, Texas

M. S. Henebry, Illinois Environmental Protection Agency, Springfield, Illinois

R. J. Huggett, Virginia Institute of Marine Science, The College of William and Mary, Gloucester Point, Virginia

G. C. Ingersoll, National Fisheries Contaminant Research Center, U.S. Fish and Wildlife Service, Columbia, Missouri

S. Jones, Department of Pharmacology, University of Mississippi, University, Mississippi

T. W. La Point, National Fisheries Contaminant Research Center, U. S. Fish and Wildlife Service, Columbia, Missouri

P. F. Landrum, Great Lakes Environmental Research Laboratory, National Oceanic and Atmospheric Administration, Ann Arbor, Michigan

G. R. Lanza, School of Public and Allied Health, East Tennessee State University, Johnson City, Tennessee

M. Lewis, Battelle Columbus Division, Columbus, Ohio

M. J. Melancon, Patuxent Wildlife Research Center, U. S. Fish and Wildlife Service, Laurel, Maryland

M. K. Nelson, National Fisheries Contaminant Research Center, U. S. Fish and Wildlife Service, Columbia, Missouri

J. T. Oris, Department of Ecology, Miami University, Oxford, Ohio

R. A. Pastorok, PTI Environmental Services, Bellevue, Washington

E. A. Power, EVS Consultants, North Vancouver, British Columbia, Canada

M. Redmond, Hatfield Marine Science Center, Science Applications International Corporation, Newport, Oregon

K. J. Scott, Science Applications, International Corporation, Narragansett, Rhode Island

R. C. Swartz, Hatfield Marine Science Center, U. S. Environmental Protection Agency, Newport, Oregon

H. Tatem, Waterways Experiment Station, U. S. Army Corps of Engineers, Vicksburg, Mississippi

W. T. Waller, Institute of Applied Sciences, University of North Texas, Denton, Texas

A. G. Westerman, Department of Environmental Protection, Kentucky Natural Resources and Environmental Protection Cabinet, Frankfort, Kentucky

C. Yoder, Ohio Environmental Protection Agency, Columbus, Ohio

G. Allen Burton, Jr. is Associate Professor and Director of the Environmental Health Sciences Program at Wright State University. He attended Ouachita Baptist University where he obtained a Bachelors of Science degree in Biology and Chemistry in 1976. Dr. Burton obtained Masters degrees in Microbiology and Environmental Science from Auburn University and the University of Texas at Dalles, respectively, in 1978 and 1980. During 1979 he conducted research at the U.S. Army's Waterways Experiment Station in Vicksburg, MS, on survival of enteric pathogens in sediments and nutrient cycling in sediment mesocosms. He obtained a Ph.D. degree in Environmental Science from the University of Texas at Dallas in 1984. From 1980 until 1985 he was a Life Scientist with the U.S. Environmental Protection Agency. He was a Visiting Fellow at the National Oceanic and Atmospheric Administration's Cooperative Institute for Research in Environmental Sciences at the University of Colorado.

Dr. Burton's research during the past 12 years has focused on detecting and evaluating toxicant effects in sediments and storm waters. He has done extensive research in the areas of metal and metalloid bioavailability, laboratory and *in situ* assay development and field validation, defining ecosystem stressors and spatial variance, interactions of toxicant mixtures, comparing indicator and indigenous species acute and chronic responses, and removal of pesticide toxicity in constructed wetlands. Ecosystem effects have been evaluated at multiple levels of biological organization, ranging from microbial activity to fish growth.

Professor Burton has been active in the development of methods for the assessment of sediment contamination and storm water effects through his research activities, the American Society of Testing and Materials, participation on state and federal technical committees, and as a consultant. During the past 4 years, he has received more than $500,000 in grants and contracts, dealing with aquatic ecosystem monitoring assessment. He has over 50 scientific publications dealing with aquatic systems.

Contents

CHAPTER 5
Evaluation of Sediment Contaminant Toxicity: The Use of Freshwater Community Structure
Thomas W. La Point and James F. Fairchild

CHAPTER 6
The Emergence of Functional Attributes as Endpoints in Ecotoxicology
John Cairns, Jr., B. R. Neiderlehner, and E. P. Smith

CHAPTER 7
The Significance of In-Place Contaminated Marine Sediments on the Water Column: Processes and Effects
Robert M. Burgess and K. John Scott

CHAPTER 8
Plankton, Macrophyte, Fish, and Amphibian Toxicity Testing of Freshwater Sediments
G. Allen Burton, Jr.

CHAPTER 9
Assessment of Sediment Toxicity to Marine Benthos
Janet O. Lamberson, Theodore H. DeWitt, and Richard C. Swartz

CHAPTER 10
Freshwater Benthic Toxicity Tests
G. A. Burton, Jr., M. K. Nelson, and C. G. Ingersoll

CHAPTER 11
Biomarkers in Hazard Assessments of Contaminated Sediments
William H. Benson and Richard T. Di Giulio

CHAPTER 12

Models, Muddles and Mud: Predicting Bioaccumulation of Sediment-Associated Pollutants

Henry Lee II

Chapter 13
Sediment Bioaccumulation Testing with Fish
Michael J. Mac and Christopher J. Schmitt

Chapter 14
Integrative Assessments in Aquatic Ecosystems
Peter M. Chapman, Elizabeth A. Power, and G. Allen Burton, Jr.

CHAPTER 15

Management Framework for Contaminated Sediments (The U.S. EPA Sediment Management Strategy)

Elizabeth Southerland, Michael Kravitz, and Thomas Wall

Chapter 16

Assessment and Management of Contaminated Sediments in Puget Sound

Thomas C. Ginn and Robert A. Pastorok

Chapter 17

The Effects of Contaminated Sediments in the Elizabeth River

R. J. Huggett, P. A. Van Veld, C. L. Smith, W. J. Hargin, Jr., W. K. Vogelbein, and B. A. Weeks

CHAPTER 1

Assessing Sediment Quality

Elizabeth A. Power and Peter M. Chapman

INTRODUCTION

This book is concerned directly with biological effects because they are critical to the assessment of sediment quality. Chemical contaminant concentrations have been used historically, and are still being used, to assess sediment quality and to regulate and make management decisions as to how to deal with contaminated sediments. When such chemical criteria are not based on known biological effects (i.e., single species, population, or community), this approach can be subject to criticism.

Background

The objective of this chapter is to introduce the field of sediment toxicity assessment. Sediments contaminated with nutrients, metals and metalloids, organics, and oxygen-consuming substances can be found in freshwater, estuarine, and marine systems throughout the world. While some of these contaminants are present in elevated concentrations as a result of natural processes, in most cases they are due to anthropogenic activities. Contaminants are introduced to aquatic ecosystems *via* many routes (e.g., effluent discharge, ocean and lake disposal, nonpoint sources, contaminant spills, airborne deposition) (Figure 1).

Traditionally, contaminated sediments have been evaluated as part of navigational dredging and disposal projects. During the 1980s the focus expanded to include sites not directly involved in dredged disposal programs (e.g., nearshore fisheries areas,

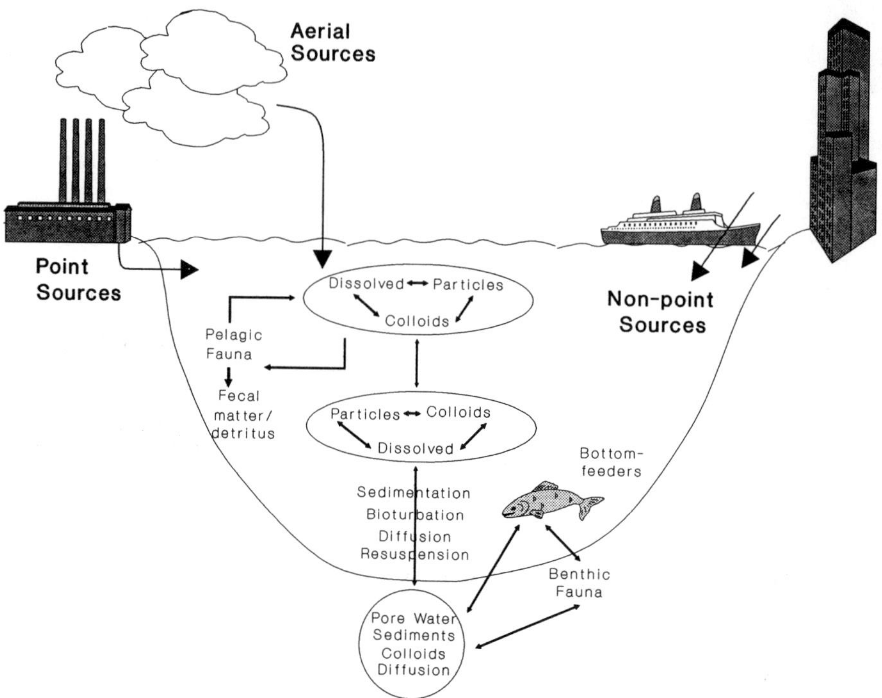

Figure 1. Schematic of fate of a contaminant in the aquatic environment, showing potential sources, routes of exposure, and movement of contaminant within various components of the ecosystem.

point-source dischargers); as a result, the large magnitude of the contaminated sediment problem has been recognized both by the scientific community and society in general.

The NRC[1] noted that contaminated marine sediments are widespread, are potentially far-reaching in their environmental and public health significance, and have emerged as an environmental issue of national and international importance. They propose a working definition: "Contaminated sediments are those that contain chemical substances at concentrations that pose a known or suspected environmental or human health threat."

In the increasingly large quantity of literature addressing contaminated sediment issues, assessment of sediment quality is consistently related to either environmental effects or human health; the focus of this book is on environmental effects. Further, since contamination per se does not imply more than the presence of substances not normally or "naturally" present in clean sediments, assessment of sediment quality includes, in this book, the process of determining whether contaminated sediments cause, or have the potential to cause, biological effects.

This introductory chapter is intended to provide the reader with a sense of the current status of sediment assessment, as well as provide an overview of some of the issues. Individual issues are discussed in detail in the following chapters.

OVERVIEW

Problem Sediments

Problem sediments typically contain toxic levels of persistent contaminants, many of which may be immediately lethal (e.g., heavy metals, chlorophenols) or have long-term deleterious effects, including reproductive impairment/birth defects (e.g., PCBs, PAH, dioxins). Reductions in or changes to sedimentary organisms, which are a major food source for other ecologically important and for commercially important trophic levels such as crabs, shrimp, and fish, have been a result of toxic sediments.[2-4] Further, bottom-dwelling organisms such as crabs and bottomfish may develop cancerous lesions as a result of contact with problem sediments.[5-8]

The above are demonstrable effects. In addition, public concern remains high that human beings eating seafood and freshwater fish from contaminated areas, or engaging in water-contact activities in these areas may suffer harm, including cancers.[9]

Chemical contaminants of toxic sediments may be the result of either past or present disposal practices. Because shutting off the source does not always solve the problem (e.g., many problem chemicals persist), addressing sediment toxicity is not an easy matter and may include dredging, capping, or upland disposal. The costs of either inaction or action are both high: the former in terms of continued environmental problems, the latter in terms of fiscal resources. Consequently, it is essential that methods for assessing toxic sediments be developed that are precise, efficient, comprehensive, and reliable.

Status of Sediment Quality Assessment

The National Oceanic and Atmospheric Administration (NOAA) National Status and Trends Program has monitored chemical concentrations in marine sediments at approximately 200 sites across the U.S. since 1984.[1] Significant contamination with potential for biological effects has been identified in major urban areas on both coasts of the continental U.S., including sites in the Hudson-Raritan estuary, Boston Harbor, western Long Island, and Oakland estuary of San Francisco Bay.[10] In Puget Sound, Washington, mounting evidence in the early 1980s of pollution problems focused attention on sediments containing chemicals of concern in urban/industrial harbors and navigation channels.[11] Because no environmental monitoring had been performed at existing dredged disposal sites, there was, initially, little actual field data to respond to this concern. Since then, considerable resources have been expended to assess sediment quality in Puget Sound as well as to develop management plans for evaluation and disposal of sediments. In many senses, developments in Puget Sound have led the field both nationally and internationally. Case examples of several marine sites with contaminated sediments are presented in Chapters 18 and 20.

The U.S. and Canada are working cooperatively on freshwater sediment issues in the Great Lakes, through the International Joint Commission (IJC). Through this collaboration, contaminated sediments have been identified in 41 of the 42 areas of concern (ARCs).[12] The Great Lakes Water Quality Control Board (GLWQCB) has focused their attention on the identification and remediation of these sediments. Further discussion of the Great Lakes as a case example is presented in Chapter 17.

As widespread as the problem of sediment contamination appears to be, there presently is no clear understanding of its ecological significance nor its extent. It is rarely possible to ascribe with certainty a particular biological effect to a given chemical concentration. Barrick et al.[13] describe two measures of reliability that the ideal sediment assessment method would fulfil completely. "Sensitivity" is the proportion of actual environmental problems that are predicted by sediment assessment as problems, and "efficiency" is the proportion of predicted problems that are actually environmental problems.

To date, no one approach to sediment assessment is wholly satisfactory (i.e., is both 100% sensitive and 100% efficient). Two separate, yet complementary, approaches are available for the biological characterization of contaminated sediments: *bioassays* and *bioassessment*. *Bioassays* are laboratory-based tests that incorporate rigorous experimental protocols and controls. Both toxicity tests and bioaccumulation studies are bioassays.[14] *Bioassessments* are field-based analyses that lack strict experimental controls.[15]

Risk assessment involves the synthesis of results from separate exposure and effects components in order to provide a scientific basis for estimating the probability of harm to the aquatic environment.[16] Exposure assessment consists of estimating the duration and intensity of contaminant exposure for potentially affected biological communities. This book focuses on the effects assessment component, which consists of estimating the biological response of these communities in terms of toxicity or bioaccumulation.

WHAT IS SEDIMENT?

Sediment is comprised of all detrital, inorganic, or organic particles eventually settling on the bottom of a body of water. Therefore, sediment is generally a matrix of materials and can be relatively heterogeneous in terms of its physical, chemical, and biological characteristics. Geologically speaking, sediments are at the end of the path for natural and anthropogenic materials, which is at the root of contaminated sediment problems.

Sediments can be thought of as having four main components. The largest volume is occupied by interstitial water, which fills the spaces between sediment particles and which usually accounts for over 50%, by volume, of surface sediments.[17] The inorganic phase includes the rock and shell fragments and mineral grains that result from the natural erosion of terrestrial materials. Organic matter occupies a low volume, but is an important component of sediment because it can regulate the sorption and bioavailability of many contaminants. Finally, anthropogenically derived materials include contaminated materials and eroded topsoil. Movement of materials into and out of sediments is controlled by chemical, biological, and physical processes. The latter includes porosity (volume of spaces between particles) and permeability (ability of water to move between, into, and out of spaces). Gravels and sands are most permeable; muds are least permeable.[18]

Sediment particles are derived from a mixture of material inputs from different sources, including eroded rocks and soils, waste particles, atmospheric fall-out, and inorganic materials produced biologically. They consist of a variety of components, including clay minerals, carbonates, quartz, feldspar, and organic solids. These particles are usually "coated" with hydrous manganese and iron oxides and organic substances.[17] Iron and manganese oxides are formed in stratified lakes, estuaries, and at the sediment-water interface in both fresh and marine waters. Under oxic conditions, organic surfaces called biofilms form on aquatic sediments from organisms such as bacteria and algae, by their breakdown and by sorption of organic matter onto sediment surfaces.

Sediments are typically categorized as to size of particles; two useful categorizations of sediment are coarse (sand and coarse material less than 62 μm in diameter) and fine (silts and clays greater than 62 μm).[19] The coarse fraction is composed primarily of stable, inorganic silicate materials that are noncohesive and generally not associated with chemical contamination. The fine fraction consists of particles with a relatively large surface area to volume ratio and, frequently, surface electric charges that cause these particles to be more chemically and biologically reactive than coarser sands and that increase the likelihood of sorption and desorption of contaminants. Problem chemical contamination is most often found in fine ("muddy") sediments, which characterize depositional areas.

The quality of a sediment is influenced by its environment as well as its own physical characteristics. For example, studies of sediments in freshwater and marine habitats have historically progressed somewhat independently of one another. In addition to differences in salinity and species, there are a host of concomitant factors (e.g., water movement, pH, organic carbon load, sediment movement, pore water residence time, water chemistry) that affect the sorption of contaminants to sediment. Consequently, marine and freshwater sediments are generally addressed in separate chapters in this book.

BIOAVAILABILITY AND TOXICITY

The bioavailability and toxicity of sediment-sorbed contaminants are linked to a host of factors, including sediment collection, storage and handling (see Chapter 3), routes of exposure, presence of other chemicals, modifying factors (including sediment characteristics), type of organisms, and feeding habits.[20] An understanding of these factors is essential when designing laboratory studies that determine sediment toxicity, because they provide the generic framework within which site-specific testing should be designed and the results interpreted.

There are three potential sources/paths for contaminants to reach benthic organisms:[21] the sediments themselves (e.g., ingestion), overlying water, and interstitial (pore) water (e.g., across respiratory surfaces and across body walls) (Figure 2). The detrimental impacts of chemicals in sediments may result from any combination of these, and the relative importance of each route may change with both the compound and a number of modifying factors.

Why Is Bioavailability Important?

The role that benthic systems play in sequestering contaminants is well recognized.[22] However, the key to sediment assessment is bioavailability; although sediments might contain relatively high concentrations of toxic compounds, this condition does not necessarily lead to adverse effects on organisms living in the sediments. The fate of contaminants in a sediment-water system is highly dependent on their sorptive behavior, which, in turn, affects bioavailability and toxicity.

The bioavailability of sediment-sorbed contaminants is of concern because sediments can serve as both sinks and sources of contaminants.[23,20] For instance, a recent news report[24] notes that the majority of PCB tissue concentrations in Lake Ontario trout are directly attributable to their presence in sediments where they are accumulated by bottom-feeding crustaceans that serve as food for alewives that are, in turn, fed upon by trout. However, the extent and magnitude of contamination does not necessarily reflect a similar level of bioeffects; even relatively high levels of contamination may be of little biological significance if bioavailability is limited.[25] The only means of measuring bioavailability is by measuring or determining a biological response, for instance, through bioassay testing. Such testing has often involved measures of bioaccumulation (the ability of an organism to accumulate contaminants in tissues). However, because bioaccumulation is a phenomenon, not an effect (and can be relatively expensive due to costly chemical analyses), emphasis has shifted to sediment toxicity tests that are effects based and relatively inexpensive.

Contaminant sorption to sediments is a complex and poorly understood process, but one that has obvious repercussions for bioavailability. Although sorption to sediments is frequently described as fast and readily reversible, true sorption equilibrium for chemicals such as nonionic organics may require days or weeks to achieve.[26]

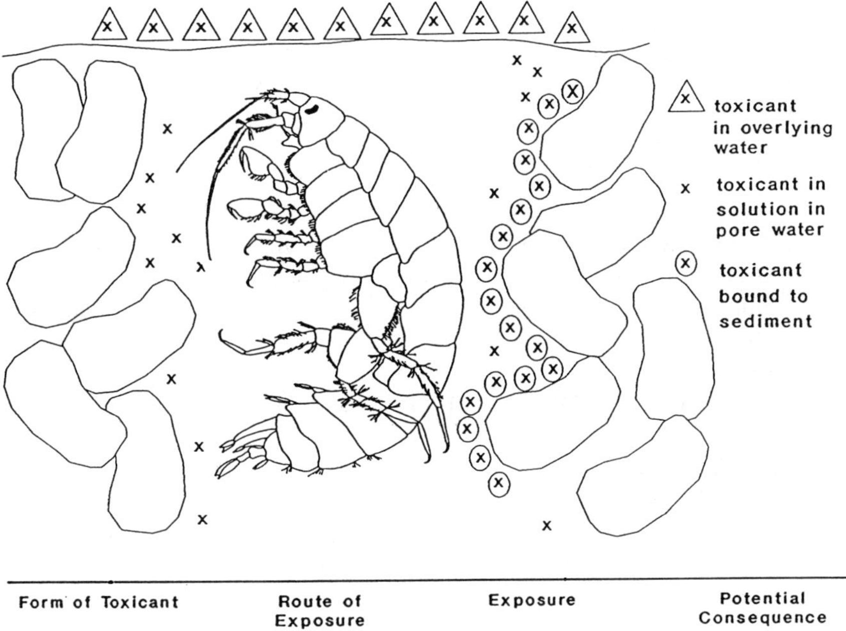

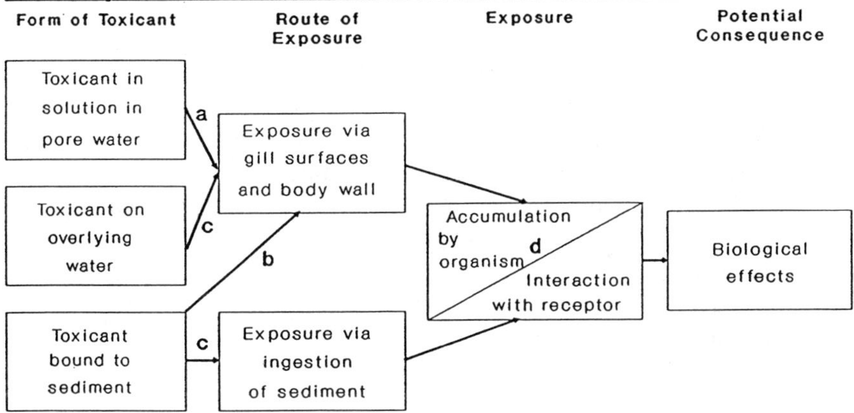

a Expected major route of uptake

b Not a major route for direct uptake; often estimated by normalization of bulk sediment contaminant data (e.g., total organic carbon [TOC] and acid volatile sulphides [AVS])

c Route of exposure; relative importance subject to debate

d Biological effects only occur through entry of toxicants into biological systems; accumulation alone is a phenomenon, not an effect

Figure 2. *Upper:* Schematic representation of benthic organism in sediment and routes of sediment exposure; *lower:* interactions of contaminated toxic sediments with benthic organism.

Sediment contaminant mixtures can produce toxic effects that are different from those of the individual contaminants. In addition, there are many circumstances when field collected sediments that are highly contaminated, based on bulk chemistry data, are not toxic. Sediment chemical measures only provide information on contamination because the toxicity of a chemical substance in sediment varies with its concentration and with conditions encountered within a specific sediment. In a study of sediment toxicity, contamination, and macrobenthic communities near a large marine sewage outfall, Swartz et al.[2] noted that Palos Verdes Shelf sediments had a relatively low acute toxicity and speculated that this may reflect a reduced bioavailability of chemical contaminants that are bound to high concentrations of organic matter.

Routes of Exposure to Benthic Organisms

Routes of organism exposure include uptake from pore water and overlying water across body walls and across respiratory surfaces, and ingestion of contaminated sediment particles (Figure 2). In all cases, if the compound is bioavailable, there is uptake by the organism and interaction with a receptor site, which may or may not elicit measurable biological effects.

If water represents the major route of exposure (pore or overlying, through the body wall or respiratory surface), then organic contaminants bound to sediment organic carbon will generally not be bioavailable. This situation would most likely occur for contaminants that sorb well to sediments with a high K_p (chemical sorption coefficient). Sediments with low levels of total volatile solids (TVS) showed more toxicity to the marine amphipod *Rhepoxynius abronius* than did sediments with higher TVS.[27] The decrease in toxicity at higher TVS levels was attributed to the increased binding of chemicals to the sediment. At low TVS levels, pore-water concentrations of chemicals were higher, corresponding with the greater bioavailability and higher toxicity of contaminant mixtures in the low TVS sediment. This example illustrates the problem of estimating sediment quality based solely on bulk chemistry concentrations.

Another example of a modifying factor for sediment toxicity, in this case for metals, is acid volatile sulfide (AVS).[28] It has been demonstrated that the toxicity (LC_{50} for cadmium) of sediments on an AVS-normalized basis is the same for sediments with over an order of magnitude difference in dry-weight-normalized cadmium toxicity. Because other toxic metals also form soluble sufides, it is likely that AVS is important in determining their toxicity in sediments as well.[28]

Any toxic effect to a benthic organism depends on its sensitivity to interstitial water and particle-bound contaminants.[3] Di Toro[29] argues that the dose-response curve for a biological effect of concern can be correlated, not to the total sediment chemical concentration, but to the interstitial water concentration. Repeatedly, the issue of bioavailability is raised when researchers are trying to link chemical and toxicological or biological community data to determine ecological significance.

In general, sediment-sorbed contaminants are more persistent, less mobile, and occur at higher concentrations than those in the overlying waters.[30] Sediment ingestion and passage through the digestive tract acts to alter the parameters affecting exposure by changing the conditions of exposure; changes in pH during digestion can affect the speciation and binding of contaminants, and their bioavailability.[31,20] In laboratory tests, the uptake of sediment-associated anthracene by the freshwater amphipod *Hyalella azteca* occurs much faster than predicted. This observation suggests that the selective feeding of *H. azteca* on smaller food particles results in a concentration of diet on fine sediments with the highest organic matter, and, hence, contaminant concentration.[32]

Not all benthic organisms ingest sediments, and the major source of organic chemicals for many is primarily interstitial water or water at the sediment-water interface.[33,34] It is important to differentiate between chemicals available because of ingestion and those that become desorbed into pore water where uptake into tissues can be rapid. Even though uptake from pore water can be faster than directly from sediments, the higher concentration of sediment may still result in this phase being, in some cases, the major source of chemicals for uptake.[35]

Exposing burrowing organisms to whole sediments is considered to be a worst-case, real-world condition. The use of aqueous elutriates provides information on the toxicity of contaminants released to the water column, as in dredging, and allows exposure of certain useful indicator organisms. Some tests use chemical means to derive test materials (i.e., chemical extracts are tested), either due to the requirements of the test system or in order to include (or exclude) certain contaminant classes from testing.

SEDIMENT QUALITY ASSESSMENT

Current research on particle-bound contaminants originated with the concept that sediments reflect the chemical, physical, and biological conditions of the overlying water.[36] The determination that water quality could be reflected in sediment quality quickly led to recognition that toxic chemicals could be characterized by their affinity for particulates. Sediment quality can be assessed by a number of methods, but they tend to fall into five main categories, as summarized in Chapter 14:

- Community structure (Chapters 4, 5, and 6)
- Sediment toxicity (Chapters 7, 8, 9, and 10)
- Tissue chemistry (Chapters 11 and 12)
- Sediment chemistry (Chapter 13)
- Pathology (discussed in several chapters)

Ideally, all five components would be utilized to assess sediment quality, providing maximal information about the ecosystem. However, the reality is that environmental managers are faced with limitations in both resources and time. Decisionmakers must optimize their resources by selecting the information that will have the greatest utility. Environmental managers strive to strike a balance between level of effort and type/quality of information needed to make effective decisions. When resources are limited or the impetus to be protective is low, a manager may be "reactive", responding when problems occur (e.g., contaminated fisheries, human health concerns, etc.). When resources are less limited or the impetus to be protective is high, a manager may be "proactive", closely monitoring the system and anticipating problems before they occur (e.g., installing an effective effluent treatment system and locating the discharge in a high-current area, establishing an effective sediment monitoring plan before required to do so by regulation, etc.).

Either approach may be effective for individual managers and specific situations. However, reactive measures tend to be short-term responses, while proactive measures are longer term, but more expensive. Because the environment is increasingly being viewed as one of the top priorities facing industry and government, obviously the proactive approach is preferable and, arguably, cost effective overall. Not only may impacts be identified before they become problematic, but the potential costs of sediment remediation should be drastically reduced.

However, as is apparent from Figure 3, as sensitivity in testing and assessment increases, the probability that the response being measured is environmentally significant decreases. At some point, only "noise" is being measured. At this point, no action is required, and this is called the "noactive" point. However, because we do not know when proactive changes to noactive, regulators generally adopt a very conservative approach. Ideally, high sensitivity should be balanced against obvious environmental effects.[13]

One of the problems faced by environmental managers, particularly for environmental chemistry, is our ability to detect a greater number of chemicals at increasingly lower detection limits, which means that allowance must be made for new information in the process of sediment quality assessment. In an editorial, MacKay[37] discusses the problem of "low, very low, and negligible concentrations" and makes the point that environmental quality criteria should be related to the effects of a compound rather than to our ability to detect them.

However, the bottom line is the determination of biological effects (summarized in Table 1). Ecosystem health can be assessed in two ways: a "top-down" (holistic) approach or a "bottom-up" (reductionist) approach. Sediment assessment has traditionally focused on bulk sediment analyses and the application of numeric criteria. Yet, it is recognized that no single method can appropriately address all assessment needs. As a result, there are a variety of rapidly developing methodologies to assess sediment quality, which are discussed in subsequent chapters.

A framework for sediment assessment was developed by Power et al.,[38] which summarizes the linkages between biological responses and levels of organization, providing perspective on data requirements and interpretation. Determination of

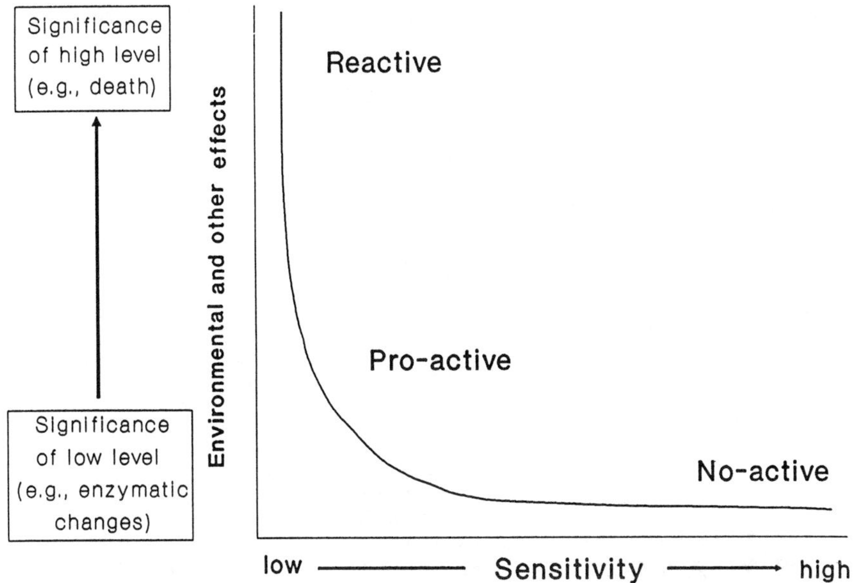

Figure 3. Schematic relating sensitivity to environmental significance. Overly sensitive measures may have no environmental significance (i.e., result in "false positives"). The term proactive means being predictive, ideally based on measurement of prealteration, bioeffects, and/or contamination.

cause and effect requires that we identify the linkages between (1) exposure and effect, (2) biochemical and whole-organism measures, and (3) laboratory and field results. These three aspects of sediment assessment are anticipated to be the major foci of future research.

Sediment Quality Criteria

In addition to developing methods for testing sediment, there is increasing public and regulatory pressure to develop sediment quality criteria to interpret sediment contaminant data and facilitate the assessment of sediment quality. Sediment quality criteria are necessary, in addition to water quality criteria, because[39]

- Various toxic contaminants found in only trace amounts in the water column accumulate in sediments to elevated levels;
- Sediments serve as both a reservoir and a source of contaminants to the water column;
- Sediments integrate contaminant concentrations over time, whereas water column contaminant concentrations are much more variable and dynamic;
- Sediment contaminants in addition to water column contaminants affect benthic and other sediment-associated organisms; and
- Sediments are an integral part of the aquatic environment, providing habitat, feeding, and rearing areas for many aquatic biota.

Table 1. Listing of Parameters used for Describing the Impacts of Chemicals[37]

Response Level	Description	Parameters	Specific Examples (where applicable)
Partitioning	Bioaccumulation	Complexation and storage	Metallothionein production
		Tissue distribution	
		Uptake kinetics	
		Metabolism and excretion	
		Bioconcentration	
Neuroendocrine	Biochemical	Neuroamine and catecholamine responses	Cortisol release
Physiological	Primary metabolic impact	Enzyme activities	Hepatic mixed function oxidase induction
		Respiration	
		Photosynthesis	
		Enzyme activities	
		Excretion	
	Primary metabolic responses	Metabolic rate	Adenylate energy charge
		Hematology	Hematocrit, leucocrit, hemoglobin
		Pigmentation	
		Osmoregulation	
		Ionoregulation	
		Hormonal changes	Changes in estradiol, testosterone
Integrators	Behavior	Sensory capacity	
		Rhythmic activities	
		Motor activity	Burrowing
		Learning/motivation	
		Avoidance/attraction	
		Reproductive behavior	
	Growth	Feeding rate/nutrition	
		Scope for growth	
		Net growth efficiency	
		Body/organ weights	Liver and spleen changes
		Developmental rate/ stages	Changes in sexual maturation
	Histopathology	Abnormal growths	Neoplasms/tumors
		Abnormal histological changes	
	Genetic	Chromosomal damage	Sister chromatid exchange
		Mutagenic/teratogenic effects	Ames testing
	Reproduction	Sexual maturation	
		Gamete viability/fertility	
		Larval development	
		Brood size/fecundity	
		Frequency of reproduction	
		Survival	

Table 1. Listing of Parameters used for Describing the Impacts of Chemicals[37] (continued)

Response Level	Description	Parameters	Specific Examples (where applicable)
Dynamics	Behavior	Recolonization/ migration	
		Aggression/predation	
		Vulnerability	
		Mating	
	Integrators	Age-class survival	
		Reproductive success	
		Density	
		Biomass	
		Diversity	
		Species richness	
		Succession	
		Nutrient cycling	
		Energy flow	

Sediment quality criteria also are necessary to provide for long-term management of contaminated sediments, including assessment of sediment quality, identification of problem areas for remedial action, and designation of "acceptable" sediments for open-water disposal. Criteria are expected, if properly designed and implemented, to assist researchers in determining appropriate chemicals for focusing laboratory and cause-effect studies. In addition, sediment quality criteria are required to establish wasteload allocations, in particular for "new" chemicals, and to design and evaluate monitoring programs. Current approaches used to develop sediment quality criteria have been reviewed by Chapman[39] and are summarized in several documents.[10,40,41]

The following factors need to be taken into consideration when evaluating sediment quality criteria:[39]

- Legal and technical defensibility
- Amount of data required and amount of data currently available
- Ability to produce both chemical-by-chemical and chemical mixture criteria, and
- The possibility of generic as well as site-specific applicability.

THE FUTURE

Scientists rarely, if ever, consider the possibility that the tasks we set for ourselves are impossible.[42] The possibility that some toxicological phenomena could be unpredictable is rarely considered by scientists. However, we must distinguish between phenomena that may always be "unpredictable" (due to their complexity, for instance) and those that we cannot predict at this time (classes of events not yet embraced by functional theory). *What is holding the field of sediment toxicity assessment back?* Rigler[42] outlined four reasons why we cannot expect to succeed in making predictions about future states of a system or about particular state variables of that system:

- The behavior of the system is random;
- The only potential methods of obtaining information about the system disturb it;
- The system gives out no information; and
- The behavior of the system is affected by theory.

In the case of sediment assessment, it is unlikely that the behavior of the system is random; researchers have already developed predictive models and demonstrated through empirical testing that at least some of these relationships are consistent. The variability of biological systems may be greater than researchers can comprehend, but this is not to be confused with unpredictability.

One of the greatest difficulties in assessing sediment quality is the problem of extracting information from the system without disturbing it. Maintaining sample integrity during removal, transport, and testing in the laboratory is extremely difficult. The sediment environment is composed of a myriad of microenvironments, redox gradients, and other physical, chemical, and biological processes. Many of these characteristics influence sediment toxicity and bioavailability to aquatic organisms. Some of these technical problems will be resolved by future developing technologies, but the problem of disturbing the system under study will likely remain an issue in the field of sediment assessment. The problems of sediment handling are discussed further in Chapters 2 and 3.

Although sediment researchers might sometimes think that a system is yielding no information, that is unlikely to be the case. We are limited by our methods, by the ecosystem's naturally high variability, and by our ability to discern between the effects of multiple causes. However, there is no question that we will continue to advance in our ability to extract useful, high quality information from the system. Clearly, a focus of future study will be how to *interpret* that information, and this is one of the most important aspects of sediment quality assessment. Methods produce data, but integrative programs provide information for decisionmaking. This point is discussed in several of this book's chapters, in particular Chapter 16, which focuses on management.

Finally, a contributor to any failure to develop accurate predictions is the reality that the system is affected by the theory; for instance, the behavior of human beings is affected by book learning and rapid acceptance of new theories. Our work as scientists is influenced by our peers, society, and by the state-of-the-art in our particular field of research. This limitation of science impinges on sediment assessment, through avenues such as societal judgements, management strategies for economic resources, and environmental regulations. The context in which we view our environment has direct implications on our perception and evaluation of that environment. For example, a regulation may place undeserved importance on a particular parameter or biological indicator, which results in the science of sediment assessment being subjected to forces outside of the objective or scientific realm. The influence the human race has on the study of the environmental phenomena it has created (i.e., contaminated sediment) is unmeasurable, but surely it is one of the most intangible influences on our assessment of sediment quality.

THE NEXT STEP

The field of toxicity assessment is undergoing consolidation and reevaluation; researchers in this field are expending considerable effort summarizing the state of the science, comparing the advantages and disadvantages of various approaches, and re-examining some basic principles. We may be on the brink of a shift in focus, hopefully to address the basic question, *"So what?"* The bottom line is, has been, and always will be the biological significance of any effects that are measured or detected, and then making a judgement about whether those effects are acceptable or not. This book, with its focus on biological effects, provides the basis for potential changes in the focus and approach into the 1990s. There are no certainties, however, other than the fact that society will have a steadily increasing influence on policy and regulation for sediment assessment and management.

SUMMARY AND CONCLUSIONS

The measurement of biological effects is an integral component of the determination of sediment quality. Contaminated sediments have come to be recognized as a significant problem contributing to degradation of environmental quality. Sediments can have both immediately lethal or long-term deleterious effects (e.g., impaired growth or reproduction, lesions). Accumulation of chemicals by body tissues (bioaccumulation) is also of concern from both ecological and public health perspectives.

However, as widespread as the problem of sediment contamination appears to be, there presently is no clear understanding of its significance at the community or ecosystem level, and the methods for its determination are new and sometimes controversial. In addition, our understanding of the mechanisms of bioavailability is limited. We can measure chemical concentrations in sediment, but presently existing approaches for predicting their bioavailability are few. While sediment chemistry is most frequently measured, direct assessments of bioavailability utilize the measurement of (1) in situ effects (community structure), (2) sediment toxicity, (3) tissue chemistry, and (4) pathology. Ideally, all five measures (sediment chemistry and the four approaches to biological assessments) would be used to assess sediment quality.

There is increasing public and regulatory pressure to develop sediment quality criteria to interpret sediment contaminant data and facilitate the assessment of sediment quality. This book examines various approaches to assessing sediment toxicity and how they can be used as tools for sediment management. Demonstration of cause and effect requires that we determine the linkages between (1) exposure and effect, (2) biochemical and whole organism measures, and (3) laboratory and field results.[38] An immediate challenge is the establishment of a perspective on ecosystem-level risk in a proactive framework. At present, until and unless breakthroughs occur, we need to continue using a burden-of-evidence approach with methods that are scientifically defensible, publicly supported, *and* reliable and sensitive.

REFERENCES

1. National Research Council. *Contaminated Marine Sediments — Assessment and Remediation* (Washington, D.C.: National Academy Press, 1989), p. 493.
2. Swartz, R. C., D. W. Schults, G. R. Ditworth, W. A. DeBen, and F. A. Cole. "Sediment Toxicity, Contamination, and Macrobenthic Communities Near a Large Sewage Outfall," in *Validation and Predictability of Laboratory Methods for Assessing the Fate and Effects of Contaminants in Aquatic Ecosystems, STP 865*, T.P. Boyle, Ed. (Philadelphia: American Society for Testing and Materials, 1985), pp. 152–175.
3. Swartz, R. C., G. R. Ditsworth, D. W. Schults, and J. O. Lamberson. "Sediment Toxicity to a Marine Infaunal Amphipod: Cadmium and Its Interactions with Sewage Sludge," *Mar. Environ. Res.* 18:133–153 (1986).
4. Chapman, P. M. "Marine Sediment Toxicity Tests," in *Chemical and Biological Characterization of Sludges, Sediments, Dredge Spoils and Drilling Muds, STP 976*, J.J. Lichtenberg, F.A. Winter, C.I. Weber, and L. Fradkin, Eds. (Philadelphia: American Society for Testing and Materials, 1988), pp. 391–402.
5. Malins, D. C., B. B. McCain, D. W. Brown, M. S. Myers, J. T. Landahl, P. G. Prohaska, A. J. Friedman, L. D. Rhodes, D. G. Burrows, W. D. Gronlund, and H. O. Hodgins. "Chemical Pollutants in Sediments and Diseases in Bottom-Dwelling Fish in Puget Sound," *Environ. Sci. Technol.* 18:705–713 (1984).
6. Baumann, P. C., and J. C. Harshbarger. "Frequencies of Liver Neoplasia in a Feral Fish Population and Associated Fish Carcinogens," *Mar. Environ. Res.* 17:324–327 (1985).
7. Mix, M. C. "Cancerous Diseases in Aquatic Animals and Their Association with Environmental Pollutants: A Critical Review," *Mar. Environ. Res.* 20:1–141 (1986).
8. Becker, D. S., T. C. Ginn, M. L. Landolt, and D. B. Powell. "Hepatic Lesions in English Sole (*Parophyrs retulas*) from Commencement Bay, Washington (U.S.A.)", *Mar. Environ. Res.* 23:153–173 (1987).
9. Black, J. J. "Aquatic Animal Neoplasm as an Indicator for Carcinogenic Hazards to Man," in *Hazard Assessment of Chemicals: Current Developments*, Vol. 3, Urban Wastes in Coastal Marine Environments, (New York: Academia Press, 1986), pp. 181–232.
10. Long, E. R., and L. G. Morgan. *The Potential for Biological Effects of Sediment-Sorbed Contaminants Tested in the National Status and Trends Program*, (Seattle, WA: NOAA Technical Memorandum NOA OMA 52, 1990), p. 175 and appendices.

11. Puget Sound Dredged Disposal Analysis (PSDDA). *Management Plan Report — Unconfined Open-Water Disposal of Dredged Material, Phase II.* EPA Region 10, Army Corps of Engineers and Washington State Department of Environment, (September 1989).
12. International Joint Commission. *Guidance on Characterization of Toxic Substances Problems in Areas of Concern in the Great Lakes Basin*, Surveillance Work Group, Windsor, Ontario, p. 179 (1987).
13. Barrick, R., H. Beller, S. Becker, and T. Ginn. "Use of the Apparent Effects Threshold Approach (AET) in Classifying Contaminated Sediments," in *Contaminated Marine Sediments — Assessment and Remediation*, NRC Committee, Eds. (Washington, D.C.: National Academy Press, 1989), pp. 64–77.
14. Chapman, P. M. "A Bioassay by Any Other Name Might Not Smell the Same," *Environ. Toxicol. Chem.* 8:557 (1989).
15. Herricks, E. E., and D. J. Schaeffer. "Compliance Monitoring — Standard Development and Regulation Enforcement Using Biomonitoring Data," in *Freshwater Biological Monitoring*, D. Pascol, and R. W. Edwards, Eds. (New York: Pergamon Press, 1984), pp. 153–166.
16. Gentile, J. H., V. J. Bierman, Jr., J. F. Paul, H. A. Walker, and D. C. Miller. "A Hazard Assessment Research Strategy for Ocean Disposal," in *Oceanic Processes in Marine Pollution, Vol. 3,* M. A. Champ, and P. K. Park, Eds. (Malabar, FL: Robert E. Krieger, 1989), pp. 199–212.
17. Förstner, U. "Sediment-Associated Contaminants — An Overview of Scientific Bases For Developing Remedial Options," *Hydrobiologia* 149:221–246 (1987).
18. Chapman, P. M. "Measurements of the Short-Term Stability of Interstitial Salinities in Subtidal Estuarine Sediments," *Estuar. Coast. Shelf Sci.* 12:67–81 (1981).
19. International Joint Commission. *Procedures for the Assessment of Contaminated Sediments Problems in the Great Lakes.* Sediment Subcommittee and its Assessment Work Group, Windsor, Ontario, p. 140 (1988).
20. Schuytema, G. S., D. F. Krawczyk, W. L. Griffis, A. V. Nebeker, M. L. Robideaux, B. J. Brownawell, and J. C. Westall. "Comparative Uptake of Hexachlorobenzene by Fathead Minnows, Amphipods and Oligochaete Worms from Water and Sediment," *Environ. Toxicol. Chem.* 7:1035–1045 (1988).
21. Boese, B. L., H. Lee, D. T. Specht, and R. Randall. "Comparison of Aqueous and Solid-Phase Uptake for Hexachlorobenzene in the Tellenid Clam *Macoma nasuta* (Conrad): A Mass Balance Approach," *Environ. Toxicol. Chem.* 9:221–231 (1990).
22. Baker, R.A. *Contaminants and Sediments* (Ann Arbor, MI: Ann Arbor Science, 1980).
23. Larsson, P. "Contaminated Sediments of Lakes and Oceans Act as Sources of Chlorinated Hydrocarbons for Release to Water and Atmosphere," *Nature* 317:347–349 (1985).
24. Anonymous. "Why Do Trout From Lake Ontario Continue to Show Levels of Polychlorinated Biphenyls that Exceed...," *Environ. Sci. Technol.* 24:1114 (1990).
25. Payne, J. F., J. Kiceniuk, L. L. Fancey, V. Williams, G. L. Fletcher, A. Rahimtula, and B. Fowler. "What is a Safe Level of Polycyclic Aromatic Hydrocarbons for Fish: Subchronic Toxicity Study on Winter Flounder (*Pseudopleuronectes americanus*)," *Can. J. Fish. Aquat. Sci.* 45:1983–1993 (1988).
26. Karickhoff, S. W., and K. R. Morris. "Sorption Dynamics of Hydrophobic Pollutants in Sediment Suspensions," *Environ. Toxicol. Chem.* 4:469–479 (1985).

27. Swartz, R. C., P. F. Kemp, D. W. Schults, and J. O. Lamberson. "Effects of Mixtures of Sediment Contaminants on the Marine Infaunal Amphipod, *Rhepoxynius abronius*," *Environ. Toxicol. Chem.* 7:1013–1020 (1988).
28. Di Toro, D. M., J. D. Mahoney, D. J. Hansen, K. J. Scott, M. B. Hicks, S. M. Mayr, and M. S. Redmond. "Toxicity of Cadmium in Sediments: The Role of Acid Volatile Sulfide," *Environ. Toxicol. Chem.* 9:1487–1502 (1990).
29. Di Toro, D. M. "A Review of the Data Supporting the Equilibrium Partitioning Approach to Establishing Sediment Quality Criteria," in *Contaminated Marine Sediments — Assessment and Remediation*, NRC Committee, Eds. (Washington, D.C.: National Academy Press, 1989), pp. 100–114.
30. Larson, L. J. "Method for Preliminary Assessment of Aquatic Contamination Sites Using Sediment Extract Toxicity Tests," *Bull. Environ. Contam. Toxicol.* 42:218–225 (1989).
31. Ekelund, R., A. Granmo, M. Berggren, L. Renberg, and C. Wahlberg. "Influence of Suspended Solids on the Bioavailability of Hexachlorobenzane and Lindane to the Deposit-Feeding Marine Bivalve, *Abra nitida* (Muller)," *Bull. Environ. Chem. Toxicol.* 38:500–508 (1987).
32. Landrum, P. F., and D. Scavia. "Influence of Sediment on Anthracene Uptake, Depuration and Biotransformation by the Amphipod *Hyalella azteca*," *Can. J. Fish. Aquat. Sci.* 40:298–305 (1983).
33. Eadie, B. J., P. F. Landrum, and W. Faust. "Polycyclic Aromatic Hydrocarbons in Sediments, Pore Water and the Amphipod *Pontoporeia hoyi* from Lake Michigan," *Chemosphere* 11:847–858 (1982).
34. Adams, W. J., R. A. Kimerle, and R. G. Mosher. "Aquatic Safety Assessment of Chemicals Sorbed to Sediments," in *Aquatic Toxicology and Hazard Assessment: Seventh Symposium, STP 854*, R. D. Cardwell, R. Purdy, and R. C. Bahner, Eds. (Philadelphia: American Society for Testing and Materials, 1985), pp. 429–453.
35. Reynoldson, T. R. "Interactions Between Sediment Contaminants and Benthic Organisms," *Hydrobiologia* 149:53–66 (1987).
36. Züllig, H. "Sedimente als Ausdruck des Zustandes eines Gewässers," *Schweiz. Z. Hydrol.* 18:7–143 (1956).
37. MacKay, D. "On Low, Very Low, and Negligible Concentrations," *Environ. Toxicol. Chem.* 7:1–3 (1988).
38. Power, E. A., K. R. Munkittrick, and P. M. Chapman. "An Ecological Impact Assessment Framework for Decision-Making Relative to Sediment Quality," in *Aquatic Toxicology and Risk Assessment: Vol. 14, STP 1124,* M. A. Mayes and M. G. Barron, Eds., (Philadelphia: American Society for Testing and Materials, 1991), pp. 48–64.
39. Chapman, P. M. "Current Approaches to Developing Sediment Quality Criteria," *Environ. Toxicol. Chem.* 8:589–599 (1989).
40. Shea, D. "Developing National Sediment Quality Criteria," *Environ. Sci. Technol.* 22:1256–1261 (1988).
41. E. P. A. *Sediment Classification Methods Compendium*, Prepared by TetraTech for the U.S. Environmental Protection Agency (1990).
42. Rigler, F. H. "Recognition of the Possible: An Advantage of Empiricism in Ecology," *Can. J. Fish. Aquat. Sci.* 39:1323–1332 (1982).

CHAPTER 2

Sediment Variability

Lars Håkanson

INTRODUCTION AND AIM

The aim of this paper is to discuss some basic factors regulating sediment variability, namely bottom dynamic conditions, sediment contamination, lake size, and lake bottom irregularity, based on empirical data from Swedish lakes. Figure 1 is meant to illustrate a fundamental problem in sediment sampling. Consider an area (a lake, a given stretch of a river, etc.) from which there are no data (all is black). One sediment sample (Figure 1A) will reveal the sediment characteristics at the sample site, and from this it is possible to make predictions about the situation near the site. But how near is near? And how many samples (n) would be required to obtain a mean, median, or characteristic value for the entire lake? And what are the most important factors that influence this number? These are fundamental questions in all contexts where recent sediments are used (e.g., in aquatic pollution control programs for metals or other types of pollutants and in matters dealing with the evaluation of sediment toxicity).

The problem is *schematically* illustrated by the "torch" in Figure 1; the knowledge of the character of the sediments is good/perfect at the actual sample site (white color); close to the sample site it is probable that the sediments are similar to the sediments at the sample site (grey shades), but far away from the sample site, predictions about the sediments are difficult (all black). The ability to make such predictions will, indeed, depend on may factors linked to either the sediment/water system and/or the measurement process.[1,2]

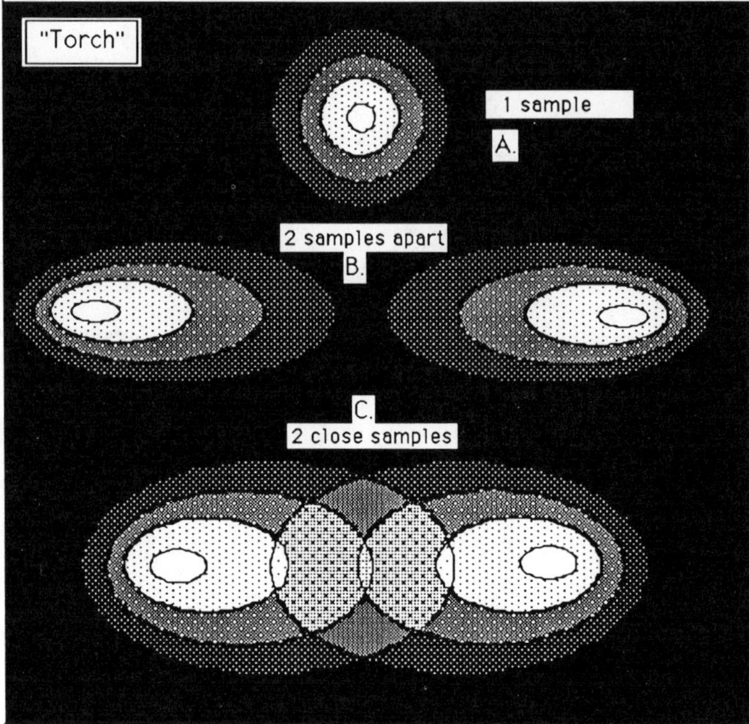

Figure 1. Illustration of the "torch" concept in sediment sampling.

Factors linked to the water system:

1. The water system. Different presuppositions exist in lakes and rivers, in shallow and deep waters, etc., with different bottom dynamic conditions (erosion, transportation, and accumulation). Areas of erosion are, in this context, characterized by hard or consolidated deposits where fine materials (i.e. materials that follow Stoke's law) are never, or rarely, deposited. Areas of transportation are often very diverse, since these are areas where fine materials are discontinuously being deposited. Areas of accumulation are almost always characterized by loose sediment, since these are areas with continuous deposition of fine materials. Table 1 illustrates the relationship between the bottom dynamic conditions and the physical and chemical status of the surficial sediments of a Swedish lake, Lake Lilla Ullevifjärden. One focus in this article is on the importance of the bottom dynamic conditions in the contexts of sediment sampling.
2. The lake size. More samples would generally be required in larger lakes than in smaller lakes to obtain the same information value. But how important is this compared to other factors?
3. The lake bottom roughness. More samples would be required in lakes with high roughness than in lakes with smooth bottoms.

Table 1. The Relationship Between Bottom Dynamic Conditions (Erosion, Transportation and Accumulation) and the Physical and Chemical Character of the Surficial Sediments of Lilla Ullevifjärden (Lake Mälaren), Sweden[a]

Number of Analyses	Erosion 15	Transportation 10	Accumulation 14
Physical Status			
Water depth, m	13.0 (5.3)	17.5 (5.4)	31.6 (8.0)
Water content, % ws	32.6 (9.0)	67.4 (9.6)	94.1 (2.3)
Bulk density, g cm^{-3}	1.72 (0.15)	1.26 (0.10)	1.03 (0.02)
Organic content, % ds	4.6 (2.2)	10.7 (4.6)	24.3 (2.5)
Chemical Status			
Nutrients and indicators of nutrient status			
Nitrogen, mg g^{-1} ds	0.6 (0.4)	3.4 (1.2)	10.7 (1.5)
Phosphorus, mg g^{-1} ds	0.8 (0.4)	2.8 (2.1)	1.6 (0.5)
Carbon, mg g^{-1} ds	0.5 (0.5)	2.7 (2.0)	10.4 (1.7)
Chlorophyll, μg g^{-1} ds	5.3 (4.2)	18.5 (9.4)	167.1 (45.5)
Not contaminating, chemically mobile elements (see also P)			
Iron, mg g^{-1} ds	24.6 (10.4)	53.5 (14.4)	41.3 (3.2)
Manganese, mg g^{-1} ds	0.8 (0.8)	3.5 (2.6)	2.5 (1.5)
Contaminating elements			
Zinc, μg g^{-1} ds	41 (19)	111 (27)	189 (17)
Nickel, μg g^{-1} ds	23 (8)	40 (8)	57 (10)
Copper, μg g^{-1} ds	18 (9)	31 (13)	59 (6)

Source: Håkanson and Jansson (1983).[1]

[a] Mean values and standard deviations (in parentheses).

4. Anthropogenic factors and type of pollution. Different contaminating substances, such as metals and organic material, appear with different types of distribution patterns.[2]
5. The chemical "climate" in the sediments. This affects the distribution of elements like P, Fe, and Mn, since the distribution and chemical form of these elements are highly influenced by the pH and Eh (= redox conditions) in the sediments. This will not be discussed in the paper more than what is stated in Table 1.[1]
6. The sediment biology, e.g., linkages to bottom fauna and bioturbation.

Factors linked to sampling and measurement:

7. The number of samples. This is in focus in this paper.
8. The type of sampling net (deterministic, stochastic, regular, grid, etc.). This has been discussed in many papers.[2,3]
9. The type of sampling equipment. This has also been discussed in many papers.[1]
10. The subsampling and sampling preparation.

11. The reliability of the laboratory analysis.
12. The interpolation and/or weighing system to determine contour lines between sample sites and area-typical mean values, e.g., kriging.[4-8]

Thus, this paper addresses some (but certainly not all) of the factors that affect the variability of sediment samples. While waiting for a scientifically tested sample formula that would include qualitative and quantitative comparisons of the relative importance of all these factors (based on a comprehensive and compatible data set), the aim of this paper is to take a few steps on that road, focusing on the role of bottom dynamic conditions, lake size, lake irregularity, and sediment contamination on sediment variability.

THE NUMBER OF SAMPLES

When one sample is taken (Figure 1A), very little information can be obtained about the real sediment variability. If two samples are taken, better predictions about the situation between the sites can be made, both when the sites are far away from each other (Figure 1B) and especially if they are close (Figure 1C).

The classical approach to studying variability would be to first estimate the number of samples necessary to obtain a certain sampling precision. This requires an estimate of the overall variance. If the data are normally distributed, then the necessary number of samples, n, needed to estimate the mean, MV, within given limits may be determined from basic statistics.[9] This also means that one has to define t, i.e., the Student's *t* at a given probability level (*p*).

Since most sediment data will probably show a skewed distribution, one may have to consider transformations in order to make the data distribution normal:

- $\mathbf{e^x}$; this transformation will maximize the influence of high values
- **log(x)**, i.e., 10-logarithm [or ln(x) = the natural logarithm]; will minimize the influence of high values
- $\mathbf{X^z}$, where z may, e.g., be = 2 or 0.5

The original skewed distribution may also be used if computing an index of clumping.[10]

It is common experience that n generally would seem very large from a practical point of view.[2,3,11] Furthermore, even if very many samples are taken, by making the preliminary grid denser and completing a very costly survey, the subsequent mapping may nevertheless create a pattern that is difficult to interpret. It may, in fact, be more effective to obtain a given precision for an area-typical mean value than for a site-typical value. The following sample formula[1] partly accounts for this. The purpose of this sample formula is to obtain a first, crude estimate of a reasonable number of samples for whole-lake estimates to be used in, for example, the context of aquatic pollution control. Thus, the idea is to obtain a general reference number and not a thorough description of the above-mentioned 12 factors that affect sample representativity:

$$\text{Sample formula 1:} \quad n' = 2.5 + 0.5\sqrt{a * F} \tag{1}$$

where a is the lake area (the formula is mainly meant to be used for lakes larger than 1 km^2); F is the shore development, a measure that is closely related to lake-bottom roughness;[1] F is defined as the relationship between the true (scale independent) shoreline length and the circumference of a circle with the same area as the lake. Sample formula 1 is graphically illustrated in Figure 2A.

Figure 2B gives a general, statistical sample formula to be discussed in this paper.[12] This formula is derived from the basic definitions of the mean value, the standard deviation, and the Student's *t* value and expresses the number of necessary samples (n'') as:

$$\text{Sample formula 2:} \quad n'' = (t * V/y)^2 + 1 \tag{2}$$

where y is the error accepted in the mean value; y = 1 implies 100% error; V is the relative standard deviation, i.e.:

$$V = 100 * SD/MV$$

Subsequently, we will determine the mean value with 95% certainty ($p = 0.05$), which gives a t value of 1.96 (or about 2). This implies that Equation 2 can be written

$$n'' = (2 * V/100 * y)^2 + 1$$

The relationship between n″, V, and y is illustrated graphically in Figure 2B. For a relative standard deviation of 25%, we can see that about 25 samples are required in order to establish a lake-typical mean value if we accept an error of 10%.

In the following sections, I first try to link these two formulas, which have been derived from different presuppositions, by quantifying the impact of bottom dynamic conditions on the V value; then we try to quantify the influence of sediment contamination on the V value and, finally, the influence of lake size and lake irregularity (F value) on the V value. The analysis is based on a set of data from Swedish lakes (all data used here emanate from Håkanson and Uhrberg[13] and Håkanson;[14] data from surficial sediments (0–1 cm) are used, except in determinations of natural background values, to express contamination factors). These exercises will also demonstrate that it may be a difficult task to get hold of a

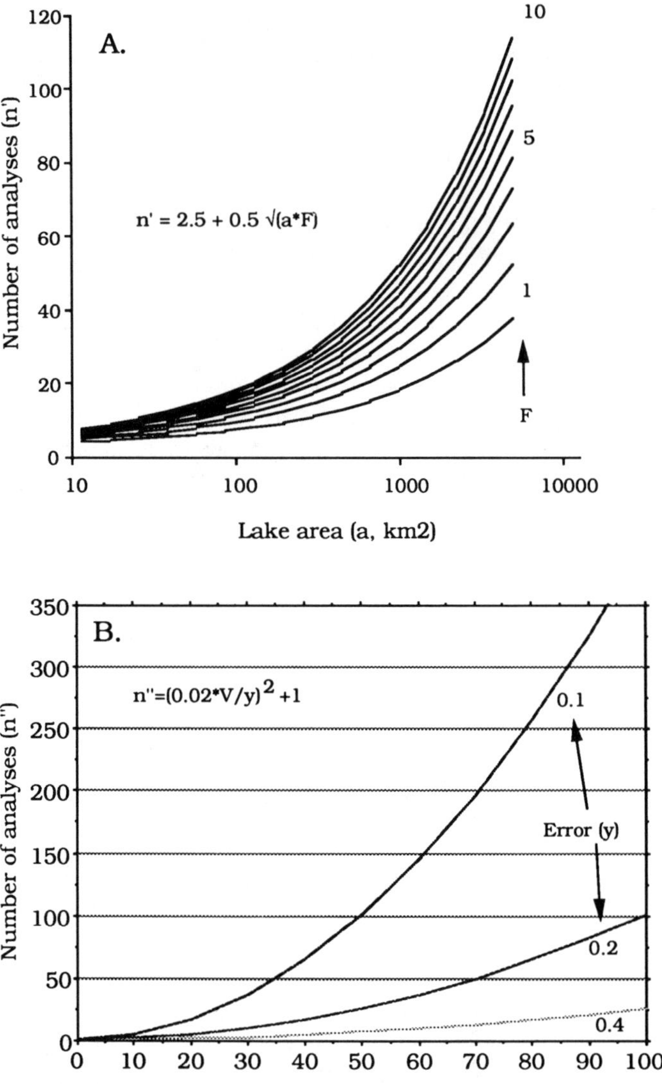

Figure 2. (A) Nomogram illustrating sample formula one. (B) Nomogram illustrating sample formula two.

comprehensive set of compatible data, and the data used here do leave room for future improvements.

The organic content, or rather the loss on ignition (IG), will be used as a key parameter. Figure 3A exemplifies that there exists a causal, and not just a statistical, link between the IG values of sediment samples, the N/C ratio of sediment samples, and the trophic and humic levels of lakes. Figure 3B exemplifies that the IG values

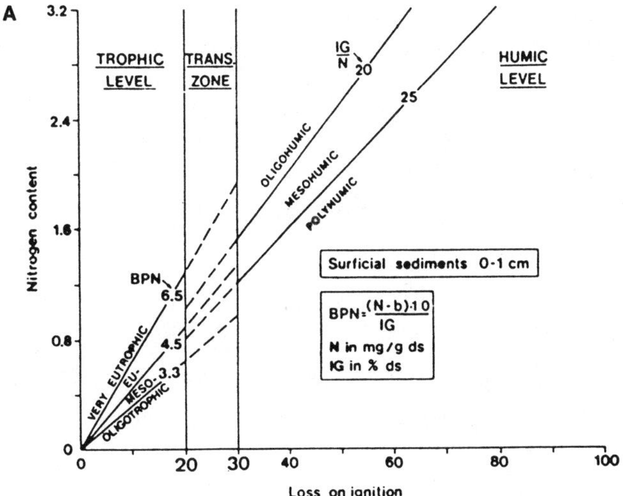

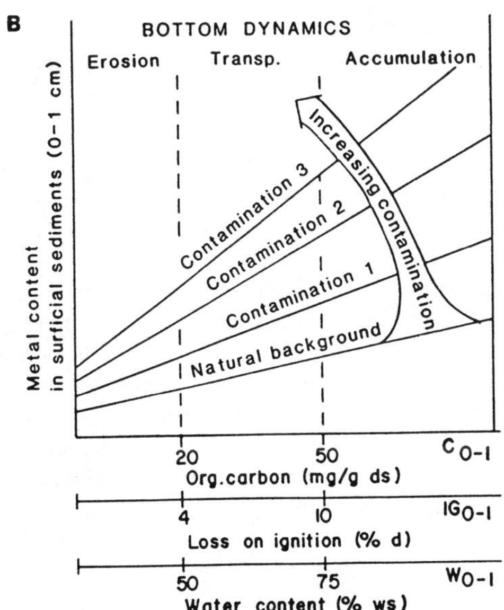

Figure 3. (A) Diagram illustrating the relationship between nitrogen concentration and loss on ignition of surficial sediments (0–1 cm) relative to lake trophic level and lake humic level. (From Håkanson, L., M. Jansson. *Principles of Lake Sedimentology* (Heidelberg: Springer-Verlag, 1983), p. 316. With permission.) (B) Diagram illustrating the linkage between bottom dynamic conditions (E, T, A) and the sediment contamination of metals as given by three regression lines between metal content and concentration of organic content, loss on ignition and water content in surficial sediments (0–1 cm). (From Håkanson, L., "Metal Monitoring in Coastal Environments," in *Metals in Coastal Environments of Latin America,* U. Seelinger et al., Eds. (Berlin: Springer-Verlag, 1988), p. 239. With permission.)

of sediment samples are also causally linked to bottom dynamic conditions and sediment contamination (of metals). The relationships in Figure 3 will be elaborated here in connection with the two sample formulas. The organic content is also causally linked to the physical character of the sediment (e.g., to water content, bulk density, and grain size).[1] The following formula expresses bulk density (D in g/cm^3) as a function of water content (W in % of the dry sediments), density of the organic materials (Dm in g/cm^3), and organic content (= loss on ignition; IG*, in % of the dry sediments):

$$D = 100 * Dm/(100 + (W + IG*)(Dm - 1)) \qquad (3)$$

So, variability in organic content (IG) would *a priori* imply variability in many, if not in most, physical, chemical, and biological sediment parameters.

BOTTOM DYNAMICS

Selected metal concentrations (as MV, SD, and V) and the loss or ignition in areas of erosion, transportation, and accumulation are presented in Table 2. The message in this table is

- The variability is generally great in areas of erosion and transportation and much smaller in areas of accumulation. This is true for IG and, hence, also for all parameters related to IG, such as pollutants and other sediment parameters, like number and species of bacteria, meio-, and macrofauna.[1]
- The relationship between the bottom dynamic conditions and the variability may also be determined quantitatively; this would have direct bearings on the number of samples necessary to determine an area-typical mean value. This is illustrated for the given metals in Table 2. From this table, we can note that the mean V value for the metals in A bottom (= accumulation areas) is about 40; it is over 60 in E and T bottoms. The mean V value for the loss on ignition is about 20 in A bottoms and three times higher in E and T bottoms. This has, as can be seen from sample formula 2 (Figure 2B), a significant affect on the number of samples that ought to be taken. These results, however, should be considered with due reservation, as they are only meant to give order of magnitude type of information of the impact of bottom dynamic conditions on these selected parameters in these lakes. These results cannot be transferred directly to other lakes and rivers, neither for these metals nor for other sediment parameters. But these data demonstrate the important characteristic of samples from areas of accumulation. These samples provide information for sediment toxicity assessments, which is superior to that provided by samples from E and T bottoms. This is also schematically illustrated in Figure 4, which is a "torch" diagram in analogy with Figure 1.

Table 2. Mean Values (MV), Standard Deviations (SD), Relative Standard Deviations (V) of Selected Metals and Loss on Ignition (IG) in Areas of Erosion, Transportation, and Accumulation in Four Swedish Lakes

	E			T			A			
	MV	SD	V	MV	SD	V	MV	SD	V	
Lake/metal										
Lilla Ullevifjärden E = 15, T = 10, A = 14; number of samples										
IG	**4.6**	**2.2**	**47.8**	**10.7**	**4.6**	**43.0**	**24.3**	**2.5**	**10.3**	
Zn	41	19	46.3	111	27	24.3	189	17	9.0	
Ni	28	13	46.4	51	13	25.5	73	7	9.6	
Cr	23	8	34.8	40	8	20.0	57	10	17.5	
Cu	18	9	50.0	31	13	41.9	59	6	10.2	
Vänern E = 11, T = 14, A = 75										
IG	**1.5**	**0.8**	**53.3**	**4.5**	**2.3**	**51.1**	**10.3**	**5.3**	**51.5**	
Cd				0.7	0.5	71.4	1.4	0.8	57.1	
Hg	0.026	0.03	107.7	0.27	0.48	177.8	1.48	2.13	143.9	
Cu	13	9	69.2	26	10	38.5	29	7	24.1	
Pb	34	17	50.0	64	36	56.3	100	34	34.0	
Zn	110	63	57.3	240	125	52.1	420	210	50.0	
Hjärmaren E = 7, T = 8, A = 20										
IG	**12**	**2.6**	**21.7**	**6.2**	**4**	**64.5**	**12.3**	**1.5**	**12.2**	
Cd	0.8	0.4	50.0	0.6	0.4	66.7	1.3	0.6	46.2	
Pb	52.3	13.2	25.2	24.4	11.5	47.1	54.1	23	42.5	
Cu	98.3	39.1	39.8	26	15.2	58.5	56.2	32.9	58.5	
Zn	297	31	10.4	103	35	34.0	265	118	44.5	
Ni	44.3	4	9.0	18.3	7.6	41.5	48.1	23.6	49.1	
Hg	0.235	0.12	51.1	0.063	0.039	61.9	0.114	0.072	63.2	
Vättern E = 6, T = 11, A = 31										
IG	**1.6**	**1.8**	**112.5**	**3.8**	**3.1**	**81.6**	**79.9**	**6.4**	**8.0**	
Ni	12	15	125.0	14.0	9	64.3	46	24.0	52.2	
Cu	10	10	100.0	10.0	8	80.0	50	24.0	48.0	
Zn	170	150	88.2	120.0	82	68.3	450	130.0	28.9	
Hg	0.04	0.05	125.0	0.2	0.23	127.8	0.129	0.1	41.9	
Pb	47	50	106.4	61.0	24	39.3	150	38.0	25.3	
	All	**Mean**	62.7			60.6			40.3	37.5 (Hg in Vänern omitted)
		SD V	36.2			34.5			28.8	
	IG	**Mean**	58.8			60.0			20.5	
		SD	38.4			16.9			20.7	

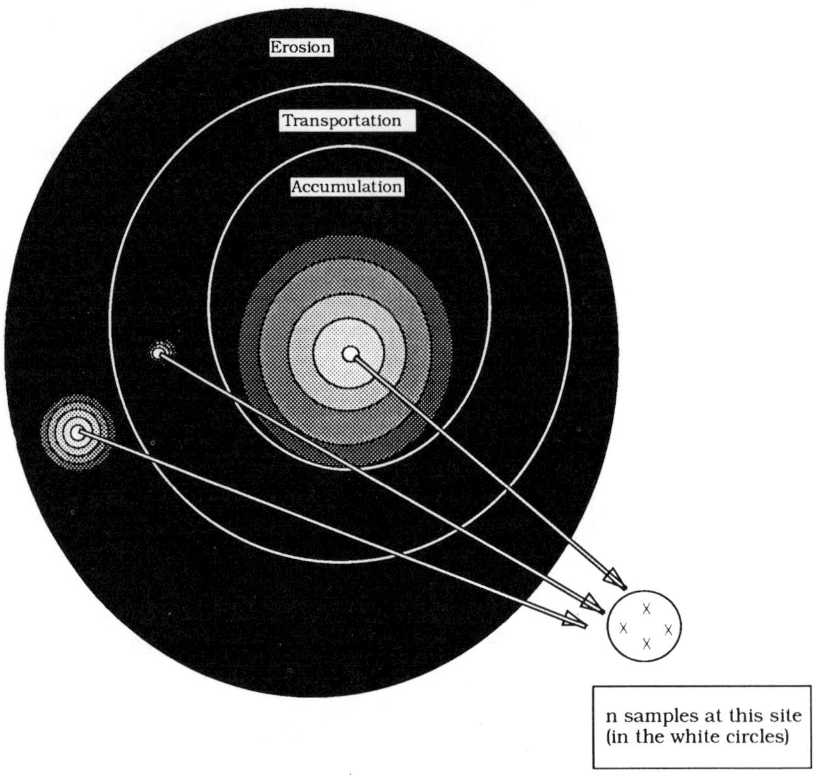

Figure 4. Illustration of the fact that the bottom dynamic conditions (E, T, A) have a profound impact on the representativity and information value of all types of sediment parameters. A few samples may reveal much if they emanate from an area of accumulation, but very little if they emanate from an area of transportation and erosion (which by definition have mixed deposits).

IMPACT OF SEDIMENT CONTAMINATION

Is it possible to obtain some quantitative indication of the role of sediment contamination on the sediment variability (in terms of organic content and selected chemical sediment parameters) in the same general way? Table 3 gives data on the number of analyses (n), mean values, standard deviations, relative standard deviations, and sediment contamination (Cf; defined according to Figure 5) for 16 Swedish lakes displaying a very wide range in contamination; from Cf = 1 to Cf = 33 (for Hg in Lake Väsman).

A better overview of the central information in Table 3 is given in Figure 6, which illustrates the relationship (regression line and r^2 value) between the Cf values and the V values. From this figure, we can note that there exists a weak, but sizeable, relationship between Cf and V; the r^2 value is between 0.18 and 0.25, depending on

Table 3. Number of Sediment Samples (n), Mean Values (MV), Standard Deviations (SD), Relative Standard Deviations (V), Contamination Factors (Cf) for Loss on Ignition (IG) and Some Metals (Hg, Cd, Pb, Cu, and Zn) in 16 Swedish Lakes

Lake		IG			Hg				Cd				Pb				Cu				Zn			
	n	MV	SD	V	MV	SD	V	Cf	MV	SD	V	Cf	MV	SD	V	Cf	MV	SD	V	Cf	MV	SD	V	Cf
Bysjön	3	11.6	1	8.6	0.34	0.07	20.6	3.8	1.4	0.3	21.4	1.7	63	12	19.0	1.3	15	2	13.3	0.7	297	58	19.5	1.4
Björken	3	12.2	4.6	37.7	0.4	0.12	30.0	4.4	2	0.2	10.0	2.4	76	23	30.3	1.6	14	5	35.7	0.7	257	45	17.5	1.2
Väsman	8	17.2	12.9	75.0	3	2.98	99.3	33.3	5.3	4.3	81.1	6.4	903	1184	131.1	19.2	85	69	81.2	4.0	1030	971	94.3	4.9
Övre Hillen	3	11.3	1.9	16.8	2.04	1.18	57.8	22.7	4.5	1.1	24.4	5.4	226	63	27.9	4.8	89	7	7.9	4.1	1270	253	19.9	6.0
Haggen	4	14.6	4.1	28.1	0.2	0.11	55.0	2.2	2.7	1	37.0	3.3	136	63	46.3	2.9	27	16	59.3	1.3	308	106	34.4	1.5
Norra Barken	6	11.2	2.2	19.6	0.95	0.51	53.7	10.6	8.7	5.8	66.7	10.5	813	448	55.1	17.3	102	62	60.8	4.7	2140	782	36.5	10.2
Södra Barken	6	8.9	1.4	15.7	0.47	0.08	17.0	5.2	5.4	1.1	20.4	6.5	260	55	21.2	5.5	45	7	15.6	2.1	1500	268	17.9	7.1
Snyten	3	9.4	3.2	34.0	0.19	0.1	52.6	2.1	0.7	0.2	28.6	0.8	65	9	13.8	1.4	32	6	18.8	1.5	203	36	17.7	1.0
Stora Aspen	5	9.9	1.6	16.2	0.47	0.26	55.3	5.2	4.8	1.7	35.4	5.8	981	499	50.9	20.9	104	37	35.6	4.8	1990	665	33.4	9.5
Lilla Aspen	3	11.5	1.6	13.9	0.39	0.15	38.5	4.3	4.3	0.6	14.0	5.2	840	254	30.2	17.9	80	11	13.8	3.7	1930	147	7.6	9.2
Amänningen	7	9.5	1.3	13.7	0.28	0.09	32.1	3.1	2.5	0.8	32.0	3.0	378	215	56.9	8.0	51	12	23.5	2.4	1610	545	33.9	7.7
Virsbosjön	3	8.5	1.4	16.5	0.3	0.08	26.7	3.3	1.8	0.4	22.2	2.2	170	33	19.4	3.6	47	10	21.3	2.2	1310	314	24.0	6.2
Gnien	3	14.0	6.6	47.1	0.31	0.14	45.2	3.4	1.9	0.5	26.3	2.3	100	9	9.0	2.1	44	15	34.1	2.0	940	143	15.2	4.5
Östersjön	11	18.2	21.4	117.6	0.63	0.37	58.7	7.0	1.9	1	52.6	2.3	140	110	78.6	3.0	130	86	66.2	6.0	1260	710	56.3	6.0
Västersjön	3	22.6	2.5	11.1	0.2	0.12	60.0	2.2	1.1	0.1	9.1	1.3	65	6	9.2	1.4	38	3	7.9	1.8	450	20	4.4	2.1
Freden	3	10.1	0.5	5.0	0.35	0.11	31.4	3.9	3	1.1	36.7	3.6	83	19	22.9	1.8	67	18	26.9	3.1	1060	159	15.0	5.0
	Mean	12.5	4.3	29.8	0.7	0.4	45.9	7.3	3.3	1.3	32.4	3.9	331.2	187.6	38.9	7.0	60.6	22.9	32.6	2.8	1097.2	326.4	28.0	5.2
	SD	3.9	5.5	29.3	0.8	0.7	20.3	8.6	2.1	1.6	19.8	2.5	341.9	308.2	31.5	7.3	34.4	26.3	22.6	1.6	649.3	307.0	21.8	3.1

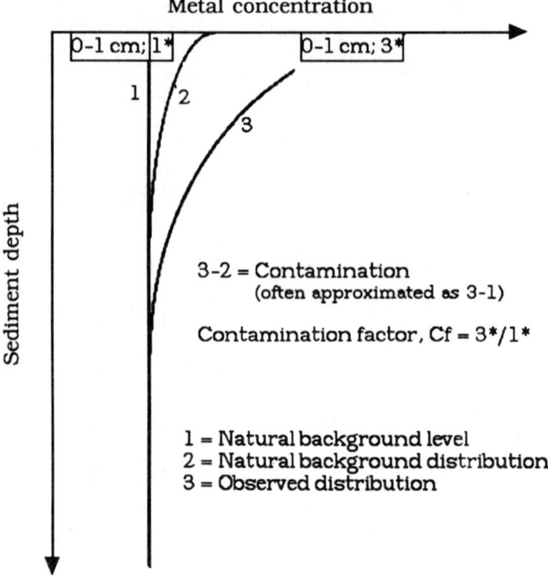

Figure 5. Illustration of natural background level, natural background distribution (which is linked to diagenetic processes in the sediments and often difficult to establish), observed distribution, and sediment contamination in a sediment core. (From Håkanson, L. and M. Jansson. *Principles of Lake Sedimentology* (Heidelberg: Springer-Verlag, 1983), p. 316. With permission.)

the number of data pairs used. This means that about 20% of the variability in V between these lakes can be linked to the variability in Cf. It also means that about 80% may be linked to other factors. One should also consider the variability due to limitations in the representativity of these empirical data.

From the two regression lines in Figure 6, we can see that, on average, an increase in Cf from 5 to 15 would mean an increase in V with about 30 to 40 units. This has bearings (Figure 2B) on the number of necessary samples. As previously stated, these results cannot be used universally; they are meant to provide order of magnitude type of information and nothing else.

INFLUENCE OF LAKE SIZE AND F

It has been difficult to obtain good data sets (that are sufficiently large and contain compatible data) on relative standard deviation, lake area, lake irregularity (F), and sediment contamination. The data given in Table 4 from nine Swedish lakes only yields initial rough estimates on this complex issue! Variability here is expressed in terms of our key parameter, loss on ignition (IG), and sediment contamination is expressed as mean Cf values for the three metals that (of all the investigated metals)

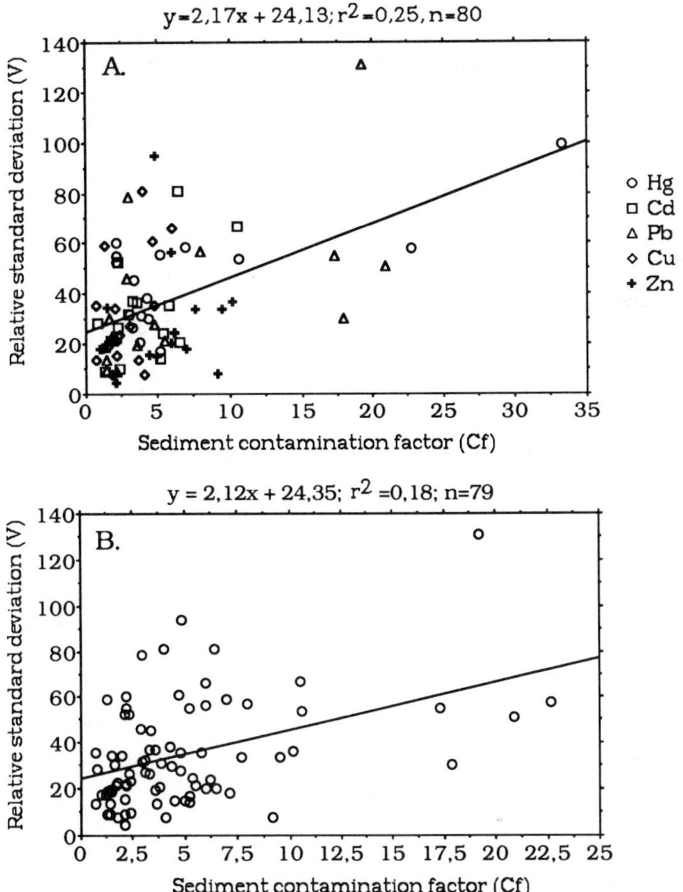

Figure 6. The relationship between sediment contamination (Cf) and relative standard deviation (V) for five metals in Swedish lakes. (A) For the entire data set (n = 80). (B) One outlier (Cf > 30) excluded.

gave the highest Cf values in each lake. The reason why I use a different set of lakes in Table 4 as compared to Table 3 is simply that there are no data available on both F and metals in the lakes included in Table 3. The Cf values in Table 4 leave, naturally, a lot to be desired; they are included here merely as reference values, since Figure 6 gives a much better indication on the role of Cf on variability (V).

Table 5 is a correlation matrix based on linear correlation coefficients, r, for the parameters given in Table 4. From this matrix, we may note that lake size seems to have a considerable influence on the given V value; the r value for log(area) is 0.74 (i.e., $r^2 = 0.55$). This means that 55% of the variability between the nine lakes in V (for IG) can be linked to lake size [log(area)]. Also F and Cf appear with positive

Table 4. Sediment Variability (V for IG), Lake Area, Lake Irregularity (F), and Lake Contamination (Mean Contamination Factor for the Three Metals in Each Lake That Have Yielded the Higher Cf Values) in Nine Swedish Lakes

Lake	V (for IG)	Area (km^2)	F	Cf (for 3 metals)
Hemfjärden	8.5	25.4	2.80	1.5
Mellanfjärden	6.4	40.1	1.88	1.6
Storhjälmaren	43.5	277.5	3.82	1.2
S. Hjälmaren	32.9	99.1	2.41	1.7
Ö. Hjälmaren	31.3	35.7	3.32	1.0
Lilla Ullevifjärden	28.7	1.9	2.74	1.1
Ekoln	24.4	18.6	2.75	5.4
Vättern	73.8	1856.0	3.00	3.3
Vänern	55.0	5650.0	7.50	3.1

Table 5. Correlation Matrix (r Values for the Parameters in Table 4)

	V (for IG)	Area	log (area)	F	Cf
V (for IG)	1.00	0.60	0.74	0.51	0.27
Area		1.00	0.79	0.92	0.31
log (area)			1.00	0.68	0.27
F				1.00	0.18
Cf					1.00

correlations to V (for IG); 26% (r = 0.51) of the variability between the V values for these lakes can, in fact, be linked to variability in F. The coupling to these Cf values were, as expected, not as marked as was obtained for the data set illustrated in Figure 6.

Figure 7 gives the linear regression between V and log(area) and between V and F. We can note that an increase in lake area from 10 to 100 km^2 will imply an average increase in V from approximately 15 to 30. An increase in F could also be linked to a probable increase in V. But for the relationship between V (for IG) and F, the spread is large ($r^2 = 0.26$); and it may not be meaningful to give a regression line, since both the slope and the intercept could be altered significantly if single lakes are omitted, or more lakes included, in the analysis. The crucial point here is not with such details, but rather with the fact that these data indicate that lake size and irregularity are important factors in sediment sampling.

It should also be noted that F and area were intercorrelated (r = 0.92). This is expected, since the definition of F is

$$F = lo/2 * \sqrt{(\pi * \text{area})}$$

CONCLUSIONS

These results may be summarized as follows:

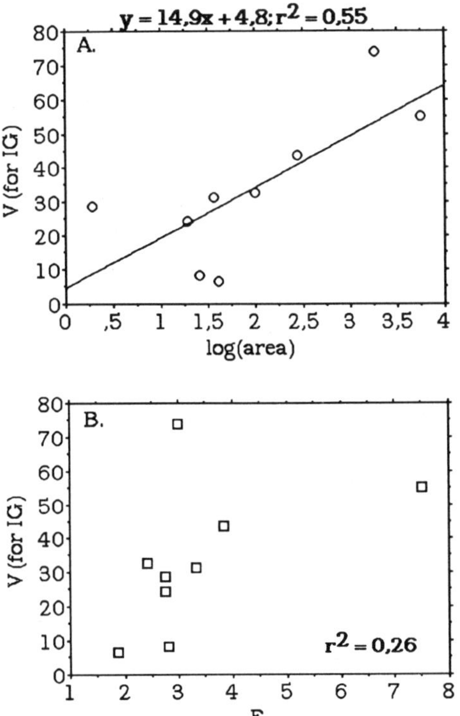

Figure 7. (A) The relationship between lake size [log(area)] and sediment variability (V for IG). (B) The relationship between lake irregularity (F) and sediment variability (V for IG).

- Using the variability of the organic content (loss on ignition = IG) of surficial sediments (0 to 1 cm) as a key parameter, one may say that the bottom dynamic conditions have a profound influence on the sediment variability and the number of samples that has to be taken to obtain area-typical mean values for entire lakes. The mean V value for IG is on the order of 20 for accumulation areas and about 60 for areas of erosion and transportation. From sample formula 2 (Figure 2B), we can note that this implies that about 140 samples should be taken from E and T bottoms (for y = 0.1); the corresponding figure for A bottoms is about 20.
- Sediment contamination also seems to influence variability. Since there exist causal links between the organic content and many sediment parameters, one may *a priori* conclude that variability in IG means variability in all parameters linked to IG. Figure 6 indicates that an increase in sediment contamination (Cf) for some selected metals (Hg, Cd, Pb, Cu, and Zn) is positively linked to an increase in the sediment variability (V) for these metals.
- Lake size influences sediment variability (V for IG) significantly. The larger the lake, the larger the potential variability in sediment status. An increase in size from 10 to 100 km^2 would, on average, increase V from 15 to 30, and n″ would then increase from about 10 to about 40 (for y = 0.1).

- Bottom roughness/shore irregularity (as expressed by F) also seems to have an influence on V (for IG).

These results are meant to provide general order-of-magnitude type of information on the relative importance of some factors on sediment variability. Other factors may also influence sediment variability, and the results derived from these Swedish lakes cannot be used uncritically on lakes in general. These results do, however, provide some information on how the considered "environmental" factors (as distinguished from methodological and statistical factors) influence the number of sediment samples that should be taken to obtain representative compatible data for research, water pollution, and sediment toxicity assessments.

REFERENCES

1. Håkanson, L., and M. Jansson. *Principles of Lake Sedimentology* (Heidelberg: Springer Verlag, 1983), p. 316.
2. Baudo, R. "Uncertainty in Description of Sediment Chemical Composition," *Hydrobiologia* 176/177:441–448 (1989).
3. Sly, P. "Statistical Evaluation of Recent Sediment Geochemical Sampling," presented at 9th International Congress of Sedimentology, Nice, 1975.
4. Chilès, J. P., and P. Chauvet. "Kriging: A Method for Cartography of the Sea Floor," *Int. Hydro. Rev.* 52:25–41 (1975).
5. Burgess, T. M., and R. Webster. "Optimal Interpolation and Isarithmic Mapping of Soil Properties. I. The Semi-Variogram and Punctual Kriging," *J. Soil Sci.* 31:315–331 (1980).
6. Burgess, T. M., and R. Webster. "Optimal Interpolation and Isarithmic Mapping of Soil Properties. II. Block Kriging," *J. Soil Sci.* 31:333–41 (1980).
7. Burgess, T. M., R. Webster, and A. B. McBratney. "Optimal Interpolation and Isarithmic Mapping of Soil Properties. IV. Sampling Strategy," *J. Soil Sci.* 32:643–659 (1981).
8. Floderus, S. "Sediment Sampling Evaluated with a New Weighting Function and Index of Reliability," *Hydrobiologia* 176/177:61–75 (1989).
9. Davis, J. C. *Statistics and Data Analysis in Geology* (New York: John Wiley & Sons, 1973), p. 550.
10. Kratochvil, B., and J. K. Taylor. "Sampling for Chemical Analysis," *Anal. Chem.* 53:924A–938A (1981).
11. Baudo, R. "Sediment Sampling, Mapping, and Data Analysis," in *Sediments: Chemistry and Toxicity of In-Place Pollutants,* R. Baudo, J. P. Giesy, and H. Muntau, Eds. (Ann Arbor, MI: Lewis Publishers, 1990), pp. 15–60.
12. Håkanson, L. "Metal Monitoring in Coastal Environments," in *Metals in Coastal Environments of Latin America,* U. Seelinger et al., Eds. (Berlin: Springer-Verlag, 1988), pp. 239–257.

13. Håkanson, L., and R. Uhrberg. "Investigations in the Water System of River Kolbäcksån. XIII. Metals in Fish and Sediments," Swedish EPA Report, (in Swedish), Uppsala (1981), p. 215.
14. Håkanson, L. "Lake Sediments in Aquatic Pollution Control Programs; Principles, Processes and Practical Examples," Swedish EPA Report 1398 (in Swedish), Uppsala,1981, p. 242.

CHAPTER 3

Sediment Collection and Processing: Factors Affecting Realism

G. Allen Burton, Jr.

INTRODUCTION

The sediment environment is a major and integral component of aquatic ecosystem functioning, yet in many ways, it is very different and separate from the overlying water environment. Spatial and temporal dynamics are often orders of magnitude different between the two environs. In many systems, fine grained sediments below the top few millimeters are relatively permanent and may never be resuspended into overlying waters. However, despite this apparent stability, there is a continual flux of physical, chemical, and biological components into and out of the surficial sediments, which may dramatically affect overlying water quality, sediment quality, and, thus, ecosystem structure and functioning.

Assessing sediment quality usually requires collection of sediment via grab, dredge, or core methods. These methods have, to varying degrees, relatively drastic effects on the sediment's integrity, disrupting physicochemical and biological gradients and their interrelationships, which may then affect toxicant partitioning, complexation, speciation, and bioavailability. Consequently, in situ methods have been proposed and shown to be useful in studies of many sediment issues, such as sediment oxygen demand,[1,2] interstitial water chemistry,[3,4] and toxicity.[5]

Obviously, in situ studies are superior, since they eliminate sampling and laboratory manipulation effects; however, for many studies and sites, they may be impractical. It is then necessary to collect and process sediments in a manner that is the least disruptive to the sample integrity and to recognize the consequences that various procedures have on sediment quality. Without this understanding, realistic extrapolations from laboratory studies to field conditions are simply impossible.

INTEGRITY AND SPATIAL/TEMPORAL PATTERNS

The holistic (as opposed to the reductionist) approach has been promoted recently[6,7] as the preferred method of studying aquatic ecosystems. Based on an abundance of sediment-related "reductionist" research on basic physical, chemical, and biological processes and considering the dynamics of stream[8-10] and lake[11,12] ecology, it is apparent that the holistic approach should be utilized when evaluating the sediment environs. Many studies have failed to recognize that a multitude of physical, chemical, and biological (micro- and macrobenthic) processes are integrated in the sediment as dynamic, yet structured, gradients often occurring on small spatial scales of microns to millimeters and temporal scales of minutes to months.[8,9] Given this reality, it is questionable whether the disruption of the sediment integrity resulting from grab sampling will allow laboratory studies of sediments that can be realistically extrapolated to in situ conditions.

Sediments in slow moving waters often tend to be fine grained, as opposed to the gravel and cobble found in fast flowing waters, and, because of their greater surface area and associated factors, accumulate higher concentrations of toxicants. These sediment environments are more complex than large-grained sediments, in part, because there are more microhabitats and interrelating redox gradients. These gradients have a particularly important role in determining the bioavailability of ionic (both metal, metalloid, and organic) toxicants. The distribution and bioavailability of nonionic toxicants (such as PCBs, dioxins, DDT, chlordane, Kepone) tend to be more associated with organic carbon,[13] which may also vary with sediment depth, but are less affected by sensitive redox gradients.

Sediments usually have steep vertical gradients (millimeters to centimeters) of some constituents that often influence partitioning processes, including oxygen, redox potential, sulfur and nitrogen species, hydrogen, methane, and labile dissolved organic compounds such as short chain fatty acid fermentation products.[14-18] Over a small range, which corresponds to the Eh (redox) profile, there is sequential consumption of different products via anaerobic respiration and methanogenesis.[14] In regards to metal partitioning (and availability), oxygen and sulfide gradients are particularly important. Another microbial product that dominates interstitial water toxicity in some sediments is ammonia.[19] This is more of a factor in organically rich, anoxic sediments receiving nutrient inputs. Interstitial water ammonia diffuses upward and may benefit epipelic algae in euphotic zones.[14] These benthic algae (periphyton, aufwuchs) create vertical distributions of oxygen in the sediment, which vary spatially (millimeters) and temporally (minutes to months).[14] In lake sediment systems where biological productivity is low, the oxygen penetrates deeper (25 mm maximum).[14] In stream systems oxygen gradients may go much deeper (centimeters) due to larger grain sizes and water exchange. These gradients can be dramatically affected by sampling, and in situ testing is recommended.[14,20]

The microbial consortia in sediments and their metabolic activity vary significantly with depth, due to gradients of natural electron acceptors.[14-18] This vertical variance may be linked to toxicant deposition patterns.[21] A contaminated stream station showing homogeneous particle size distributions of sand and silt and water content, was sampled on both 0.04 m^2 and 1.0 m^2 grids, using a hand corer and Ekman dredge. Indigenous sediment microbial activity (β-galactosidase, β-glucosidase, alkaline phosphatase, and dehydrogenases), and *Daphnia magna* and *Ceriodaphnia dubia* mortality were not significantly different in subsamples; however, large spatial differences were observed on small and large scales, both horizontally (20 cm to 10 m) and vertically (4 cm intervals to 20 cm depth).[21]

Some stream benthic macroinvertebrate community distributions appear to be determined by hydraulics.[22] Small-scale sampling is more apt to define meiofaunal patches than is large-scale sampling, which homogenizes patchiness,[23] thus ameliorating significant differences. The number of replicate samples needed to obtain a given precision decreases with increased density and sampler size, and the optimal sampler size (considering cost and precision) depends on mean density.[24]

The influence of storm events and watershed characteristics on chemical element dynamics is poorly understood, particularly because some are lumped into operationally defined units such as dissolved or total organic carbon.[25] Significant heterogeneity (62 to 100%) has been observed between sediment cores in concentrations of organic matter, water, and total phosphorus.[26] Some heterogeneity is likely due to invertebrate bioturbation,[27] sediment transport by currents, and small-scale variations in bottom profiles.[26,28]

The surficial layer (upper few centimeters) tends to be the ecosystem active portion, while deeper sediments are passive and more permanently "in place." The deeper layers are primarily of interest as historical records of ecosystem activity, but also may be reintroduced into the active portion of the ecosystem via dredging activities, bioturbation, and severe storm or hydrogeological events. There is a continual flux of inorganic[29] and organic compounds[30,31] through the sediment-water interface. These processes are accelerated by biological activity (planktonic and benthic), and, thus are seasonally linked,[31-33] and also by other physical disturbances (e.g., flow-induced resuspension, dredging) whose temporal relationships are more chaotic in nature.[34,35] Though the surficial sediment layer ecosystem is more "active," it is significantly more permanent (or less active) than the overlying waters and, therefore, often serves as a better record of recent watershed activities (disturbances) than the water column. This realization has led to increased sediment monitoring of contaminant concentrations and benthic macroinvertebrate communities in recent years by federal and state regulatory agencies.[36,37]

Acid volatile sulfide (AVS) concentrations in sediments also varied seasonally, peaking in late summer/early fall and spring, which appears to correlate with productivity inputs.[38] This would be expected, based on microbial activity studies where peak activity in sediments correlated with seasonal plankton blooms/organic carbon inputs.[31,33] Microbial processing contributed significantly to carbon cycling in the summer, but 95% of the primary production went to the benthos in the winter.[32]

TOXICANT DYNAMICS

In comparison to nonpolar organic contaminants, metal and metalloid dynamics between sediments and interstitial and overlying waters are particularly complex. Their movement, availability, and possible toxicity are influenced by chemical and physical reactions and factors such as oxygen/redox gradients, pH, temperature, adsorption, sedimentation, complexation, precipitation, and grain size.[39] In addition, a variety of common sediment bacterial communities can metabolize and alter metal/metalloid valence states via oxidation-reduction reactions, thereby altering chemical fate and toxicity.[40,41] Copper, iron, and zinc appear to be particularly sensitive to sample handling.[42] The release of metals from sediments to water has primarily been investigated in dredging effect-related studies. There is little release of metals from reduced sediments in oxygenated waters during dredging operations.[43] The release of metals is more likely after the sediments have been redeposited.[43] However, clearly dominating mechanisms do not appear to control availability in all freshwater systems. Salomons et al.[44] observed Cd release from anoxic marine sediments into oxic water, but metals were sorbed from oxic waters to freshwater sediments. Water concentrations of some metals have been shown to decrease by four orders of magnitude within 1 hr of dredging.[43] Any metals released from anoxic dredged sediments tend to adsorb onto freshly precipitated Fe/Mn oxyhydroxides.[39,45] This oxidation and release of sulfide-bound cadmium and subsequent readsorption to oxyhydroxides has been reported to occur in less than an hour.[46,47] Similarly, Jenne and Zachara[48] found that a significant portion of dissolved metals are irreversibly adsorbed to solids within several hours. These observations suggest that studies on the bioavailable reversible fraction should be conducted after this initial period. However, with hydrophobic organics, it may take days to reach equilibrium between the water and sediment phases.[49]

Exchange rates, such as desorption, for many metals (Fe^{+3}, Zn^{+2} Ni^{+2}, Cu^{+2}, As^{+3}, Cd^{+2}) are slow under undisturbed conditions.[40,43,50,51] The maximum remobilization rate of cadmium after dredging occurred in 2 to 4 weeks, with maximum release after 1 month.[46] Remobilization (desorption) is stimulated by bacterial decomposition of substrates/ligands and the formation of soluble organic compounds.[39] Peak concentrations of metals in the water column have been shown to correlate with organic matter decomposition during the year, low-flow conditions, and initial stormwater flushing.[39] Higher levels of Cd have been observed in oxic sediments overlying anoxic sediments, possibly due to sulfide removal.[52] This observation supports the acid volatile sulfide partitioning results with cadmium, recently shown by Di Toro et al.[53]

Metals are partitioned in sediments in many forms, as soluble free ions, soluble organic (low-molecular-weight humic) and inorganic complexes, easily exchangeable ions, precipitates of metal hydroxides, precipitates with colloidal ferric and manganic oxyhydroxides, insoluble organic complexes, insoluble sulfides, and residual forms.[43] The residual fraction serves as the matrix vehicle[54] and is associated with labile components (e.g., carbonates, amorphous aluminosilicates, organic matter)[55] that are coated with iron/manganese oxides and organic matter. This variable coating serves as an active sorption site for toxicants.[39] Free metal (e.g., Cu^{+2}) is generally thought to possess the greatest toxicity to aquatic organisms,[56,57] so it is important to understand binding dynamics (such as rates) controlling conditions (such as pH, Eh), and sorption/desorption properties.

In some sediments, sorption of metals is driven by amorphic oxides of Fe, Mn, and reactive particulate organic carbon.[45] Amorphic oxides of Si and Al, and clay and zeolite minerals, are particularly important sorbants for anionic metals and metalloids.[48] The use of total organic carbon to determine sorption potential for metals is inappropriate, since the aluminosilicate or carbonate coatings may isolate portions of the particulate organic carbon from the aqueous phase, thus making it "nonreactive".[48] Manganese oxides are highly reactive, strongly sorb many ions, are involved in many redox reactions, and are common in sediments, thereby influencing the mobility and fate of many pollutants.[58] Mn^{+4} and Fe^{+3} adsorb As^{+3} and are primary electron acceptors in its oxidation to As^{+5}, a less toxic form. Mn^{+4} can oxidize As^{+3} and As^{+5} within 48 hr.[51] Higher soluble As concentrations in soils were related to soluble Fe^{+2}.[59] Another factor controlling partitioning is pH. Adsorption of cadmium is easily affected between pH 7 and 9.[50] pH and Eh changes can alter iron solubility by three orders of magnitude, but rarely change the valence states.[43] Fe and Mn can desorb faster than can Cu and Zn, which are insensitive to pH and oxygen changes.[47]

Biological processes influence the toxicity and fate of some metals and metalloids. Arsenic and selenium are readily biomethylated and demethylated[41,60] to more or less toxic/available forms. Good correlations have been observed in bivalve tissue between As/Fe and Pb/Fe ratios and readily extractable Fe from sediments.[61,62]

Predicting the partitioning of toxicants (and, thus, bioavailability) is difficult[39,48,49,63] due to the myriad of possible undefined processes that may simultaneously reduce and increase availability. For example, reduced sediments have shown release of Pb, Cu, and Fe while Zn and Hg decreased,[43,64,65] and oxidized sediments released Cd, Cu, Pb, and Zn while Fe decreased.[43,66] The concentration of metal observed in interstitial waters is dependent, to a large degree, on sorption/precipitation processes. The process depends on the metal and the environment. Adsorption is complicated, being related to the solid type, concentration, adsorption species, and surface property changes resulting from interactions such as coagulation. In addition, there is sorption-site competition and reaction kinetics of constituents in mixtures that are unknown. Calcium reduces cadmium sorption by amorphic iron oxides, yet zinc is unaffected.[67] High dissolved organic matter concentrations enhance the solubility and complexation of metals.[50] Currently, organic ligand effects cannot be predicted.

Some suggest that the solution to solids ratio is critical in determinations of Kp (sediment/water partition coefficient).[68,69] Sharp increases in sorptions at low sediment concentrations (400 mg/L) may be due to disaggregation of sediment particles, thereby increasing the exposed surface area; however, others suggest this is an experimental artifact.[70,71] In sulfidic environments, Cu, Cd, and Zn concentrations are controlled by the precipitation-dissolution process and are independent of solids concentration.[50] In anoxic environments without sulfides, As and Cr concentrations are controlled by adsorption/desorption reactions and are solids concentration dependent. Partitioning coefficients vary one to three orders of magnitude in low particle concentration environments as a result of these interactive and nondeterministic effects.[63]

When resuspension events occur, predicting metal remobilization may be possible in site-specific studies; however, remobilization is dependent on particle residence time in the water column, which varies between sites, storms, and ecosystems.[39] In most systems, however, remobilization of metals from resuspended sediments is likely to be insignificant due to the slow reaction rates.[46] Suspended solids altered the bioavailability and acute toxicity of organic chemicals when the dose-response curve was steep. However, most organics do not have K_p's greater than 10^4 and are unlikely to sorb sufficiently to suspended or bed sediments to significantly affect their bioavailability.[72] Rogers et al.[72] questioned whether sediment organic carbon content explains the variability observed in sorption coefficients. Chemicals with high sorption coefficients were usually not acutely toxic below their solubility concentration. Chlordane availability to *Daphnia magna* was reduced by montmorillonite clay (0% total organic carbon [TOC]), and sediments (1.7% TOC) but were not a factor in controlling bioavailability if suspended solids were greater than 200 to 300 mg/L.[73]

Because of the heterogeneity of sediments, sediment sorption partition coefficients cannot be fully normalized using one sediment characteristic.[49] For nonionic organics with low solubilities, nonspecific van der Waals' interactions of the solute with the organic fraction of the sediment dominate.[49] With some compounds this relationship has shown that valid K_p's can be determined.[74] Equilibrium rates and bioavailability interactions are still a point of debate.[75] Ionic organic compound partitioning, like metals and metalloids, is more complicated, being influenced by numerous charge measures such as cation exchange capacity, pH, and Eh.[49] Sorption of cationic pesticides was shown to occur on negatively charged clay and organic matter sites.[76]

Jenne and Zachara[48] state that there are three critical areas where lack of data limit our ability to quantitatively describe mobility (thus bioavailability) of toxic elements in aquatic systems: (1) equilibrium and kinetics of sorption to solids; (2)

thermodynamics of metal-dissolved organic carbon complexation, species formation, and solids solubility; and (3) kinetics of dissolution and precipitation reactions. Therefore, predictions of metal sorption have orders of magnitude uncertainty and major discrepancies with field data.[48]

ECOSYSTEM INTERACTIONS

Understanding of community ecology in lotic and lentic systems has progressed significantly in recent years.[8-10] "Biotic dynamics and interactions are intimately and inextricably linked to variation in abiotic factors,"[77] and lotic systems are not in equilibrium due to natural disturbances that may occur frequently or infrequently.[10] Disturbance can be defined as a discrete event that alters the community/population structure and function and changes the physical environment and resource availability. Disturbances vary in type, frequency, and severity, both among and within ecoregions. The frequency of and intensity of disturbances cannot be predicted. Intermediate levels of disturbance maximize species richness.[10] This relationship to diversity has been described in a dynamic equilibrium model[78] whereby the frequency of disturbance controls whether competitive species and long-life-cycle species exist. Equilibrium conditions will tend to occur if disturbances are infrequent, thus excluding opportunistic species.[79] In stream ecology, disturbance is the dominating organizing factor, having a "major impact on productivity, nutrient cycling and spiralling, and decomposition."[10] Disturbances such as storm events (or the presence of toxicants) can eliminate biota.[80] Recovery and succession of these systems between disturbances is typified by recurrent or divergent patterns.[9,10]

In some systems, "bottom-up" effects have been observed where algal succession or community composition alterations affect zooplankton grazers.[81-83] Interruption of microbial cycling/decomposition processes reduced ecosystem productivity and altered sediment redox, produced anoxia, and increased H_2S production and acidity.[82,83] When macro-and meiobenthic invertebrate and protozoan cropping of bacteria was removed due to contaminant-induced lethality, the sediments serve as a carbon sink,[82,83] so organic carbon and nutrients necessary for secondary productivity are unavailable, and food web alterations are likely.[82,83] The sediment bacterial community is available to sediment ingesters, and bacterial production is transferred to fish through a single intermediate consumer, such as *Chironomus,* relatively efficiently.[84] In streams, bacterial productivity may be greater from sediments than from total primary production.[85] Benthic shredders (amphipods) could differentiate between fungal species colonizing detritus. So a change in the fungal community could alter organic matter processing rates.[86]

If one is concerned with energy flow, then the "bottom-up" approach dominates; however, predator removal (top-down) will alter community structure, productivity, or biomass.[81] This reemphasizes the complex and simultaneous functioning of ecosystem relationships.

The major role that natural and anthropogenic disturbances have on aquatic ecosystems pushes a high level of spatial and temporal variance under equilibrium conditions to even higher levels. Spatial and temporal dimensions span 16 orders of magnitude in stream ecology.[8,9] Significant spatial variance in sediments is common.[21,28] Each level of the system has different dimensions that have different variances associated with it, and is interacting simultaneously with other ecosystem levels and their respective dimensions/variances. This complex reality is difficult, if not impossible, to define accurately, but must be considered in all assessments of sediment quality or ecosystem health.

Orians[87] stated that one of the greatest challenges in ecology (and ecotoxicology) is bridging the conceptual gap between micro- and macroecology. Aquatic systems can be considered as a mosaic of patches.[9] "A patch is a spatial unit that is determined by the organism and problems in question."[9] The heterogenous environment has highly clumped distributions (patches) of organisms whose spatial/temporal patterns and relationships change seasonally and due to factors such as food (resource) patterns;[23] therefore, this poses severe sampling problems. The appropriate sampling scale will depend on the organism size, density, distribution, life cycle, and question being asked,[9] which, unfortunately, are often not considered. Aquatic ecosystems are open nonequilibrium systems[9,11] where patches are in transitory equilibrium with other patches.[88] Different life histories and variable interactions between species may prevent equilibrium.[11]

SAMPLING AND PROCESSING

Given the discussions in the preceding sections, it is apparent that sampling of sediments and subsequent assays of toxicity will be difficult if the researcher intends to maintain sample integrity and define in situ conditions. Few studies have focused on sampling and sample manipulation (e.g., spiking) effects on measured toxicity responses.[89,90] A standard guide was recently published by the American Society for Testing and Materials,[91] concerning collection, storage, characterization, and manipulation of sediments for toxicological testing. Other regulatory guidance documents exist that are concerned, in part, with sediment collection and characterization procedures.[92;93]

Sampling

Disrupting the sensitive sediment environment is of major concern when collecting samples for toxicity studies, since the bioavailability and toxicity can change significantly when in-place sediments are disturbed. A number of sampling related factors can contribute to the loss of the sample's integrity, including sampler-induced pressure waves, washout of fine-grained sediments during retrieval, compaction due to sampler wall friction, and sampling vessel- or person-induced disturbance of surficial layers. Choosing the most appropriate sediment sampler for a study will depend on the sediment's characteristics, the volume and efficiency required, and the study's objective (Table 1). Several references are available that discuss the various collection devices.[28,91,93,94,96-98] The efficiency of benthic collection samplers has been compared, and in general, the grab samplers are less efficient collectors than the corers.[91]

The disadvantages of dredge (grab) samplers vary with their designs. Common problems include shallow depth of penetration and the presence of a shock wave that results in the loss of the fine surface sediments. Dredge samplers that quantitatively sample surface sediments have been described.[99] The depth profile of the sample may be lost in the removal of the sample from the sampler. Dredge sampling promotes the loss of not only fine sediments, but also of water-soluble compounds and volatile organic compounds present in the sediment.[91]

Studies of macroinvertebrate sampling efficiency with various grab samplers have provided useful information for sampling in sediment toxicity and sediment quality evaluations. The Ekman dredge is the most commonly used sampler for benthic investigations.[97] The Ekman's efficiency is limited to less-compacted, fine-grained sediments, as are the corer samplers. However, these are usually the sediments of greatest concern in toxicity assessments. The most commonly used corer is the Kajak-Brinkhurst, or hand corer. In more resistant sediments the Petersen, Ponar, Van Veen, and Smith-McIntyre dredges are used most often.[97] Based on studies of benthic macroinvertebrate populations, the sediment corers are the most accurate samplers, followed by the Ekman dredge, in most cases.[97] For consolidated sediments, the Ponar dredge was identified as the most accurate, while the Petersen was the least effective.[97] A comparison of sampler precision for macrobenthic purposes showed the Van Veen sampler to be the least precise; the most precise were the corers and Ekman dredge.[97] The Smith-McIntyre and Van Veen samplers are more commonly used in marine studies, due to their weight. Shipek samplers are also used in marine investigations, but may lose the top 2 to 3 μm of sediment fines from washout.[96]

Many of the problems associated with dredge samplers are largely overcome with the corers. The best corers for most sediment studies are hand-held polytetrafluoroethylene plastic, high-density polyethylene, glass corers (liners), or large box corers.

If used correctly, box corers can maintain the integrity of the sediment surface while collecting a sufficient depth for most toxicity studies. Conventional gravity corers compress the sediment, as evidenced by altered pore-water alkalinity gradients, and box coring was superior for studies of in situ gradients.[100] The box core can be subcored or sectioned at specific depth intervals, as required by the study. Unfortunately, the box corer is large and cumbersome; thus, it is difficult to use and usually requires a lift capacity of 2000 to 3000 kg. Box cores typically require fine-grained sediments of at least a 30-cm depth. Other coring devices that have been successfully used include the percussion corer[101] and vibratory corers.[102-104]

Table 1. Popular Recommended Sediment Samplers: Strengths and Weaknesses

Sampler	Strengths	Weaknesses
Hand and gravity corers	Maintains sediment layering of inner core. Fine surficial sediments retained. Replicate samples efficiently obtained. Removable liners. Inert liners may be used. Quantitative sampling allowed.	Small sample volume. Liner removal required for repetitive sampling. Not suitable in large grain or consolidated sediments.
Box corer	Maintains sediment layering of large volume of sediment. Surficial fines retained relatively well. Quantitative sampling allowed.	Size and weight require power winch; difficult to handle and transport. Not suitable in consolidated sediments.
Vibratory corers	Samples deep sediments for historical analyses. Samples consolidated sediments.	Expensive and requires winch. Outer core integrity slightly disrupted.
Ekman or box dredge	Relatively large volume may be obtained. May be subsampled through lid. Lid design reduces loss of surficial sediments as compared to many dredges. Usable in moderately compacted sediments of varying grain sizes.	Loss of fines may occur during sampling. Incomplete jaw closure occurs in large-grain sediments or with large debris. Sediment integrity disrupted. Not an inert surface.
Ponar	Commonly used. Large volume obtained. Adequate on most substrates. Weight allows use in deep waters.	Loss of fines and sediment integrity occurs. Incomplete jaw closure occurs occasionally. Not an inert surface.
Van Veen or Young grab	Useful in deep waters and on most substrates. Young grab coated with inert polymer. Large volume obtained.	Loss of fines and sediment integrity occurs. Incomplete jaw closure possible. Van Veen has metal surface. Young grab is expensive. Both may require winch.
Peterson	Large volume obtained from most substrates in deep waters.	Loss of fines and sediment integrity. Not an inert surface. Incomplete jaw closure may occur. May require winch.
Orange-peel	Large volume obtained from most substrates. Efficient closure.	Loss of fines and sediment integrity. Not an inert surface. Requires winch.
Shipek	Adequate on most substrates.	Small volume. Loss of fines and sediment integrity. Not an inert surface.

Corer samplers also have limitations in some situations.[91] Most corers do not work well in sandy sediments; dredge samplers, or diver-collected material remain the only current alternatives. In general, corers collect less sediment than do dredge samplers which may provide inadequate quantities for some toxicity studies. Small corers tend to increase bow (pressure) waves (disturbance of surface sediments) and compaction, thus altering the vertical profile. However, these corers provide better confidence limits and spatial information when multiple cores are obtained.[23,97,105-107] As shown by Rutledge and Fleeger[108] and others, care must be taken in subsampling from core samples, since surface sediments might be disrupted in even hand-held core collection. They recommend subsampling in situ or homogenizing core sections before subsampling. Slowing the velocity of entry of coring equipment also reduces vertical disturbance.[109] Samples are frequently of a mixed depth, but a 2-cm sample is recommended[76] and the most common depth obtained, although depths up to 40 ft have been used in some dredging studies. For dredging, remediation, and/or historical pollution studies, it is sometimes necessary to obtain cores of depths up to several meters. This often requires the use of vibracores that are somewhat destructive to sediment integrity, but are often the only feasible alternative for deep- or hard sediment sampling. For some studies it has been advantageous or necessary to composite or mix single sediment samples. Composites usually consisted of three to five grab samples.[91]

Subsampling, compositing, or homogenizing of sediment samples is often necessary, and the optimal methods will depend on the study objectives. Important considerations that may influence sediment test results and interpretations include loss of sediment integrity and depth profile; changes in chemical speciation via oxidation and reduction or other chemical interactions; disruption of chemical equilibrium resulting in volatilization, sorption, or desorption; changes in biological activity; completeness of mixing; and sampling container contamination. In most studies of sediment toxicity, it is advantageous to subsample the inner core area (not contacting the sampler), since this area is most likely to have maintained its integrity and depth profile and not be contaminated by the sampler. Subsamples from the depositional layer of concern, for example, the top 1 or 2 cm, should be collected with a nonreactive sampling tool, such as a polytetrafluoroethylene-lined calibration scoop.[110] Subsamples are placed in a nonreactive container and mixed until the texture and color appear uniform.

Due to the large volume of sediment that is often needed for toxicity or bioaccumulation tests and chemical analyses, it might not be possible to use subsampled cores because of sample size limitations. In those situations the investigator should be aware of the above considerations and their possible affect on test results as they relate to in situ conditions.

Once sediment samples are collected, it is important, in most situations, to maintain an anoxic environment. The majority of fine-grained sediments, which are of concern in toxicity assessments, are anaerobic below the top few surface millimeters,[14] and any introduction of oxygen will likely alter the valence state of many ionic chemicals and result in sediment toxicant(s) bioavailability and toxicity levels that differ from in situ conditions. To protect sediments from oxygenation, the use of a glove box with an inert gas supply for subsampling and processing, e.g., preparation of sediments for centrifugation, may be necessary.

Safety

Many study sediments are severely contaminated or contain unknown chemicals and unknown concentrations of potentially hazardous substances. Therefore, it is recommended that all sampling and sample processing be conducted in thoroughly ventilated areas, to protect workers from exposure. Steps must also be taken to insure that the sample is not inadvertently contaminated with laboratory contaminants. Work areas must be free from potential contaminants such as combustion engine exhausts, lubricants, metals, and surfactants, and other synthetic substances that may be inert.

Reference and Control Sediment

Assessment of in situ sediment toxicity is aided by the collection and testing of reference and control samples.[91] A reference sediment may be defined as a sediment possessing similar characteristics to the test sediment, but without anthropogenic contaminants. Sediment characteristics, such as particle size distribution and percent organic carbon, should simulate, as closely as possible, that of the test sediment.[91] In some situations the reference sediment might also show toxicity due to naturally occurring chemical, physical, or biological properties. For this reason it is important to also test the toxicity of control sediments. Control sediments have been successfully used in toxicity evaluations.[13] A control sediment might consist of natural or artificially prepared sediments of known composition and of consistent quality that have been used in prior sediment toxicity tests or culturing that has been shown to be nontoxic.[91]

Characterization

The characteristics that have been most often measured in sediments are moisture content, organic carbon or volatile matter content, and particle size. When characterizing a sediment, quality assurance should always be addressed.[93-95,112,113] Sediments, by their nature, are very heterogenous; they exhibit significant temporal and spatial heterogeneity in the laboratory and in situ. Replicate samples should be analyzed to determine the variance in sediment characteristics and analytical methods. The type of sediment characterization needed will depend on the study objectives and the contaminants of concern; however, a minimum set of parameters should be included that are known to influence toxicity and will aid data interpretation. Recommended measurements include the following: in situ temperature, particle size-surface area distribution, moisture or interstitial water content, ash free weight, organic carbon (determined by titration or combustion), pH, Eh, acid volatile sulfides, ammonia, and cation exchange capacity. Many of the characterization methods have been based on analytical techniques for soils and waters, and the literature should be consulted for further information.[98,109,114-116]

The moisture content of sediments is measured by drying the sediments at 50° to 105°C to a constant weight.[98]

Volatile matter content is often measured instead of, and in some cases in addition to, organic carbon content as a measure of the total amount of organic matter in a sample. This measurement is made by ashing the sediments at high temperature and reporting the percent ash free dry weight.[117-119] Although the exact method for ashing the sample is often not specified, the normally accepted temperature is 550 ± 50°C.[95,98]

Carbon fractions that may be of importance in determining toxicant fate and bioavailability include total organic carbon,[95,120-122] dissolved organic carbon,[123] dissolved inorganic carbon, sediment carbonates, and reactive particulate carbon.[124] Reactive particulate carbon is that portion that equilibrates with the aqueous phase. The organic carbon content of sediments has been measured by wet oxidation, which is also useful for the determination of the organic carbon content of water.[125] Organic carbon analyses have also been conducted by titration,[126] modification of the titration method,[127] or combustion after removal of carbonate by the addition of HCl and subsequent drying.[128]

Particle sizing of sediments can be measured by numerous methods,[114,129] dependent on the particle properties of the sample.[130] Particle size distribution is often determined by wet sieving.[95,112,114] Particle size classes might also be determined by the hydrometer method,[131,132] the pipet method,[114] settling techniques,[133] X-ray absorption,[134] and laser light scattering.[135] The pipet method may be superior to the hydrometer method.[136] To obtain definite particle sizes for the fine material, a Coulter (particle size) counter method might be employed.[137] This method gives the fraction of particles with an apparent spherical diameter. Another potential method for determining the particle size distribution of a very fine fraction is through the use of electron microscopy.[138] The collection technique for the very fine materials can result in aggregation to larger colloidal structures.[138,139] Comparisons of particle sizing methods have shown that some produce similar results, and others do not. These differences are expected, given the fact that different techniques measure different particle properties. For example, the Malvern Laser Sizer and Electrozone Particle Counter are sizing techniques, while the hydrophotometer and SediGraph determine sedimentation diameter based on particle settling.[130,140-142] It is preferable to use a method that incorporates particle settling as a measure, as opposed to strictly sediment sizing.

Various methods have been recommended to determine the bioavailability of metals in sediments.[46,143-145] One extraction procedure, cation exchange capacity, provides information relevant to metal bioavailability studies.[109] Amorphic oxides of iron and manganese, and reactive particulate carbon, have been implicated as the primary influences on metal sorption potential in sediments.[54,145,146-148] Measurement of acid volatile sulfide (AVS) and divalent metal concentrations associated with AVS extraction provides insight into the metals availability in anaerobic sediments.[143,149] Easily extractable fractions are usually removed with cation-displacing solutions: for example, neutral ammonium acetate, chloride, sodium acetate, or nitrate salts.[150] Extraction of saltwater or calcareous sediments, however, is often complicated by complexation effects or dissolution of other sediment components.[46,151] The advantages and limitations of other extractants have been recently discussed.[46,149,152,153] EDTA or HCl extractants have been used successfully in evaluations of trace metals in nondetrital fractions of sediments.[46,154,155] Metal partitioning in sediments might be determined by using sequential extraction procedures that fractionate the sediments into several components, such as interstitial water, ion exchangeable, easily reducible organic, and residual sediment components.[153,156-158] Unfortunately, at this time no one method has been clearly demonstrated as being superior to the others.[152] This may be due, in part, to site-specific characteristics that influence bioavailability, such as desorption and equilibration processes.

The pH of sediments is an important factor determining the fate and effect of many ionic chemicals. This parameter can be measured directly[98] or in a 1 to 1 mixture of sediment/soil to water.[159]

Redox (Eh) is a measure of the oxidation-reduction status of the sediment and is a particularly important factor controlling metal speciation and determining the extent of sediment oxidation. Redox gradients in sediments often change rapidly over a small depth and are easily disturbed. When measuring Eh, care must be taken in probe insertion and in allowing equilibration to occur. These measurements are potentiometric and are measured with a platinum electrode relative to a standard hydrogen electrode.[98]

Biochemical oxygen demand and chemical oxygen demand might provide useful information in some cases.[98] Sediment oxygen demand might also be a useful descriptor; however, a wide variety of methods exist.[1,2,160-162]

Analysis of toxicants in sediments is generally performed by standard methods such as those of the EPA.[98,112,117] Soxhlet extraction is generally best for removal of organics, but depends on extraction parameters.[163,164] Concentrations are generally reported on a dry weight basis; however, this can lead to misleading interpretations. Toxicant levels should always be based on particle size and/or particle surface area distributions.

Storage

Drying, freezing, and storage temperature all affect toxicity.[91] Often the storage time of sediments used in toxicity tests has not been specified or has ranged from a few days[89] to one year.[165] Storage of sediments after arrival at the laboratory was generally by refrigeration at 4°C.[91] Significant changes in metal toxicity to cladocerans and microbial activity have been observed in stored sediments.[90] Recommended limits for storage of metal-spiked sediments have ranged from within 2 days[90] to 5 days[166] to 7 days[76] to less than 2 weeks.[91,167] A study of sediments contaminated with nonpolar organics found that interstitial water storage time did not affect toxicity to polychaetes when samples were frozen.[168] AVS is a reactive solid-phase sulfide pool that apparently binds some metals, thus reducing toxicity.[143] Cadmium toxicity in sediments has been shown to be related to AVS complexation.[143,169] When anoxic sediments were exposed to air, AVS were volatilized. If a study intends to investigate metal toxicity and the sediment environment is anoxic, then exposure to air might reduce or increase toxicity due to oxidation and precipitation of the metal species or loss of acid volatile sulfide complexation. It is generally agreed that sediments used for toxicity testing should not be frozen[170] and should be stored at 4°C with no air space or under nitrogen.[76]

Although risking changes in sediment composition, several investigations elected to freeze samples.[168,171-173] Fast-freezing of sediment cores has been recommended for chemical analyses; however, this alters the sediment structure and distorts the sample profile.[174] Freezing and thawing appeared to increase the release of soluble organic carbon.[168] Freezing in anoxic atmospheres has been reported to inhibit oxidation of reduced iron and manganese compounds.[148] It has also been recommended for stored sediments that are to be analyzed solely for organics and nutrients.[175]

Storage conditions have been shown to change pore water chemistry. Interstitial water chemistry changed significantly after 24-hr storage, even when stored at in situ temperatures.[176,177] Coagulation and precipitation of the humic material was noted when interstitial water was stored at 4°C for more than 1 week.[123] Oxidation of reduced arsenic species in pore water of stored sediments was unaffected for up to 6 weeks when samples where acidified and kept near 0°C without deoxygenation. When samples were not acidified, deoxygenation was necessary.[178] The above studies highlight the importance of timely analyses of sediment and interstitial water samples.

Interstitial Water Collection

Numerous reviews and studies have demonstrated that commonly used methods in pore water investigations produced experimental artifacts.[179-183] Collection of sediment interstitial water has been accomplished by several methods: centrifugation, squeezing, suction, and equilibrium dialysis.[91] In general, methods for recovery of relatively large volumes of interstitial water from sediments are limited to either centrifugation[184] or squeezing.[185] Other methods, such as suction,[186] gas pressurization,[110] in situ samplers,[187] and equilibration by using dialysis membrane or a fritted glass sampler,[188-191] do not produce large quantities of interstitial water.

The in situ methods are superior to others, since they minimize interferences associated with sample collection and processing. Dialysis chambers, often called

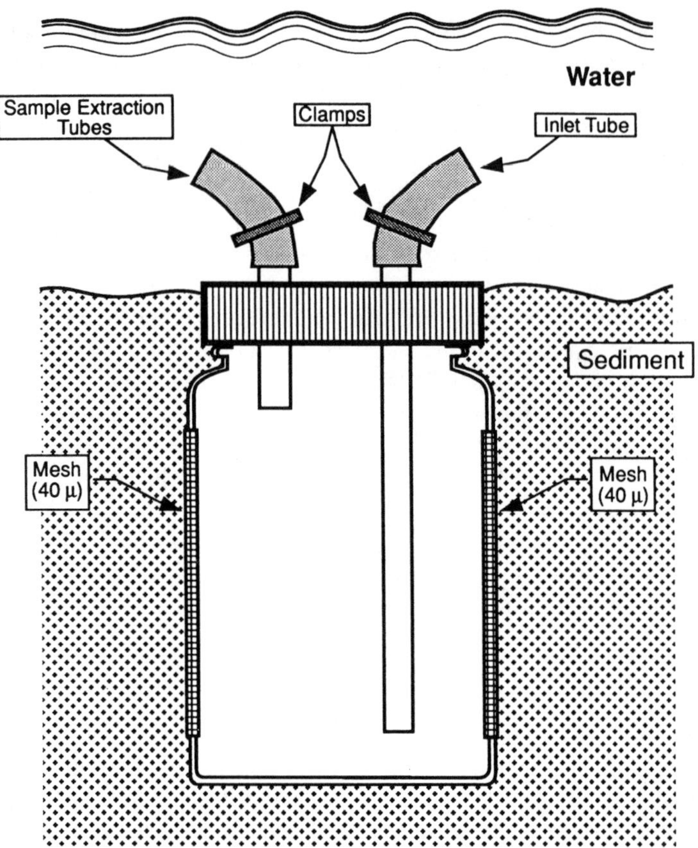

Figure 1.

"peepers," were first described by Hesslein[189] and Mayer.[190] Modifications of dialysis chambers[158,181,192] have been superior to the original cellulose-based membrane chambers. Typically, chambers are placed in situ and allowed to equilibrate for periods of one week, but have ranged from three days to a month.[179] Many of the artifact problems unique to dialysis samplers have been discussed[3,4] and highlight the extremely sensitive and tenuous nature of pore water investigations.

Recently, pore water toxicity has been measured in situ using plastic chambers covered with various polymer screens with mesh sizes of 0.15 to 0.50 mm (Figure 1). Short-term chronic studies of 7 days, using *Pimephales promelas* larvae, and 48-hr acute studies with *Daphnia magna,* demonstrated significantly higher survival in situ than in laboratory exposures to centrifuged pore water.[193] In many noncontaminated fine-grained sediments, low dissolved oxygen levels and high ammonia concentrations were primarily responsible for the observed toxicity. In larger-grained sediments (e.g., sand), realistic measures of sediment contamination with in situ methods could be conducted efficiently, avoiding sampling- and processing-induced errors. The test system was superior to dialysis chambers in that pore waters with the chamber equilibrated within 24 hr.

Some pore water constituents (e.g., dissolved organic carbon or dimethylsulfide) can be significantly affected by the collection method.[45,187] Other constituents, such as salinity, dissolved inorganic carbon, ammonia, sulfide, and sulfate, might not be affected by collection methods, providing oxidation is prevented.[187] Given the fact that most sediments are anoxic, all steps involved in sample processing may require an inert atmosphere (argon, nitrogen, helium, or carbon dioxide) to prevent the oxidation of reduced chemicals.[187,194,195]

If interstitial water is collected by centrifugation and filtration, potential effects on the interstitial chemistry (and thus toxicity) need to be considered.[91] Centrifugation followed by 2-μm filtration yielded similar metal concentrations to dialysis methods.[4] However, filtration with glass fiber or plastic filters is not appropriate in some cases and has been shown to remove nonpolar organics.[196] Centrifugation at 7600 × *g* with glass contact only was shown to be a superior separation method.[196] Others have produced contrary results, recommending filtration with polycarbonate filters.[186] Filtration is normally conducted to remove particles with a 0.45-μm pore size; however pore sizes of 0.20 μm have been recommended.[4,48] Centrifugation speeds may not be sufficiently high to remove dispersible clays. Since trace metals or organics concentrate on solids, this limitation may have significant effects on sorption studies.[48] The effects of centrifugation speed, filtration, and oxic conditions on some chemical concentrations in pore water have been documented in numerous studies.[179,197-199]

Sediment Dilution

In order to obtain concentration-effect information in solid phase sediment toxicity evaluations, different concentrations of the test sediment should be used.[200,201] Currently there is little information available regarding the most appropriate methodology and potential artifacts that may occur when diluting test sediments to obtain a graded contaminant concentration. A noncontaminated sediment may be used as the "diluent," which has physicochemical characteristics (e.g., organic matter/carbon, particle size) similar to the test sediment, but does not contain elevated (above background) levels of the toxicants of concern.[91,200,201] Others have diluted test sediments with water,[202] clean sand,[200] or diluted pore water.[201] In all dilution methods, both the effect of the contact time of the interstitial water and sediment (i.e., equilibrium) and the effect of disrupting the sediment's integrity on toxicant availability must be considered.

Elutriates

Many studies of sediment toxicity have been conducted on the elutriate or water-extractable phase.[92,203] This method was developed to assess the effects of dredging operations on water quality. Sediments are typically shaken in site or reconstituted water (1:4 volume-to-volume ratio) for 30 min. The water phase may then be separated from the sediment by centrifugation, followed by filtration of the supernatant through a 0.45-µm filter when conducting some tests, such as algal growth assays. The solids-to-solution ratio affects sorption rates[48,49] and should be considered when doing sediment extractions, but may not be a critical factor in whole-sediment tests[21] where there is no mixing.

CONCLUSIONS

The complicated nature of sediments and their fragile integrity, which, if disrupted, likely influences their chemical composition and toxicity, make the process of sample collection and handling an extremely critical component of any ecosystem assessment. Historical practices of using a grab (e.g., Peterson dredge) sampler to collect one mixed sample per site, allowing aeration, loss of surficial fines and temperature elevation to occur, and then storing sediments for weeks to months, are crude at best and do not allow contamination zones and effects to be accurately assessed. The state-of-the-practice now requires more resources and care be placed in the initial study phases (design, collection, and processing) so that meaningful data will be produced, improving the chances of an accurate assessment of the sediment's quality.

REFERENCES

1. Davis, W. S., L. A. Fay, and C. E. Herdendort. "Overview of USEPA/CLEAR Lake Erie Sediment Oxygen Demand Investigations During 1979," *J. Great Lakes Res.* 13:731–737 (1987).
2. Davis, W. S., T. M. Brosnan, and R. M. Sykes. "Use of Benthic Oxygen Flux Measurements in Wasteload Allocation Studies. Chemical and Biological Characterization of Sludges, Sediments, Dredge Spoils, and Drilling Muds," STP 976 (Philadelphia, PA: American Society for Testing and Materials, 1988), pp. 450–462.
3. Carignan, R. "Interstitial Water Sampling by Dialysis: Methodological Notes," *Limnol. Oceanogr.* 29:667–670 (1984).
4. Carignan, R., F. Rapin, and A. Tessier. "Sediment Porewater Sampling for Metal Analysis: A Comparison of Techniques," *Geochim. Cosmochim. Acta* 49:2493–2497 (1985).
5. Sasson-Brickson, G., and G.A. Burton, Jr. "*In Situ* and Laboratory Sediment Toxicity Testing with *Ceriodaphnia dubia*," *Environ. Toxicol. Chem.* 10:201–207 (1991).
6. Harris, H. J., P. E. Sager, H. A. Regier, and G. R. Francis. "Ecotoxicology and Ecosystem Integrity: The Great Lakes Examined," *Environ. Sci. Technol.* 24:598–603 (1990).
7. Chapman, P. M., E. A. Power, and G. A. Burton, Jr. Chapter 14, this volume.

8. Minshall, G. W. "Stream Ecosystem Theory: A Global Perspective," *J. N. Am. Benthol. Soc.* 7:263–288 (1988).
9. Pringle, C. M., R. J. Naiman, G. Brelschko, J. R. Karr, M. W. Oswood, J. R. Webster, R. L. Welcomme, and M. J. Winterbourn. "Patch Dynamics in Lotic Systems: The Stream as a Mosaic," *J. N. Am. Benthol. Soc.* 7:503–524 (1988).
10. Resh, V. H., A. V. Brown, A. P. Covich, M. E. Gurtz, H. W. Li, G. W. Minshall, S. R. Reice, A. L. Sheldon, J. B. Wallace, and R. C. Wissmar. "The Role of Disturbance in Stream Ecology," *J. N. Am. Benthol. Soc.* 7:433–455 (1988).
11. Carpenter, S. R., J. F. Kitchell, and J. R. Hodgson. "Cascading Trophic Interactions and Lake Productivity," *Bioscience* 35:634–639 (1985).
12. Carpenter, S. R., Ed. *Complex Interactions in Lake Communities.* (New York: Springer-Verlag, 1988).
13. Adams, W. J., R. A. Kimerle, and R. G. Mosher. "Aquatic Safety Assessment of Chemicals Sorbed to Sediments," in *Aquatic Toxicology and Hazard Assessment*, 7th Symposium, STP 854 (Philadelphia: American Society for Testing and Materials, 1985), pp. 429–453.
14. Carlton, R. G., and M. J. Klug. "Spatial and Temporal Variations in Microbial Processes in Aquatic Sediments: Implications for the Nutrient Status of Lakes," in *Sediments: Chemistry and Toxicity of In-Place Pollutants*, R. Baudo, J. Giesy, and H. Muntau, Eds., (Boca Raton, FL: Lewis Publishers, 1990), pp. 107–130.
15. Capone, D. G., and R. P. Kiene. "Comparison of Microbial Dynamics in Marine and Freshwater Sediments: Contrasts in Anaerobic Carbon Catabolism," *Limnol. Oceanogr.* 33:725–749 (1988).
16. Cooke, J. G., and R. E. White. "Spatial Distribution of Denitrifying Activity in a Stream Draining an Agricultural Catchment," *Freshwater Biol.* 18:509–519 (1987).
17. King, G. M. "Characterization of β-Glucosidase Activity in Intertidal Marine Sediments," *Appl. Environ. Microbiol.* 51:373–380 (1986).
18. Novitsky, J. A. "Microbial Growth Rates and Biomass Production in a Marine Sediment: Evidence for a Very Active but Mostly Nongrowing Community," *Appl. Environ. Microbiol.* 53:2368–2372 (1987).
19. Ankley, G. T., A. Katko, and J. W. Arthur. "Identification of Ammonia as an Important Sediment-Associated Toxicant in the Lower Fox River and Green Bay, Wisconsin," *Environ. Toxicol. Chem.* 9:313–322 (1990).
20. Hesslein, R. H. "An In Situ Sampler for Close Interval Pore Water Studies," *Limnol. Oceanogr.* 21:912 (1976).
21. Stemmer, B. L., G. A. Burton, Jr., and G. Sasson-Brickson. "Effect of Sediment Spatial Variance and Collection Method on Cladoceran Toxicity and Indigenous Microbial Activity Determinations," *Environ. Toxicol. Chem.* 9:1035–1044 (1990).
22. Statzner, B., J. A. Gore, and V. H. Resh. "Hydraulic Stream Ecology: Observed Patterns and Potential Applications," *J. N. Am. Benthol. Soc.* 7:307–360 (1988).
23. Findlay, S. E. "Small-Scale Spatial Distribution of Meiofauna on a Mud- and Sandflat," *Estuar. Coast. Shelf Sci.* 12:471–484 (1981).
24. Morin, A. "Variability of Density Estimates and the Optimization of Sampling Programs for Stream Benthos," *Can. J. Fish. Aquat. Sci.* 42:1530–1534 (1985).
25. Meyer, J. L., W. H. McDowell, T. L. Bott, J. W. Elwood, C. Ishizaki, J. M. Melack, B. L. Peckarsey, B. J. Peterson, and P. A. Rublee. "Elemental Dynamics in Streams," *J. N. Am. Benthol. Soc.* 7:410–432 (1988).

26. Downing, J. A., and L. C. Roth. "Spatial Patchiness in the Lacustrine Sedimentary Environment," *Limnol. Oceanogr.* 33:447–458 (1988).
27. Reynoldson, T. B. "Interactions Between Sediment Contaminants and Benthic Organisms," *Hydrobiologia* 149:53–66 (1987).
28. Baudo, R. "Sediment Sampling, Mapping, and Data Analysis," in *Sediments: Chemistry and Toxicity of In-Place Pollutants,* R. Baudo, J. Giesy, and H. Muntau, Eds. (Boca Raton, FL: Lewis Publishers, 1990), pp. 15–60.
29. Mortimer, C. H. "The Exchange of Dissolved Substances Between Mud and Water in Lakes," *J. Ecol.* 30:147–201 (1942).
30. Meyer-Reil, L.-A. "Measurement of Hydrolytic Activity and Incorporation of Dissolved Organic Substrates by Microorganisms in Marine Sediments," *Mar. Ecol. Prog. Ser.* 31:143–149 (1986).
31. Meyer-Reil, L.-A. "Seasonal and Spatial Distribution of Extracellular Enzymatic Activities and Microbial Information of Dissolved Organism Substrates in Marine Sediments," *Appl. Environ. Microbiol.* 53:1748–1755 (1987).
32. Nielsen, T. G., and K. Richardson. "Food Chain Structure of the North Sea Plankton Communities: Seasonal Variations of the Role of the Microbial Loop," *Mar. Ecol. Prog. Ser.* 56:75–87 (1989).
33. van Duyl, F. C., and A. J. Kop. "Seasonal Patterns of Bacterial Production and Biomass in Intertidal Sediments of the Western Dutch Wadden Sea," *Sea Mar. Ecol. Prog. Ser.* 59:249–261 (1990).
34. Pool, R. "Is It Chaos, or Is It Just Noise?" *Science* 243:25–27 (1989).
35. Pool, R. "Ecologists Flirt with Chaos," *Science* 243:310–313 (1989).
36. "An Overview of Sediment Quality in the United States," Office of Water Regulations and Standards, U.S. Environmental Protection Agency, Washington, D.C. and Region 5, Chicago, IL, EPA-905/9-88-002 (1987).
37. Southerland, E., T. Wall, and M. Kravitz. Chapter 15, this volume.
38. Di Toro, D. M., C. Zarba, D. Hansen, W. Berry, R. Swartz, C. Cowan, S. Pavlou, H. Allen, N. Thomas, and P. Paquin. "Technical Basis for Establishing Sediment Quality Criteria for Nonionic Organic Chemicals by Using Equilibrium Partitioning." *Environ. Toxicol. Chem.* 10:1541–1586 (1991).
39. Förstner, U. "Inorganic Sediment Chemistry and Elemental Speciation," in *Chemistry and Toxicity of In-Place Pollutants,* R. Baudo, J. Giesy, and H. Muntau, Eds. (Boca Raton, FL: Lewis Publishers, 1990), pp. 61–106.
40. Wood, J.M. "Biological Processes Involved in the Cycling of Elements Between Soil or Sediments and the Aqueous Environment," *Hydrobiologia* 149:31–42 (1987).
41. Drotar, A., G. A. Burton, Jr., J. E. Tavernier, and R. Fall. "Widespread Occurrence of Bacterial Thiol Methyltransferases and the Biogenic Emission of Methylated Sulfur Gases," *Appl. Environ. Microbiol.* 53:1626–1631 (1987).
42. Mudroch, A., and R. A. Bourbonniere. "Sediment Preservation, Processing, and Storage," in *CRC Handbook of Techniques for Aquatic Sediment Sampling*, A. Mudroch, and S. D. MacKnight, Eds. (Boca Raton, FL: CRC Press, 1991), pp. 131–169.
43. Gambrell, R. P., R. A. Khalid, and W. H. Patrick, Jr. "Physicochemical Parameters that Regulate Mobilization and Immobilization of Toxic Heavy Metals," in *Proc. Specialty Conference on Dredging and Its Environmental Effects,* Mobile, AL (New York: American Society of Civil Engineering, 1976).
44. Salomons, W., N. J. de Rooij, H. Keedijk, and J. Bril. "Sediments as a Source for Contaminants?" *Hydrobiologia* 149:13–30 (1987).

45. Lee, G. F., and R. A. Jones. "Water Quality Significance of Contaminants Associated with Sediments: An Overview," in *Fate and Effects of Sediment-Bound Chemicals in Aquatic Systems,* K.L. Dickson, A.W. Maki, and W.A. Brungs, Eds. (New York: Pergamon Press, 1984), pp. 1–34.
46. Kersten, M., and U. Förstner. "Cadmium Associations in Freshwater and Marine Sediments," in *Cadmium in the Aquatic Environment,* J.O. Nriagu, and J. B. Sprague, Eds., (New York: Wiley, 1987).
47. Hirst, J. M., and S. R. Aston. "Behavior of Copper, Zinc, Iron and Manganese During Experimental Resuspension and Reoxidation of Polluted Anoxic Sediments," *Estuar. Coast. Shelf Sci.* 16:549–558 (1983).
48. Jenne, E. A., and J. M. Zachara. "Factors Influencing the Sorption of Metals," in *Fate and Effects of Sediment-Bound Chemicals in Aquatic Systems,* K. L. Dickson, A. W. Maki, and W. A. Brungs, Eds. (New York: Pergamon Press, 1987), pp. 83–98.
49. Podoll, R. T., and W. R. Mabey. "Factors to Consider in Conducting Laboratory Sorption/Desorption Tests," in *Fate and Effects of Sediment-Bound Chemicals in Aquatic Systems,* K. L. Dickson, A. W. Maki, and W. A. Brungs, Eds. (New York: Pergamon Press, 1984), pp. 83–98.
50. Salomons, W. "Sediments and Water Quality," *Sci. Technol. Lett.* 6:315–326 (1985).
51. Oscarson, D. W., P. M. Huang, C. Defosse, and A. Herbillon. "Oxidative Power of Mn(IV) and Fe(III) Oxides with Respect to As(III) in Terrestrial and Aquatic Environments," *Nature* 291:50–51 (1981).
52. Emerson, S., R. Jahnke, and D. Heggie. "Sediment-Water Exchange in Shallow Water Estuarine Sediments," *J. Mar. Res.* 42:709–730 (1984).
53. Di Toro, D. M., J. D. Mahony, D. J. Hansen, K. J. Scott, M. B. Hicks, S. M. Mayr, and M. S. Redmond. "Toxicity of Cadmium in Sediments: The Role of Acid Volatile Sulfide," *Environ. Toxicol. Chem.* 9:1487–1502 (1990).
54. Jenne, E. A. "Trace Element Sorption by Sediments and Soil-Sites and Processes,"In *Symposium on Molybdenum in the Environment,* Vol. 2, (New York: Marcel Dekker, 1977), pp. 425–553.
55. Martin, J. M., P. Nirel, and A. J. Thomas. "Sequential Extraction Techniques: Promises and Problems," *Mar. Chem.* 22:313–342 (1987).
56. Cairns, M. A., A. V. Nebeker, J. N. Gakstatter, and W. L. Griffis. "Toxicity of Copper-Spiked Sediments to Freshwater Invertebrates," *Environ. Toxicol. Chem.* 3:435–445 (1984).
57. Sunda, W., and R. R. L. Guillard. "The Relationship Between Cupric Ion Activity and the Toxicity of Copper to Phytoplankton," *J. Mar. Res.* 34:511–529 (1976).
58. Oscarson, D. W., P. M. Huang, U. T. Hammer, and W. K. Liaw. "Oxidation and Sorption of Arsenite by Manganese Dioxide as Influence by Surface Coatings of Iron and Aluminum Oxides and Calcium Carbonate," *Water, Air, Soil Pollut.* 20:233–244 (1983).
59. Deuel, L. E., and A. R. Swoboda. "Arsenic Solubility in a Reduced Environment," *Soil Sci. Soc. Am. Proc.* 36:276–278 (1971).
60. Holm, T. R., M. A. Anderson, R. R. Stanforth, and D. C. Iverson. "The Influence of Adsorption of the Rates of Microbial Degradation of Arsenic Species in Sediments," *Limnol. Oceanogr.* 25:23–30 (1980).
61. Luoma, S. N., and G. W. Bryan. "Factors Controlling the Availability of Sediment-Bound Lead to the Estuarine Bivalve *Scrobicularia plana*," *J. Mar. Biol. Assoc.* 58:793–802 (1978).

62. Langston, W. J. "Arsenic in U.K. Estuarine Sediments and Its Availability to Benthic Organisms," *J. Mar. Biol. Assoc.* 60:869–881 (1980).
63. Honeyman, B. D., and P. H. Santschi. "Metals in Aquatic Systems-Predicting Their Scavenging Residence Times for Laboratory Data Remains a Challenge," *Environ. Sci. Technol.* 22:862–871 (1980).
64. Brooks, B. B., J. J. Presley, and I. R. Kaplan. "Trace Elements in the Interstitial Waters of Marine Sediments," *Geochim. Cosmach. Acta* 32:397–414 (1968).
65. Windom, H. L. "Investigations of Changes in Heavy Metals Concentrations Resulting From Maintenance Dredging of Mobile Bay Ship Channel, Mobile Bay, Alabama," U.S. Army Corps of Engineers, Mobile Bay, AL, Contract No. DACW01-73-C-0136 (1973).
66. Holmes, C. W., A. S. Elizabeth, and C. J. McLerran. "Migration and Redistribution of Zinc and Cadmium in Marine Estuarine System," *Environ. Sci. Technol.* 8:254–259 (1974).
67. Demsey, B. A., and P. C. Singer. "The Effect of Calcium on the Adsorption of Zinc by MnOx(S) and $Fe(OH)_3$ Cams," in *Contaminants and Sediments,* Vol. 2, R. A. Baker, Ed. (Ann Arbor, MI: Ann Arbor Science, 1980), pp. 333–354.
68. Di Toro, D. M., and L. M. Horzempa. "Reversible and Resistant Components of PCB Adsorption-Desorption Isotherms," *Environ. Sci. Technol.* 16:594 (1982).
69. O'Connor, D. J., and J. P. Connolly. "The Effect of Concentration of Adsorbing Solids on the Partition Coefficients," *Water Res.* 14:1517 (1980).
70. Curl, R. L., and G. A. Keolelan. "Implicit-Adsorbate Model for Apparent Anomalies with Organic Adsorption on Natural Adsorbents," *Environ. Sci. Technol.* 18:916 (1984).
71. Gschwen, P. M., and S. Wu. "On the Constancy of Sediment-Water Partition Coefficients of Hydrophobic Organic Pollutants," *Environ. Sci. Technol.* 19:90 (1985).
72. Rodgers, J. H., Jr., K. L. Dickson, F. Y. Saleh, and C. A. Staples. "Bioavailability of Sediment-Bound Chemicals to Aquatic Organisms — Some Theory, Evidence and Research Needs," in *Fate and Effects of Sediment-Bound Chemicals in Aquatic Systems,* K. L. Dickson, A. W. Maki, and W. A. Brungs, Eds. (New York: Pergamon Press, 1984), pp. 245–266.
73. Hall, W. S., K. L. Dickson, F. Y. Saleh, and J. H. Rogers, Jr. "Effects of Suspended Solids on the Bioavailability of Chlordane to *Daphnia magna,*" *Arch. Environ. Contam. Toxicol.* 15:509–534 (1986).
74. Adams, W. J. "Bioavailability of Neutral Lipophilic Organic Chemicals Contained in Sediment: A Review," in *Fate and Effects of Sediment-Bound Chemicals in Aquatic Systems,* K. L. Dickson, A. W. Maki, and W. A. Brungs, Eds. (New York: Pergamon Press, 1984), pp. 219–244.
75. Landrum, P. F., and J. A. Robbins. "Bioavailability of Sediment-Associated Contaminants to Benthic Invertebrates," in *Sediments Chemistry and Toxicity of In-Place Pollutants,* R. Baudo, J. Giesy, and H. Muntau, Eds. (Boca Raton, FL: Lewis Publishers, 1990), pp. 237–264.
76. Anderson, J., W. Birge, J. Gentile, J. Lake, J. Rodgers, Jr., and R. Swartz. "Biological Effects, Bioaccumulation, and Ecotoxicology of Sediment-Associated Chemicals," in *Fate and Effects of Sediment-Bound Chemicals in Aquatic Systems,* K. L. Dickson, A. W. Maki, and W. A. Brungs, Eds. (New York: Pergamon Press, 1984), pp. 267–296.
77. Power, M. E., R. J. Stout, C. E. Cushing, P. P. Harper, F. R. Hauer, W. J. Matthews, P. B. Moyle, B. Statzner, and I. R. Wars de Badgen. "Biotic and Abiotic Controls in River and Stream Communities," *J. N. Am. Benthol. Soc.* 7:456–479 (1988).

78. Huston, M. "A General Hypothesis of Species Diversity," *Am. Nat.* 113:81–101 (1979).
79. Minshall, G. W., and R. C. Peterson. "Towards a Theory of Macroinvertebrate Community Structure in Stream Ecosystems," *Archiv. Hydrobiol.* 104:49–76 (1985).
80. Power, M. E., R. J. Stout, C. E. Cushing, P. P. Harper, F. R. Hauer, W. J. Matthews, P. B. Moyle, B. Statzner, and I. R. Wars de Badgen. "Biotic and Abiotic Controls in River and Stream Communities," *J. N. Am. Benthol. Soc.* 7:456–479 (1988).
81. Crowder, L. B., W. Drenner, W. C. Kerfoot, K. J. McQueen, E. L. Mills, U. Sommer, C. N. Spencer, and M. J. Vanni. "Food Web Interactions in Lakes," in *Complex Interactions in Lake Communities,* S. R. Carpenter, Ed. (New York: Springer-Verlag, 1987), pp. 141–160.
82. Griffiths, R. P. "The Importance of Measuring Microbial Enzymatic Functions While Assessing and Predicting Long-Term Anthropogenic Perturbations," *Mar. Pollut. Bull.* 14:162–165 (1983).
83. Griffiths, R. P., B. A. Caldwell, W. A. Broich, and R. Y. Morita. "Long-Term Effects of Crude Oil on Microbial Processes in Subarctic Marine Sediments," *Mar. Pollut. Bull.* 13:273–278 (1982).
84. Porter, K. G., H. Paerl, R. Hodson, M. Pace, J. Priscu, B. Riemann, D. Scavia, and J. Stockner. "Microbial Interactions in Lake Food Webs," in *Complex Interactions in Lake Communities,* S. R. Carpenter, Ed. (New York: Springer-Verlag, 1987), pp. 209–227.
85. Bott, T. L., and L. A. Kaplan. "Bacterial Biomass, Metabolic State, and Activity in Stream Sediments: Relation to Environmental Variables and Multiple Assay Comparisons," *Appl. Environ. Microbiol.* 50:508–522 (1985).
86. Bärlocher, F. "The Role of Fungi in the Nutrition of Stream Invertebrates," *Bot. J. Linn. Soc.* 91:83–94 (1985).
87. Orians, G. H. "Micro and Macro in Ecological Theory," *Bioscience* 30:79 (1980).
88. Sheldon, A. L. "Colonization Dynamics of Aquatic Insects," in *Ecology of Aquatic Insects,* V. H. Resh, and D. M. Rosenberg, Eds. (New York: Praeger, 1984), pp. 401–429.
89. Malueg, K. W., G. S. Schuytema, and D. F. Krawczyk. "Effects of Sample Storage on a Copper-Spiked Freshwater Sediment," *Environ. Toxicol. Chem.* 5:245–253 (1986).
90. Stemmer, B. L., G. A. Burton, Jr., and S. Leibfritz-Frederick. "Effect of Sediment Test Variables on Selenium Toxicity to *Daphnia magna,*" *Environ. Toxicol. Chem.* 9:381–389 (1990).
91. "Standard Guide for Collection, Storage, Characterization, and Manipulation of Sediments for Toxicological Testing," (Philadelphia: Standard No. E 1391, American Society for Testing and Materials, 1991) in press.
92. "Ecological Evaluation of Proposed Discharge of Dredged Material into Ocean Waters," Environmental Effects Laboratory, U.S. Army Engineer Waterways Experiment Station, U.S. Environmental Protection Agency, Corps of Engineers, Vicksburg, MI (1977).
93. "Handbook for Sampling and Sample Preservation of Water and Wastewater," Environmental Monitoring and Support Lab, U.S. Environmental Protection Agency, Cincinnati, OH, EDA-600/4-82-029 (1982).
94. "Sampling Protocols for Collecting Surface Water, Bed Sediment, Bivalves, and Fish for Priority Pollutant Analysis," Office of Water Regulations and Standards Monitoring and Data Support Division, U.S. Environmental Protection Agency, Washington, D.C., Final Draft Report (1982).

95. "Recommended Protocols for Conducting Laboratory Bioassays on Puget Sound Sediments," Puget Sound Estuary Program, prepared by TetraTech, Inc. and E.V.S. Consultants, Inc., U.S. Environmental Protection Agency, Region 10, Seattle, WA, Final Report TC-3991-04 (1986).
96. Mudroch, A., and S. D. MacKnight. "Bottom Sediment Sampling," In *CRC Handbook of Techniques for Aquatic Sediments Sampling,* A. Mudroch, and S. D. MacKnight, Eds. (Boca Raton, FL: CRC Press, 1991), pp. 29–95.
97. Downing, J. A. "Sampling the Benthos of Standing Waters, a Manual on Methods for the Assessment of Secondary Productivity in Freshwaters," in *IBP Handbook 12,* 2nd ed., J. A. Downing, and F. H. Rigler, Eds. (Boston: Blackwell Scientific, 1984), pp. 87–130.
98. Plumb, R. H. "Procedures for Handling and Chemical Analysis of Sediment and Water Samples," Environmental Protection Agency/Corps of Engineers Technical Committee on Criteria for Dredged and Fill Material, Contract EPA-4805572010, EPA/CE-81-1, 478, p. (1981).
99. Grizzle, R. E., and W. E. Stegner. "A New Quantitative Grab for Sampling Benthos," *Hydrobiologia* 126:91–95 (1985).
100. Lebel, J., N. Silverberg, and B. Sundby. "Gravity Core Shortening and Pore Water Chemical Gradients," *Deep Sea Res.* 29:1365 (1982).
101. Gilbert, R., and J. Glew. "A Portable Percussion Coring Device for Lacustrine and Marine Sediments," *J. Sed. Petrol.* 55:607–608 (1985).
102. Fuller, J. A., and E. P. Meisburger. "A Simple, Ship-Based Vibratory Corer," *J. Sed. Petrol.* 52:642–644 (1982).
103. Imperato, D. "A Modification of the Vibracoring Technique for Sandy Sediment," *J. Sed. Petrol.* 57:788–789 (1987).
104. Lanesky, D. E., B. Logan, R. G. Brown, and A. C. Hine. "A New Approach to Portable Vibracoring Underwater and on Land," *J. Sed. Petrol.* 49:654–657 (1979).
105. Blomqvist, S. "Reliability of Core Sampling of Soft Bottom Sediment — An *in situ* Study," *Sedimentology* 32:605–612 (1985).
106. Elliott, J. M., and C. M. Drake. "A Comparative Study of Seven Grabs for Sampling Benthic Macroinvertebrates in Rivers," *Freshwater Biol.* 11:99–120 (1981).
107. Fleeger, J. W., W. Sikora, and J. Sikora. "Spatial and Long-Term Variation of Meiobenthic-Hyperbenthic Copepods in Lake Pontchartrain, Louisiana," *Estuar Coast. Shelf Sci.* 16:441–453 (1983).
108. Rutledge, P. A., and J. Fleeger. "Laboratory Studies on Core Sampling with Application to Subtitle Meiobenthos Collection," *Limnol. Oceanogr.* 33:274–280 (1988.)
109. Black, C. A., Ed. "Methods of Soil Analysis," American Society of Agronomy, Agronomy Monograph No. 9, Madison, WI (1965).
110. Long, E. R., and M. F. Buchman. "An Evaluation of Candidate Measures of Biological Effects for the National Status and Trends Program," NOAA Tech. Memorandum NOS OMA 45, Seattle, WA (1989).
111. Reichert, W. L., B. Le Eberhart, and U. Varanasi. "Exposure of Two Species of Deposit-Feeding Amphipods to Sediment-Associated [^{3}H]benzoapyrene: Uptake, Metabolism and Covalent Binding to Tissue Macromolecules," *Aquat. Toxicol.* 6:45–56 (1985).
112. Burton, G. A., Jr. "Quality Assurance Issues in Assessing Receiving Waters," in *Proc. Conf. on Effects of Urban Runoff on Receiving Systems* (New York: Engineering Foundation, 1992), in press.

113. "Guidance for Contracting Biological and Chemical Evaluations of Dredged Material," U.S. Army Engineer Waterways Experiment Station, Vicksburg, MS, Tech Rep. D-90-10.
114. Guy, H. P. "Techniques of Water-Resources Investigations of the U.S.G.S.," chap. C1, in *Laboratory Theory and Methods for Sediment Analysis; Book 5, Laboratory Analysis, U.S.G.S.* (Arlington, VA: U. S. Geological Survey, 1969) p. 58.
115. "Water Environmental Technology," in *ASTM Annual Book of Standards,* Vol. 11.02 (Philadelphia: American Society for Testing and Materials, 1989).
116. Page, A. L., R. H. Miller, and D. R. Keeney, Eds. "Methods of Soil Analysis," Parts 1 and 2, American Society of Agronomy, Madison, WI (1982).
117. "Standard Methods for the Examination of Water and Wastewater," 16th ed., American Waterworks Association and Water Pollution Control Federation, American Public Health Association, Washington, D.C. (1985).
118. Keilty, T. J., D. S. White, and P. F. Landrum. "Sublethal Responses to Endrin in Sediment by *Limnodrilius hoffmeisteri* (Tubificidae) and in Mixed-Culture with *Stylodrilius heringianus* (Lumbriculidae)," *Aquat. Toxicol.* 13:251–270 (1988).
119. McLeese, D. W., C. D. Metcalfe, and D. S. Pezzack. "Uptake of PCB's from Sediment by *Nereis virens* and *Crangon septemspinosa,*" *Arch. Environ. Contam. Toxicol.* 9:507–518 (1980).
120. Kadeg, R. D., and S. Pavlou. "Reconnaissance Field Study Verification of Equilibrium Partitioning: Nonpolar Hydrophobic Organic Chemicals," submitted by Battelle Washington Environmental Program Office, Washington, D.C., for U. S. Environmental Protection Agency, Criteria and Standards Division, Washington, D.C. (1987).
121. Kadeg, R. D., S. Pavlou, and A. S. Duxbury. "Sediment Criteria Methodology Validation, Work Assignment 37, Task II: Elaboration of Sediment Normalization Theory for Nonpolar Hydrophobic Organic Chemicals," prepared by Envirosphere Co., Inc., for Battelle Pacific Northwest Laboratories and U. S. Environmental Protection Agency, Criteria and Standards Division (1986).
122. Pavlou, S., R. Kadeg, A. Turner, and M. Marchlik. "Sediment Quality Criteria Methodology Validation: Uncertainty Analysis of Sediment Normalization Theory for Nonpolar Organic Contaminants," prepared by Envirosphere Co., Inc., for Environmental Protection Agency, Criteria and Standards Division, Washington, D.C. Submitted by Battelle, Washington Environmental Program Office, Washington, D.C. (1987).
123. Landrum, P. F., S. R. Nihart, B. J. Eadie, and L. R. Herche. "Reduction in Bioavailability of Organic Contaminants to the Amphipod *Pontoporeia hoyi* by Dissolved Organic Matter of Sediment Interstitial Waters," *Environ. Toxicol. Chem.* 6:11–20 (1987).
124. Cahill, R. A., and A. D. Autrey. "Improved Measurement of the Organic Carbon Content of Various River Components," *J. Freshwater Ecol.* 4:219–223 (1987).
125. Menzel, D. W., and R. F. Vaccaro. "The Measurement of Dissolved Organic and Particulate Carbon in Sea Water," *Limnol. Oceanogr.* 9:138–142 (1964).
126. Walkley, A., and I. A. Black. "An Examination of the Degtjareff Method for Determining Soil Organic Matter and a Proposed Modification of the Chromic Acid Titration Method," *Soil Sci.* 37:29–38 (1934).

127. Yeomans, J. C., and J. M. Bremmer. "A Rapid and Precise Method for Routine Determination of Organic Carbon in Soil," *Commun. Soil Sci. Plant Anal.* 19:1467–1476 (1985).
128. Wood, L. W., G. Y. Rhee, B. Bush, and E. Barnard. "Sediment Desorption of PCB Congeners and Their Bio-Uptake by Dipteran Larvae," *Water Res.* 21:873–884 (1987).
129. Allen, T. "Particle Size Measurement," (New York, NY: John Wiley & Sons, 1975), p. 452.
130. Singer, J. K., J. B. Anderson, M. T. Ledbetter, I. N. McCave, K. P. N. Jones, and R. Wright. "An Assessment of Analytical Techniques for the Size Analysis of Fine-Grained Sediments," *J. Sed. Petrol.* 58:534–543 (1988).
131. Day, P. R. "Particle Fractionation and Particle-Size Analysis," Section 43-5, Hydrometer Method of Particle Size Analysis, Monograph No. 9, American Society of Agronomy, Madison, WI, pp. 562–566, (1965).
132. Patrick, W. H. Jr. "Modification of Method Particle Size Analyses," *Proc. Soil Sci. Soc. Am.* 4:366–367 (1958).
133. Sanford, R. B., and D. J. P. Swift. "Comparisons of Sieving and Settling Techniques for Size Analysis, Using a Benthos Rapid Sediment Analyzer," *Sedimentology* 17:257–264 (1971).
134. Duncan, G. A., and G. G. Lattaie. *Size Analysis Procedures Used in the Sedimentology Laboratory, NWRI Manual,* National Water Research Institute, Canada Centre for Inland Waters (1979).
135. Cooper, L. R., R. Haverland, D. Hendricks, and W. Knisel. "Microtrac Particle Size Analyzer: An Alternative Particle Size Determination Method for Sediment and Soils," *Soil Sci.* 138:138–146 (1990).
136. Sternberg, R. W., and J. S. Creager. "Comparative Efficiencies of Size Analysis by Hydrometer and Pipette Methods," *J. Sed. Petrol.* 31:96–100 (1961).
137. McCave, I. M., and J. Jarvis. "Use of the Model T Coulter® Counter in Size Analysis of Fine to Coarse Sand," *Sedimentology* 20:305–315 (1973).
138. Leppard, G. G., J. Buffle, R. R. De Vitore, and D. Pereet. "The Ultrastructure and Physical Characteristics of a Distinctive Colloidal Iron Particulate Isolated from a Small Eutrophic Lake," *Arch. Hydrobiol.* 113:405–424 (1988).
139. Leppard, G. G. "The Fibrillar Matrix Component of Lacustrine Biofilms," *Water Res.* 20:697–702 (1986).
140. Kaddah, M. T. "The Hydrometer Method for Detailed Particle-Size Analysis: 1. Graphical Interpretation of Hydrometer Readings and Test of Method," *Soil Sci.* 118:102–108 (1974).
141. Stein, R. "Rapid Grain-Size Analyses of Clay and Silt Fraction by Sedigraph 5000D: Comparison with Coulter Counter and Atterberg Methods," *J. Sed. Petrol.* 55:590–615 (1985).
142. Welch, N. H., P. B. Allen, and D. J. Galinds. "Particle-Size Analysis by Pipette and Sedigraph," *J. Environ. Qual.* 8:543–546 (1979).
143. Di Toro, D. M., J. D. Mahony, D. J. Hansen, K. J. Scott, M. B. Hicks, S. M. Mayr, and M. S. Redmond. "Toxicity of Cadmium in Sediments: The Role of Acid Volatile Sulfide," draft report to U.S. Environmental Protection Agency, Criteria and Standards Division, Washington, D.C. (1989).
144. Chao, T. T., and L. Zhou. "Extraction Techniques for Selective Dissolution of Amorphous Iron Oxides from Soils and Sediments," *Soil Sci. Soc. Am. J.* 47:225–232 (1983).

145. Crecelius, E. A., E. A. Jenne, and J. S. Anthony. "Sediment Quality Criteria for Metals: Optimization of Extraction Methods for Determining the Quantity of Sorbents and Adsorbed Metals in Sediments," Battelle Report to U.S. EPA, Criteria and Standards Division, Washington, D.C. (1987).
146. Salomons, W. "Adsorption processes and hydrodynamic conditions in estuaries," *Environ. Technol. Lett.* 1:356–365 (1980).
147. Jenne, E. A. "Controls on Mn, Fe, Co, Ni, Ca, and Zn Concentrations in Soils and Water: The Significant Role of Hydrous Mn and Fe Oxides," *Adv. Chem.* 73:337–387 (1968).
148. Jenne, E. A. "Sediment Quality Criteria for Metals: II. Review of Methods for Quantitative Determination of Important Adsorbants and Sorbed Metals in Sediments," Battelle Report to U.S. EPA, Criteria and Standards Division, Washington, D.C. (1987).
149. Di Toro, D. M., J. D. Mahony, D. J. Hansen, K. J. Scott, M. B. Hicks, S. M. Mayr, and M. S. Redmond. "Toxicity of Cadmium in Sediments: The Role of Acid Volatile Sulfide," *Environ. Toxicol. Chem.* 9:1487–1502 (1990).
150. Lake, D. L., P. Kirk, and J. Lester. "Fractionation, Characterization, and Speciation of Heavy Metals in Sewage Sludge and Sludge-Amended Soils: A Review," *J. Environ. Qual.* 13:175–183 (1984).
151. Maher, W. A. "Evaluation of a Sequential Extraction Scheme to Study Associations of Trace Elements in Estuarine and Oceanic Sediments," *Bull. Environ. Contam. Toxicol.* 32:339 (1984).
152. O'Donnel, J. R., B. M. Kaplan, and H. E. Allen. "Bioavailability of Trace Metals in Natural Waters," Aquatic Toxicology and Hazard Assessment: 7th Symposium, STP 854 (Philadelphia: American Society for Testing and Materials, 1985) pp. 485–501.
153. Salomons, W., and U. Förstner. "Trace Metal Analysis on Polluted Sediment. II. Evaluation of Environmental Impact," *Environ. Technol. Lett.* 1:506–517 (1980).
154. Fiszman, M., W. C. Pfeiffer, and L. Drude de Lacerda. "Comparison of Methods Used for Extraction and Geochemical Distribution of Heavy Metals in Bottom Sediments from Sepetiba Bay," *Environ. Technol. Lett.* 5:567–575 (1984).
155. Malo, B. A. "Partial Extraction of Metals from Aquatic Sediments," *Environ. Sci. Technol.* 11:277–282 (1977).
156. Engler, R. M., J. Brannon, J. Rose, and G. Bigham. "A Practical Selective Extraction Procedure for Sediment Characterization," in *Chemistry of Marine Sediments* (Ann Arbor, MI: Ann Arbor Science, 1977), pp. 163–171.
157. Khalid, R.A., R. Gambrell, and W. Patrick, Jr. "Chemical Availability of Cadmium in Mississippi River Sediment," *J. Environ. Qual.* 10:523-528 (1981).
158. Tessier, A., D. Campbell, and M. Bisson. "Sequential Extraction Procedure for the Speciation of Particulate Trace Metals," *Anal. Chem.* 51:844–851 (1979).
159. Jackson, M. L. *Soil Chemical Analysis* (Englewood Cliffs, NJ: Prentice-Hall, 1958).
160. Andersen, F. O., and W. Helder. "Comparison of Oxygen Microgradients, Oxygen Flux Rates and Electron Transport System Activity in Coastal Marine Sediments," *Mar. Ecol. Progr. Ser.* 37:259–264 (1987).
161. Bott, T. L. "Benthic Community Metabolism in Four Temperate Stream Systems: An Inter-Biome Comparison and Evaluation of the River Continuum Concept," *Hydrobiologia* 123:3–45 (1985).

162. Urchin, G. G., and W. K. Ahlert. "*In Situ* Sediment Oxygen Demand Determination in the Passaic River (NJ) During the Late Summer/Early Fall 1983," *Water Res.* 19:1141–1144 (1985).
163. Haddock, J. D., P. F. Landrum, and J. P. Giesy. "Extraction Efficiency of Anthracene from Sediments," *Anal. Chem.* 55:1197–1200 (1983).
164. Sporstoel, S., N. Gjøs, and G. E. Carlberg. "Extraction Efficiencies for Organic Compounds Found in Aquatic Sediments," *Anal. Chim. Acta* 151:231–235 (1983).
165. Bailey, H. C., and D. H. Liu. "*Lumbriculus variegatus,* A Benthic Oligochaete, as a Bioassay Organism," in *Aquatic Toxicology*, STP 707 (Philadelphia: American Society for Testing and Materials, 1980), pp. 205–215.
166. Swartz, R. C., W. A. DeBen, J. K. Jones, J. O. Lamberson, and F. A. Cole. "Phoxocephalid Amphipod Bioassay for Hazard Assessment," 7th Symposium, STP 854, R. D. Cardwell, R. Purdy, and R. C. Bahner, Eds. (Philadelphia: American Society for Testing and Materials, 1985), pp. 284–307.
167. Nebeker, A. V., M. A. Cairns, J. H. Gakstatter, K. W. Malueg, G. S. Schuytema, and D. F. Krawczyk. "Biological Methods for Determining Toxicity of Contaminated Freshwater Sediments to Invertebrates," *Environ. Toxicol. Chem.* 3:617–630 (1984).
168. Carr, R. S., J. W. Williams, and C. T. B. Fragata. "Development and Evaluation of a Novel Marine Sediment Pore Water Toxicity Test with the Polychaete *Dinophilus gyrocilatus,*" *Environ. Toxicol. Chem.* 8:533–543 (1989).
169. Forstner, U., and G. T. W. Wittmann. *Metal Pollution in the Aquatic Environment* (New York: Springer-Verlag, 1979).
170. Schuytema, G. S., A. V. Nebeker, W. L. Griffis, and C. E. Miller. "Effects of Freezing on Toxicity of Sediments Contaminated with DDT and Endrin," *Environ. Toxicol. Chem.* 8:883–891 (1989).
171. Prater, B. L., and M. A. Anderson. "A 96-h Bioassay of Duluth and Superior Harbor Basins (Minnesota) Using *Hexagenia limbata, Asellus communis, Daphnia magna,* and *Pimphales promelas* as Test Organisms," *Bull. Environ. Contam. Toxicol.* 18:159–169 (1977).
172. Hoke, R. A., and B. L. Prater. "Relationship of Percent Mortality of Four Species of Aquatic Biota from 96-Hour Sediment Bioassays of Five Lake Michigan Harbors and Elutriate Chemistry of the Sediments," *Bull. Environ. Contam. Toxicol.* 25:394–399 (1980).
173. Muir, D. C. G., G. P. Rawn, B. E. Townsend, W. L. Lockhart, and R. Greenhalgh. "Bioconcentration of Cypermethrin, Deltamethrin, Fenvalerate and Permethrin by *Chironomus tentans* Larvae in Sediment and Water," *Environ. Toxicol. Chem.* 4:51–61 (1985).
174. Rutledge, P. A., and J. Fleeger. "Laboratory Studies on Core Sampling with Application to Subtidal Meiobenthos Collection," *Limnol. Oceanogr.* 33:274–280 (1988).
175. Rochon, R., and M. Chevalier. "Sediment Sampling and Preservation Methods for Dredging Projects," Conservation and Protection-Environment Canada, Quebec Region (1987).
176. Hulbert, M. H., and M. P. Brindle. Effects of Sample Handling on the Composition of Marine Sedimentary Pore Water," *Geol. Soc. Am. Bull.* 86:19–110 (1975).
177. Watson, P. G., P. Frickers, and C. Goodchild. "Spatial and Seasonal Variations in the Chemistry of Sediment Interstitial Waters in the Tamar Estuary," *Estuar Coast. Shelf Sci.* 21:105–119 (1985).

178. Aggett, J., and M. R. Kreigman. "Preservation of Arsenic (III) and Arsenic (V) in Samples of Sediment Interstitial Water," *Analyst* 112:153–157.
179. Adams, D. D. "Sediment Pore Water Sampling," in *CRC Handbook of Techniques for Aquatic Sediments Sampling*, A. Mudroch, and S. D. MacKnight, Eds. (Boca Raton, FL: CRC Press, 1991), pp. 131–169.
180. Manheim, F. T. "Interstitial Waters of Marine Sediment," in *Chemical Oceanography,* Vol. 6, J. P. Riley, and R. Chester, Eds. (New York: Academic Press, 1976), p. 115.
181. Kriukov, P. A., and F. T. Manheim. "Extraction and Investigative Techniques for Study of Interstitial Waters of Unconsolidated Sediments: A Review," in *The Dynamic Environment of the Ocean Floor,* K. A. Fanning, and F. T. Manheim, Eds. (Washington, D.C.: Lexington Books, 1982), p. 3–26.
182. Froelich, P. M., G. P. Klinkhammer, M. L. Bender, N. A. Luedtke, G. R. Heath, D. Cullen, P. Dauphin, D. Hammond, B. Hartmann, and V. Maynard. "Early Oxidation of Organic Matter in Pelagic Sediments of the Eastern Equatorial Atlantic: Suboxic Diagenesis," *Geochim. Cosmochim. Acta* 43:1075–1090 (1979).
183. Malcolm, S. J., P. J. Kershaw, M. B. Lovett, and B. R. Harvey. "The Interstitial Water Chemistry of ^{239,240}Pu and ^{241}Am in the Sediments of the Northeast Irish Sea," *Geochim. Cosmochim. Acta* 54:29 (1990).
184. Edmunds, W. M., and A. H. Bath. "Centrifuge Extraction and Chemical Analysis of Interstitial Waters," *Environ. Sci. Technol.* 10:467–472 (1976).
185. Bender, M., W. Martin, J. Hess, F. Sayles, L. Ball, and C. Lambert. "A Whole-Core Squeezer for Interfacial Pore-Water Sampling," *Limnol. Oceanogr.* 32:1214–1225 (1987).
186. Knezovich, J. P., and F. L. Harrison. "A New Method for Determining the Concentrations of Volatile Organic Compounds in Sediment Interstitial Water," *Bull. Environ. Contam. Toxicol.* 38:937–940 (1987).
187. Howes, B. L., J. W. H. Dacey, and S. G. Wakeham. "Effects of Sampling Technique on Measurements of Porewater Constituents in Salt Marsh Sediments," *Limnol. Oceanogr.* 30:221–227 (1985).
188. Bottomley, E. Z., and I. L. Bayly. "A Sediment Porewater Sampler Used in Root Zone Studies of the Submerged Macrophyte, *Myriophyllum spicatum,*" *Limnol. Oceanogr.* 29:671–673 (1984).
189. Hesslin, R. H. "An In Situ Sampler for Close Interval Pore Water Studies," *Limnol. Oceanogr.* 21:912–914 (1976).
190. Mayer, L. M. "Chemical Water Sampling in Lakes and Sediments with Dialysis Bags," *Limnol. Oceanogr.* 21:909–911 (1976).
191. Pittinger, C. A., V. C. Hand, J. A. Masters, and L. F. Davidson. "Interstitial Water Sampling in Ecotoxicological Testing: Partitioning of a Cationic Surfactant," *Toxicol. Haz. Assess.* 10:138–148 (1988).
192. Carignan, R., and R. J. Flett. "Postdepositional Mobility of Phosphorus in Lake Sediments," *Limnol. Oceanogr.* 26:361 (1981).
193. Skalski, C., R. Fisher, and G. A. Burton, Jr. "An *In Situ* Interstitial Water Toxicity Test Chamber," Abstr. Annu. Meet. Soc. Environ. Toxicol. Chem. P058, Arlington, VA, p. 132, (1990).
194. Bray, J. T., O. P. Bricker, and B. M. Troup. "Phosphate in Interstitial Waters of Anoxic Sediments: Oxidation Effects During Sampling Procedure," *Science* 180:1362–1364 (1973).

195. Lyons, W. B., J. Gaudette, and G. Smith. "Pore Water Sampling in Anoxic Carbonate Sediments: Oxidation Artifacts," *Nature* 277:48–49 (1979).
196. Word, J. Q., J. A. Ward, L. M. Franklin, V. I. Cullinan, and S. L. Kiesser. "Evaluation of the Equilibrium Partitioning Theory for Estimating the Toxicity of the Nonpolar Organic Compound DDT to the Sediment Dwelling Amphipod *Rhepoxynius abronius,*" Battelle/Marine Research Laboratory Report, Sequim, WA, Task 1, WA56 (1987).
197. Adams, D. D., D. A. Darby, and R. J. Young. "Selected Analytical Techniques for Characterizing the Metal Chemistry and Geology of Fine-Grained Sediments and Interstitial Water," in *Contaminants and Sediments*, Vol. 2, R. A. Baker, Ed. (Ann Arbor, MI: Ann Arbor Science, 1980), pp. 3–28.
198. Klinkhammer, G. P. "Early Diagenesis in Sediments from the Eastern Equatorial Pacific. II. Pore Water Metal Results. *Earth Planet. Sci. Lett.* 49:81–101 (1980).
199. Simon, N. S., M. M. Kennedy, and C. S. Massoni. "Evaluation and Use of a Diffusion Controlled Sampler for Determining Chemical and Dissolved Oxygen Gradients at the Sediment-Water Interface," *Hydrobiologia* 126:135–141 (1985).
200. Landrum, P. F., V. N. Tsymbal, M. K. Nelson, C. G. Ingersoll, D. C. Gossiax, G. A. Burton, and G. Sasson-Brickson. "Sediment Associated Contaminant Toxicity Assessment by Dilution Experiments," Abstr. Ann. Meet. Soc. Environ. Toxicol. Chem., Arlington, VA, No. 435, p. 107, (1990).
201. Giesy, J. P., C. J. Rosieu, R. L. Graney, and M. G. Henry. "Benthic Invertebrate Bioassays with Toxic Sediment and Pore Water," *Environ. Toxicol. Chem.* 9:233–248 (1990).
202. Dave, G. "Sediment Toxicity in Lakes Along the River Kolbäcksån, Central Sweden," Abstr. 5th Intern. Symp., The Interactions Between Sediment and Water, Uppsala, Sweden, No. 175 (1990).
203. Munawar, M., R. L. Thomas, H. Shear, P. McKee, and A. Mudroch. "An Overview of Sediment-Associated Contaminants and Their Bioassessment," Department of Fisheries and Oceans, Burlington, Ontario (1984).

CHAPTER 4

Ecosystem Assessment Using Estuarine and Marine Benthic Community Structure*

Robert J. Diaz

INTRODUCTION

This chapter is an introduction to the use of benthic community structure as a means of assessing ecosystem response to contaminated sediments in estuarine and marine ecosystems. Since contaminants that enter estuarine and marine ecosystems eventually bind to sediment particles and are deposited on the bottom,[1] emphasis on benthic organisms (bottom dwelling invertebrates and vertebrates) as a primary means of assessing ecosystem response is warranted. Of particular importance are the macrobenthic invertebrates because of their basic longevity, sedentary life styles, proximity to sediments, influence on sedimentary processes, and trophic importance (as reviewed in Diaz and Schaffner[2]).

Community Structure Concepts

The conceptual underpinning of community structure theory has always been central to the framework of ecological theory. Species diversity is perhaps the best known (and when expressed as a derived index, it is also the most controversial and least understood) community structure parameter. Other parameters include species composition (taxonomic identity, the simplest form of species diversity), relative

* Contribution No. 1719 of the Virginia Institute of Marine Science.

abundance (distribution of individuals among species), resemblance (similarity or dissimilarity of species or relative abundances), spatial structure or pattern, size structure (biomass properties), functional group composition, and trophic complexity (energy-flow pathways or ascendancy). What trajectory a community follows during its development (successional vs. deterministic and stocastic processes) and the sources of spatial and temporal variation have received the most attention from ecologists (see Diamond and Case[3] and Menge and Sutherland[4] as examples).

In the marine environment Petersen[5] was one of the first to identify community types. But it was the work of researchers like Sanders,[6] Copeland,[7] and Johnson[8] that focused attention on community structure as a means of assessing the impacts of pollutants and other types of disturbances in estuarine and marine environments. They proposed models to explain the characteristic values for the various community structure parameters in different communities. Pollution researchers then postulated that stress (toxic or enrichment) applied to a community would change, usually lowering, the value of community structure parameters and allow for the measurement of impacts. A large mass of literature has accumulated based on this postulate, but much of this literature demonstrates a lack of understanding of community dynamics and just how complex community structure can be.

Central to using community structure parameters as a means of testing ecological theories or assessing toxic impacts is the definition of community. Current concepts of community have grown to reflect an integrated, holistic view that combines all of the above-mentioned community structure parameters. For example, Mills[9] succinctly defined a community as a group of organisms in a particular environment, presumably interacting with each other and with the environment, and separable by means of ecological survey from other groups. Twenty years later Drake[10] defined a community as the ensemble of species in some area, whose limits are determined by the practical extent of energy flow, including indirect and cascading effects: or more simply, all species at all trophic levels at a particular location in a given habitat. Use of the term community, as defined by Drake,[10] then implies a high degree of complexity even in the simplest of habitats. Rarely are communities evaluated at this level in applied ecological studies. In general, assessments of the effects of pollution on an ecosystem are focused on only a very narrowly defined community or subset of a community and are usually based on the size of the organisms. For example, sampling gear and sieve size are generally matched to effectively collect a single size category of organisms (micro-, meio-, macro-, megabenthos). Often even these size-defined communities are then subsampled taxonomically (heterotrophic bacteria, nematodes, polychaetes, crabs, etc.).

Community Structure Applied to Assessment of Pollution

The application of community structure parameters to the assessment of pollution and toxicity effects mandates that the ecological framework surrounding the definition of community be considered. Even though it is necessary to adopt an operational (and possibly artificial) definition of the community to be studied, this definition must have an ecological basis.

All too often the species or habitats chosen to define the community are dictated by the resources available for sample processing and by the availability of taxonomic expertise. Often an overwhelming emphasis is placed on the importance of common species, and rare species are typically ignored or overlooked. This is not a particularly ecologically realistic approach, since most species in a community (as defined by Mills,[9] Drake,[10] and others) are rare. While some analytical approaches are improved by dropping rare species (see later section), unbiased evaluation of a community structure needs to include information contained in rarer species.

The place of community structure in impact assessment hinges on the assumptions that

1. It is possible to measure the intrinsic properties and magnitude of any community structure parameter used, including an estimation of natural variation;
2. A cause and effect relationship exists between community structure and the impact; although cause and effect may not need to be proven in every study.[11]

If, for the purposes of impact assessment, these assumptions are met, the final decision as to the "meaning" of collected data has to be ascribed by a management strategy. Basically, how far from its undisturbed natural value must a community structure measure be in order to be particularly important and set in motion remedial or punitive action? These types of decisions are too important to make without data from properly thought-out and executed field designs.

This chapter attempts to present the strengths and weaknesses of a community structure approach to the assessment of contaminated sediment, relative to the ecological framework that defines communities. Basic guidelines for the appropriate use of community structure data in impact assessment are presented. The references selected are intended to be illustrative and not a complete review of the literature.

COMMUNITY STRUCTURE PARAMETERS

Species Composition

Implicit in the definition of community is the individual identity of the species that compose the community. Unfortunately, their identity can only be determined through classical taxonomic studies that require specialists and much time and that managers (who so love the simplicity of indices) are generally unwilling to support. Poole[12] argued that the number of species in a defined community is the simplest and likely the only truly objective measure of community diversity. It is certainly the least ambiguous description of community structure, but by no means the simplest to determine (see Hutchinson,[13] Sanders,[6] Woodwell and Smith,[14] and Hurlbert[15] for examples).

Evenness of Individuals Among Species

How individuals in a community are distributed among the species is an integral part of a community structure. Next to the number of species, the evenness in the distribution of individuals among the species constitutes a basic characteristic of a community. Minimum evenness occurs when one species numerically dominates a community, with all other species being represented by few individuals. Maximum evenness occurs when all species have the same abundance.

Diversity Indices

Diversity is the most commonly used community structure measure, either as "number of species" (S) or as a composite index (such as H′). Diversity indices are formulated to combine two independent components of community structure, the number of species (or species richness) and the relative abundance of individuals among species (or evenness), into a single derived variable.[16] Since their first formulation, diversity indices have been used to explain and simplify the community concept. The widely used information theory-based indices were first used in community studies by Margalef.[17] The application of diversity to applied studies followed with the work of Wilhm and Dorris.[18]

Implicit in the use of diversity for assessing disturbance is that a given community has a characteristic level of diversity. While commonly used to measure change in a community structure, diversity indices tend to be ambiguous, with no clearly definable relationship between the community and its functional processes. Green[19] very succinctly summarizes the strengths and weaknesses of diversity indices. Green's cautions on diversity must be given serious consideration prior to the use of any diversity index. The selection of any particular diversity index needs to be justified on the empirical value of the index for the specific study purpose, and not on theoretical grounds.[19] Because of the sensitivity of any diversity index to sample size, sieve size, and taxonomic level, it makes little sense to argue one index over another on theoretical grounds. Sample diversity is a biased estimator of community diversity, which is not a serious problem for comparisons that are made between stations within a single study.[19] A summary of popular diversity indices used in assessing impacts is given by Perkins.[20]

Resemblance Indices

Similarity and dissimilarity indices are similar to diversity indices in that most formulations use the same components of community structure (species richness and evenness for quantitative, and only species richness for qualitative indices). However, resemblance exists only as a comparative measure between communities or among samples from a single community. A single community or a single sample from a community has no intrinsic resemblance. As the term resemblance implies, one sample resembles another. The most familiar resemblance measure is the Pearson product moment correlation coefficient.

There are many formulations for resemblance (each of which makes certain assumptions about the species data) used in a wide variety of ways.[16,21] A resemblance matrix in itself typically has limited interpretive value because of the large number of stations compared and the complicated interrelationships of species associations. They are most often used by multivariate techniques for defining community pattern,[22-25] but can be used as a statistic for assessing the overall strength of species associations.[26] Spatial and niche overlap studies also use resemblance to quantify patterns.[27] In applied community studies, resemblance was initially used to characterize change along gradients.[28,29]

Biomass and Trophic Levels

Biomass is an integral part of community structure, being the basic medium for energy flow. At the species level, biomass contains a great deal of information about community structure that is directly related to community processes. The distribution of biomass among species may be a conservative property of a community[30,31] and diagnostic of the type of disturbance to which a community is exposed. Pearson and Rosenberg[32] found that organic enrichment predictably changed the size distribution of biomass. Changes in the size distribution of biomass between[33] and within[34] species, due to pollution, have been documented. There is also evidence that growth and productivity can be altered by pollution stress.[35,36]

The quantitative relationships between energy flow, biomass size distribution, and stress on communities still need to be clearly established. Attempts to link pollution impacts to changes in the functional aspects of communities (in particular, energetics and trophic levels) is useful to resource-yield-oriented management strategies. To address these relationships, Luckenbach et al.[37] evaluated benthic biomass as a total resource for higher trophic levels, and environmental stress. Elliott and Taylor[38] assessed benthic energy flow within an ecosystem and related the effects of both environmental variables and pollution to energy flow.

Trophic complexity of a community expresses the levels at which energy flows and recycles through an ecosystem. This complexity is directly related to an ecosystem's overhead and ascendancy. Taken together they constitute a systems capacity for development.[39] As an ecosystem is stressed, there could be a restructuring of its trophic base, which may result in a lower capacity for community development. Basically, stressed systems tend to have simpler energy-flow pathways,[40] which may be manifested through an undesirable series of changes in community structure.

Life History and Functional Characteristics

Typically, life history characters of estuarine and marine benthic species include

• Life span	Grassle and Grassle[41]
• Maximum body size	Schwinghamer[30]
• Reproduction	Fredette and Diaz[42]
• Feeding mode	Fauchald and Jumars[43]
• Defecation mode	Fauchald and Jumars[43]
• Mobility	Fauchald and Jumars[43]
• Dwelling structure	Schaffner[26]
• Sediment processing	Diaz and Schaffner[2]
• Burrowing depth	Pearson and Rosenberg[44]
• Recruitment mode	Grassle and Grassle[41]
• Recruitment time	Diaz[45]
• Habitat preference	Schaffner and Diaz[46]
• Growth	Bansa and Mosher[47]
• Productivity	Diaz and Schaffner[2]
• Genetic plasticity	Grassle and Grassle[41]

Communities are composed of different species, each of which has a unique set of life history characteristics. For this reason species that compose communities are not uniformly distributed across the surface of the sediment or with depth in the sediment. The way species utilize their habitat reflects the plasticity and overlap of life history and functional characteristics.[2] Species typically range from generalists or opportunists, occurring over a wide range of habitats, to specialists or equilibrium species, being limited to narrower ranges.[29] Opportunistic species were found to be common in stressed ecosystems, while equilibrium species were typically absent.[29,41,48] Odum[49] contrasted community structure, community function, and species life history characteristics, for the development of ecosystems. Additionally, he presented several models to explain the general nature of community development and why there is a succession of species types as a community develops or is disturbed (naturally or anthropogenically).

Models based on life history and functional characteristics for predicting the response of marine benthos to stress have been developed for organic enrichment[32] and physical disturbance.[50]

Indicator Species

The concept of indicator species developed early in our attempts to assess the environmental impacts of pollution[51,52] and is founded on the premise that an organism's presence or absence in a particular disturbed habitat is an indicator of the habitat quality and, by extension, the health of the total ecosystem. Because organisms integrate their response through time and react to all synergistic and antagonistic effects of combined pollutants and physical stresses,[53] they do provide convenient full-time monitors of total habitat quality.

The first, and probably best, accommodation of any species to toxic stress is through taking advantage of, or varying certain of its life history and not gaining tolerance to, stress.[54] In using an individual species population as an indicator of pollution, it is then necessary to have a good understanding of their life history and functional characteristics. Through their quantitative and qualitative presence or absence, species carry information on the actual responses of the ecosystem, rather than predicted responses from physical measurements or bioassays. These latter predicted responses may not model the real world as well as needed. Problems arise, however, in the clear interpretation and extrapolation of a single indicator's response because of

1. The integrating nature of the species response to all variables (pollutants and natural)
2. The representativeness of the particular species response relative to the total community response from which it came
3. The population-specific (intraspecific) variation of species responses to disturbance or stress (i.e., *Capitella capitata,* the sibling species of Grassle and Grassle[48])

It must be recognized that indicator species have a limited ability to ascribe cause and effect from changes observed in applied pollution studies. Indicator species are only part of a large array of variables that should be considered in environmental assessment.

Recent trends in indicator species have been away from population-level responses, to the examination of individual organism, or cellular, response at the physiological level: the biomarker concept.[55] This refocusing on the organism requires that increased attention be paid to systematic studies. As aptly pointed out by Grassle and Grassle,[56] the species name is the only retrievable way of entering information (physiological, toxicological, population) about an organism into the scientific literature. Misidentification of species means that efforts expended gaining such information have been totally wasted.

APPROACHES TO ASSESSMENT OF ESTUARINE AND MARINE ECOSYSTEMS USING COMMUNITY STRUCTURE

Given the range and complexity of benthic communities and pollutants, it is not possible to give a shopping list of techniques to apply. There are many innovative applications of community structure in the literature, which need to be researched for transferability to a specific need. An excellent example of the diversity of approaches that can be applied is the work of the Group of Experts on the Effects of Pollutants (GEEP), which studied the same pollution gradient, complemented by experimental exposures, at all levels of biological organization, from the cell to the community.[57] In this section, I will present examples of how community structure parameters have been used in pollution studies.

Indicator Species

The work of Grassle and Grassle[56] with the capitellid polychaete *Mediomastus ambiseta* is a good example of how a single species can be used as a reliable indicator of environmental stress. The effects of No. 2 fuel oil and organic enrichment were evaluated in laboratory mesocosms and both impacted and unimpacted field populations. Their laboratory studies demonstrated two basic responses of *Mediomastus ambiseta* populations: (1) populations of this species were severely reduced by oil that reached the sediment, and recovery did not begin within one year after oil additions ceased; (2) this species responded rapidly to increased food supply by the enhanced recruitment and survival of postlarvae, rapid growth to maximum size, increased fecundity, and elevated survival of over-wintering individuals. Knowledge of the life history characteristics of *Mediomastus ambiseta* helped to explain the observed dynamics of the field populations. When densities of *Mediomastus ambiseta* increased rapidly under disturbance conditions (i.e., after an oil spill, around a sewage outfall, following natural defaunation), the primary cause was a response to increased food quality and quantity. Even though No. 2 fuel oil was found to reduce populations, populations increased when both oil and organic enrichment occurred together. Grassle and Grassle[56] suggest that the positive effects of higher levels of food temporarily outweighed the negative effects of the oil. When the source of enrichment persists, as around a sewage outfall, the dominance of *Mediomastus ambiseta* may be expected to persist. Where the source declines, as following an oil spill, populations fluctuate and also decline through time.

The work of Grassle and Grassle[56] points out several features of an indicator species that must be carefully evaluated. (1) At what point does a species become a predictor of disturbance from pollution? (2) Is increased abundance or absence the key response? (3) Are the population changes observed out of the natural range of variation for the species? An interesting feature of *Mediomastus ambiseta* as an indicator is that it started out as the harbinger of acute disturbance or pollution.[56] Now, in at least parts of Narragansett Bay and Buzzards Bay, its populations have gradually increased over the last several decades to the point that it is a dominant species, possibly in response to a gradual eutrophication.[56] A sudden reduction in abundance of *Mediomastus ambiseta* could now be the most sensitive indicators of acute disturbance or impact from pollution.

Diversity

The best use of diversity indices is to combine them with other uni- or multivariate methods to increase the robustness of an analytical approach. One of the biggest problems in relying solely on a diversity index to assess change is that sampling is often inadequate to characterize true community diversity. Sanders[6] aptly pointed this problem out in the presentation of his rarefaction method, which dealt with the rate at which species were added to a collection as the area sampled of a habitat increased. If used, diversity should be treated with all the cautions proposed by Green[19] and in a manner similar to Boesch.[58] Boesch[58] thoroughly described the patterns of diversity for a geographical area of interest, by major habitat type. This set the stage for the informed use of diversity in assessing change by defining the expected range of diversity. Boesch[58] pointed out that diversity can be a valuable indicator of pollution effects, but must be evaluated in the context of the considerable variation that typifies most systems both between and within habitats.

To determine the degree to which communities were responding to pollution, Diaz[59] combined an ordination of physical parameters with diversity (H′) by plotting H′ in the ordination space of the first two axes. This approach related community level changes to pollution levels. Also, this approach aided in the interpretation of cluster analysis by providing an independent separation of the effects of natural environmental variables from pollution.

Species-Abundance-Biomass

The interrelationships of the total number of species (S), their total abundance (A), and their total biomass (B), for the most part, predictably vary in response to organic enrichment[32,60] and, possibly, can be used to assess other sediment contaminants.[33] The values of S, A, and B can be plotted along a gradient of pollution or used to form ratios of A/S, the abundance ratio, and B/A, the size ratio.[61,62] The attractiveness of these community statistics is that they are easy to comprehend. The A/S ratio is maximized in highly enriched areas, and the B/A ratio is maximized in unenriched areas.[62]

Warwick[33] also developed a graphical method, the abundance-biomass comparison (ABC) curve, for assessing pollution stress using S, A, and B. He proposed that if both cumulative percent B and A were plotted against S by rank on a log scale, unpolluted sites would have the B line above the A line, with the reverse situation at polluted sites. At moderately polluted sites, the two lines would overlap. The advantage of the ABC curve over the A/S and B/A curves is that an entire pollution gradient does not need to be sampled. ABC curves can be generated from a single sample.

The exact shape of these curves varies from site to site (see Warwick[33] for examples) and requires good baseline data for proper interpretation. Natural interspecific and intraspecific station variation accounts for a certain amount of the change in the shape of the curves. How useful these curves will be in assessing the effects of toxic compounds still needs to be evaluated. They do possess the very desirable characteristic of summarizing community structure changes in a way that is less ambiguous than diversity indices.

Multivariate Analysis

The strength of multivariate analytical techniques in assessing sediment contamination is that they can provide spatial and temporal descriptions of the patterns between stations, based on community structure data. These biologically delineated patterns can then be compared to spatial or temporal patterns in pollutant distribution or physical environmental parameters.[25] As with any correlative statistical approach, however, the tendency to ascribe cause-effect should be avoided. Smith et al.[25,63] provide an excellent introduction to the use and interpretation of multivariate techniques in assessing pollution impacts on benthic communities. They conclude that the use of several multivariate techniques, combined with selected univariate techniques, provides more sensitivity and power than is obtained from individual use of either multivariate or univariate analyses. In particular, Smith et al.[25] recommend the use of cluster analysis[22] and ordination[23] to describe community pattern, and discriminant analysis[64,65] to describe correlation patterns among environmental variables.

Cluster and ordination analyses provide a distinct description of community pattern, using all of the species information, which can be interpreted at various spatial or temporal scales. Diversity indices also use all of the species information, but they condense it into a single number, and individual species patterns are lost. Descriptions of community pattern based on cluster or ordination permit evaluation of the strength of environmental correlations in a much more detailed way than diversity indices. If the data are generated from a properly designed study,[19] these analyses can identify clear limits to the environmental correlations and can lead to the generation of hypotheses for further resolution of sediment contaminant issues and the development of management strategies.

Trophic Level Analysis

Growth and secondary production are the community processes that are most easily understood in a management strategy. Because of the trophic link between benthic invertebrates and many fisheries species,[2] this may be the most relevant level to address the impacts of sediment contaminants. Trophic level studies of invertebrates, however, are very difficult to conduct, and most effort has been spent on aspects of community structure that are easier to measure. For bacteria and other microheterotroph communities, the reverse of this situation is the case, where growth and production studies are common.[66,67]

Elliott and McLusky[36] and Gray[68] proposed the use of invertebrate secondary production as a means of assessing sublethal effects of contaminated sediments. However, in areas where organic enrichment[32] or other nontoxic stress occur,[69] production is likely to increase and offset any negative effect of sediment contamination. Therefore, a change in productivity from what is expected in a particular habitat may signal that a problem exists.

Growth and annual production of the tellinid bivalve *Macoma balthica* was evaluated by Elliott and McLusky[36] in the heavy-metal–contaminated Forth estuary. They found both growth and production to be mainly influenced by sediment physical and chemical factors and tidal height. Highest production was found in areas with muddy-fine sand substrates at midtide levels. While there was a great deal of variation in the production at various stations within the Forth estuary, comparisons of *Macoma balthica* production to reference areas suggest the Forth estuary population is under stress.[36] The reduced productivity in the Forth estuary cannot be completely explained by the environmental variables measured, but may be caused by other abiotic factors, including levels of sediment contamination.

STRATEGIES AND GUIDELINES FOR USING COMMUNITY STRUCTURE IN ASSESSING POLLUTION IMPACTS

Management Strategy

The key to successful use of community structure in assessing toxic impacts is to clearly define the problem area and then delineate the community for which a management strategy should be developed. This is not a simple task, since most polluted areas have a variety of toxic compounds in the sediments, which are delivered to the environment in a variety of ways and at varying concentrations through time.

An approach to the definition of the problem, similar to that taken by Elliott and O'Reilly[70] with marine benthic communities or with fish communities,[71] needs to be considered. These authors clearly define the communities of concern and then attempt to characterize levels of natural variation within the communities, through appropriately scaled sampling of the communities. This places them in a good position to make informed decisions about what may be causing change (either natural variation or pollution impacts) and the implications of any change to the ecosystem or fisheries.

Chapman et al.[72] and Chapman,[73] after having evaluated field impacts of pollution by solely evaluating community data, bioassay data, or sediment contamination data, developed an approach known as the sediment quality triad (see Chapter 14). The triad approach incorporates synoptically collected data from all three of these disciplines, which are then evaluated for correspondence of trends observed at the site in question.

As pointed out by Mahon and Smith[71] and earlier in this chapter it is necessary to adopt an operational approach to defining a community for management purposes, but it is very important to consider the ecological basis for community definition. After all, it is the combination of natural and anthropogenic factors that continually shape communities in space and through time. Any attempt to understand what in particular is shaping a community requires an understanding of how these factors work in concert on the community. A nebulously defined community or a lack of comprehension as to how a community functions can only lead to nebulous management decisions.

Field Designs

The regular long-term (greater than one year) monitoring of a community is superior to aperiodic or short-term sampling, because the concentration of pollutants and intensity of other stressors that influence communities vary with time. Communities will likely have a threshold response to a stressor or may show a delayed response because of some seasonal element, such as recruitment or food availability or temperature. Long-term monitoring, however, is often impractical for use by managers because of time or funding constraints.

Prior to designing any field study basic information is needed on

1. Where in the environment will (or have) the toxic compounds settled
2. Which aspect of ecosystem function will be impacted
3. A working definition of "community" for assessing impact.

After consideration is given to these three points, a coherent plan of action can be developed. Before planning actually begins, it is a good idea to have a clear statement of the objectives. A list of questions the study is intended to answer or hypotheses about the expected causal effects would be a good starting point. An important factor, often overlooked in most studies, is the magnitude of change from "normal", for any of the parameters to be measured, that will be particularly important to the objectives of the study. Basically, you have to consider the size of the effect and form some judgement as to its importance. Statistically, the magnitude of change considered important should match the minimum detectable difference used for estimating the number of replicates. This will insure the execution of the most statistically powerful sampling design.

The design of an effective field study is not simple and should follow good experimental design principles, such as those of Green[79] or Cochran.[74] Additional issues that need to be considered include the taxonomic ability to resolve species to the level where their change is important to the ecosystem and cost efficiency of the sampling design. Design features that need special attention in impact studies are

1. The duration and extent of impact. The simplest is a one-time event that effects all members of a population equally. In field situations, however, the level and duration of exposure, in some way, varies from individual to individual. It may not be possible to do any detailed measurement on individuals or individual samples, so we simply have to compare impacted and unimpacted, or group data, roughly by equal impact level. In field studies it is best to have replicates for each level of impact that can be discerned, because there is an ever present danger in any field study that some source of bias that

affects comparisons has not been rendered unimportant (statistically accounted for or randomized) by the design.

2. Reference or unimpacted areas. After identification of the impacted area, suitable comparison areas are needed to judge the extent of impact. The term "control area" should be avoided in describing these areas because it implies an ability to remove bias. Reference areas should be subject to all forces (that are not themselves possible consequences of the impact) known to affect the impacted area,
3. Quality control or assurance. In addition to obvious factors such as station location and taxonomic verification, quality control should be extended to all parameters measured.
4. Timing of Measurements. Timing generally makes or breaks a study when pollutants are expected to create some response that lasts into the future. Selection of collection dates should then focus on the original questions asked and expected recovery rate of the defined community being sampled.

Statistical Considerations

Simply because you can measure something does not mean it will be good for assessing environmental quality. This is the first caveat to consider in designing a field or laboratory study for assessing sediment contamination. The second caveat is do not measure anything unless you already know how the data will be analyzed. Statistical preplanning is essential and can make or break a study. Clarke and Green[75] provide an excellent summary of what must be considered in a biological-effects study. They list six main statistical stages:

1. Purpose: reason for survey, and definition of hypotheses and expected endpoints
2. Survey design: criteria for selection of samples and replication levels
3. Preprocessing: initial examination and possible transformation of data for conformity with statistical test assumptions
4. Tests of null hypotheses: of "no biological differences" between sites
5. Descriptive and explanatory analyses: displaying the relationships between responses at each site and relation of those changes to contaminants
6. Retrospective assessment: of the sensitivity of various response measures and analyses employed as part of the continuous cycle of improving subsequent designs.

If sufficient baseline data exist, it may be possible to streamline a sampling design by developing a sequential decision plan (see Jackson and Resh 1988[76]) that will eventually do two things:

1. Reduce the sampling effort, thereby reducing cost and time to an answer
2. Allow for refinement of the management strategy by producing data in a more timely manner.

If a sequential decision plan philosophy is adopted, care must be taken not to sacrifice statistical power[77] for expedience.

SUMMARY

Limitations and Strengths

Because benthos are such good integrators of all the environmental factors to which they are exposed, more then one community structure parameter needs to be measured in order to extract the effects of sediment-associated pollutants. Thus, for community structure to be meaningfully applied in impact assessment, more then one aspect must be evaluated. The fact remains that no single community structure parameter can reliably be used to predict an ecological event or response to a toxicant. The main reason being that community structure does respond predictably to variation in ecological processes (disturbance, competition, predation), but the importance of ecological processes in shaping the community varies in response to the levels of environmental stress.

Diversity alone is weak for assessing any impact or ecosystem change; but when combined with life history and multivariate analyses, diversity can provide information on how community structure changes. Less cryptic presentation of community structure data that are easily interpreted may be best, such as the abundance-biomass comparison (ABC) curves of Warwick.[33]

The size of the study area and the duration of the study limit what can be seen of a community.[78] However, when different methods and conceptual approaches are combined, the view one gets of the ecosystem's response to pollutants is broadest. The biggest problem in looking at pollution effects, at the management level, is that changes in community structure parameters have little connection with a common driving management concern: commercial species yield. Environmental scientists need to develop assessment strategies that rely on testable hypotheses with quantitative data.

Future Directions

For community structure to be of practical use in assessing toxic impacts links need to be established between

1. Management goals and the definition of community
2. Aspects of community structure measured and community function
3. How pollutants alter this function
4. Characteristic values of community structure in different communities from which change is measured.

Once established, these links between ecosystem community structure and pollutants will lead us to the development of long-term risk assessment.[79,80]

Despite all the problems associated with community structure as a means of assessing toxic effects, it still provides the best indication, when data are properly planned and collected, of the consequences of pollution for management purposes.

REFERENCES

1. Olsen, C. R., N. H. Cutshall, and I. L. Larsen. "Pollutant-Particle Associations and Dynamics in Coastal Marine Environments: A Review," *Mar. Chem.* 11:501–533 (1982).
2. Diaz, R. J., and L. C. Schaffner. "The Functional Role of Estuarine Benthos," in *Perspectives on the Chesapeake Bay, 1990. Advances in Estuarine Sciences,* M. Haire, and E. C. Krome, Eds. (Gloucester Pt., VA: Chesapeake Research Consortium, Rpt. No. CBP/TRS41/90, 1990), pp. 25–56.
3. Diamond, J., and T. J. Case, Eds. *Community Ecology* (New York: Harper and Row, 1986), p. 665.
4. Menge, B. A., and J. P. Sutherland. "Community Regulation: Variation in Disturbance, Competition, and Predation in Relation to Environmental Stress and Recruitment," *Am. Nat.*130:730–757 (1987).
5. Peterson, C. G. "On the Animal Communities of the Sea Bottom in the Skagerrak, the Christiania Fjord and the Danish waters," *Rep. Dan. Biol. Stat.* 23:3–28 (1915).
6. Sanders, H. L. "Marine Benthic Diversity: A Comparative Study," *Am. Nat.* 102:243–282 (1968).
7. Copeland, B. J. "Estuarine Classification and Responses to Disturbances," *Trans. Am. Fish. Soc.* 99:826–835 (1970).
8. Johnson, R. G. "Conceptual Models of Benthic Marine Communities," in *Models in Paleobiology,* T. J. M. Schopf, Ed. (San Francisco: Freeman and Cooper, 1972), pp. 149–159.
9. Mills, E. L. "The Community Concept in Marine Zoology: A Review," *J. Fish. Res. Bd. Can.* 26:1415–1428 (1969).
10. Drake, J. A. "Communities as Assembled Structures: Do Rules Govern Pattern?" *Trends Ecol. Evol.* 5:159–164 (1990).
11. Underwood, A. J., and C. H. Peterson. "Towards an Ecological Framework for Investigating Pollution," *Mar. Ecol. Prog. Ser.* 46:227–234 (1988).
12. Poole, R. W. *An Introduction to Quantitative Ecology* (New York: McGraw-Hill, 1974), p. 532.
13. Hutchinson, G. E. "Homage to Santa Rosalia or Why are There so Many Kinds of Animals?," *Am. Nat.* 93:145–159 (1959).
14. Woodwell, G. M., and H. H. Smith. "Diversity and Stability in Ecological Systems," *Brookhaven Symposia in Biology No. 22,* (New York: Upton, 1969), p. 264.
15. Hurlbert, S. H. "The Nonconcept of Species Diversity: A Critique and Alternative Parameters," *Ecology* 52:577–586 (1971).
16. Krebs, C. J. *Ecological Methodology.* (New York: Harper and Row, 1989), p. 654.
17. Margalef, R. "Information Theory in Ecology," *Gen. Syst.* 3:36–71 (1958).
18. Wilhm, J. L., and T. C. Dorris. "Biological Parameters for Water Quality Criteria," *Bioscience* 18:447–481 (1968).
19. Green, R. H. *Sampling Design and Statistical Methods for Environmental Biologists* (New York: Wiley, 1979), p. 257.
20. Perkins, J. L. "Bioassay Evaluation of Diversity and Community Comparison Indexes," *J. Water Poll. Cont. Fed.* 55:522–530 (1983).
21. Bloom, S. A. "Similarity Indices in Community Studies: Potential Pitfalls," *Mar. Ecol. Prog. Ser.* 5:125–128 (1981).

22. Boesch, D. F. "Application of Numerical Classification in Ecological Investigations of Water Pollution," U.S. Environmental Protection Agency, Ecol. Res. Ser. EPA-600/3-77-033 (1977).
23. Gauch, H. G. *Multivariate Analysis in Community Ecology* (New York: Cambridge University Press, 1982), p. 298.
24. Jongman, R. H., C. J. F ter Braak, and O. F. R. van Tongeren. *Data Analysis in Community and Landscape Ecology* (The Netherlands: Pudoc Wageningen, 1987), p. 299.
25. Smith, R. W., B. B. Bernstein, and R. L. Cimberg. "Community Environmental Relationships in the Benthos: Applications of Multivariate Analytical Techniques," in *Marine Organisms as Indicators,* D. F. Soule. and G. S. Kleppel, Eds. (New York: Springer-Verlag, 1988), pp. 247–326.
26. Schaffner, L. C. "Small-Scale Organism Distributions and Patterns of Species Diversity: Evidence for Positive Interactions in an Estuarine Benthic Community," *Mar. Ecol. Prog. Ser.* 61:107–117 (1990).
27. Hurlbert, S. H. "The Measurement of Niche Overlap and Some Relatives," *Ecology* 59:67–77 (1978).
28. Terborgh, J. "Distribution on Environmental Gradients: Theory and a Preliminary Interpretation of Distributional Patterns in the Avifauna of the Cordillera Viacabamba, Peru," *Ecology* 52:23–40 (1971).
29. Boesch, D. F. "A New Look at the Zonation of Benthos Along the Estuarine Gradient," in *Ecology of Marine Benthos,* B. C. Coull, Ed. (Columbia, SC: University of South Carolina Press, 1977), pp. 245–266.
30. Schwinghamer, P. "Generating Ecological Hypotheses from Biomass Spectra Using Causal Analysis: A Benthic Example," *Mar. Ecol. Prog. Ser.* 13:151–166 (1983).
31. Schwinghamer, P. "Influence of Pollution Along a Natural Gradient and in a Mesocosm Experiment on Biomass-Size Spectra of Benthic Communities," *Mar. Ecol. Prog. Ser.* 46:199–206 (1988).
32. Pearson, T. H., and R. Rosenberg. "Macrobenthic Succession in Relation to Organic Enrichment and Pollution of the Marine Environment," *Oceangr. Mar. Biol. Ann. Rev.* 16:229–311 (1978).
33. Warwick, R. M. "A New Method for Detecting Pollution Effects on Marine Macrobenthos Communities," *Mar. Biol.* 92:557–562 (1986).
34. Dauer, D. M., and W. G. Connor. "Effects of Moderate Sewage Input on Benthic Polychaete Populations," *Estuar. Coast. Mar. Sci.* 10:335–346 (1980).
35. Warwick, R. M., I. R. Joint, and P. J. Radford. "Secondary Production of the Benthos in an Estuarine Environment," in *Ecological Processes in Coastal Environments,* R. L. Jeffries, and A. J. Davy, Eds. (Oxford: Blackwell Scientific, 1979), pp. 429–450.
36. Elliott, M., and D. S. McLusky. "Invertebrate Production Ecology in Relation to Estuarine Quality Management," in *Estuarine Management and Quality Assessment,* W. Halcrow, and J. G. Wilson, Eds. (New York: Plenum Press, 1985), pp. 85–104.
37. Luckenbach, M. W., R. J. Diaz, E. C. Zobrist, and C. H. Hutton. "Evaluation of the Benthic Resource Value of Impounded and Non-Impounded Tidal Creeks in Virginia, USA," *Ocean Shore Manage.* 14:35–50 (1990).
38. Elliott, M., and C. J. L. Taylor. "The Production Ecology of the Subtidal Benthos of the Forth Estuary, Scotland," *Sci. Mar.* 53:531–541 (1989).
39. Baird, D., and R. E. Ulanowicz. "The Seasonal Dynamics of the Chesapeake Bay Ecosystem," *Ecol. Monogr.* 59:329–364 (1989).

40. Odum, H. T. *Environment, Power, and Society* (New York: Wiley-Interscience, 1971), p. 331.
41. Grassle, J. F., and J. P. Grassle. "Opportunistic Life Histories and Genetic Systems in Marine Benthic Polychaetes," *J. Mar. Res.* 32:253–284 (1974).
42. Fredette, T. J., and R. J. Diaz. "Life History of *Gammarus mucronatus* Say (Amphipoda: Gammaridae) in Warm Temperate Estuarine Habitats, York River, Virginia," *J. Crust. Biol.* 6:57–78 (1986).
43. Fauchald, K., and P. A. Jumars. "The Diet of Worms: A Study of Polychaete Feeding Guilds," *Oceanogr. Mar. Biol. Ann. Rev.* 17:193–284 (1979).
44. Pearson, T. H., and R. Rosenberg. "Feast and Famine: Structuring Factors in Marine Benthic Communities," in *Organization of Communities, Past and Present. 27th Symposium British Ecological Society, Aberystwyth, 1986,* J. H. R. Gee, and P. S. Giller, Eds. (Oxford: Blackwell Scientific, 1987), pp. 373–395.
45. Diaz, R. J. "Short Term Dynamics of the Dominant Annelids in a Polyhaline Temperate Estuary," *Hydrobiologia* 115:153–158 (1984).
46. Schaffner, L. C., and R. J. Diaz. "Disribution and Abundance of Overwintering Blue Crabs, *Callinectes sapidus,* in the Lower Chesapeake Bay," *Estuaries* 11:68–72 (1988).
47. Bansa, K., and S. Mosher. "Adult Body Mass and Annual Production/Biomass Relationships of Field Populations," *Ecol. Monogr.* 50:353–379 (1980).
48. Grassle, J. P., and J. F. Grassle. "Sibling Species in the Marine Pollution Indicator *Capitella* (Polychaeta)," *Science* 192:567–569 (1976).
49. Odum, E. P. "The Strategy of Ecosystem Development," *Science* 164:262–270 (1969).
50. Rhoads, D. C., and J. D. Germano. "Interpreting Long-Term Changes in Benthic Community Structure: A New Protocol," *Hydrobiologia* 142:291–308 (1986).
51. Thomas, W. A., G. Goldstein, and W. H. Wilcox, Eds. *Biological Indicators of Environmental Quality* (Ann Arbor, MI: Ann Arbor Science, 1974), p. 254.
52. Soule, D. F., and G. S. Kleppel. *Marine Organisms as Indicators.* (New York: Springer-Verlag, 1988), p. 342.
53. Wass, M. L. "Biological and Physiological Basis of Indicator Organisms and Communities. Section II — Indicators of Pollution," in *Pollution and Marine Ecology,* T. A. Olson, and F. J. Burgess, Eds. (New York: Interscience, 1967), pp. 271–283.
54. Gray, J. S. "Why Do Ecological Monitoring?" *Mar. Pollut. Bull.* 11:62–65 (1980).
55. McCarthy J. F. and L. R. Shugart, Eds., Biomarkers of Environmental Contamination (Boca Raton: Lewis Publ., 1990), p. 457.
56. Grassle, J. P., and J. F. Grassle. "The Utility of Studying the Effects of Pollutants on Single Species Populations in Benthos of Mesocosms and Coastal Ecosystems," in *Concepts in Marine Pollution Measurements,* H. H. White, Ed. (College Park MD University of Maryland Sea Grant, 1984), UM-SG-TS-84-03, pp. 621–642.
57. Bayne, B. L., R. F. Addison, J. M. Cappuzzo, K. R. Clarke, J. S. Gray, M. N. Moore, and R. M. Warwick. "An Overview of the GEEP Workshop," *Mar. Ecol. Prog. Ser.* 46:235–243 (1988).
58. Boesch, D. F. "Species Diversity of Marine Macrobenthos in the Virginia Area," *Chesapeake Sci.* 13:206–211 (1972).
59. Diaz, R. J. "Pollution and Tidal Benthic Communities of the James River Estuary, Virginia," *Hydrobiologia* 180:195–211 (1989).
60. Gray, J. S., and T. H. Pearson. "Objective Selection of Sensitive Species Indicative of Pollution-Induced Change in Benthic Communities. 1. Comparative Methodology," *Mar. Ecol. Prog. Ser.* 9:111–119 (1982).

61. Pearson, T. H., G. Duncan, and J. Nuttall. "The Loch Eil Project: Population Fluctuations in the Macrobenthos," *J. Exp. Mar. Biol. Ecol.* 56:305–321 (1982).
62. Pearson, T. H. "Benthic Ecology in an Accumulating Sludge Disposal Site," in *Ocean Processes in Marine Pollution, Vol. 1: Biological Processes and Wastes in the Ocean,* J. M. Cappuzzo, and D. R. Kester, Eds. (Malabar, FL: Robert E. Krieger, 1987) pp. 195–200.
63. Smith, E. P., K. W. Pontasch, and J. Carins. "Community Similarity and the Analysis of Multispecies Environmental Data: A Unified Statistical Approach," *Water Res.* 24:507–514 (1990).
64. Bernstein, B. B., J. H. Hessler, R. Smith, and P. A. Jumars. "Spatial Dispersion of Benthic Foraminifera in the Abyssal Central North Pacific," *Limnol. Oceanogr.* 23:401–416 (1978).
65. Shin, P. K. S. "Multiple Discriminant Analysis of Macrobenthic Infaunal Assemblages," *J. Exp. Mar. Biol. Ecol.* 59:39–50 (1982).
66. Montagna, P. A. "In situ Measurement of Microbenthic Grazing Rates on Sediment Bacteria and Edaphic Diatoms," *Mar. Ecol. Prog. Ser.* 5:151–156 (1984).
67. Koepfler, E. T., and H. I. Kator. "Ecotoxicological Effects of Creosote Contamination on Benthic Microbial Populations in an Estuarine Environment," *Tox. Assess.* 1:465–485 (1986).
68. Gray, J. S. "Detecting Pollution-Induced Changes in Communities Using the Log-Normal Distribution of Individuals Among Species," *Mar. Pollut. Bull.* 12:173–176 (1981).
69. Rhoads, D. C., P. L. McCall, and J. Y. Yingst. "Disturbance and Production on the Estuarine Seafloor," *Sci. Am.* 66:577–586 (1978).
70. Elliott, M., and M. G. O'Reilly. "The Variability and Prediction of Marine Benthic Community Parameters," in *Coasts and Estuaries, Spatial and Temporal Intercomparisons*, M. Elliott and J. P. Ducrotoy, Eds. (Fredensborg, Denmark: Olson and Olson, 1991).
71. Mahon, R., and R. W. Smith. "Demersal Fish Assemblages on the Scotian Shelf, Northwest Atlantic: Spatial Distribution and Persistence," *Can. J. Fish. Aquat. Sci.* 46(Suppl. 1):134–152 (1989).
72. Chapman, P. M., R. N. Dexter, and E. R. Long. "Synoptic Measures of Sediment Contamination, Toxicity and Infaunal Community Composition (the Sediment Quality Triad) in San Francisco Bay," *Mar. Ecol. Prog. Ser.* 37:75–96 (1987).
73. Chapman, P. M., Chapter 14, this volume.
74. Cochran, W. G. *Planning and Analysis of Observational Studies* (New York: Wiley, 1983), p. 145.
75. Clarke, K. R., and R. H. Green. "Statistical Design and Analysis for a 'Biological Effects' Study," *Mar. Ecol. Prog. Ser.* 46:213–226 (1988).
76. Jackson, J. K., and V. H. Resh. "Sequential Decision Plans in Monitoring Benthic Macroinvertebrates: Cost Savings, Classification Accuracy, and Development of Plans," *Can. J. Fish. Aquat. Sci.* 45:280–286 (1988).
77. Green, R. H. "Power Analysis and Practical Strategies for Environmental Monitoring," *Environ. Res.* 50:195–205 (1989).
78. Wiens, J. A., J. F. Addicott, T. J. Case, and J. Diamond. "Overview: The Importance of Spatial and Temporal Scale in Ecological Investigations," in *Community Ecology,* J. Diamond, and T. J. Case, Eds. (New York: Harper and Row, 1986), pp. 145–153.

79. Carins, J. "Biological Monitoring, Part IV, Future Needs," *Water Res.* 15:941–952 (1981).
80. Carins, J., and D. R. Orvos. "Developing an Ecological Risk Assessment Strategy for the Chesapeake Bay," in *Perspectives on the Chesapeake Bay, 1990. Advances in Estuarine Sciences,* M. Haire, and E. C. Krome, Eds. (Gloucester Pt., Virginia: Chesapeake Research Consortium, Rpt. No. CBP/TRS41/90, 1990), pp. 83–98.

CHAPTER 5

Evaluation of Sediment Contaminant Toxicity: The Use of Freshwater Community Structure

Thomas W. La Point and James F. Fairchild

INTRODUCTION

Ecological assessments of sediment toxicity often use resident biota as indicators of sediment quality. For such an assessment to be successful, closely integrated biological, chemical, and physical data are required. Sediments integrate historical water quality, and thus, the spatial and temporal distribution of resident organisms can reflect the degree to which chemicals in the overlying water have been toxic. Field surveys of algae, fishes, and invertebrates provide an essential component of biological assessments of toxicity associated with contaminated sediments. Incorporating field surveys in such biological assessments has at least three advantages: (1) indigenous benthic organisms complete all or most of their life cycles in the aquatic environment and serve as continuous monitors of sediment quality by integrating spatial and temporal fluctuations in contaminant exposure; (2) a field assessment of natural populations can be used to screen potential sediment contamination[1] (if a spatial gradient of contamination exists, extrapolations of interspecies sensitivity, water quality differences, or chemical interaction [additivity, antagonism, or synergism] are not necessary); and (3) results of an assessment of indigenous populations are biologically interpretable, which should quantify resource damage in a manner more easily understood by managers, regulators, and the general public.

This chapter describes various approaches to the use of indigenous aquatic flora and fauna in assessing the toxicity of contaminated sediments in freshwater aquatic environments. We first discuss the importance of experimental design and planning in conducting an assessment. This is followed by a discussion of concepts applicable to biomonitoring in aquatic ecosystems in general. Next, we present a review of the techniques and measures known to be sensitive to changes in biota in lotic and in lentic waters, followed by a recommended set of approaches and methods. We conclude the chapter by presenting a set of research needs we think are important to advancing knowledge about the use of biotic community structure as an assessment technique for contaminated sediments.

BIOASSESSMENT OF CONTAMINATED SEDIMENTS IN LENTIC AND LOTIC HABITATS: COMMON CONCEPTS

Importance of Experimental Design

A sound experimental design is critical to assessing biotic responses to contaminated sediments. Ideally, a sound monitoring design requires an understanding of the complexity of aquatic ecosystems such that confounding factors (e.g., current velocity, depth, light penetration, variance in substrate size, organic matter, nutrients, seasonality) are controlled or accounted for in sample and site comparisons. Each of these factors can profoundly affect sediment characteristics and benthic invertebrate distribution in the absence of chemical contamination.[2] Resh[3] has stressed the critical importance of sediment size and substrate surface area in determining the distribution of species and life stages within species. La Point et al.[4] showed that within apparently homogeneous stretches of rivers, sediment quality can vary considerably.

In a faunal survey designed to assess contaminant effects, reference sites are critical.[5] Reference sites are chosen from areas in which the sediments have substantially less[6] contamination than sites affected by contaminants. In choosing a reference site, one should ensure that the chemical and physical characteristics of the reference and contaminated sites are sufficiently similar. Doing so lessens the probability of making an erroneous conclusion about the effect of contaminants on community composition.[7] The degree of heterogeneity in contaminant distribution within and among sample sites increases the variance in species distribution. Hence, unless the gradient in contaminant concentration is strong, a faunal survey has to be supplemented by screening bioassays and supportive information on the chemical and physical characteristics of the sediments.

Results from sampling programs have led investigators to infer causality from observed changes in the numbers of individuals, species richness, species distributions, and other parameters.[8] To optimize the information from the sampling effort expended, an appropriate experimental design must be developed before initiating a study; mistakes in the study design cannot be statistically corrected after the sampling is concluded. Green[9] discusses sampling design, sample numbers, statistical assumptions, and statistical testing, in the context of an optimal study design. Davis and Lathrop[5] recommend a minimum of five separate benthos samples per site. However, they note that the number of samples taken should be sufficient for the among-sample coefficient of variation (for the total number of invertebrates collected) to be less than 50%. Hurlbert[10] also provides an excellent discussion on the benefits of appropriate sample design and the pitfalls of using inappropriate statistical analyses. *Standard Methods*[11] provides helpful, generalized suggestions for sampling periphyton, macrophyton, and macroinvertebrates. Finally, consideration must be given to the taxonomic resolution necessary to detect shifts or alterations in a biological community. Identification to species requires more expertise than identifications to order, family, or genus. The degree of taxonomic resolution required depends on the degree of environmental contamination, the intensity of effect, the type of study (whether a broad, preliminary survey or more intensive monitoring), and, too often, the level of financial resources available for the assessment. However, taxonomic expertise is becoming widely available. Identification of taxa to the species level should no longer be seen as a hindrance to sediment bioassessments.

Measures of Community Structure

Most quantification of contaminant effects on freshwater biota in lentic or lotic environments has involved changes in benthic invertebrate community structure. Matthews et al.[12] defined community structure, referencing Odum,[13] as the composition of the biological community, including species, numbers, biomass, life histories, and distribution in space of populations; the quantity and distribution of the abiotic (nonliving) materials, such as nutrients and water; and the range or gradient of conditions of existence, such as temperature and light. In this section we discuss both direct and derived measures of community structure. Many early studies on the response of indigenous fauna to contaminated sediments were of a qualitative nature. Recently, the use of biotic indices as quantitative measures resurged, as evidenced by Hellawell,[14] Plafkin et al.,[7] Jackson and Resh,[15] and Hilsenhoff.[16] Quantification of biotic responses will greatly enhance the ability to assess the influence of contaminated sediments on aquatic biota.

Various direct or derived measures of community structure, such as species richness or diversity, are sensitive to contaminants of sediments. Investigators have studied the effects of anthropogenic influences on aquatic biota for over 100 years. The Saprobian Index[17] was a biologically oriented approach developed in the early 1900s. That index, based on the tolerance of different macroinvertebrate species to organic pollution, and other indices based on indicator species concepts,[18,19] have remained the basis for water quality investigations for over 50 years.

Species Richness and Relative Abundance

Species richness (the number of species in a community) and relative abundance (the number of individuals in any given species, relative to the total number of individuals in the community) are common concepts to structural measures of periphytic, planktonic, benthic, and fish communities. Estimates of the relative abundance or species richness can yield readily interpretable information on the degree of contamination of an aquatic habitat.[20-22] The degree of success in using invertebrate community structure and spatial distributions to assess ecological damage depends upon statistical considerations of sufficient sample numbers,[23,24] adequate substrate characterization,[25,26] and an adequate characterization of the suite of water quality characteristics that influence invertebrate distributions independently of sediment-associated contaminants.[4,27] Species richness and composition are important because the loss of a particular species to an ecosystem can be critical when that species plays a critical role in community or ecosystem functions, such as predation,[28] grazing,[29] or competition.[30]

Measures of species richness and relative abundance are taken by sampling known substrate areas or volumes. Taxonomic identification has not always been taken to the species level; this is especially true in monitoring invertebrate communities. Quite often taxonomic, fiscal, and time restraints predicate the need for rapid bioassessment[7,16] using taxonomic identifications to the genus and family levels. However, wide tolerance ranges to contaminants can exist among species within genera.[31] Hence, taxonomic and life history studies at the species level remain a pressing need.

Indicator Species

The presence or absence of indicator species is commonly used to assess damage to ecological communities.[7,16,32,33] The concept originally derived from the saprobian system in which certain species and species groups were used to characterize stream and river reaches subjected to organic wastes. Increasing anthropogenic organic matter in aquatic habitats serves to fill the energy requirements of tolerant species while, at the same time, reducing the numbers of sensitive species that respond negatively to competition, predation, or decreased oxygen levels.[34,35] Westman[36] summarizes three characteristics that impart value to species used as indicators of habitat condition: (1) the species should be sufficiently numerous and widespread to occur in most sites in a study area; (2) the species should possess a sufficiently narrow environmental tolerance to the contaminant so changes in abundance reflect changes in concentration; and (3) the species should have a relatively short generation time, leading to a quick response time.

Experience has shown the indicator species concept lacks broad applicability to all types of contaminated sediments. Sheehan[35] indicates that communities do not respond to organic wastes and toxic chemicals in the same manner. Organic sewage stimulates certain species by increasing their food supply; other species consequently diminish as a result of interspecific interactions. Organic contaminants may affect benthic species as much by clogging interstitial spaces as by toxic effects.[37] The wastes reduce oxygen levels and water flow and exclude the habitat to burrowing invertebrates. Habitat exclusion, unassociated with toxic effects, for benthic species also occurs from the deposition of fine inorganic sediments.[38]

Indicator species also respond, to a greater or lesser degree, to toxic chemicals dissolved in the overlying water; they may not be indicative of sediment contamination. The degree of binding to sediments influences the degree of bioavailability of the contaminants to benthic organisms.[39,40] Furthermore, in aquatic habitats chronically polluted with low levels of contaminants, genetic selection may occur over sufficiently long time periods.[41] In such instances, certain sensitive species may appear to be quite tolerant.[42] Classification of organisms into different tolerance categories can be subjective because the measure of species response may be related to the investigator's arbitrary classification of responses to be measured.[43]

However, the indicator species concept can be applied to the assessment of ecological damage if sufficient care limits how broadly the results are applied. In such cases, species may be found upstream, near the contaminant source, or in habitats known to be unaffected by deposition and storage of contaminated sediments.[44] The indicator species concept is presently used in assessments of hazardous effluents[45] and contaminanted sediments.[46] In a modification of the indicator species concept, Barton[26] studied the occurrence of 20 common groups of benthic macroinvertebrates in Lake Erie. His results indicate that substrate composition and seasonality are primary factors controlling species distribution. Larvae of another indicator species, *Cricotopus bicinctus,* are resistant to copper[20] and to electroplating wastes containing high levels of chromium, cyanide, and copper.[47] Hence, the species becomes dominant under such effluent regimes as more sensitive species decrease. In a similar approach, although at lower taxonomic resolution, the total numbers of insects in the orders Ephemeroptera, Plecoptera, and Trichoptera are counted and referred to as the number of EPT individuals.[7] Many species of these three orders are sensitive to metals and other inorganic contaminants and, hence, provide an index of ecological effects.

Derived Measures

Frequently, taxonomic and abundance data are mathematically reduced to single, calculated numerical indices to simplify data interpretation. Indices derived from direct measures of taxa presence have been extensively developed, reviewed, and critiqued.[9,35,48,49] Indices used in assessing integrated community responses to contaminants are derived variables.[9,50] They are derived in the sense that they are typically proportions, based on the sum of individuals in one taxon (or community) relative to the sum of individuals among the remaining taxa (or other community). Indices (Table 1) can be classified among several types: evenness (how equitably individuals in a community are distributed among the present taxa), relative abundance (the abundance of individuals in one taxon relative to the total abundance of individuals in all other taxa), similarity (comparing likeness of community composition among two sites[51]), and biotic indices that examine the relative importance of key sensitive and insensitive taxa.[14,16,23,45,52]

Although indices can aid in data reduction, they may actually limit interpretability. When indices are calculated, information on individual species occurrence is lost even though money and time were spent on the original identifications.[43] Reliance on reductive indices may actually have hindered progress in the area of indicator organism research. Thus, reliance on indices has occurred at the expense of studies on taxonomy and life history. Furthermore, relying upon a single index, such as the Shannon-Weiner Index, can yield misleading results. For example, a few individuals evenly distributed among several species could give a relatively high index of diversity even though a habitat is grossly polluted.[9] Indices are also criticized for their statistical recalcitrance.[9,50] Statistical assumptions of independence, normal distribution, and homogeneous variances are frequently invalid for these derived, proportional measures. Hence, when indices are used, statistical transformations (e.g., arc-sine) or rank-order statistics[53,54] are recommended.

A final consideration for methods to assess changes in community structure from sediment contamination is to use changes in guild structure or functional groupings of taxa in affected communities. Functional groups or guilds refer to the manner in which organisms obtain their required nutrient and energy resources. Invertebrates can be classified among such functional groups as collector-gatherers, piercers, predators, scrapers, and shredders.[55-57] Fish have also been classified by functional feeding guilds (piscivore, insectivore, herbivore, omnivore).[7]

Shifts in community guild structure reflect changes in the trophic-dynamic status of an aquatic ecosystem. For example, contaminant influences may drastically reduce herbivorous insects (scraper guild), which concomitantly increases algal standing crop.[21,58] Changes may also occur in a guild, such as when a contaminant alters the level of competition between two species that compete for a common resource.[59] Generally, measures of guild structure have low sensitivity (because of high sample variances) and high redundancy;[60] hence, changes in guild structure require fairly strong effects before being detected.

METHODS AND INDICES FOR AQUATIC ECOSYSTEMS

Introduction

Freshwater lentic and lotic systems present different sampling demands for assessing the effects of contaminated sediments on resident biota. In flowing water, a unidirectional downstream movement of potential contaminants is in a somewhat predictable gradient. However, in standing waters, exposure is less predictable because contaminants move more slowly and in multiple directions via wind mixing, diffusion, and upwelling. Assessments in standing waters are further complicated by heterogenous distributions of plankton, nekton, and fishes; thus, other techniques, such as on-site toxicity bioassays, standard acute and chronic laboratory bioassays with invertebrates and fish, and sediment bioassays with benthic invertebrates,[61-63] are often useful.

Under appropriate situations (e.g., harbors, bays, sites with mapped contaminant distribution or known seepage rates), sampling techniques exist that are sensitive enough to demonstrate adverse effects on the resident biota. As previously discussed, consideration must be given to the comparability among samples and sampling methods. For instance, the variance associated with microinvertebrate samples is dependent upon the mean population density and sample volume.[64] Thus, given the necessity of showing that a difference exists with a given probability, the necessary number and size of samples to quantify differences among sites should be ascertained previous to the study. Algorithms for such calculations are readily available.[9,50] For epibenthic plankton, Vollenweider[65] notes it is probably better to take a large number of samples and only count a few subsamples from each, than to take fewer samples and completely enumerate each. The chosen techniques depend on the limnological characteristics of the reference and contaminated sites (such as the depth, surface area, trophic status, and general morphoedaphic characteristics). Some commonly used and accepted sampling techniques are described in the following section.

Periphyton

Periphyton communities can be very sensitive to changes in lotic environments from contamination.[66-70] Monitoring may involve sampling either natural or artificial substrates. Sampling natural substrates yields a wider variance in taxonomic composition and relative abundance than does sampling from artificial substrates, although the variance can be reduced by carefully selecting specific microhabitats with similar physical and chemical characteristics (e.g., substrate type, current velocity, depth, ambient light, etc.).[69] On solid substrates (e.g., tiles, slides, cobble), algal abundance, biomass, and species composition can be obtained by removing the substrate and isolating the flora by scraping or brushing a known area into a separate container. Alternatively, the desired sampling area can be isolated or enclosed by using a chamber sealed to the substrate, with the material removed by suction into a vial.[71] Collecting algae from soft sediments is much more laborious because it involves vacuum suctioning to remove the soft organic surficial sediment layer and then sorting through the debris for enumeration.[69]

Artificial substrates for periphyton colonization and sampling have been widely applied in environmental assessments. Material for artificial substrates includes granite slabs, plastic strips, tiles, and glass slides. Diatometers, consisting of frosted glass slides placed into a holding frame and immersed into the water, have been widely used[66,68,72] and are broadly accepted. Diatometers are known to be somewhat

Table 1. Indices of Species Diversity, Evenness, and Community Similarity[a]

Index	Reference
Species Richness: The number of collected species per sample, assumed to represent the number of species in the community of interest.	Fisher et al.;[116] Lloyd and Ghelardi[117]
Diversity Indices[b]:	
$SI = \frac{[n_i(n_i - 1)]}{N(N-1)}$	Simpson[118]
$D = \frac{S-1}{\ln N}$	Margalef[119]
$H' = -p_i \log_2 p_i$	Shannon and Weaver[120]
$H = \frac{1}{N} \log \left[\frac{N!}{\overset{s}{\underset{i=1}{}} n_i} \right]$	Lloyd et al.[121]
$SCI = \frac{R}{N'}$	Cairns et al.[122]
$E(S_n) = \left[1 - \frac{\begin{matrix} N - n^i \\ n \end{matrix}}{\begin{matrix} N \\ n \end{matrix}} \right]$	Hurlbert[123]
Evenness Indices	
$J = \frac{H'}{\log_2 S}$	Pielou[124,125]
$e = \frac{S}{S'}$	Lloyd and Ghelardi[117]
Community similarity indices	
$I = \frac{2c}{a+b}$	Sorensen[126]
$I = \frac{c}{a+b-c}$	Jaccard[127]

Table 1. Indices of Species Diversity, Evenness, and Community Similarity[a] (continued)

$PS = 100\left(1 - 0.5\left[p_{ij} - p_{ik}\right]\right)$	Whittaker and Fairbanks[128]
$BC = \dfrac{\left[n_{1i} - n_{2i}\right]}{\left[n_{1i} + n_{2i}\right]}$	Bray and Curtis[129]

[a] Compiled from Peet[48] and Sheehan.[35]

[b] Notations: N = Total number of individuals (in all samples or in community). S = Total number of species (in all samples or in community). S′ = Total number of species according to McArthurs's broken stick model. n^i = Number of individuals in the ith species. p_i = n^i/N; the proportion of total number of individuals in the ith species. p_{ij} = the proportion of the ith species in the jth sample or community. p_{ik} = the proportion of the ith species in the jth sample or community. N! = Total number of individuals factorial. R = Number of changes in species per scan of the sample. N′ = Total number of individuals scanned in the sample $\overset{S}{i=1} = n_1!\cdot n_2!\cdot n_3!\cdots n_5!$ n_{1i} = Number of individuals in the ith species of the first community. n_{2i} = Number of individuals in the ith species of the second community. a = Number of species in the first sample or community. b = Number of species in the second sample or community. c = Number of species common to both samples a and b.

more species-selective than are natural substrates. However, this disadvantage is offset by gains in sampling convenience and replicability (similarities in surface texture, surface area, colonization time, and microenvironmental conditions). Additional detail on diatometers and methods of their use are in Patrick,[66] Gale,[72] and APHA.[11]

After the periphyton samples are collected, they may be analyzed for taxonomic composition (cell number, species richness, relative abundance, and taxa evenness) and standing crop (chlorophyll *a* or biomass per unit area).[11,65] One common caution for algal surveys is that a sufficient number of cells should be counted to ensure that rare species are quantified. For example, Stevenson and Lowe[69] recommend counting 200 cells from each sample to ensure complete enumeration of dominants, 500 cells to ensure counting uncommon taxa, and 1000 cells to adequately record rare species. Alternatively, they suggest counting until fewer than one new species is encountered for each additional 100 algal cells counted.

Community indices (diversity, community similarity; see Table 1) can be derived from the collected data. In addition, standing crop biomass, as ash free dry weight (AFDW), is used to calculate an Autotrophic Index and other productivity-related indices.[11,70] However, the structure and function of epilithic periphyton are probably more indicative of changes in water quality than in sediment quality. In any case, separation of the relative effects of sediment- and water-derived contaminants on epilithic microflora is necessary.

Plankton

Many devices are available for sampling phytoplankton and zooplankton. They are categorized as four types: closing samplers, nets, pumps, and traps.[73-75] Closing samplers are bottles or tubes lowered into the water to a particular depth and closed by a drop-weight messenger.[74,76] Plankton nets for quantitative zooplankton counts are typically greater than or equal to 60-μm pore size. However, a mesh of this size does not retain ultra- and nannoplankton; hence, nets are not recommended for phytoplankton studies or any plankton studies in which corresponding measures of ultra- and nannoplankton are collected.[74,77] Also, primary and secondary production measures, taken in conjunction with plankton population turnover estimates, need to be sampled with closing samplers or tubes that take several liters of water to obtain adequate sample size and include all size classes of organisms.

Integrating tube samplers have been used with great success in shallow water systems such as ponds and limnocorrals (cf., papers in Bloesch[78]). These types of samplers capture a known volume of water by extending a tube down through the water column. The sampled water cores typically vary in length and diameter (from 1 to several meters long and 1 to several centimeters in diameter), depending on the experimental conditions. These collect all plankton in the water within the tube; hence, they cannot be used to measure zooplankton stratification. If the interest is in plankton in a particular layer, pumps or closed-tube (Van Dorn) samplers are more appropriate.[79]

Upon collection, appropriate taxonomic expertise is needed to identify the taxa for estimates of species richness; measures of diversity, evenness, and dominance; and measures of community similarity. Note that planktonic communities have a turnover in production measured in hours or days.[77,79] Community structural measures may be enhanced with measures of planktonic production and respiration, although these measures are not commonly used in routine contaminant biomonitoring.

Benthic Invertebrates

Benthic invertebrates represent the most common fauna used in ecological assessments of contaminants in aquatic environments. Many excellent references exist dealing with the collection, identification, and analyses of benthic invertebrate populations.[7,11,57,80,81] Most published biomonitoring and research with benthic invertebrates have dealt primarily with macroinvertebrate populations. Macroinvertebrates are operationally defined as those invertebrates retained by screens of mesh size greater than 0.2 mm.[82] Larger mesh sizes (such as the 0.595 mm, U.S. Standard No. 30[11]) have been accepted as standard. Microinvertebrates (rotifers, nematodes, gastrotrichs) are of particular ecological interest; however, their taxonomy is much less known. Hence, they have not been routinely monitored in environmental assessments.

Macroinvertebrates can be collected from lotic environments with a variety of techniques.[11,80,83] Good discussions of sampling methods and associated problems and remedies are in Peckarsky[81] and *Standard Methods.*[11] In any given effects study, careful consideration must be given to the comparability of samples among stations. Not only must the type of chosen sampling device be appropriate, but the sampling effort per habitat must be uniform among different stations.[9,80] High spatial heterogeneity of toxic constituents in sediments presents a serious sampling problem when trying to link contaminant exposure to invertebrate distribution and abundance.[84] In certain cases different sampling devices are necessary to isolate the effects of contaminanted sediments on benthic macroinvertebrates from variation in distribution and abundance from normal chemical and physical effects. Only very rarely does one type of device sufficiently sample all the microhabitats and associated fauna in a biomonitoring study. Mapping local stretches of habitat enables a focused approach on segments with similar substrate composition;[85] this approach minimizes the confounding influences of sediment texture on species presence.

As in contaminant effects studies with periphyton, macroinvertebrates can be collected and quantified by sampling either standardized or natural substrates. Standardized sampling substrates placed into lotic environments can be made of artificial components, such as tempered hardboard plates (Hester-Dendy samplers), or natural materials, such as gravel or cobble in wire baskets, to semiquantitatively collect macroinvertebrates.[11,57,86] Colonization substrates rely upon macroinvertebrate behavior to successfully collect individuals. Colonized substrates may not reflect sediment exposure; rather, they may reflect habitat availability.[130] Therefore, caution must be used to insure the validity of data for a sediment contaminant assessment. Recommendations for the use of colonized substrates are described in *Standard Methods*.[11] The optimum time for colonization of substrate samplers before collection is six weeks.[87] As with any sampling program, care should be taken to insure uniformity in colonization time, depth, light penetration, temperature, and current velocity. The benefit in using colonization samplers is their comparability among sites and relative ease of sampling. The principal drawbacks to their use are their relative selectivity in types and numbers of collected invertebrates and their potential for not reflecting sediment exposure.

Various nets for collecting dislodged individuals are the second type of typical sampling devices to quantify the abundance and distribution of benthic macroinvertebrates. Two very commonly used devices are the Surber and Hess samplers.[81,83] These samplers are similar because each encloses a defined area of substrate. Substrate within the confines of the sampler is disturbed and mixed by hand or by a stake; dislodged invertebrates are carried downstream into the net by current flow. Large rocks in the sampled area are manually lifted from the substrate and brushed or scrubbed at the mouth of the sampler to dislodge attached or clinging individuals.

Specific techniques are required for sampling larger rivers and sections of lotic habitats with soft substrates. Surber and Hess samplers generally do not work in these situations because current velocity is needed to dislodge and wash invertebrates into the sampler net.[88] Furthermore, if the water is too deep, it will flow over the top of the sampler. Hence, in these types of habitats, recommended macroinvertebrate samplers include corers,[11,64] dredges,[11,64] and drags.[75] A useful handbook on these techniques is by Lind.[89] Large river sampling presents problems much akin to those in lakes and reservoirs. In these habitats, corers (e.g., Kajak-Brinkhurst) are recommended for soft substrates such as silts or clays. Corers consist of long, open tubes that rely on gravity to penetrate the substrate. Various closure methods are used to seal the tube prior to retrieving and removing it from a fixed area of sediment. Dredges (e.g., Eckman and Peterson dredges) sample an area of the substrate defined by the surface area of the open dredge. Dredges typically incorporate more of the deeper sediments than do drag samples; however, this becomes more important if the number of invertebrates are compared by unit volume of sediment rather than by unit area.[90] Drags (Usinger and Needham types) are pulled over a defined distance, taking a relatively shallow sample of the substrate and benthos. Drags perform well in gravel areas, but do not sample soft muds. A quantitative sampling program with such drags must incorporate precise measures of pulled distance, but may still be more qualitative than other methods.

Once sampled, invertebrates should be preserved, isolated, and identified to the desired taxonomic unit. To ensure comparability of data from different stations, similar expertise and effort should be applied in sorting and isolating species from associated debris. For taxonomic identification, a list of useful invertebrate keys is found in *Standard Methods*.[11] Typical measurements include abundance, species (or taxon) richness, relative abundance, and biomass per unit area. From taxonomic identifications to species, trophic guild structure can be determined.[57,91] From the direct measures, indices of diversity, evenness, and community similarity can be calculated (cf., Table 1).

Fishes

Quantifying responses by fish populations remains an important need for aquatic resource managers. Fishes have been recommended for biomonitoring for several reasons: (1) regulators and the general public understand and accept the implications of the effects of pollution on fishes better than on other trophic groups; (2) fisheries have economic, recreational, and aesthetic values; (3) the identification of fishes is relatively easy (compared with algae and invertebrates); (4) the environmental requirements of fishes are better known; and (5) fishes are perceived as integrators of effects at lower trophic levels.[92] Difficulties in using fishes as sediment contaminant biomonitors stem from associating fishes with particular sediments or habitats and in determining their residence time in or over contaminated areas. Their absence in a water body may more directly reflect water quality.

Two excellent chapters outlining the capture of fishes have been written by Lagler[93] and Hendricks.[92] As Lagler[93] notes, the key to successfully assessing the response of fisheries populations to changes in environmental conditions is to know the number of fishes, the sex and year-class distribution, and the rates of mortality and recruitment. However, the size, distribution, and response of freshwater fish populations can be difficult to quantify because of large variances in spatial distribution and in year class.[93] In addition, quantification of fish populations can be difficult because of the selectivity and efficiency of the gear.[92,94] However, consideration of these factors can lead to a successful biomonitoring program with fishes.

Three techniques proven to function well in lotic environments are electrofishing, shore seines, and poisoning. In large rivers and in lakes (see below), most data on fish abundance and distribution are provided by electrofishing or netting (gill, trammel, and fyke types). All such fishing gear is selective by species and size of fish.[93] The degree to which the gear is selective has to be determined by comparison among techniques; this is important whether sampling large or small river ecosystems. Lagler[93] and Southwood[75] describe an approach to determining the sampling selectivity of the gear. Marked individuals are released into the population and subsequently recaptured with the gear; differences in the proportions of different length groups recaptured by any particular gear provides a direct measure of its selectivity.[93] In streams (up to about the sixth order), both upstream and downstream approaches can be blocked with seines or nets placed across the streambed to prevent fish movement into the sampling area. In these instances, repeated sampling with either electrofishing or seining yields robust estimates of fish species presence and abundance.[80]

The types of analyses of collected fish data include calculating fish species richness, total abundance per unit area, and changes in year-class structure and in sex ratio. In a contaminant effects assessment in which the fish community is repeatedly sampled, population size can be estimated with a maximum-likelihood estimation technique, the Zippin method following multiple-step removal-depletion sampling, or other methods.[80,94]

Karr[33] developed an index of biological integrity (IBI) to determine the effects of decreased habitat quality on fish communities of Midwestern streams. IBI, unlike traditional diversity indices, is weighted by individual species tolerances for water quality and habitat conditions. The index is composed of 12 individual metrics divided into areas of species composition/richness, trophic composition, abundance, and condition. Scores of each metric are classified as best, average, or worst (each class has a numerical weighting), according to information from published or competent ichthyological sources for streams of a given size or geographical area.[7,95] Typically, electrofishing or seining is used to determine the species composition and relative abundance of the fish community in selected habitats. After scoring each, the sum of the metrics is obtained to reveal the score ranging from 12 (poorest conditions) to 60 (best conditions).

Many researchers have tested or applied the IBI in various situations across the U.S. IBI, adjusted for differences in zoogeography and stream size, accurately reflected fish community response to degraded habitat conditions in Midwestern streams;[95] generalized pollution;[96] siltation;[97] chlorine and ammonia;[97] generalized pollution in small, cool-water streams in the Appalachian Plateau;[98] and in a ninth-order Northwestern river.[99] Thus, IBI represents a potential fish community index to augment assessment of large-scale habitat degradation and land use influences on watersheds.[100] The listed studies collectively indicate that IBI responds to influences other than contaminants in sediments. Hence, the relative effects of sediments and waterborne contaminants must be isolated by parallel single-species bioassays and by incorporating appropriate reference sites into the biomonitoring program.

As supportive measures in surveys of fish populations and community structure, the percentage of tumors,[101,102] vertebral anomalies,[103] disease and parasites,[104] and fin erosion[105] have been proposed as biological indicators of contaminant effects. Leonard and Orth[98] urge caution in relying on these features because of the mobility of fishes, statistical errors in inferences, differential species sensitivity, and subjectivity in observations. However, these observations may serve to elucidate causal factors in a survey. In fact, percentage anomalies in fishes is used as one of the12 metrics in IBI of Karr.[33]

INTEGRATION OF METHODS: CASE STUDIES OF MULTIVARIATE APPROACHES FOR CONTAMINANT BIOASSESSMENT

Great Lakes

Sediments of the Great Lakes were extensively examined with a sediment-triad approach[106-109] (cf., Chapter 14) that uses laboratory toxicity assessments of collected sediment samples, a chemical determination of potentially toxic constituents, and a survey of the benthic macroinvertebrate fauna. The laboratory tests were generally successful in predicting which sediments should have a depauperate benthos. The correspondence was highest in sites where the sediments are grossly contaminated. For example, Indiana Harbor in Lake Michigan has sediments ranging from oily sludges to sandy gravel.[110] All toxicity tests and the benthic collections[111] indicated the severe nature of the pollution in the inner segments of the harbor. Concordance was less between laboratory testing and the results of a field benthic community survey for stations marginally contaminated, because the physical characteristics and heterogeneity of the sediment and flow regimes become more important in determining faunal presence than contaminant effects. For example, in the Saginaw River in Lake Huron, researchers used artificial samplers and Ekman dredges to describe the invertebrate fauna.[130] The types of collected animals of the two systems are vastly different; oligochaetes dominate the dredge samplers and amphipods dominate the artificial substrates. This study points out the need for information on habitat limitation and the role it and sediment exposure play in determining the presence of benthos.

Riverine

Fewer studies have incorporated the triad approach (sediment toxicity, measures of faunal distribution and abundance, and chemistry) in rivers. The Trinity River study (Chapter 19, this volume) used the triad approach to determine the influences of many industrial and municipal wastes from the Dallas and Fort Worth, TX area. More often, riverine research and monitoring have sought to identify the effects of aqueous contaminants, rather than those bound to sediments. However, some studies integrated the response of the fauna, using invertebrate drift, rates of litter decomposition, and density estimates as a function of sediment type.[112-114] However, this series of studies examined the effects of waterborne methoxychlor on headwater streams in the Coweeta Hydrologic Laboratory. The Coweeta study showed the importance of understanding the indirect effects of contaminants and the direct toxic effects. The strong secondary effects of methoxychlor not only changed the community structure (a shift from insect-dominated communities to oligochaete dominance), but drastically increased the concentration of suspended fine particulate materials exported downstream. These studies are valuable because they point out the complex interactions influencing the response of a benthos to a contaminant. Although direct toxic effects are more easily measured and, hence, understandable, more understanding of the ecosystem is required before we can predict the complete effect of pollutants on the aquatic biota.

RECOMMENDATIONS

An aquatic field survey designed to assess changes in community structure as a result of contaminated sediments should be conducted in a tiered, stepwise approach. The approach should complement a series of sediment toxicity bioassays.[62] An initial checklist should require a basic site evaluation consisting of physical habitat evaluations and descriptions (e.g., Platts et al.,[80] Hughes et al.[115]). In lotic environments the site descriptions should be used to select appropriate reference and affected sites for a rapid bioassessment with macroinvertebrates or fishes.[7] If suitable reference sites are unavailable, consideration should be given to using screening bioassays for comparisons. One must always keep in mind that substantial differences in community structure often stem from substrate characteristics totally unrelated to contaminant influences.

Initial screening tests in standing waters are more difficult because potential gradients may not be as easily identified as in flowing waters. In these situations a bioassay with water, sediments, or caged organisms may be more useful.

The results of preliminary site surveys determine subsequent steps in the assessment sequence. The advantages and limitations of using macroinvertebrates, algae, or fishes have been discussed. Because survey requirements may vary in individual assessment situations, the recommendation of one trophic component for study is difficult. However, in most instances, major emphasis should be placed on the structural analysis of the macroinvertebrate community because of inherent problems in using fishes (mobility, sampling problems) or periphyton (confounding influence of nutrient enrichment, turbidity, etc.). Quantification of the macroinvertebrate community allows maximum flexibility (in terms of expertise, expense, and protocol development) in determining zones of contamination and recovery to the potentially wide variety of sediment-associated contaminants.

Specific instances occur when periphyton or fish assessments are important. Macroinvertebrates and fishes may not respond directly to sediment-bound nutrients or herbicides; in these situations, an assessment of periphyton community structure with artificial substrates[11,66,72] is recommended. These situations are evident from site history and preliminary observations. An assessment of the fish community may be needed in several instances: for example, when recreational or economic values are important. Fishes bioaccumulate nonpolar organic contaminants from sediments (Mac and Schmitt, Chapter 12, this volume) and offer sufficient biomass to measure contaminant bioavailability. This is important when emission of contaminants is low; in such situations the contaminants may not be toxic, but can bioaccumulate to levels that exceed limits for human consumption, thus creating health, recreational, aesthetic, or economic problems.

RESEARCH NEEDS

1. Research is needed to further establish basic ecological relationships between biota and their physical and chemical requirements. Such information assists in understanding changes in community structure along gradients (e.g., stream size, distance, substrate differences) that can confound perceived cause and effect relationships for biota and sediment contaminants.
2. Rapid bioassessment procedures must be further developed, implemented, and validated to screen the effects of bound contaminants on resident biota. Such procedures are needed to further identify and prioritize assessments.
3. Further development of complementary biochemical and physiological indicators of contaminant stress are needed. These indices should be studied in relation to observed population-level responses in both field and laboratory organisms. Successful development of these techniques will provide sensitive early warning indicators of contaminant stress.
4. Further research should be conducted to develop the indicator or sentinel species concepts. Recognition of species sensitive to contaminants will allow a further integration of single species laboratory testing, in situ bioassay methods, and ecological assessments.
5. Additional studies that measure both structural and functional responses at the community and ecosystem level are needed to further quantify the effects of contaminants on aquatic resources. Such studies will increase our understanding of the effects of contaminants on aquatic ecosystems and will serve as useful paradigms for explaining the implications of observed changes in structure to managers, regulators, and the general public.

REFERENCES

1. "Sediment Classification Methods Compendium," U.S. EPA Draft Final Report, (1989).
2. Pettigrove, V. "The Importance of Site Selection in Monitoring the Macroinvertebrate Communities of the Yarra River, Victoria," *Environ. Monitor. Assess.* 14:297–313 (1990).
3. Resh, V. H. "Sampling Variability and Life History Features: Basic Considerations in the Design of Aquatic Insect Studies," *J. Fish. Res. Bd. Can.* 36:290–311 (1979).
4. La Point, T. W., S. M. Melancon, and M. K. Morris. "Relationships Among Observed Metal Concentrations, Criteria, and Benthic Community Structural Responses in 15 Streams," *J. Water Pollut. Contr. Fed.* 56:1030–1038 (1984).
5. Davis, W. S., and J. E. Lathrop. "Freshwater Benthic Macroinvertebrate Community Structure and Function," Chap. 7, in *Sediment Classification Methods Compendium,* U.S. EPA Draft Final Report, (1989).
6. Chapman, P. M. "Sediment Quality Criteria from the Sediment Quality Triad: An Example," *Environ. Toxicol. Chem.* 5:957–964 (1986).
7. Plafkin, J. L., M. T. Barbour, K. D. Porter, S. K. Gross, and R. M. Hughes. *Rapid Bioassessment Protocols for Use in Streams and Rivers: Benthic Macroinvertebrates and Fish,* EPA 444/4-89-001 (Washington, D.C.: U. S. Environmental Protection Agency, 1989), pp. 190 .
8. Barton, D. R. "Some Problems Affecting the Assessment of Great Lakes Water Quality Using Benthic Invertebrates," *J. Great Lakes Res.* 15:611–622 (1989).
9. Green, R.H. *Sampling Design and Statistical Methods for Environmental Biologists* (New York, NY: John Wiley & Sons, 1979), pp. 257.
10. Hurlbert, S. H. "Pseudo-Replication and the Design of Ecological Field Experiments," *Ecol. Monogr.* 54:187–211 (1984).
11. American Public Health Association. *Standard Methods for the Examination of Water and Wastewater* (Washington, D.C.: American Public Health Association, 1985), pp. 1268.
12. Matthews, R. A., A. L. Buikema, Jr., J. Cairns, Jr., and J. H. Rodgers, Jr. "Biological Monitoring Part IIa: Receiving System Functional Methods, Relationships, and Indices," *Water Res.* 16:129–139 (1982).
13. Odum, E. P. "Relationship Between Structure and Function in Ecosystems," *Japan J. Ecol.* 12:108–118 (1962).
14. Hellawell, J. M. *Biological Indicators of Freshwater Pollution and Environmental Management* (London: Elsevier, 1986), pp. 546.
15. Jackson, J. K., and V. H. Resh. "Sequential Decision Plans in Monitoring Benthic Macroinvertebrates: Cost Savings, Classification Accuracy and Development of Plan," *Can. J. Fish. Aquat. Sci.* 45:280–286 (1988).
16. Hilsenhoff, W. L. "Rapid Field Assessment of Organic Pollution with a Family-Level Biotic Index," *J. N. Am. Benthol. Soc.* 7:65–68 (1988).
17. Kolkwitz, R., and M. Marsson. "Grundsatz fur die Biologische Beunteilung des Wassers Nach Seiner Flora und Fauna," *Mitteilungen PrufAnst. Wasser Versorgung Abwasserbeseit. Berlin* 1:33–72 (1902).

18. Gaufin, A. R., and C. M. Tarzwell. "Aquatic Invertebrates as Indicators of Stream Pollution," *Publ. Health. Rep.* 67:57–64 (1952).
19. Patrick, R. "A Proposed Biological Measure of Stream Conditions, Based on a Survey of the Conestoga Basin, Lancaster County, Pennsylvania," *Proc. Acad. Nat. Sci. Philadelphia* 101:277–341 (1949).
20. Sheehan, P. J., and R. W. Winner. "Comparison of Gradient Studies in Heavy-Metal Polluted Streams," in *Effects of Pollutants at the Ecosystem Level,* P. J. Sheehan, D. R. Miller, G. C. Butler, and P. Bourdeau, Eds. (New York: John Wiley & Sons, 1984), pp. 255–271.
21. Lamberti, G. A., and V. H. Resh. "Distribution of Benthic Algae and Macroinvertebrates Along a Thermal Stream Gradient," *Hydrobiologia* 128:13–21 (1985).
22. Van Hassel, J. H., D. S. Cherry, J. C. Hendricks, and W. L. Specht. "Discrimination of Factors Influencing Biota of a Stream Receiving Multiple-Source Perturbations," *Environ. Pollut.* 55:271–287 (1988).
23. Chutter, M. "A Reappraisal of Needham and Usinger's Data on the Variability of a Stream Fauna when Sampled with a Surber Sampler," *Limnol. Oceanogr.* 17:139–141 (1972).
24. Morin, A. "Variation of Density Estimates and the Organization of Sampling Programs for Stream Benthos," *Can. J. Fish. Aquat. Sci.* 42:1530–1534 (1985).
25. Reice, S. R. "Predation and Substratum: Factors in Lotic Community Structure," in *Dynamics of Lotic Ecosystems,* T. D. Fontaine, III, and S. M. Bartell, Eds. (Ann Arbor, MI: Ann Arbor Science , 1983), pp. 325–345.
26. Barton, D. R. "Distribution of Some Common Benthic Invertebrates in Nearshore Lake Erie, with Emphasis on Depth and Type of Substratum," *J. Great Lakes Res.* 14:34–43 (1988).
27. Wiederholm, T. "Responses of Aquatic Insects to Environmental Pollution," in *The Ecology of Aquatic Insects,* V. H. Resh, and D. M. Rosenberg, Eds. (New York,: Praeger, 1984), pp. 508–557.
28. Paine, R.T. "A Note on Trophic Complexity and Community Stability," *Am. Nat.* 103:91–93 (1969).
29. Giesy, J. P., Jr., H. J. Kania, J. W. Bowling, R. L. Knight, S. Mashburn, and S. Clarkin. "Fate and Biological Effects of Cadmium Introduced into Channel Microcosms," U.S. Environmental Protection Agency, Athens, GA, EPA 600/3-79-039 (1979).
30. Matczak, T. Z., and R. J. Mackay. "Territoriality in Filter-feeding Caddisfly Larvae: Laboratory Experiments," *J. N. Am. Benthol. Soc.* 9:26–34 (1990).
31. Chagnon, N. L., and S. I. Guttman. "Differential Survivorship of Allozyme Genotypes in Mosquitofish Populations Exposed to Copper or Cadmium," *Environ. Tox. Chem.* 8:319–326 (1989).
32. Hynes, H. B. N. *The Biology of Polluted Waters* (Liverpool: Liverpool University Press, 1966), pp. 202 .
33. Karr, J. R. "Assessment of Biotic Integrity Using Fish Communities," *Fisheries* 6:21–27 (1981).
34. Gaufin, A. R. "The Effects of Pollution on a Midwestern Stream," *Ohio J. Sci.* 58:197–208 (1958).
35. Sheehan, P. J. "Effects on Community and Ecosystem Structure and Dynamics," in *Effects of Pollutants at the Ecosystem Level,* P. J. Sheehan, D. R. Miller, G. C. Butler, and P. Bourdeau, Eds. (New York: John Wiley & Sons, 1984), pp. 51–99.

36. Westman, W. E. *Ecology, Impact Assessment, and Environmental Planning,* (New York: John Wiley & Sons, 1985), pp. 532 .
37. Warwick, W. W. "Chironomidae (Diptera): Response to 2800 Years of Cultural Influence; A Paleoecological Study with Special Reference to Sedimentation, Eutrophication and Contamination Processes," *Can. Entomol.* 112:1193–1238 (1980).
38. Dahm, C. N., E. H. Trotter, and J. R. Sedell. "Role of Anerobic Zones and Processes in Stream Ecosystem Productivity," in *Chemical Quality of Water and the Hydrologic Cycle,* R. C. Averett, and D. M. McKnight, Eds. (Chelsea, MI: Lewis Publishers, 1987), p. 157.
39. Chant, L., and R. J. Cornett. "Measuring Contaminant Transport Rates Between Water and Sediments Using Limnocorrals," *Hydrobiologia* 159:237–245 (1988).
40. Winsor, M., B. L. Boese, H. Lee, II, R. C. Randall, and D. T. Specht. "Determination of the Ventilation Rates of Interstitial and Overlying Water by the Clam *Macoma nasuta,*" *Environ. Toxicol. Chem.* 9:209–213 (1990).
41. Gillespie, R. B., and S. I. Guttman. "Effects of Contaminants on the Frequencies of Allozymes in Populations of the Central Stoneroller," *Environ. Toxicol. Chem.* 8:309–317 (1988).
42. Hersh, C. M., and W. G. Crumpton. "Determination of Growth Rate Depression of Some Green Algae by Atrazine," *Bull. Environ. Contam. Toxicol.* 39:1041–1048 (1987).
43. Herricks, E. E., and J. Cairns, Jr. "Biological Monitoring Part III: Receiving System Methodology Based on Community Structure," *Water. Res.* 16:141–153 (1982).
44. Bargos, T., J. M. Mesanza, A. Basaguren, and E. Orive. "Assessing River Water Quality by Means of Multifactorial Methods Using Macroinvertebrates. A Comparative Study of Main Water Courses of Biscay," *Water. Res.* 24:1–10 (1990).
45. Courtemanch, D. L., and S. P. Davies. "A Coefficient of Community Loss to Assess Detrimental Change in Aquatic Communities," *Water Res.* 21:217–222 (1987).
46. Blodgett, K. D., and R. E. Sparks. "Documentation of a Mussel Die-Off in Pools 14 and 15 of the Upper Mississippi River," in *Die-Offs of Freshwater Mussels in the United States,* Workshop Proceedings, R. J. Neves, Ed., Davenport, IA (1987), pp. 76–88.
47. Surber, E. W. "*Cricotopus bicinctus*, a Midge Fly Resistant to Electroplating Wastes," *Trans. Am. Fish. Soc.* 88:111–116 (1959).
48. Peet, R. K. "The Measurement of Species Diversity," *Ann. Rev. Ecol. Syst.* 5:285–307 (1974).
49. Pielou, E. C. *Mathematical Ecology* (New York: John Wiley & Sons, 1977), p. 385.
50. Sokal, R. R., and F. J. Rohlf. *Biometry* (San Francisco: W. H. Freeman, 1981), p. 859.
51. Whittaker, R. H. "Evolution and Measurement of Species Diversity," *Taxon* 21:213–251 (1972).
52. Hellawell, J. M. "Change in Natural and Managed Ecosystems: Detection, Measurement and Assessment," *Proc. R. Soc. London, Ser. Biol. Sci.* 197:31–56 (1977).
53. Siegel, S. *Nonparametric Statistics for the Behavioral Sciences* (New York: McGraw-Hill , 1956), p. 312 .
54. Hoaglin, D. C., F. Mosteller, and J. W. Tukey. *Exploring Data Tables, Trends, and Shapes* (New York: John Wiley & Sons, 1985), p. 527.
55. Cummins, K. W., and M. J. Klug. "Feeding Ecology of Stream Invertebrates," *Ann. Rev. Ecol. Syst.* 10:147–172 (1979).

56. Bruns, D. A., and G. W. Minshall. "Macroscopic Models of Community Organization: Analyses of Diversity, Dominance, and Stability in Guilds of Predacious Stream Insects," in *Stream Ecology: Application and Testing of General Ecological Theory,* J. R. Barnes, and G. W. Minshall, Eds. (New York: Plenum Press, 1983), pp. 231–264.
57. Merritt, R. W., and K. W. Cummins, Eds. *An Introduction to the Aquatic Insects of North America* (Dubuque, IA: Kendall/Hunt, 1984), p. 441.
58. Yasuno, M., S. Fukushima, J. Hasegawa, F. Shioyama, and S. Hatakeyama. "Changes in Benthic Fauna and Flora After Application of Temephos to a Stream on Mt. Tsukuba," *Hydrobiologia* 89:205–214 (1982).
59. Petersen, R. C. "Population and Guild Analysis for Interpretation of Heavy Metal Pollution in Streams," in *Community Toxicity Testing,* J. Cairns, Jr., Ed., STP 920 (Philadelphia: American Society for Testing and Materials, 1986), pp. 180–198.
60. Wallace, J. B. "Structure and Function of Freshwater Ecosystems: Assessing the Potential Impact of Pesticides," in *Using Mesocosms to Assess the Aquatic Ecological Risk of Pesticides: Theory and Practice,* J. R. Voshell, Jr., Ed., Miscellaneous Publ. of the Entomological Society of America 75:4–17 (1989).
61. Burton, G. A., Jr., and B. L. Stemmer. "Evaluation of Surrogate Tests in Toxicant Impact Assessments," *Tox. Assess.* 3:255–269 (1988).
62. Burton, G. A., Jr. "Evaluation of Seven Sediment Toxicity Tests and Their Relationships to Stream Parameters," *Tox. Assess.* 4:149–159 (1989).
63. American Society for Testing and Materials. *Standard Guide for Conducting Sediment Toxicity Tests with Freshwater Invertebrates*, ASTM E 1383, in press (1991).
64. Downing, J. A. "Sampling the benthos of standing waters," in *A Manual on Methods for the Assessment of Secondary Productivity in Fresh Waters, IBP Handbook 17,* J. A. Downing, and F. H. Rigler, Eds. (Oxford: Blackwell Scientific, 1984), pp. 87–130.
65. Vollenweider, R. A. *Primary Production in Aquatic Environments, IBP Handbook No. 12* (Oxford: Blackwell Scientific, 1969), p. 213.
66. Patrick, R. "Use of Algae, Especially Diatoms, in the Assessment of Water Quality," in *Biological Methods for the Assessment of Water Quality,* J. Cairns, Jr., and K. L. Dickson, Eds., STP 528 (Philadelphia: American Society for Testing and Materials, 1973), pp. 76–95.
67. Cairns, J., Jr., Ed. *Community Toxicity Testing,* STP 920 (Philadelphia: American Society for Testing and Materials, 1986), p. 350.
68. Lewis, M. A., M. J. Taylor, and R. J. Larson. "Structural and Functional Response of Natural Phytoplankton and Periphyton Communities to a Cationic Surfactant with Considerations on Environmental Fate," in *Community Toxicity Testing,* J. Cairns, Jr., Ed., STP 920 (Philadelphia: American Society for Testing and Materials, 1986), pp. 241–268.
69. Stevenson, R. J., and R. L. Lowe. "Sampling and Interpretation of Algal Patterns for Water Quality Assessments," in *Rationale for Sampling and Interpretation of Ecological Data,* B. G. Isom, Ed. STP 894 (Philadelphia: American Society for Testing and Materials, 1986), pp. 118–149.
70. Crossey, M. J., and T. W. La Point. "A Comparison of Periphyton Community Structural and Functional Responses to Heavy Metals," *Hydrobiologia* 162:109–121 (1988).
71. Hamala, J. A., S. W. Duncan, and D. W. Blinn. "A Portable Pump Sampler for Lotic Periphyton," *Hydrobiologia* 80:189–191 (1981).

72. Gale, W. F., A. J. Gurzynski, and R. L. Lowe. "Colonization and Standing Crops of Epilithic Algae in the Susquehanna River, Pennsylvania," *J. Phycol.* 15:117–123 (1979).
73. Welch, P. S. *Limnological Methods* (New York: McGraw-Hill, 1948), p. 382.
74. Tonolli, V. "Zooplankton," in *Secondary Productivity in Fresh Waters, IBP Handbook No. 17,* W. T. Edmondson, and G. G. Winberg, Eds. (Oxford: Blackwell Scientific, 1971), pp. 1–25.
75. Southwood, T. R. E. *Ecological Methods* (New York: John Wiley & Sons, 1978), p. 524
76. Herman, D., N. K. Kaushik, and K. R. Solomon. "Impact of Atrazine on Periphyton in Freshwater Enclosures and Some Ecological Consequences," *Can. J. Fish. Aquat. Sci.* 43:1917–1925 (1986).
77. Bender, M., K. Grande, K. Johnson, J. Marra, P. J. L. Williams, J. Sieburth, M. Pilson, C. Langdon, G. Hitchcock, J. Orchardo, C. Hunt, P. Donaghay, and K. Heinemann. "A Comparison of Four Methods for Determining Planktonic Community Production," *Limnol. Oceanogr.* 32:1085–1098 (1987).
78. Bloesch, J., Ed. *Mesocosm Studies* (Dordrecht, The Netherlands: W. Junk Publishers, 1988), Hydrobiologia 159 (3):221–313.
79. Edmondson, W. T., and G. G. Winberg. *A Manual on Methods for the Assessment of Secondary Productivity in Fresh Waters, IBP Handbook 17* (Oxford: Blackwell Scientific, 1971), p. 358.
80. Platts, W. S., W. F. Megahan, and G. W. Minshall. "Methods for Evaluating Stream, Riparian, and Biotic Conditions," USDA General Technical Report, INT-138, Ogden, UT, p. 70 (1983).
81. Peckarsky, B. L. "Sampling the Stream Benthos," in *A Manual on Methods for the Assessment of Secondary Productivity in Fresh Waters, IBP Handbook 17,* J. A. Downing, and F. H. Rigler, Eds. (Oxford: Blackwell Scientific, 1984), pp. 131–160.
82. Hynes, H. B. N. "Benthos of Flowing Qater," in *Secondary Productivity in Fresh Waters, IBP Handbook No. 17,* W. T. Edmondson, and G. G. Winberg, Eds. (Oxford: Blackwell Scientific, 1971), pp. 66–80.
83. Merritt, R. W., K. W. Cummins, and V. H. Resh. "Collecting, Sampling, and Rearing Methods for Aquatic Insects," in *An Introduction to the Aquatic Insects of North America,* R. W. Merritt, and K. W. Cummins, Eds. (Dubuque, IA: Kendall/Hunt, 1984), pp. 11–26.
84. Stemmer, B. L., G. A. Burton, Jr., and G. Sasson-Brickson. "Effect of Sediment Spatial Variance and Collection Method on Cladoceran Toxicity and Indigenous Microbial Activity Determinations," *Environ. Toxicol. Chem.* 9:1035–1044 (1990).
85. Minshall, G. W., J. T. Brock, and T. W. La Point. "Characterization and Dynamics of Benthic Organic Matter and Invertebrate Functional Feeding Group Relationships in the Upper Salmon River, Idaho," *Int. Rev Gesampten Hydrobiol.* 67:793–820 (1982).
86. Rosenberg, D. M., and V. H. Resh. "The Use of Artificial Substrates in the Study of Freshwater Benthic Macroinvertebrates," in *Artificial Substrates,* J. Cairns, Jr., Ed. (Ann Arbor, MI: Ann Arbor Science, 1982), pp. 175–236.
87. American Public Health Association. *Standard Methods for the Examination of Water and Wastewater* (Washington, D.C. : American Public Health Association, 1985), pp. 1268.
88. Minshall, G. W. "Aquatic Insect-Substratum Relationships," in *The Ecology of Aquatic Insects,* V. H. Resh, and D. M. Rosenberg, Eds. (New York: Praeger, 1984), pp. 358–400.

89. Lind, O. T. *Handbook of Common Methods in Limnology* (St. Louis: C.V. Mosby, 1979), p. 199.
90. American Society for Testing and Materials. *Standard Guide for Collection, Storage, Characterization, and Manipulation of Sediments for Toxicological Testing,* ASTM E 1391, in press (1991).
91. Cummins, K. W. "Trophic Relations of Aquatic Insects," *Ann. Rev. Entomol.* 18:183–206 (1973).
92. Hendricks, M. L., C. H. Hocutt, and J. R. Stauffer, Jr. "Monitoring of Fish in Lotic Habitats," in *Biological Monitoring of Fish,* C. H. Hocutt, and J. R. Stauffer, Jr., Eds. (Lexington, MA: Lexington Books, 1980), pp. 205–233.
93. Lagler, K. F. "Capture, Sampling and Examination of Fishes," in *Fish Production in Fresh Waters, IBP Handbook No. 3,* T. Bagenal, Ed. (London: Blackwell Scientific, 1978), pp. 7–47.
94. Ricker, W. E. *Computation and Interpretation of Biological Statistics of Fish Populations,* Bulletin 191, (Ottawa, Canada: Department of the Environment, Fisheries and Marine Service, 1975), p. 382.
95. Fausch, K. D., J. R. Karr, and P. R. Yant. "Regional Application of an Index of Biotic Integrity Based on Stream Fish Communities," *Trans. Am. Fish. Soc.* 113:39–55 (1984).
96. Berkman, H. E., C. F. Rabeni, and T. P. Boyle. "Biomonitors of Stream Quality in Agricultural Areas: Fish Versus Invertebrates," *Environ. Manage.* 10:413–419 (1985).
97. Karr, J. R., R. C. Heidinger, and E. H. Helmer. "Effects of Chlorine and Ammonia from Wastewater Treatment Facilities on Biotic Integrity," *J. Water Poll. Contr. Fed.* 57:912–915 (1985).
98. Leonard, P. M., and D. J. Orth. "Application and Testing of an Index of Biotic Integrity in Small, Coolwater Streams," *Trans. Am. Fish. Soc.* 115:401–414 (1986).
99. Hughes, R. M., and J. R. Gammon. "Longitudinal Changes in Fish Assemblages and Water Quality in the Willamette River, Oregon," *Trans. Am. Fish. Soc.* 116:196–209 (1987).
100. Steedman, R. J. "Modification and Assessment of an Index of Biotic Integrity to Quantify Stream Quality in Southern Ontario," *Can. J. Fish. Aquat. Sci.* 45:492–501 (1988).
101. Baumann, P. C. "Cancer in Wild Freshwater Fish Populations with Emphasis on the Great Lakes," *J. Great Lakes Res.* 10:251–253 (1984).
102. Baumann, P. C., W. D. Smith, and W. K. Parland. "Tumor Frequencies and Contaminant Concentrations in Brown Bullheads from an Industrialized River and a Recreational Lake," *Trans. Am. Fish. Soc.* 116:79–86 (1987).
103. Bengtsson, B. E. "Vertebral Damage in Fish Induced by Pollutants," in *Sublethal Effects of Toxic Chemicals on Aquatic Animals,* J. H. Koeman, and J. J. Strik, Eds. (Amsterdam: Elsevier, 1975), pp. 48–70.
104. Overstreet, R. M., and H. D. Howse. "Some Parasites and Diseases of Estuarine Fishes in Polluted Habitats of the Mississippi," *Ann. N.Y. Acad. Sci.* 298:427–462 (1977).
105. Sherwood, M. J., and A. J. Mearns. "Environmental Significance of Fin Erosion in Southern California Demersal Fishes," *Ann. N. Y. Acad. Sci.* 298:177–189 (1977).
106. Burton, G. A., Jr., B. L. Stemmer, K. L. Winks, P. E. Ross, and L. C. Burnett. "A Multitrophic Level Evaluation of Sediment Toxicity in Waukegan and Indiana Harbors," *Environ. Toxicol. Chem.* 8:1057–1066 (1989).

107. Giesy, J. P., Jr., C. J. Rosiu, and R. L. Graney. "Benthic Invertebrate Bioassays with Toxic Sediment and Pore Water," *Environ. Toxicol. Chem.* 9:233–248 (1990).
108. Ingersoll, C. G., and M. K. Nelson. "Testing Sediment Toxicity with *Hyalella azteca* (Amphipoda) and *Chironomus riparius* (Diptera)," in *Aquatic Toxicology and Risk Assessment: Thirteenth Volume,* W. G. Landis, and W. H. van der Schalie, Eds., STP 1096 (Philadelphia: American Society for Testing and Materials, 1990), pp. 93–109.
109. Chapman, P. M., R. N. Dexter, and E. R. Long. "Synoptic Measures of Sediment Contamination, Toxicity, and Infaunal Community Composition (the Sediment Quality Triad) in San Francisco Bay," *Mar. Ecol. Prog. Ser.* 37:75–96 (1987).
110. Ross, P., J. Jones, L. C. Burnett, E. Jockusch, and C. Kermode. *Biological and Toxicological Investigations of Sediment Samples from Indiana Harbor and Canal and Adjacent Lake Michigan,* J. B. Risatti, and P. Ross, Eds., U. S. Army Corps of Engineers, Final Report, pp. 29–46 (1989).
111. Wetzel, M. *Aquatic Macroinvertebrates from Indiana Harbor and Canal and Adjacent Lake Michigan*, J. B. Risatti, and P. Ross, Eds., U. S. Army Corps of Engineers, Final Report, pp. 47–67 (1989).
112. Cuffney, T. F., J. B. Wallace, and J. R. Webster. "Pesticide Manipulation of a Headwater Stream: Invertebrate Responses and Their Significance for Ecosystem-Level Processes," *Freshwater Invert. Biol.* 3:153–171 (1984).
113. Wallace, J. B., D. S. Vogel, and T. F. Cuffney. "Recovery of a Headwater Stream from an Insecticide-Induced Community Disturbance," *J. N. Am. Benthol. Soc.* 5:115–126 (1986).
114. Wallace, J. B., G. J. Lughart, T. F. Cuffney, and G. A. Schurr. "The Impact of Intensive Insecticidal Treatments of Drift and Benthos of a Headwater Stream," *Hydrobiologia* 179:135–147 (1989).
115. Hughes, R.M., D.P. Larsen, and J.M. Omernik. "Regional Reference Sites: A Method for Assessing Stream Potentials," *Environ. Manage.* 10:629–635 (1986).
116. Fisher, R. A., A. S. Corbet, and C. B. Williams. "The Relation Between the Number of Species and the Number of Individuals in a Random Sample of an Animal Population," *J. Anim. Ecol.* 12:42–58 (1943).
117. Lloyd, M., and R. J. Ghelardi. "A Table for Calculating the Equitability Component of Species Diversity," *J. Anim. Ecol.* 33:217–225 (1964).
118. Simpson, E. H. "Measurement of Diversity," *Nature* 163:688 (1949).
119. Margalef, R. 1951. "Species Diversity in Natural Communities," *Publnes. Inst. Biol. Appl. Barcelona* 6:59–72 (1951).
120. Shannon, C. E., and W. Weaver. *The Mathematical Theory of Communications* (Chicago: University of Illinois Press, 1963), p. 117.
121. Lloyd, M., J. H. Zar, and J. R. Karr. "On the Calculation of Information-Theoretical Measures of Diversity," *Am. Midl. Nat.* 79:257–272 (1968).
122. Cairns, J., Jr., D. W. Albaugh, F. Busey, and M. D. Chaney. "The Sequential Comparison Index: A Simplified Method for Non-Biologists to Estimate Relative Differences in Biological Diversity in Stream Pollution Studies. *J. Water Pollut. Contr. Fed.* 40:1607–1613 (1968).
123. Hurlbert, S. H. "The Nonconcept of Species Diversity: A Critique and Alternative Parameters," *Ecology* 52:577–586 (1971).
124. Pielou, E. C. "Shannon's Formula as a Measure of Species Diversity: Its Use and Misuse," *Am. Nat.* 100:463–465 (1966).

125. Pielou, E.C. "The Measurement of Diversity in Different Types of Biological Collections," *J. Theor. Biol.* 13:131–144 (1966).
126. Sorenson, T. "A Method of Establishing Groups of Equal Amplitude in Plant Sociology Based on Similarity of Species Content and Its Application to Analysis of the Vegetation on Danish Commons," *Biol. S. Kr.* 5:1–34 (1948).
127. Jaccard, P. "The Distribution of Flora in the Alpine Zone," *New Phytol.* 11:37–50 (1912).
128. Whittaker, R. H., and C. W. Fairbanks. "A Study of Plankton Copepod Communities in the Columbia Basin, Southeastern Washington," *Ecology* 39:46–65 (1958).
129. Bray, J. R., and J. T. Curtis. "An Ordination of the Upland Forest Communities of Southern Wisconsin," *Ecol. Monogr.* 27:325–349 (1957).
130. Swift, M. Personal communication, Wright State University, Dayton.

CHAPTER 6

The Emergence of Functional Attributes as Endpoints in Ecotoxicology

John Cairns, Jr., B. R. Niederlehner, and E. P. Smith

INTRODUCTION

When a person has a medical checkup, assessments of his or her health are focused mainly on the successful functioning of various systems (e.g., heart rate, blood pressure, liver function, etc.). In contrast, assessments of ecosystem health have focused primarily on structural attributes rather than on functional attributes. For most of the world population and its representatives, the most important argument for the restoration, maintenance, and preservation of natural ecosystems is the protection of the services they provide. These services include the maintenance of the global atmosphere in a condition supportive of life, including humans; the transformation of noxious wastes; and the improvement of human health through models for complex chemicals that serve as a basis for many drugs. Most of these services are functional attributes and must be protected for the ecosystem to be sustained. Assessments of functional responses to stress seem an obvious approach to assuring the continuation of the services provided by ecosystems. This chapter is divided into three components that address functional attributes important in assessing the ecological consequences of contamination: (1) an introductory discussion of functional vs. structural assessment of ecosystem condition; (2) some illustrative examples of assessments based on functional attributes; and (3) some illustrative examples of statistical methods appropriate to evaluate rate processes.

Table 1. Examples of Common Structural and Functional Endpoints Used in Assessment of Health in Aquatic Ecosystems

Structural Endpoints
Species richness
Species abundance
Similarity
Biomass
Functional Endpoints
Rate of biomass production
Rate of primary productivity
Rate of community respiration
Rate of nutrient uptake/regeneration
Rate of decomposition
Biological regulation

STRUCTURAL VS. FUNCTIONAL ASSESSMENT OF ECOSYSTEM CONDITION

While ecosystem structure is studied through the assessment of the numbers and kinds of species and other component parts at one point in time, ecosystem function describes the dynamics or changes in the system over time (Table 1). Examples of common measures of ecosystem structure are species abundance and diversity, and assessments of standing crop, organic carbon, chlorophyll, oxygen, and particle-size profiles. Measures of ecosystem function are usually rates describing changes over time, or, occasionally, ratios of pool sizes from which rate changes can be inferred. These rates commonly describe energy flow and material cycling or biological regulation of the environment. Examples of functional endpoints are production rates, respiration rates, mineralization rates, and nutrient regeneration rates.

Appreciation for the importance of the functional attributes of ecosystems is currently increasing due to some well-publicized environmental issues. The basic process now called the "greenhouse effect" has been recognized for over 100 years.[1] The great debate over risk from global climate change now focuses on the ability of natural systems to offset the effects of increased greenhouse gases by increased biological activity, thus keeping the atmosphere in a condition favorable to present life on earth. This is a functional attribute (for a more elaborate discussion see works by Schneider[2] or Rosswall[3]). Even modest changes in global temperature, with concomitant changes in the rainfall pattern, etc., could have drastic effects on an already overstrained agricultural system and natural ecosystems. Nor is the regulation of the atmospheric gaseous composition the only functional role of natural ecosystems. Ehrlich and Ehrlich[4] and Wilson[5] have discussed at length a variety of ecosystem services provided to mankind. Some of these services are functional, such as the transformation of sewage and other wastes; others are structural in the sense that many drugs are modeled after chemicals found in plants and animals (but the resulting drugs usually cure diseases through functional changes in human physiology).

Although many ecosystem functions are easily described, others are not. Even the ones easily described are not, as yet, well quantified. The understanding of ecosystem function will probably progress through a hierarchy of sizes and complexity. The functions of small and simple systems, most likely microbial, will be described and understood first by scientists, and the beginnings of an understanding of ecosystem function will occur at the regional level before the global level.

There are three possible relationships between structure and function in natural systems responding to toxicants:[6]

1 Structure and function are so intimately associated that a change in one attribute inevitably leads to a corresponding, closely linked change in the other attribute;
2. Functional attributes are more sensitive than structural attributes because functional capabilities of organisms can diminish without killing them, thus changing the functional characteristic before the structural characteristic is affected;
3. Structural characteristics are more sensitive than are functional characteristics because functional redundancy exists in most complex natural systems. Several species perform each functional activity; thus, species can be lost without changing the functional performance. The abundance of the remaining organisms in a particular functional niche would expand to replace the function lost when a more sensitive species was eliminated.

It is not clear which of these relationships between structure and function is most typical. Some data suggest that the responses of structure and function to chemical stresses are not so closely tied as to yield identical estimates of safe concentrations.[7-11] In some situations, disturbance leads to increases in species richness. Odum and colleagues[12] hypothesized that there is a subsidy-stress gradient and suggested that low levels of disturbance often act as "subsidies" to ecosystems. Evidence supports this hypothesis in the case of the addition of biologically useful materials, such as inorganic nutrients or carbon, to the ecosystem. More interesting are the observations that toxic stresses can also result in an apparent subsidy.[13] For example, in a microbial microcosm, low levels of atrazine (32 μg/L) enhanced species richness while impairing production, as monitored by oxygen concentrations.[14] It is difficult to understand how a general herbicide (photophosphorylation inhibitor) such as atrazine might subsidize the ecosystem structure. Rather, enhanced species numbers may be a result of the loss in populations of certain regulating species, thereby allowing other taxa to proliferate. In an experimental system containing several hundred species, keystone taxa that might exert these controlling effects are extraordinarily difficult to identify. This type of finding suggests that enhancement of species numbers cannot always be interpreted as "good," since the response may, in fact, be due to the removal of certain critical members of the community and should be viewed as an early warning that more severe, chronic effects are likely if the stress is increased. In the same study, species richness, biomass, chlorophyll concentrations, and equilibrium species numbers were enhanced at low concentrations (3 to 32 μg/L), but severely reduced at higher concentrations (337 μg/L).

If a nonlinear response to stress is common, "enhancement" may be a first sign of stress, preceding a final observation of severe impoverishment.[15] Basing an assessment of risk exclusively on customary structural measurements, in this case, could be misleading. However, a suite of responses, including both structural and functional attributes, would provide the additional evidence necessary to clearly identify stress rather than subsidy, even at low concentrations. This combination of both structural and functional information is currently being used for efficient field surveys to characterize environmental health.[16]

Several factors may discourage scientific interest in functional attributes or exploration of the relationship between structure and function. First, structural methods are well developed for studying systems (richness, diversity, similarity, etc.), while methods for studying function are still developing. Second, another consequence of less experience with functional measures is that the normal operating ranges (NOR) for structural attributes (e.g., tolerance ranges for populations of different species, indicator species, typical species richness for various ecoregions) have been studied more thoroughly and have been better described than those for functional attributes. As such, until background information on the functioning of healthy ecosystems is available, deviations from a healthy state are more easily pinpointed using structural measures. Third, while ecosystems do undergo successional processes, with one set of species replacing another, this process is relatively slow (except for microorganisms). In contrast, functional attributes often fluctuate strikingly even in a space of 24 hr. Therefore, structural attributes may appear much more constant than functional attributes. For example, any attribute associated with photosynthesis (e.g., pH, oxygen concentration) changes markedly from light to dark periods, but the species carrying out this activity remain constant. Fourth, some scientists perceive a problem in communicating the relevance and importance of functional measures: i.e., who cares if changes occur in function? However, communicating the importance of functional attributes is no more difficult than communicating the importance of structural attributes. A deliberately exaggerated example would compare perceptions of relevance and importance of global temperature maintenance (a functional attribute) to the maintenance of populations of the Houston toad (a structural attribute). Fifth, determining the relationship between the structure and function of natural ecosystems in the assessment of pollution has not had high priority in agencies funding environmental research. Yet, because the relationship between structure and function has not been defined, inferring adequate protection of ecosystem function from the more common tests on ecosystem structure is problematic.

EXAMPLES OF FUNCTIONAL MEASURES IN BIOASSESSMENT OF SEDIMENTS

Sediments are the site of great biological activity in aquatic systems. Concentrations of bacteria are up to seven orders of magnitude higher in the surface sediments than in the water column.[17] This concentration of microbial activity makes sediments the most active site for transformations of organic carbon, nitrogen, phosphorus, magnesium, and sulfur. If the processes of decomposition, mineralization, and nutrient regeneration are disturbed by contaminants, the nature of the ecosystem will be changed. Despite the importance of these and other processes concentrated in the sediments, relatively few assessments have focused on functional changes attributable to sediment-associated contaminants (Table 2). The focus of most studies of environmental effects of contaminated sediments has been on indicator or surrogate organisms and community structure. This is in contrast to assessments of contaminant fate in sediments, which often focus on dynamics. A discussion of some rate processes in sediments important to the health of aquatic systems and endpoints that could be used to monitor their sensitivity to toxic stress follows.

Energy Flow

The most commonly measured functional endpoints in aquatic ecosystems are production (P), respiration (R), and their ratio (P/R). Common methods for assessing productivity include determinations of biomass accumulation over time, oxygen uptake over time, and [^{14}C] incorporation over time.[36,37] Bacterial production in streambed sediments has been assessed by monitoring [^{3}H] thymidine incorporation,[38] and the same technique has been applied to the assessment of the toxicity of waterborne chemicals.[39] Research with contaminated sediments suggests that primary productivity can be adversely affected.[18-20]

Microbial community oxygen demand is the sole functional measure included in a list of endpoints recommended for the future hazard assessment of contaminated sediments.[40] Methods used to monitor respiration of sediments are reviewed by Medine and colleagues.[21] Common methods include oxygen uptake or CO_2 evolution in the absence of primary production and measurements of electron transport system (ETS) enzyme activity. Medine and colleagues[21] found that sediment oxygen demand was lower in laboratory microcosms exposed to low levels of heavy metals. Lake acidification had little effect on sediment oxygen demand.[22] Oxygen uptake of communities in the overlying water was depressed in microcosms with Cu-amended sediments.[20] Electron transport system activity of microbial communities in the sediment was stimulated in harbor sediments with many contaminants[9] and in a metal-contaminated stream,[23] but inhibited in sediments contaminated by drilling fluids.[24] Electron transport system activity did not distinguish field sites influenced by a coking effluent from reference sites.[25]

Much decomposition of organic carbon occurs in the sediment-litter compartment of aquatic systems. The effects of contaminants on rates of decomposition and degradation have been assessed. Loss of mass in leaf packs has often been used to evaluate the effects of waterborne contaminants on decomposition. For example, Andersson[26] found no differences in the loss of mass in leaf packs placed in contact with sediments in acidified and unacidifed lake waters. Another approach has been

Table 2. Selected Functional Responses of Sediment Communities to Pollution or of Other Receptor Organisms to Sediment-Borne Contamination[a]

Study	Test System	Results
Ross and colleagues[18]	Laboratory tests exposing a green alga to sediment elutriates	P(^{14}C uptake) decreased with increasing elutriate concentration
Munawar and Munawar[19]	Laboratory tests exposing indigenous phytoplankton to whole sediment and elutriates	P(^{14}C uptake) decreased with increasing elutriate or sediment concentration
Scanferlato and Cairns[20]	Naturally derived sediment-water microcosms amended with Cu	P(O_2 production), R(O_2 demand), P/B, and R/B were significantly reduced in microcosms with 1 mg Cu/g, while AR increased; P/R showed no differences
Medine and colleagues[21]	Naturally derived sediment-water microcosms amended with Zn, Cr, Cd, Pb, Hg mixture	R(O_2 demand) decreased by 8 and 28% relative to control
Gahnstrom[22]	Activity of sediment cores from acidified lakes	R(O_2 demand) in acidified lakes not different from that in unacidified lakes
Burton and colleagues[9]	Activities of indigenous microbes in contaminated harbor sediments	R(ETS), phosphatase, glucosidase, and galactosidase activity in contaminated sediments higher than that in reference sediments
Burton and colleagues[23]	Activities of indigenous sediment microbes in a metal-impacted stream	Galatosidase, glucosidase, protease, and R(ETS) activity were negatively correlated with in situ species richness; Amylase, phosphatase, and arylsulfatase activity were not significantly correlated
Portier[24]	Activities of indigenous sediment microbes after a drilling-fluids spill	R(ETS) and phosphatase decreased in impact zone relative to distant sites
Sayler and colleagues[25]	Activities of indigenous sediment microbes in a stream receiving coking effluent	Phosphatase; mineralization of glucose, lipid, phenanthrene, and naphthalene; and nitrogen fixation were useful in discriminating control and impact zones. R(ETS) and methanogensis were not
Andersson[26]	Naturally derived sediment-water microcosms and in situ determinations in acidified lakes	No differences in leaf mineralization in acidified and unacidified lake waters. Presence of detritivores more important in determining rate
Burton and Stemmer[27]	Activities of indigenous sediment microbes in seven contaminated stream and landfill sites	Galactosidase correlated to in situ chemical and biological responses in 80% of studies; glucosidase and R(ETS) in 60%. Microbial activity was inhibited or stimulated

Table 2. Selected Functional Responses of Sediment Communities to Pollution or of Other Receptor Organisms to Sediment-Borne Contamination[a] (continued)

Study	Test System	Results
Gahnstrom and Fleischer[28]	Activities of indigenous sediment microbes in acidified lakes	Glucose mineralization was as high in acidified as in unacidified lakes
Mills and Cowell[29]	Activity of naturally derived sediment microbes exposed to metals	Glucose oxidation inhibited at 10 and 100 mg/L of Cr, Pb, Cd, Hg, and Co, individually
Broholm and colleagues[30]	Activity of naturally derived sediment microbes exposed to trichloroethanes	Methane degradation inhibited by trichloroethanes
Jones and Hood[31]	Activity of naturally derived sediment microbes exposed to pesticides	Ammonium oxidation completely inhibited after 14-d exposure to methyl parathion, parathion, fonofos, and guthion at 10 mg/L
Perez and colleagues[32]	Naturally derived sediment-water microcosms amended with phthalate ester	Ammonia flux reduced at phthalate treatments of 100 μg/L (6166 μg/kg sediment), but there was corresponding decrease in density of benthic species
Bruns and colleagues[33]	Naturally derived sediment-water microcosms amended with HNO_3 and H_2SO_4	Increased denitrification was the predominant mechanism for neutralizing additions of nitric acid, while cation exchange was more important for sulfuric acid
Sayler and colleagues[34]	Activity of indigenous sediment microbes and naturally derived sediment-water microcosms amended with oil	Phosphatase activity depressed in impact zone of stream, but unaffected in microcosms 2 months after oil treatment
Hansen and Tagatz[35]	Naturally derived sand-water microcosms	Numbers and kinds of organisms present after 8–16 weeks of colonization depressed by PCB, toxaphene, PCPs, and drilling mud

[a] Abbreviations are used for production (P), respiration (R), and electron transport system activity (ETS).

to monitor the activity of the enzymes involved in decomposition. Of the enzymes studied, galactosidase activity in sediment microbial communities seems to be most consistently affected by pollution and most strongly related to more conventional indices of effect (e.g., diatom diversity).[27] Galactosidase and glucosidase activity were stimulated in contaminated harbor sediments.[9] Galactosidase, glucosidase, and protease activity were stimulated in a metal-impacted stream, but amylase and arylsulfatase were unaffected.[23] The effects of contaminants on heterotrophic activity of sediment microbial communities, measured as a breakdown of radiolabelled glucose or other substrates, have also been assessed. Stream segments receiving coking effluent had higher mineralization rates for glucose, napthalene, and phenanthrene.[25] Glucose mineralization rates were similar for sediment microbial communities from acidified and unacidified lakes.[28] Glucose mineralization in Chesapeake Bay sediments was inhibited by addition of various metals.[29] Methanogenesis or methane oxidation rates in sediments have been used to assess the effects of contamination on carbon cycling and energy transfer. Methanogenesis was highly variable over site and sample date in a stream receiving a coking effluent.[25] Methane degradation was inhibited by addition of trichloroethanes.[30]

Nutrient Cycling

Aquatic systems depend on the microbial communities concentrated in the sediments to remove toxic or excessive nutrients from the system, to store limiting nutrients (thus preventing their export downstream), and to regenerate nutrients in forms useful to organisms, in general, to facilitate local production.[41] Nutrient cycling has been identified as an important endpoint in describing the effects of disturbance in stream ecosystems.[42] Evidence of the value of nutrient cycling to society can be found in the fact that artificial wetlands (systems designed to maximize microbial processes normally associated with the sediment) are constructed specifically to process the nutrients in anthropogenic wastes. Nutrient removal efficiencies for these artificial wetlands can be greater than 95%.[43]

Laboratory assays to monitor various pathways in the nitrogen cycle (nitrification, denitrification, nitrogen fixation) have been developed (see review by Prichard and Bourquin[44]) and can be used to assess changes in function due to contaminant stress. Jones and Hood[31] used these assays to evaluate the effects of toxic substances on nitrogen cycling in sediment microbial communities. They found that all seven pesticides tested inhibited nitrification by estuarine sediment microbial communities. Sayler and colleagues[25] found that nitrogen fixation in sediment microbial communities was useful in discriminating control and impact zones in an effluent-influenced stream. Nitrogen cycling has also been monitored in situ. Observations of nutrient flux across the sediment-water interface have been made by enclosing areas of sediments in chambers and monitoring nutrient levels or radiolabelled nutrient content over time.[45,46] Perez and colleagues[32] found that phthlate ester contamination reduced sediment ammonia flux in microcosms. Uptake rates for added nitrogen over stream lengths have also been monitored to determine the cycling efficiency and spiraling length.[47] Although this holistic approach does not isolate the sediments in any way, it is the sediments, detritus, and associated microbes on the stream bottom that are probably responsible for nutrient spiralling through sorption, retention, and turnover. Likens and colleagues[48] found disturbances in the nitrogen cycle in a deforested watershed treated with herbicides. Changes were attributed to increased microbial nitrification in the disturbed system. Also, increased rate of denitrification along with sulfate reduction and ion exchange in the sediments are important mechanisms for neutralizing pH changes from acid precipitation.[33,49]

The effects of disturbance on phosphorous cycling also have been monitored. One mechanism for regeneration of phosphate (alkaline phosphatase activity) has been monitored in several contaminated sediments. Phosphatase activity was stimulated in contaminated harbor sediments,[9] but depressed by a coking effluent[34] and by drilling fluids.[24] Phosphatase activity did not distinguish control from impact zones in a metal-impacted stream,[23] or control from oil-treated microcosms.[34] In a summary of five stream studies, phosphatase activity in sediment microbial communities showed significant differences with pollution stress in two systems.[27] More holistic determinations of the effects of stress on phosphorous cycling have also been made. Uptake rates for phosphate have been monitored in laboratory microcosms,[50] benthic enclosures,[45,46] littoral enclosures,[51] and stream segments of control and perturbed watersheds.[52] Uptake of phosphate was slower in streams disturbed by clearcuts of the surrounding watershed than in reference streams, perhaps due to differences in the structural characteristics of the litter-sediment compartment.[52]

Biological Interactions Over Time

Functional endpoints also include rates of change in structural endpoints over time. Patterns over time are sometimes difficult to interpret, but some expected patterns of change at the community level have been used to evaluate pollution stress. Colonization and succession describe predictable changes in the numbers and kinds of organisms making up the community over time and integrate the effects of toxic substances on immigration, emigration, competition, and predation. Colonization reflects the ability of the community to replicate and organize itself and is somewhat analogous to reproduction in the single species. Colonization of sand substrate by benthic estuarine organisms was limited by PCB, toxaphene, PCPs, barite, and drilling mud.[35] Colonization of artificial substrates by freshwater microbial communities was limited by exposure to leachates from contaminated soils.[53] Additional approaches for measuring the integrity of biological interactions are possible, but these approaches have not been specific foci of assessments of sediment contamination. Odum,[54] Rapport and colleagues,[55] and others suggest that toxic stress can reverse the successional process. Comparisons of community similarity through time over a stress gradient might be used to test this idea. In addition to colonization and succession, important evidence of biological regulation of the community can be found in a lack of change over time. The integrity of homeostatic mechanisms of the community can be assessed by measurements of inertia and elasticity.[56] Exposure to stress may affect homeostatic mechanisms and the severity of effects from subsequent stress.[57] The inherent variability of a structural endpoint over time can also change with stress. For example, the amplitude of fluctuations over time in the numbers of important fish species in the Great Lakes has become wider with increasing stress.[55]

STATISTICAL CONSIDERATIONS IN USING FUNCTIONAL MEASURES

Any indicator of ecosystem health, either structural or functional, must be reliable and have reasonably low variance if it is to be useful in hazard assessment. It has been suggested that functional endpoints are more variable than structural endpoints and, as such, are less sensitive indicators of adverse changes to communities. However, this concern may be related to the lack of wide experience with functional endpoints. Variability may not necessarily be greater for functional measures,[10,11] and variability can often be controlled and/or accommodated in the design of the study.[8,10]

Variability involves a number of components.[10] An endpoint is repeatable when similar values are obtained when the same experimental unit (station, tank, microcosm) is sampled at different times or locations. For example, the productivity of a single microcosm can be measured four times. Factors that can decrease repeatability and increase variability are succession and changes in environmental conditions over time, environmental patchiness, and methodological variation. An endpoint is replicable if similar values are obtained when different experimental units treated similarly are sampled at one point in time and space. For example, in a toxicity test four microcosms may receive the same amount of contaminated material. Other than the amount of contaminant added, these four microcosms have no more in common than any other four microcosms, regardless of the amount of contaminant added. One could measure productivity once in each of these microcosms, sampling at the same time and at a similar location within each microcosm. Factors that decrease replicability and increase variability are environmental patchiness on a larger scale and methodological variation.

In contrast to designed toxicity tests, field studies of pollution effects rarely have true replication,[58] and consequently, replicability cannot be assessed. A consequence of the lack of replication in field studies is that no clear assessment of within-treatment variation can be made, and conclusions about the cause of any observed differences between stations may be confounded.[58] For example, differences between stations could be due to differences in habitat rather than pollution. This is one reason that the most convincing evaluations of environmental impact have three types of information: (1) assessment of environmental effects in the field, (2) chemical data, and (3) toxicity data. A field survey provides the most realistic, relevant, and convincing evidence that environmental integrity has been harmed, but it cannot establish a causal link to a contaminant. A chemical survey can confirm the presence of the contaminant in the natural system. Toxicity tests with true replication can establish a link between an adverse effect and the presence of a contaminant, but such tests sacrifice realism. Although this combination of information is equally important for field assessments of effects of pollution in all media, this approach has been a notable design feature of some studies of sediment contamination.[59]

In developing standard methods for the use of functional endpoints in the assessment of ecosystem health, a crucial consideration is which types of variance are important to measure and which types are important to control. In general, the variance should reflect the natural variability in the system of interest, as altered by the pollution. Depending on the purpose of the study, the researcher may not choose to control variance, since a reduction in variance can also be seen as a loss of realism.[10] A discussion of some sources of variance and ways to accommodate them follows, using an example from Crossey and La Point.[8]

Crossey and La Point[8] found that variability in measures of gross primary productivity made it impossible to distinguish between field sites affected by heavy metals, even though mean values appeared fairly different. Several sources of variance were controlled in this study. Artificial substrates were used to control environmental patchiness of the substrate. Successional age was controlled by colonizing all substrates for the same length of time. However, several additional techniques might improve the ability to detect differences; some were suggested by the authors. Methodological variation might be reduced by using fixation of ^{14}C, rather than oxygen uptake, to determine productivity, since carbon fixation is a more precise method. Another methodological change might be the use of an expression of production weighted to remove variance from structural differences of the community. For example, an assimilative ratio (AR) weights production by chlorophyll concentration. This ratio focuses particularly on the functional integrity of the autotrophic component of the community and is better able to distinguish stations in the Crossey and LaPoint study. Variance from changes in environmental conditions over time might be eliminated by simultaneous sampling or by using time as a blocking factor in statistical analyses. The production measurements in the study were made over a period of 10 days. If each site was measured each day, the day of determination could be used as a blocking factor in the statistical analysis, reducing the error term and increasing sensitivity. Variance from environmental patchiness (differences in incident light, current velocity, temperature, etc.) can also be accommodated by blocking. For example, the samples at the control site might be taken at different incident light intensities or current velocities. By repeating this sampling approach at the other sites, light intensity could be used as a blocking factor in the statistical analysis, reducing error and increasing sensitivity. If blocking would require too many determinations, environmental patchiness could be accommodated by compositing smaller samples from different locations within a site before making the determination of function. Several smaller substrates, colonized at different locations within a site, could be combined in a chamber for determinations of production. Thus, compositing is a relatively inexpensive way to reduce variance. Obviously, production measured in model streams, using controlled lighting and flow regimes, will have less patchiness and lower variance than production measured in natural systems. However, variation in these factors contributes to the realism of the assessment.

The sensitivity of functional endpoints can be improved by simply increasing replication, although this approach is limited by the cost and time required to obtain replicate measures. Power analysis[60] can determine how much replication is needed to identify biologically important degrees of impairment, given an estimate of the variance of the endpoint. A power of 0.8 (an 80% probability that differences are found between treatment groups when they, in fact, exist) is often used to estimate the number of replications necessary. Suppose, for example, that the researcher wishes to detect a difference in production of 0.4 mg $O_2/m^2/day$ between two station means, and the between-group variance is 0.05. At least seven replicates from each treatment would be required. If the variance is 0.10 rather than 0.05, then the number of replicates required roughly doubles. In the Crossey and LaPoint[8] study, variance for measures of productivity apparently fell between these two values. As such, increases in sample size alone may not be a practical approach to increasing sensitivity. However, by using a combination of method changes (^{14}C uptake, blocking, compositing, expression of results) and sample size adjustment, the variance of productivity measures could be reduced and sensitivity improved.

Experience, codified into standard methods, often resolves problems with insensitivity due to excessive variance by successfully identifying and accommodating the major sources of variation. Standard measurement methods for functional endpoints can be developed most easily using controlled and replicable systems such as microcosms or artificial streams. In these systems, repeatability and replicability can be characterized and optimized, and a cause-effect relationship between pollution and functional change can be established. Methods can then be extended to more realistic, but uncontrolled, in situ tests and field surveys. Standard sampling methods for functional endpoints in natural systems are then developed.

Following the development of standard methods, the utility of functional endpoints in distinguishing biological integrity must be validated by comparison to more accepted measures. One approach to validating the use of functional measures in evaluations of environmental health has been to compare the responses of the functional parameters to concurrent structural measures. The statistical methods used to assess the relationship between structure and function have been based on inference or on correlation. For example, Crossey and La Point[8] used analysis of variance and Kruskal-Wallis tests and compared the ability of structural and functional endpoints to distinguish between field sites. In another study, Burton and colleagues[9,23] used correlation to examine relationships between functional and structural endpoints in contaminated sediments at several field sites.

Another approach to the validation of functional endpoints is the meta-analysis approach.[61,62] In this approach, the information in many studies is combined to make conclusions about effects and to model relationships. The individual studies that make up the analysis are from different regions and may involve different toxicants and different species. Meta-analysis can be very effective at removing individual site effects associated with many environmental studies.

FUTURE RESEARCH

Despite the potential importance of the functional impairment of ecosystems, functional measures are not now a common component of field studies. Functional endpoints are displaced by structural endpoints that are currently better defined, more decisive, and measured with more consistency and confidence. Additional background information is needed to make functional endpoints useful additions to structural measures in routine assessments of ecosystem health.

Designed experiments should systematically screen functional endpoints and identify those with attributes most useful for general assessments of ecosystem health. These experiments can address the questions of the sensitivity of the endpoint, its ability to distinguish along a gradient of contamination, its generality or specificity to types of contamination, and redundancy of the endpoint compared to others commonly monitored. If, after appropriate method development, an endpoint is still unable to distinguish grossly impaired from unimpaired communities in replicated and controlled experiments or provides no additional insight into the environmental state, its usefulness in biological monitoring will be limited. Insensitive or inconsistent endpoints, all or none responses, and redundant endpoints should be dropped from development in favor of more promising ones. Endpoints that respond very generally to all toxicants are useful for screening when the causative agent is not identified, while endpoints that respond specifically to a narrow range of contaminants are useful for diagnosis. Both are worthy of development, but will be used in different situations. Normal operating ranges and unacceptable levels may be more difficult to characterize for endpoints that are both stimulated and inhibited by pollution. Endpoints that are not intuitively important to the general public and their representatives must be empirically related to one that is. For example, increases in enzyme activity may not be a cause for concern to the general public, but when related to the malodorous result of oxygen-depleted water, changes in enzyme activity could result in support for management action.

Information about sources of variation is needed in order to develop optimal standard methods for functional endpoints. Functional measures will continue to lack power if important sources of variation are not accommodated or controlled. The sensitivity of these endpoints will be improved by appropriate attention to measurement method, sampling, blocking, and compositing. Of equal importance is attention to the method for expressing rate changes. For example, the production-to-respiration ratio is the most common functional measure reported, but seems particularly insensitive and uninformative compared to its component parts. The distribution of ratios such as P/R can be skewed and widely variable when the values for component parts are not highly correlated. Log transformations of the ratio are sometimes, but not invariably, used to compensate. Expressing overall community metabolism as net daily metabolism (gross community production less community respiration) eliminates the problem and maintains the units in linear scale. Areal or volume expressions of rate changes are most common and are useful for holistic evaluations (such as the net daily metabolism of an entire stream length). However, alternative expressions may be more useful when there are concomitant structural changes and an interest in efficiency of function. For example, an assimilation ratio removes some of the structural basis for a change in primary production and, instead, looks at the integrity of the rate process. Similarly, enzyme activity is sometimes weighted by a biomass measure. While respiration measures responded variably to contamination, Odum[54] and Rapport and colleagues[55] have suggested that the ratio of respiration to biomass should consistently increase under stressful conditions as one indication of the decreased efficiency of a stressed system. Increased attention to the expression of rates may clarify the type of change, facilitate comparisons between communities, and help to identify generalized functional responses to stress. In some cases, a functional change will be due to a change in the community composition. In others, the integrity of the rate process itself will have been changed.

Once adequate methods have been developed, decisiveness, or the ability to specify a permissible concentration, is no more troublesome for functional endpoints than for structural ones. In both cases, decisiveness depends on statistical determinations of difference or dose-response curves supplemented by subjective decisions about what is and is not an acceptable consequence of toxic stress. In designed experiments, statistical comparisons to the control make the determination of permissible concentrations relatively easy. Lowest-observable-effects concentrations are calculated and used as evidence of impact. Data from systems with a contamination gradient can also define a concentration-response curve. Some degree of change, a compromise between natural variation and unacceptable impairment, is used as an approximation of an acceptable concentration (e.g., IC20, a concentration impairing response by 20% relative to reference). Obviously, the degree of change that is acceptable requires some subjective input based on knowledge of the normal operating ranges for the endpoint and on the magnitude of changes observed in unacceptably impacted systems. However, these evaluations of the biological significance of the magnitude of the changes observed are advisable even when relying on lowest-observable-effect concentrations as estimates of safe concentrations. Field studies will probably rely on regionally appropriate scoring systems for functional endpoints derived from normal operating ranges in the same manner as those for fish and macroinvertebrate community structure.[16] To develop such a scoring system, differences in the normal operating range for an endpoint across system type, season, photoperiod, and ecoregion must be described and used to define an unacceptable impact, based on distributions of representative systems and on professional judgment.

In addition to the inherent value of protecting ecosystem integrity, society depends on a variety of ecosystem functions. Global temperature maintenance, degradation of sewage and other wastes, and storage and regeneration of nutrients for production of food are functional attributes and must be protected for the global ecosystem to be sustained. Assessments of functional responses seem to be an obvious approach to assuring the continuation of these services provided by ecosystems.

ACKNOWLEDGMENTS

We are indebted to Darla Donald for editorial assistance, to Teresa Moody for word processing, and to reviewers for their thoughtful suggestions.

REFERENCES

1. Ehrlich, P. R. "Carbon Dioxide and the Human Predicament: An Overview," in *On Global Warming,* J. Cairns, Jr., and P. F. Zweifel, Eds. (Blacksburg, VA: Virginia Polytechnic Institute and State University, 1989), pp. 5–11.
2. Schneider, S. H. *Global Warming* (San Francisco: Sierra Club Books, 1989).
3. Rosswall, T. "Global Change: Research Challenge and Policy Dilemma," in *On Global Warming,* J. Cairns, Jr., and P. F. Zweifel, Eds. (Blacksburg, VA: Virginia Polytechnic Institute and State University, 1989), pp. 13–23.
4. Erhlich, P. R., and A. H. Ehrlich. *The Population Explosion* (New York: Simon and Schuster, 1990).
5. Wilson, E. O., Ed. *Biodiversity* (Washington, D.C.: National Academy Press, 1988).
6. Cairns, J., Jr., and J. R. Pratt. "On the Relation Between Structural and Functional Analyses of Ecosystems," *Environ. Toxicol. Chem.* 5:785–786 (1986).
7. Schindler, D. W. "Detecting Ecosystem Responses to Anthropogenic Stress," *Can. J. Fish. Aquat. Sci.* 44:6–25 (1987).
8. Crossey, M. J., and T. W. La Point. "A Comparison of Periphyton Community Structural and Functional Responses to Heavy Metals," *Hydrobiologia* 162:109–121 (1988).
9. Burton, G. A., Jr., B. L. Stemmer, K. L. Winks, P. E. Ross,and L. C. Burnett. "A Multitrophic Level Evaluation of Sediment Toxicity in Waukegan and Indiana Harbors," *Environ. Toxicol. Chem.* 8:1057–1066 (1989).
10. Giesy, J. P., and P. M. Allred. "Replicability of Aquatic Multispecies Test Systems," in *Multispecies Toxicity Testing,* J. Cairns, Jr., Ed. (New York: Pergamon Press, 1985), pp. 187–247.
11. Rodgers, J. H., Jr., K. L. Dickson, and J. Cairns, Jr. "A Review and Analysis of Some Methods used to Measure Functional Aspects of Periphyton," in *Methods and Measurements of Periphyton Communities: A Review,* STP690, R. L. Weitzel, Ed. (Philadelphia: American Society for Testing and Materials, 1979), pp. 142–167.
12. Odum, E. P., J. T. Finn, and E. H. Franz. "Perturbation Theory and the Subsidy-Stress Gradient," *Bioscience* 29:349–352 (1979).
13. Stebbing, A. R. J. "Hormesis: the Stimulation of Growth by Low Levels of Inhibitors," *Sci. Tot. Environ.* 22:213–234 (1982).
14. Pratt, J. R., N. J. Bowers, B. R. Niederlehner, and J. Cairns, Jr. "Effects of Atrazine on Freshwater Microbial Communities," *Arch. Environ. Contam. Toxicol.* 17:449–457 (1988).
15. Meier, R. L. "Communications Stress," *Ann. Rev. Ecol. Syst.* 3:289–314 (1972).
16. Plafkin, J. L., M. T. Barbour, K. D. Porter, S. K. Gross, and R. M. Hughes. *Rapid Bioassessment Protocols for Use in Streams and Rivers: Benthic Macroinvertebrates and Fish* (Springfield, VA: National Technical Information Service, EPA 444/4-89-001, 1989).

17. Wetzel, R. G. "Sediments and Microflora," in *Limnology, 2nd Ed.* (New York: Saunders College Publishing, 1983), pp. 591–614.
18. Ross, P., Jarry, V., and H. Sloterdijk. "A Rapid Bioassay Using the Green Alga *Selenastrum capricornutum* to Screen for Toxicity in St. Lawrence River Sediment Elutriates," in *Functional Testing of Aquatic Biota for Estimating Hazards of Chemicals,* STP988, J. Cairns, Jr., and J. R. Pratt, Eds. (Philadelphia: American Society for Testing and Materials, 1988), pp. 68–73.
19. Munawar, M., and I. F. Munawar. "Phytoplankton Bioassays for Evaluation Toxicity of In Situ Sediment Contaminants," *Hydrobiologia* 149:87–105 (1987).
20. Scanferlato, V. S., and J. Cairns, Jr. "Effects of Sediment-Associated Copper on Ecological Structure and Function of Aquatic Microcosms," *Aquat. Toxicol.* 18:23–34 (1990).
21. Medine, A. J., D. B. Porcella, and V. D. Adams. "Heavy-Metal and Nutrient Effects on Sediment Oxygen Demand in Three-Phase Aquatic Microcosms," in *Microcosms in Ecological Research,* J. P. Giesy, Jr., Ed. (Springfield, VA: National Technical Information Service, CONF-781101, 1980), pp. 279–303.
22. Gahnstrom, G. "Sediment Oxygen Uptake in the Acidified Lake Gardsjon, Sweden," *Ecol. Bull. (Stockholm)* 37:276–286 (1985).
23. Burton, G. A., Jr., A. Drotar, J. M. Lazorchak, and L. L. Bahls. "Relationship of Microbial Activity and *Ceriodaphnia* Responses to Mining Impacts on the Clark Fork River, Montana," *Arch. Environ. Contam. Toxicol.* 16:523–530 (1987).
24. Portier, R. J. "Evaluation of Specific Microbiological Assays for Constructed Wetlands Wastewater Treatment Management," in *Constructed Wetlands for Wastewater Treatment: Municipal, Industrial, and Agricultural,* D. A. Hammer, Ed. (Chelsea, MI: Lewis Publishers, 1989), pp. 515–524.
25. Sayler, G. S., T. W. Sherrill, R. E. Perkins, L. M. Mallory, M. P. Shiaris, and D. Pedersen. "Impact of Coal-Coking Effluent on Sediment Microbial Communities: A Mulitvariate Approach," *Appl. Environ. Microbiol.* 44(5):1118–1129.
26. Andersson, G. "Decomposition of Alder Leaves in Acid Lake Waters," *Ecol. Bull. (Stockholm)* 37:293–299 (1985).
27. Burton, G. A., Jr., and B. L. Stemmer. "Evaluation of Surrogate Tests in Toxicant Impact Assessments," *Tox. Assess.* 3:225–269 (1988).
28. Gahnstrom, G., and S. Fleischer. "Microbial Glucose Transformation in Sediment from Acid Lakes," *Ecol. Bull. (Stockholm)* 37:287–292 (1985).
29. Mills, A. L., and R. R. Colwell. "Microbiological Effects of Metal Ions in Chesapeake Bay Water and Sediment," *Bull. Environ. Contam. Toxicol.* 18:99–103 (1977).
30. Broholm, K., B. K. Jensen, T. H. Christensen, and L. Olsen. "Toxicity of 1,1,1-Trichloroethane and Trichloroethane on a Mixed Culture of Methane-Oxidizing Bacteria," *Appl. Environ. Microbiol.* 56:2488–2493 (1990).
31. Jones, R. D., and N. A. Hood. "The Effects of Organophosphorous Pesticides on Estuarine Ammonia Oxidizers," *Can. J. Microbiol.* 26:1296–1299 (1980).
32. Perez, K. T., E. W. Davey, N. F. Lackie, G. E. Morrison, P. G. Murphy, A. E. Soper, and D. L. Winslow. "Environmental Assessment of a Phthalate Ester, Di (2-Ethylhexyl) Phthalate (DEHP), Derived from a Marine Microcosm," in *Aquatic Toxicology and Hazard Assessment: Sixth Symposium,* STP802, W. E. Bishop, R. D. Cardwell, and B. B. Heidolph, Eds. (Philadelphia: American Society for Testing and Materials, 1983), pp. 180–191.

33. Bruns, D. A., T. P. O'Rourke, and G. B. Wiersma. "Acid Neutralization in Laboratory Sediment-Water Microcosms from a Rocky Mountain Subalpine Lake (USA)," *Environ. Toxicol. Chem.* 9:197–203 (1990).
34. Sayler, G. S., M. Puziss, and M. Silver. "Alkaline Phosphatase Assay for Freshwater Sediments: Application to Perturbed Sediment Systems," *Appl. Environ. Microbiol.* 38:922–927 (1979).
35. Hansen, D. J., and M. E. Tagatz. "A Laboratory Test for Assessing Impacts of Substances on Developing Communities of Benthic Estuarine Organisms," in *Aquatic Toxicology,* STP707, J. G. Eaton, P. R. Parrish, and A. C. Hendricks, Eds. (Philadelphia: American Society for Testing and Materials, 1980), pp. 40–57.
36. Vollenweider, R. A. *A Manual on Methods for Measuring Primary Production in Aquatic Environments, 2nd ed., IBP Handbook No. 12,* (London: Blackwell Scientific, 1974).
37. American Public Health Association, American Water Works Association, and Water Pollution Control Federation. *Standard Methods for the Examination of Water and Wastewater, 17th ed.* (Washington, D.C.: American Public Health Association, 1989).
38. Kaplan, L. A., and T. L. Bott. "Assessment of the Thymidine Technique for Determination of the Bacterial Productivity in Streambed Sediments," *Bull. N. Am. Benthol. Soc.* 7:146 (1990).
39. Riemann, B., and P. Lindgaard-Jorgensen. "Effects of Toxic Substances on Natural Bacterial Assemblages Determined by Means of [^{3}H] Thymidine Incorporation," *Appl. Environ. Microbiol.* 56:75–80 (1990).
40. Anderson, J., W. Birge, J. Gentile, J. Lake, J. Rodgers, Jr., and R. Swartz. "Synopsis of Discussion Session 3: Biological Effects, Bioaccumulation, and Ecotoxicology of Sediment-Associated Chemicals," in *Fate and Effects of Sediment-Bound Chemicals in Aquatic Systems,* K. L. Dickson, A. W. Maki, and W. A. Brungs, Eds. (New York: Pergamon Press, 1987), pp. 267–298.
41. Forsberg, C. "Importance of Sediments in Understanding Nutrient Cycling in Lakes," *Hydrobiologia* 176/177:263–277 (1989).
42. Resh, V. H., A. V. Brown, A. P. Covich, M. E. Gurtz, H. W. Li, G. W. Minshall, S. R. Reice, A. L. Sheldon, J. B. Wallace, and R. C. Wissmar. "The Role of Disturbance in Stream Ecosystems," *J. N. Am. Benthol. Soc.* 7:433–455 (1988).
43. Faulkner, S. P., and C. J. Richardson. "Physical and Chemical Characteristics of Freshwater Wetland Soils," in *Constructed Wetlands for Wastewater Treatment: Municipal, Industrial, and Agricultural,* D. A. Hammer, Ed. (Chelsea, MI: Lewis Publishers, 1989), pp. 41–72.
44. Prichard, P. H., and A. W. Bourquin. "Microbial Toxicity Studies," in *Fundamentals of Aquatic Toxicology,* G. M. Rand, and S. R. Petrocelli, Eds. (New York: Hemisphere, 1985), pp. 177–217.
45. Hale, S. S. "The Role of Benthic Communities in the Nitrogen and Phosphorus Cycles of an Estuary," in *Mineral Cycling in Southeastern Ecosystems,* F. G. Howell, J. B. Gentry, M. H. and Smith, Eds. (Washington, D.C.: U.S. Energy Research and Development Administration, CONF-740513, 1975), pp. 291–308.
46. Paul, B. J., K. E. Corning, and H. C. Duthie. "An Evaluation of the Metabolism of Sestonic and Epilithic Communities in Running Waters Using an Improved Chamber Technique," *Freshwater Biol.* 21:207–215 (1989).

47. Elwood, J. W., J. D. Newbold, R. V. O'Neill, and W. Van Winkle. "Resource Spiraling: An Operational Paradigm for Analyzing Lotic Ecosystems," in *Dynamics of Lotic Ecosystems,* T. D. Fontaine, III, and S. M. Bartell, Eds. (Ann Arbor, MI: Ann Arbor Science, 1983), pp. 3–27.
48. Likens, G. E., F. H. Bormann, N. M. Johnson, D. W. Fisher, and R. S. Pierce. "Effects of Forest Cutting and Herbicide Treatment on Nutrient Budgets in the Hubbard Brook Watershed-Ecosystem," *Ecol. Monogr.* 40:23–47 (1970).
49. Likens, G. E. "Acid Rain and Its Effects on Sediments in Lakes and Streams," *Hydrobiologia* 176/177:331–348 (1989).
50. Heath, R. T. "Holistic Study of an Aquatic Microcosm: Theoretical and Practical Implications," *Int. J. Environ. Stud.* 13:87–93 (1979).
51. Brazner, J. C., S. J. Lozano, M. L. Knuth, L. J. Heinis, D. A. Jensen, K. W. Sargent, S. L. O'Halloran, S. L. Bertelsen, D. K. Tanner, and E. R. Kline. *A Research Design for Littoral Enclosure Studies.* (Duluth, MN: U.S. Environmental Protection Agency, Environmental Research Laboratory, 1987).
52. Webster, J. B., D. J. D'Angelo, and G. T. Peters. "Nitrate and Phosphate Uptake in Streams at Coweeta Hydrologic Laboratory," *Verh. Int. Verein. Limnol.* 24, in press (1991).
53. Pratt, J. R., P. V. McCormick, K. Pontasch, and J. Cairns, Jr. "Evaluating Soluble Toxicants in Contaminated Soils," *Water Air Soil Pollut.* 37:293–307 (1988).
54. Odum, E. P. "Trends Expected in Stressed Ecosystems," *Bioscience* 35:419–422 (1985).
55. Rapport, D. J., H. A. Regier, and T. C. Hutchinson. "Ecosystem Behavior Under Stress," *Am. Nat.* 125:617–640 (1985).
56. Sheehan, P. J. "Ecotoxicological Considerations," in *Controlling Chemical Hazards: Fundamentals of the Management of Toxic Chemicals,* R. P. Cote, and P. G. Wells, Eds. (London: Unwin Hyman, 1991), pp. 79–118.
57. Cairns, J., and B. R. Niederlehner. "Adaptation of Microbial Communities to Toxic Stress," *Bull. N. Am. Benthol. Soc.* 7:140 (1990).
58. Hurlbert, S. H. "Pseudoreplication and the Design of Ecological Field Experiments," *Ecol. Monogr.* 54:187–211 (1984).
59. Long, E. R., and P. M. Chapman. "A Sediment Quality Triad: Measures of Sediment Contamination, Toxicity and Infaunal Community Composition in Puget Sound," *Mar. Pollut Bull.* 16:405–415 (1985).
60. Zar, J. H. *Biostatistical Analysis* (Englewood Cliffs, NJ: Prentice Hall, 1984).
61. Mann, C. "Meta-Analysis in the Breech," *Science* 249:476–480 (1990).
62. Suter, G. W., II, D. S. Vaughn, and R. H. Gardner. "Risk Assessment by Analysis of Extrapolation Error: A Demonstration of Effects of Pollutants on Fish," *Environ. Toxicol. Chem.* 2:369–378 (1983).

CHAPTER 7

The Significance of In-Place* Contaminated Marine Sediments on the Water Column: Processes and Effects

Robert M. Burgess and K. John Scott

INTRODUCTION

In the brief history of marine environmental toxicology, most research has been media oriented. A great deal of scientific and regulatory effort in aquatic toxicology has concentrated on characterizing the toxic effects in the water column environment resulting from contaminants of terrestrial origin.[1] Likewise, the hazards posed by contaminated sediments also have been recognized.[2-6] Now it is necessary for environmental scientists, managers, and regulators to expand the scope of their respective media-oriented research. This broadening of research scope must acknowledge the potential for interaction between the water column and sediment environments. This interaction, as a function of several chemical, physical, and biological processes, may result in the eventuality of detrimental effects in one medium (i.e., the water column) as a result of the contaminated status of another (i.e., the sediments). Understanding these processes and the resulting detrimental effects will provide us a greater foundation for studying, managing, and regulating the entire integrated aquatic environment.

---

* "In-Place" is emphasized in order to exclude dredge materials and other forms of sediment that have been grossly altered relative to their natural state, other than being contaminated.

Despite the role of marine sediments as the ultimate sink for many classes of anthropogenic contaminants, little research has examined their significance as a source of adverse impact to the water column environment. In contrast, the geochemical processes involved in the release (mobilization) and transport of pollutants from contaminated sediments have been intensively investigated,[7-11] and it has been well demonstrated that contaminated sediments act as a source of pollutants to the oceans and atmosphere.[12-14] Some consider the sediments to be a source of aquatic contamination that is as significant as other nonpoint sources, such as urban and agricultural runoff, solid-waste leachate, mining spoils, and atmospheric deposition.[15] Recently, the U.S. Environmental Protection Agency Science Advisory Board designated contaminated sediments as a nonpoint source of pollution (see Chapter 16).

The significance of sediment as a source of contaminants is particularily germane when one considers that, during the last decade, environmental regulation has resulted in the reduction of loading to the marine environment from atmospheric and terrestrial sources.[16] However, sediment reservoirs have the potential to release, in a chronic fashion, decades worth of accumulated contaminants (Figure 1). Data from several sites suggest that bioaccumulated contaminants probably originate from the sediments[17-20] via exposure to dissolved and particulate phases present in the water column, as well as through trophic transfer. Therefore, water column exposure has been demonstrated, but the degree to which this exposure causes adverse effects requires further study.

Future research and regulatory efforts must address the potentially adverse role contaminated sediments may have on the water column environment. In this context, several preliminary questions posed in this chapter address this potential problem. They are

- What processes may cause a problem, e.g., water column toxicity, to occur?
- Can we measure the effects of such a problem?
- Does a problem actually exist?

The first of three sections discusses chemical, physical, and biological processes that result in the release and transport of contaminants from sediments into the water column. The second section reviews the toxicological effects that have been observed during field surveys, laboratory, in situ organism testing, and microcosm/mesocosm studies. The final section draws conclusions from the available information relative to the questions above.

PROCESSES

The processes involved in the release of contaminants from marine sediments include physical, chemical, and biological components. Individually or in

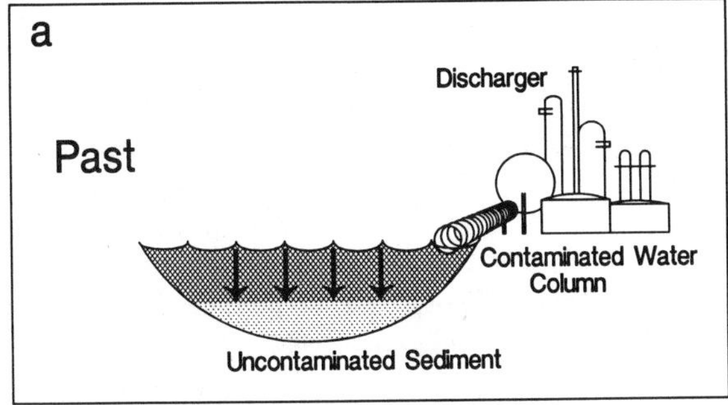

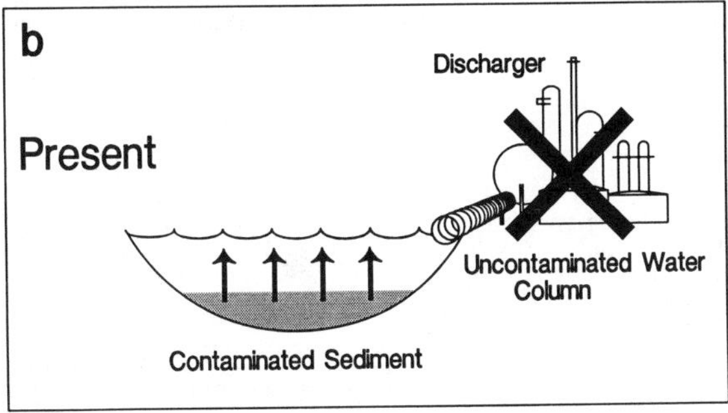

Figure 1. (a) Historical contamination of sediments from anthropogenic sources; (b) contemporary release of contaminants into the water column from contaminated sediments.

combination these processes contribute to the transport of potentially available contaminants from sediments to the water column (also see Chapters 1, 3, and 15).

Chemical

The chemical form of a pollutant determines its potential bioavailability to marine organisms. The toxicological literature indicates that "free" and/or dissolved forms of chemicals are often the most bioavailable and toxic.[21-27] Conversely, reduced bioavailability has been associated with nonionic organic compounds that are bound to sediments or colloidal matter (dissolved organic carbon).[21,25-27] Further, the toxicity of metals are observed to decrease when bound to organic ligands or sulfide precipitates.[22,23,28] For this discussion the "free" form of a given contaminant will be considered the one of greatest toxic potential to water column organisms.

Lee,[7] in a thorough discussion of processes of exchange between sediment and water column, reviews several of the chemical reactions that may result in the release of contaminants and nutrients. The primary processes include a variety of common reactions. Among these are acid-base reactions that affect pollutant speciation and sorption as a function of water column and interstitial pH, as well as precipitation reactions that result in sequestration of some types of pollutants by carbonates, hydrous oxides, and sulfides. Also included are oxidation-reduction reactions that are mediated by the addition or loss of electrons in the sediment environment. These reactions, and others, can be affected by microorganisms that metabolize organic complexes, resulting in the release of associated contaminants and the reduction of oxygen concentrations in the sediments (discussed below).

In the 20 years since Lee's review, several insights concerning these exchange processes have been made. Investigations into the chemistry of contaminants in marine sediments have evolved to the point that partitioning behavior can be predicted for several chemical classes, based on geochemical models. The most commonly modeled and best-described contaminants are nonionic organics and transition metals.

Nonionic Organics

Currently, the environmental partitioning behavior of nonionic organic contaminant is the best understood of any class of organic chemical.[29,30] Several geochemical processes control the form of organic pollutants when in the sediment environment with sorption dominating.[31,32] Sorption processes consist of a myriad of chemical reactions, including hydrophobic sorption, oxidation/reduction, complexation, hydrogen bonding, and ion exchange.[32] A convenient, although not well-understood, result of all these reactions affecting the sorption behavior, and therefore bioavailability, is that certain traits have been found among chemicals. These common traits allow all of the thousands of synthetic chemicals that have been manufactured to be catagorized into a few broad classes as a function of their partitioning behavior. Nonionic organic pollutants, exemplified by polychlorinated biphenyls (PCB), polycyclic aromatic hydrocarbons (PAH), and some pesticides, like DDT and Chlordane, will behave in a somewhat similiar manner, but not in the same fashion as ionic compounds like the phenols, carboxylic acids, and amines.[33] Once associated with the sediments, the members of one class of organic chemicals will partition differently from members of the other classes.

The dominant sediment characteristic affecting the partitioning behavior of many nonionic organics has been shown to be the fraction of organic carbon associated with the sediments and interstitial waters.[30,34,35] Prediction of the environmental fate of chemicals like PCBs and some PAHs have utilized partitioning models that are based on the strong relationship between organic carbon and nonionic organic contaminants. This approach has led to the development of the three-phase partitioning model, expressed in mass balance form:

$$C_T = C_d + mf_{oc}C_d + m_{DOC}K_{DOC}C_d \tag{1}$$

where C_T is the total concentration of a nonionic organic contaminant associated with the sediment (μg/kg), C_d is the concentration of the free chemical (μg/L), m the

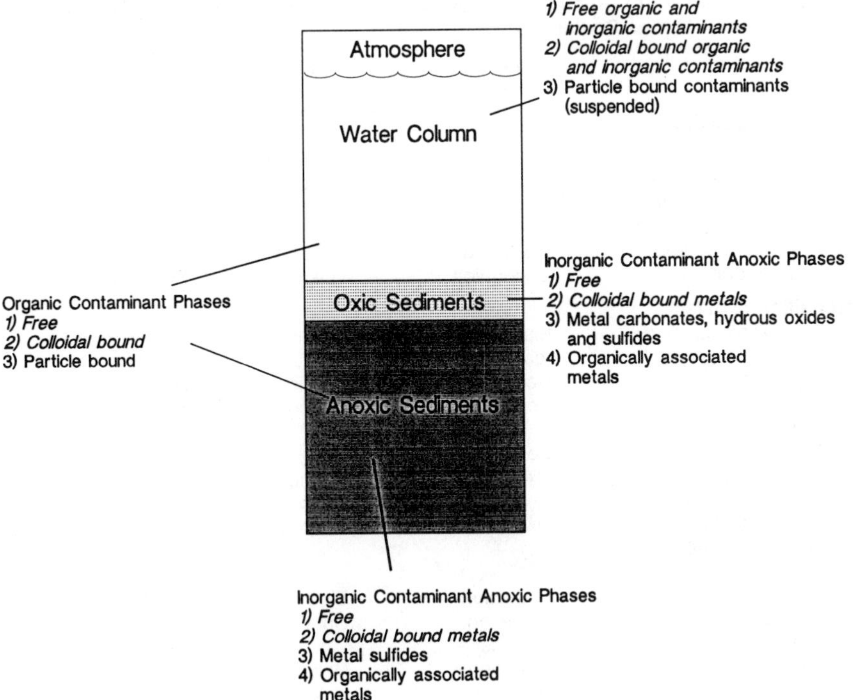

Figure 2. Cross-section of the sediment/water column interface and the prevalent forms of contaminants. Forms presented in italics are considered the most mobile and therefore are the most likely to be released into the water column.

particle concentration (kg/L), f_{OC} the fraction of organic carbon (kg organic carbon/kg), m_{DOC} the colloid concentration (kg/L), and K_{DOC} the partitioning coefficient for colloids. The model predicts that while associated with the sediment, a given nonionic organic pollutant will be partitioned into three possible compartments: sediment-sorbed chemical; organic-carbon-sorbed chemical (colloidal); and free chemical.[30,36-38] It is this free chemical, and to some extent the colloidal form, that is "chemically mobile" and can be transported from the sediment matrix into the water column where it potentially can affect sensitive organisms (Figure 2).[11,39-43]

The release of nonionic organic compounds from contaminated sediments has been inferred, at several field sites, to explain the presence of these chemicals in unusually high concentrations in the water column. Generally, this work has acknowledged that flux does occur, but few studies have investigated the forms in which pollutants are released and, thus, their ultimate bioavailability and toxicity. Pavlou and Dexter[44] concluded that the sediments in Puget Sound, Washington act as the predominant sink for PCBs; however, when sediments are disturbed they may act as a source of PCBs to the water column. Larsson[12] concludes, based on freshwater studies,[45,46] that contaminated sediments are a source of PCBs to the water

Table 1. Summary of Studies That Have Demonstrated the Release of Contaminants from Marine Sediments

Contaminant	Release Process	Form Released	Source
Nonionic Organics			
PCB Mixtures	Desorption	Unknown	Wildish et al.[51]
Naphthalene, 1-methyl Naphthalene, 2-methyl Naphthalene, biphenyl	Desorption	Unknown	Pruell and Quinn[52]
Benz[a]anthracene	Desorption	Unknown	McElroy et al.[53]
PCB congeners	Desorption	Unknown	Burgess[54]
Transition Metals			
Copper, nickel, zinc	Biologically reduced pH	Unknown	Presley et al.[65]
Copper, zinc	Reduction	Unknown	Duinker et al.[66]
Cadmium, cobalt, copper, lead, zinc	Competing ions	Unknown	Muller and Forstner[68]
Zinc	Competing ions	Free	Troup and Bricker[69]
Cadmium, copper, lead, nickel, zinc	Oxidation	Unknown	Lu and Chen[70]
Arsenic, cadmium, copper, mercury, lead, zinc	Oxidation	Unknown	Brannon, Plumb, and Smith[9]
Mercury	Reduction	Elemental	Bothner et al.[67]
Cadmium, copper, lead	Oxidation	Unknown	Hunt and Smith[71]
Cadmium, lead, zinc	Oxidation	Unknown	Baeyens et al.[72]
Copper	Oxidation	Unknown	Burgess[54]

columns of lakes as well as oceans. The presence of PCBs in the waters of New Bedford Harbor, Massachusetts are directly correlated with the highly contaminated sediments of the harbor.[47,48] Bioaccumulation of several PCB congeners by mussels deployed in the harbor correlated best with the dissolved form (free and colloidal) of the specific pollutants.[49] Similiarily, mussels exposed to contaminated sediments from Providence River, Rhode Island in laboratory exposures accumulated the dissolved form of PCBs and PAHs.[50] These observations suggest the importance of the form that a contaminant has assumed when released from the sediments, to the prediction of its availability to water column organisms.

In laboratory studies, release of nonionic organic pollutants from contaminated sediments into the water column also has been observed (Table 1). These studies demonstrate the mobility of PCB mixtures (Aroclors) and congeners, benz[a]anthracene, naphthalene, 1-methyl naphthalene, 2-methyl naphthalene, and biphenyl.[51-54] Because of operational difficulties in distinguishing between phases, i.e., free vs. bound forms, none of the studies determined the actual form of the released pollutant.

The quantity of the free form of a nonionic organic chemical must be established to predict possible adverse effects. Given this requirement, the amount of the particulate form, which can be suspended into the water column due to turbulent diffusive events, e.g., bioturbation and waves, cannot be dismissed as entirely insignificant. The particulate form cannot be dismissed because of the potential for the ingestion of the particle-sorbed form of contaminants by water column organisms. But more important to this discussion, the partitioning of particle-sorbed contaminants, both organic and metals, into the dissolved phase has been shown to increase as suspended particle concentrations increase.[55-57] For example, when contaminated sediments are suspended into the water column, an increase in water column concentrations of dissolved contaminants occurs. How and why this release occurs is not well understood, and factors under consideration range from increases in the concentration of colloids, macromolecules, and dissolved ligands, to particle aggregation resulting in reduced particle surface area, to kinetic interactions.[30] Regardless of the cause, the chemical behavior resulting from this phenomenon has been observed and indicates that, when contaminated sediments are suspended into the water column, not only is the particulate form of the pollutant accessible to water column organisms, but an increase in the concentration of dissolved and colloidal forms also occurs (Figure 3).

Ionic Organics

The behavior of ionic organic pollutants, including surfactants, in marine sediments has not been as extensively studied.[58] As with the nonionic organic chemicals, organic carbon appears to be a critical factor in the partitioning behavior of the ionic organics in sediment systems. The critical micelle concentration also has been identified as being important in these predictions. The partitioning behavior of organic acids such as 2,4-dinitro-o-cresol, 2-(2,4,5-trichlorophenoxy)propanoic acid (Silvex), and pentachlorophenol are being investigated, and the organic carbon fraction and pH appear to be dominating factors.[59-61]

Transition Metals

The fate of transition metals in aquatic systems also is a function of several geochemical processes. These processes include physical-chemical sorption effects caused by long-range van der Waals forces acting between charges present on the surfaces of particles and dissolved species. Chemical processes can include ion exchange, organic complexation, precipitation and coprecipitation, solid-state diffusion, and isomorphic substitution. Dominant processes in the marine environment are ion exchange reactions that may be altered by changes in salinity. Also dominant are precipitation and coprecipitation reactions (metallic hydrous oxides, sulfides, and carbonates) that are prone to dissolution, resulting in the release of metals based on pH and oxidation-reduction-state conditions.[8,62,63]

To date, the factor(s) that control the form in which metals exist in sediment systems have not been clearly delineated, suggesting that one master factor (i.e., like organic carbon for nonionic organics) does not exist for all conditions, but that

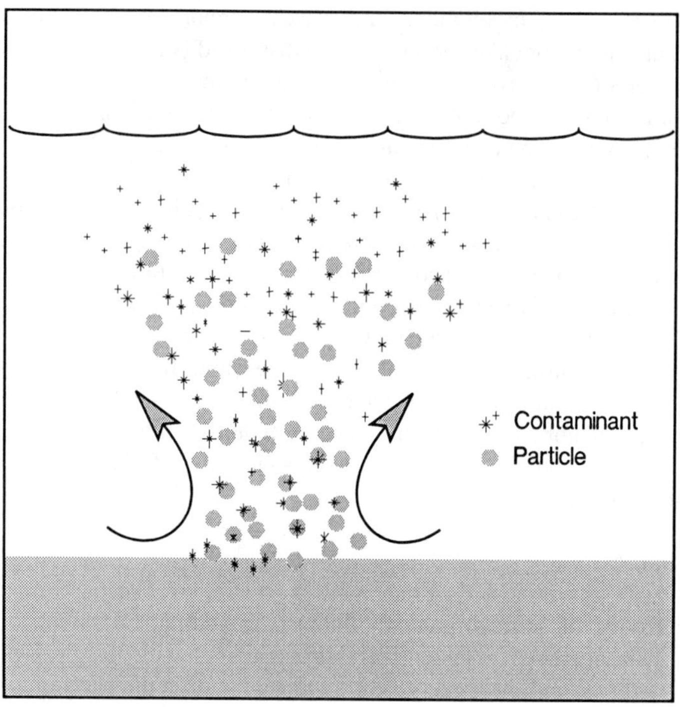

Figure 3. Conceptual representation of the release of contaminants from suspended sediments.

several are active.[64] This lack of understanding of metal speciation exasperates the difficulty involved in determining which processes are most important in the release of metals into the water column. Despite these limitations, Forstner and Wittman[8] list four types of chemical processes that may cause metals to occur in forms that are amenable to transport out of the sediment matrix. These processes include (1) the elevation of salinity, resulting in alkali (e.g., Na^+, K^+) and alkaline earth cations (e.g., Mg^{2+}, Ca^{2+}) competing with metal ions sorbed onto particle surfaces; (2) changes in the oxidation-reduction state of the sediments toward reduced conditions upon which iron and manganese hydroxides begin to dissolve, releasing associated transition metals; (3) reduction of pH, causing dissolution of carbonates and hydrous oxides, and increased competition with hydrogen ions, resulting in transition metal release; and (4) the presence of natural and synthetic complexing agents that compete with metals for sorption sites on sediments. The first three of these processes result in the release of mobile and free metal; the fourth process occasionally releases an organically bound, but still mobile, metal (Figure 2).[42] Other processes that are less well defined include biochemical transformations of the sediment environment by benthic bacteria.[8]

The partitioning of metals to acid volatile sulfide (AVS) has recently been shown to be a major factor controlling the availability of toxic transition metals in anoxic sediments.[28] According to this theory, once the molar concentration of a metal(s) exceeds the binding capacity of the sediment acid volatile sulfide pool, transition metals are present in bioavailable forms:

$$\frac{C_{Me}}{C_{AVS}} > 1 = \text{Free Metal} \tag{2}$$

where C_{Me} is the molar concentration of a transition metal, and C_{AVS} is the molar concentration of acid volatile sulfide.

Several field and laboratory studies have investigated the release of metals from contaminated sediments (Table 1). Mobilization of the metals arsenic, cadmium, cobalt, copper, lead, mercury, nickel, and zinc have been associated with several processes, including lowered pH,[65] reducing conditions,[66,67] competing seawater ions,[68,69] changes in the oxidation-reduction condition of the sediments,[9,54,70,71] and the degradation of organic complexes.[72]

Analytical measurements of the total concentration of metal in a sample are relatively simple and commonplace; however, the actual or predicted degree of metal speciation is far more useful for determining transition metal behavior in marine systems.[73] Unfortunately, methods for making these types of analytical measures have seldom been applied under field conditions.[74,75]

Physical

Free and mobile contaminants often do not remain in the sediment matrix; rather, several physical processes transport the contaminants vertically into the water column. Previous work has identified three major physical processes that facilitate contaminant transport: advection, diffusion, and convection. These processes result in the bulk transport of interstitial water (including colloids) and particles from the sediment to the overlying water column.

Advection is defined as the flow or movement of interstitial water or particles. An important quantitative component of advection, as well as diffusion and convection, is flux. Flux is the rate of transfer of material from one reservoir to another, such as from the sediment to the water column. Advective fluxes can be described simply by

$$F_a = \varnothing CU \tag{3}$$

where F_a is the advective flux (mass/surface area-time), C is the concentration of the contaminant, U is the velocity of flow, and $\varnothing$ is the porosity of the sediment.[76-78] Porosity is the relative pore volume of the sediment and is equal to the areal porosity.

Conceptually, porosity can be thought of as the percentage of a given section of the sediment-water column interface, that consists of "holes" through which interstitial water can flow into the water column.[79]

Diffusion is defined as the random motion of objects, and consists of two components. The first component, molecular diffusion, is the thermally induced random motion of molecules, atoms, or ions by intermolecular forces in gases, fluids, or solids. The second component, turbulent or eddy diffusion, involves a medium in which relatively larger objects or volumes, like particles and interstitial water, are distributed. The two types of diffusion ultimately result in the dispersal of material throughout a medium in different ways. Molecular diffusion operates as a function of chemical potential acting to decrease different types of gradients. Turbulent diffusion is based on the dissipation of kinetic energy. Another distinction between molecular and turbulent diffusion is scale. Molecular diffusion operates most effectively on the scale of atoms, molecules, and small quantities of materials; turbulent diffusion operates on the scale of lakes, oceans, and the atmosphere. Molecular diffusional flux, in one dimension, is described simply in the following equation:

$$F_d = \varnothing\,\theta\,(dC/dz) \tag{4}$$

where F_d is the diffusion flux (mass/surface area-time), θ is the diffusion coefficient, $\varnothing$ is the porosity of the sediment, and dC/dz describes a vertical concentration gradient of the contaminant or particle of interest, with C equaling the concentration of contaminant, and z the depth of sediment.[76-78] Turbulent diffusive events (i.e., waves, storms, tides) are not as easily expressed.

Convection has also been indicated as a possible physical process contributing to the release of contaminants from sediments into the water column.[16,78,80] Convection, like advection, is a mode of fluid flow. A form of convection, forced convection, involves the vertical movement of fluid, i.e., interstitial water, as a result of mechanical forces. Thibodeaux and Boyle[80] demonstrate the transport of dye-injected interstitial water into the water column caused by a current flowing over an irregular sediment surface. In general, convection occurs in high-energy environments where sedimentary features (for example, sand ripples[81]) are likely to form and where significant sediment deposition is not very likely to occur. Based on the low probability of contaminated sediment deposition in such environments, convection will not be discussed further as an important transport process.

The emphases of this discussion are the processes that lead to the release of contaminants from the sediment matrix into the water column. Some consideration of the processes that resulted in the contamination of the sediments is necessary. The physical process of deposition, together with chemical adsorption, causes contaminants present in the water column to associate with particles in suspension, which eventually accumulate as sediments,[82-87] generally expressed as:

$$D_f = -\varnothing SC \tag{5}$$

where D_f is the depositional flux on to the sediment surface (mass/surface area-time) and S is the deposition rate (∅ and C as defined previously).

This combination of adsorption and deposition has led to the contamination of a multitude of coastal marine and freshwater locations. As mentioned earlier, the water column is now less influenced by contaminants from terrestrial sources, and thus, the sediments that accumulate on the sea floor presently are not as contaminated as those deposited in the past. One may suppose that these newly deposited sediments may form a cap over the historically contaminated sediments, sealing them from further contact with the water column environment. While it can be assumed that this layer may retard release processes, it likely does not stop these processes.

The combined fluxes of interstitial water and particles, and the deposition of particles across the sediment-water interface due to the effects of advective, diffusive, and depositional processes, can be described by:

$$F_t = \varnothing\left[\theta(dC/dz) + UC - SC\right] \tag{6}$$

where F_t is the total flux of contaminants from the sediment.[76-78,88]

The processes described above are assumed to occur continually over time. The potential for wave action, atmospheric phenomena, tides, currents, and biotic interactions to cause sediment resuspension and fluid transport which augment turbulent diffusive and advective processes are critical considerations that are not fully understood.[89,90] Terms that describe sporadic events and processes can be added to the equation, but they are not as easily applied.

Several "real-world" events occur in the marine environment, that are based on the described processes and result in the transport of pollutants to the water column from sediments.[14,91-94] An example of advective transport is the upward flow of interstitial water, caused by hydrostatic pressure gradients formed by terrestrial ground water aquifers adjacent to coastal regions. Molecular diffusion, which causes interstitial water to flow from the sediments into the water column, operates as a function of salinity, pollutant concentration, and density gradients. Turbulent diffusion is observed as the transport of interstitial water and particles by currents, wave action, tides, and bioturbation. In terms of relative magnitude, diffusive processes are considered to be dominant, with advective processes estimated to be only 1% of diffusive contribution in marine systems.[93]

Biological

The chemical and physical processes described above are the primary abiotic forces resulting in the transport of pollutants from the sediments to the water column.

However, biological processes on both the microscale and macroscale can accelerate the abiotic processes. Microbenthic processes include microbial activity that results in alterations of the sediment chemical environment. These changes often sustain the cycling of contaminants through form alterations, which increases their ability to be released.[7] Macrobenthic effects are illustrated by the construction and ventilation of borrows and the swimming of larger benthic and epibenthic organisms, which result in advective and turbulent diffusive events, resulting in fluid transport and sediment suspension into the water column. Together, the microbenthic and macrobenthic processes can be placed into an ecological succession with stages that have specific physical and chemical properties.[95,96]

Microbenthic Processes

The significance of microorganisms in aquatic ecosystems is well documented and includes mediation of the cycling of many elements and compounds.[97] Of particular interest to this discussion is their contribution to benthic respiration of oxygen and the metabolism of complexes that include contaminants and organic carbon.[98] This activity can result in contaminant mobilization.[7] A second aspect of this process is the formation of low molecular weight organic compounds (colloids) to which contaminants readily partition.[8,42] Along with the metabolism of organic carbon complexes, benthic bacteria induce changes to sediment oxidation-reduction and pH conditions. This type of reaction can be exemplified by:

$$MeS + 8Fe^{3+} + 4H_2O \rightarrow Me^{2+} + SO_4 + 8H^+ + 8Fe^{2+} \qquad (7)$$

where Me^{2+} is a transition metal in which the bacteria metabolize metal sulfides, releasing metal ions and causing an excess of hydrogen ions, which reduces the pH to the optimal bacterial conditions. This reduction in pH will cause metal hydrous oxides, carbonates, and sulfides, as well as iron and manganese coprecipitates, to dissolve, releasing more metals. Conversely, bacteria can also enzymatically convert transition metals into metal organic complexes. This process is exemplified by the conversion of divalent mercury to the toxic form, methyl mercury (Figure 4). The methylated form of mercury is very mobile and is also a neurotoxin to higher organisms.[99,100]

The formation of nitrogen and methane gas bubbles by microorganisms and then their vertical movement (ebullition) through the sediment constitute a novel enhancement of pollutant transport (Figure 5). Semivolatile pollutants partition from the sediments and interstitial waters into the nitrogen and methane bubbles. Once the bubble has reached the water column, some amount of the pollutants partition to the water phase, while the remaining pollutants move into the atmosphere.[101]

Macrobenthic Processes

The significance of macrobenthic biological processes on the benthic environment has been summarized previously.[95,102-106] Also, several mathematical models of

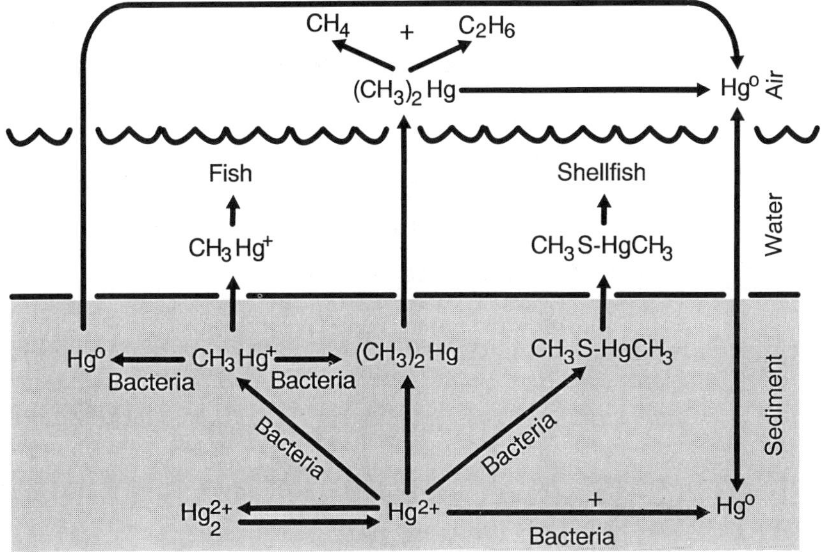

Figure 4. The biological cycle of mercury indicating the potential for bacteria to enhance the transport of mercury from the sediments into the water column. (Based on Wood.[99])

bioturbation exist, with varying levels of sophistication, that address biotic resuspension of particles and the transport of interstitial waters.[107] This body of work demonstrates that, while microbenthic organisms generally impact chemical processes, the presence and activity of macrobenthic infaunal and epifaunal organisms alter physical processes that mediate contaminant transport. From the point of view of effects on the sediment physical properties, organisms can modify sediment stratigraphy and alter sediment particle size, shear strength, and porosity, which then have various effects on diffusional and advectional fluxes. Other effects, such as increases in turbidity, changes in the sediment oxygen content, pH, and penetration of the reduction-oxidation horizon, all resulting from borrowing activity via the introduction of seawater, will affect the form of many contaminants in the interstitial waters.

Six biological processes have been identified as important in the cycling of pollutants in benthic systems. These processes are bioaccumulation, trophic transfer, migration, biodegradation, biodeposition, and bioturbation.[104,105] For this discussion, the most pertinent of these processes are migration and bioturbation, both of which can result in the transport of contaminants from the sediments into the water column.

Migration is defined as the "spatial movement of animals entering or leaving areas of sediment contamination on seasonal, daily, or other temporal intervals."[105] These migrations include the movement of infaunal and epifaunal organisms into the water column. Many of these migrating organisms have bioaccumulated contaminants from the sediments. Observations of such phenomena include deep sea scavenging amphipods feeding in the water column, benthic organisms with pelagic larvae, and the ascent of lipid-rich particles associated with decomposing animals or feeding

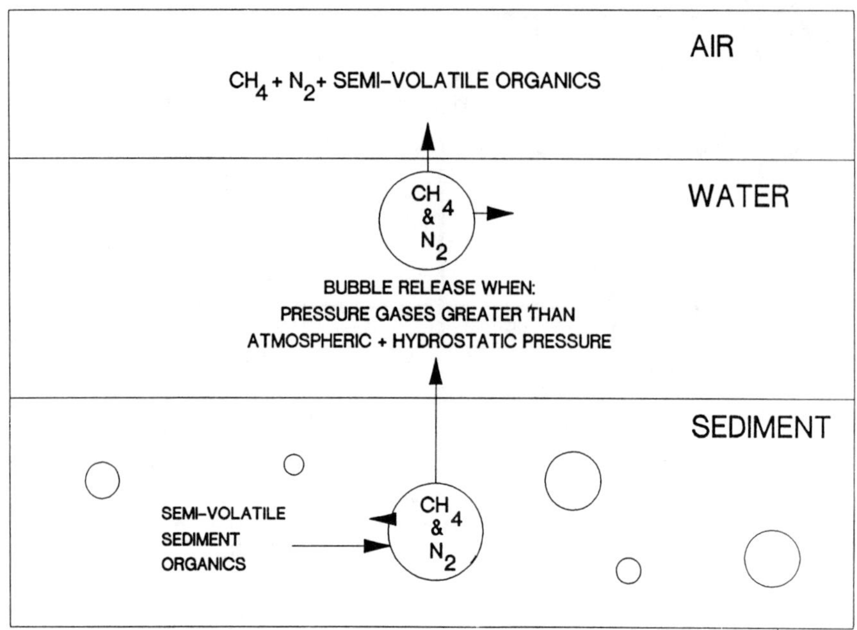

Figure 5. Suggested pathway for contaminants transported from sediment by gas bubble ebullition. (From Adams, D. D., N. J. Fendinger, and D. E. Glotfelty. "Biogenic Gas Production and Mobilization of In-Place Sediment Contaminants by Gas Ebullition," in *Sediments: Chemistry and Toxicity of In-Place Pollutants* (Chelsea, MI: Lewis Publ., 1990), p. 215. With permission.)

activities.[108] The significance of these types of processes have often been overlooked and considered unimportant to contaminant transport.

Bioturbation is the primary biogenic activity that can intensify physical processes that result in fluid and particle transport (Table 2; Figure 6). Generally, fluid transport is 10 to 100 times greater on a weight basis than particle transport.[95] Further, bioturbation can cause destabilization of the sediment through increase in sediment water content, disaggregation of the sediment organic matrix, and modification of the surface topography.[104] Disaggregation of the organic matrix, like elevation of sediment water content, increases the porosity and, in turn, increases both diffusional and advective fluxes. Similiarily, microtopography increases the potential for turbulent diffusion at the sediment surface, which may result in resuspension.

Classification of particular bioturbation activities allows linkage to specific physical processes. For example, the building of burrows and reworking of the sediment enhances diffusion processes. Biogenic reworking of the sediment surface results in particle transport into the water column caused by turbulent diffusion.[109] Infaunal tube-building organisms create systems of burrows that act as networks of diffusive cylinders that effectively increase sediment porosity and allow for

Table 2. Effects of the Activities of Benthic Marine Invertebrates on the Water Column

	Feeding Parameter			
Class Zone	**Suspension Feeders**		**Deposit Feeders**	
	Filter	**Raptorial**	**Surface**	**Subsurface**
	(bottom into water column)		**(surface to –3 cm)**	**(surface to –30 cm)**
Effect				
Feeding	0[a]	0	+	++
Burrowing	0	0	+	+
Excavation	0	0	+	+
Biodeposition	+[b]	+	+	0
Pelletization	+	0	+	+
Tubes	0	0	+	0
Fluid transport	++[c]	0	+	++

Source: Lee and Swartz.[104]

[a] 0 Process absent or of minor importance.
[b] + Process present and is moderately or highly significant.
[c] ++ Process present and has major effects .

enhanced release of contaminated interstitial waters into the overlying water column.[109]

Upon completion of the burrow system, the infauna initiate activity that results in advective processes. Irrigation of the burrows with uncontaminated overlying water enhances diffusion by increasing the magnitude of the concentration gradient between the interstitial water and the burrow water. This process increases diffusion into the burrows of contaminants that are then advected into the water column. The resulting fluid transport may significantly increase the exchange of pollutants from the sediments to the overlying water column.[96,109]

The magnitude of the increase in the diffusional and advective transport of materials between the sediments and water column, due to the presence of organisms, is significant. Generally, an increase in the flux of materials out of the sediment of about two to ten times occurs, as compared to that observed for purely abiotic diffusion and advection physical processes.[53,110] The flux of silica from marine sediments into a field-deployed flux chamber decreased significantly during the winter when biological activity was reduced naturally or when benthic organisms were killed by asphyxiation (Table 3).[110]

Ecological Succession

An approach for defining the ecological status of a benthic community is by placing it on a successional continuum. Rhoads and Boyer [95] and Rhoads and Germano [111] describe a multistage ecological succession for temperate marine soft-sediment communities (Figure 7A). The pioneering stages (e.g., stage 1) generally consist of small, opportunistic, tube-dwelling polychaetes and amphipods.

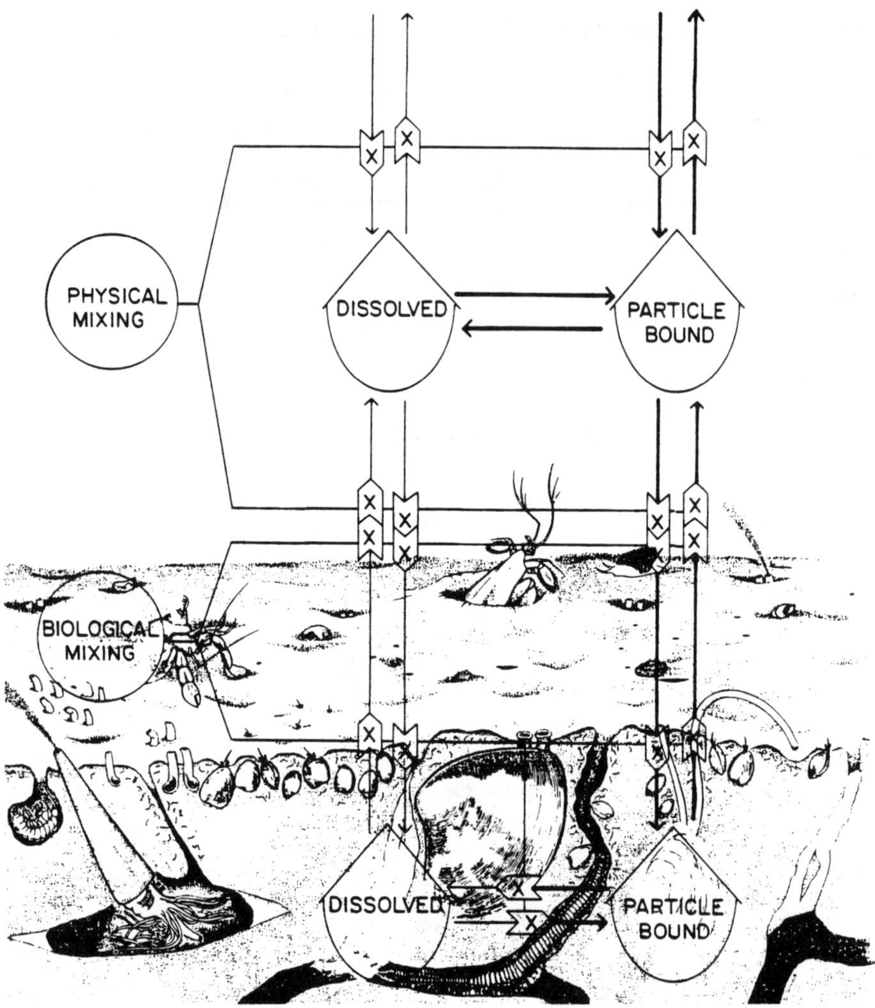

Figure 6. Diagramatic view of the activity of macrobenthic organisms. (From Davis, W. R. and J. C. Means. "A Developing Model of Benthic-Water Contaminant Transport in Bioturbated Sediment," in *Proc. 21st EMBS* (Gdańsk, Poland: Polish Acad. Sci.-Inst. Oceanology, 1989), p.215. With permission.)

Characteristically, these organisms feed from the water column and surficial sediments. The biota of this stage have several minor effects on the sediment-water column interface: (1) tube aggregations alter surface microtopography and roughness, (2) fluid and particle bioturbation results in the transport of water in and out of the sediment, and (3) deposit and suspension feeding causes the accumulation of fecal pellets. While pioneering stages have only minor impacts on the sediment-water column interface, equilibrium stages dramatically alter the interface. Equilibrium stages (e.g., stage 3) are associated with both tubiculous and free-living

Table 3. Summary of the Effects of Abiotic and Biological Processes on the Transport of Silica from the Sediments to the Water Column

	Magnitude of Flux (mg Silica per m²/d)		
Season	**Summer**	**Fall**	**Winter**
Water Temperature (C°)	**14**	**14**	**–1**
Abiotic processes only	15.7	10.1	5.5
Biological-enhanced processes	161.0	64.6	5.5

Source: Van der Loeff et al.[110]

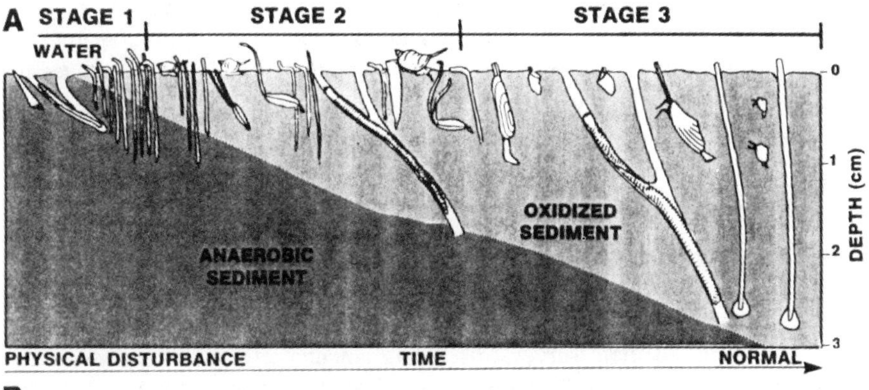

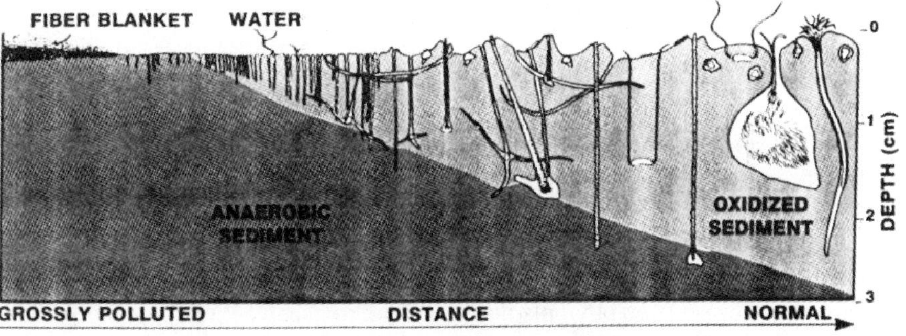

Figure 7. Schematic representation of the marine benthic successional continuum: (A) continuum in a pristine setting in which physical disturbances, e.g., storms, may result in disruption of ecological succession; (B) continuum along a gradient in sediment contamination, which demonstrates a disruption in ecological succession in the heavily contaminated portion of the gradient. (From Rhoads, D. C., and J. D. Germano. "Interpreting Long-Term Changes in Benthic Community Structure: A New Protocol," *Hydrobiologia* 142:291–308 (1986). With permission.)

infaunal deposit feeders, exemplified by "head-down" conveyer-belt feeders. Generally, these stages are found in deeply oxygenated sediments with reduced conditions at depths of greater than 10 cm. This zone of transition between oxidized and reduced conditions is correlated with high microbenthic productivity.[112]

Equilibrium stages are also associated with significant vertical particle and interstitial water exchange. Effects on the sediment/water column interface include (1) transfer of overlying water and particles to sediment depths of 10 to 20 cm, (2) homogeneous mixing of the sediment particles, (3) formation of void spaces in the sediment profile, and (4) alteration of the surface by feeding pits and fecal and excavation mounds.

Utilization of this ecological paradigm allows for some speculation as to the impacts biotic processes may have at a site of a given stage. For example, would we expect greater exchange of contaminants to the water column from sediments with an equilibrium stage community relative to a pioneering community because of the greater biological activity? Furthermore, what factors, i.e., chemical, physical, biological, determine the benthic successional stage, or in the case of contaminated sediments, do the toxicants "lock" the benthic community into an eternal pioneering stage, as has been suggested (Figure 7B)?[113]

EFFECTS

The following sections discuss the adverse impacts that have been observed to occur in the water column which have been directly associated with in-place contaminated sediments. Four areas of interest where sediments may have an impact have been identified: field surveys discussing potential effects on planktonic and nekton populations, laboratory studies using traditional acute and sublethal toxicity testing methods, in situ testing organisms in field settings, and microcosm/mesocosm studies.

Field Surveys

While the toxic effects of sediment contaminants on field surveyed marine benthic populations and communities have received considerable attention and attempts at identifying a relationship between the two are on-going (also see Chapter 4),[114,115] the toxic effects of these same contaminants when transported into the water column has not been reported (Table 4). Therefore, indications of the direct impacts of contaminated sediments on phytoplankton, zooplankton, and nekton populations and communities do not exist. Based on the limits of our ability to measure significant differences in population and community parameters between stations on relatively small areal scales, it is not surprising that effects have not been readily observed. The natural heterogeneity and patchiness of planktonic populations and communities has been reported and commented upon as a major hindrance in defining biological effects, of any sort, in planktonic communities.[116] Analysis of large-scale data sets over extensive temporal periods and geographic areas may detect differences in populations and communities, but this work has yet to be performed and correlated with the spatial distribution of contaminated sediments.

Therefore, we do not know to what degree shifts in community diversity, composition, or other ecological parameters occur in the waters above contaminated

Table 4. Summary of the Toxicological Effects that Contaminated Sediments have Elicited in Water Column Exposures

Exposure Species	Site	Effect	Source
Field Surveys			
Phytoplankton	—	Unknown	
Zooplankton	—	Unknown	
Nekton	—	Unknown	
Laboratory			
Macroalga			Burgess[54]
Champia parvula	Providence River, RI	Increased necrotic tissue and reduced survival and reproduction	
	Black Rock Harbor, CT	Increased necrotic tissue and reduced survival and reproduction	
	New Bedford Harbor, MA	Increased necrotic tissue and reduced survival and reproduction	
Sea Urchin			Burgess[54]
Arbacia punctulata	Black Rock Harbor, CT	Reduced fertilization and development	
	New Bedford Harbor, MA	Reduced fertilization	
Mysid			Burgess[54]
Mysidopsis bahia	Black Rock Harbor, CT	Reduced survival, growth, and fecundity	
	New Bedford Harbor, MA	Reduced survival, growth, and fecundity	
Spot			Hargis et al.[134]
Leiostomus xanthurus	Elizabeth River, VI	Increased lesions, eye damage, and fin erosion	
In Situ Testing of Organisms			
Mussel			Nelson and Hansen[48], Bergen[49]
Mytilus edulis	New Bedford Harbor, MA	Accumulation of contaminants and reduced metabolic function	
Microcosm/Mesocosm			
Phytoplankton	Providence River, RI	Initial reduction in diversity	Oviatt et al.[151]

sediments. Exposure to contaminants could have several potential consequences. From a population vantage point, for example, one would expect severe alterations of the populations of zooplankton and nekton if they have life stages that maintain intimate contact with heavily contaminated sediments, e.g., copepod resting eggs. We would then expect, in an ecological perspective, alterations in the phytoplankton populations, which result from reductions in copepod predation. Other potential

Table 5. Net Flux of Oxygen, Ammonia, and Phosphate Measured at the Sediment-Water Interface in Various Coastal Marine Systems During Summer

	Flux		
System	O_2 (mg/m^2-hr)	NH^+_4 (μg/m^2-hr)	PO_4 (μg/m^2-hr)
Lock Ewe (Scotland)	25.6–49.6	360–1440	—
Buzzards Bay (Massachusetts)	59.2	2250	–1425
Eel Pond (Woods Hole, U.S.)	44.8	1530	1520
Narragansett Bay (Rhode Island)	75.2	3600	2850–4750
Long Island Sound (U.S.)	—	900–3600	475–1900
New York Bight (New York)	35.2	450	190
Patuxent River Estuary (Maryland)	126.4	12780	4560
Pamlico River Estuary (North Carolina)	—	810	—
South River Estuary (North Carolina)	65.6	4500	1615
Cap Blanc (West Africa)	—	4230	4750
Vostok Bay (U.S.S.R.)	44.8	2700	1900
Maizura Bay (Japan)	—	234–576	—
Kaneohe Bay (Hawaii)	19.2	972	285
La Jolla Bight (California)	—	720	570

Source: Nixon.[122]

adverse effects would be manifested as changes in phytoplankton, zooplankton, and nekton diversity, abundance, and composition. Scott [115] describes some of the changes in benthic community structure that may potentially result from sediment contamination. Such changes include increases in the abundance of opportunistic species and decreases in long-lived species. We may expect similiar community shifts in the water column overlying contaminated sediments. Other described benthic impacts include reduced total taxa abundance, diversity, and major taxa abundance.[113,117-119]

The significance of benthic-pelagic interactions has been described for the cycling of nutrients through marine ecosystems.[120-122] Table 5 presents the net flux of several essential nutrients from marine sediments into the water column, for a variety of coastal marine systems.[122] This flux is proposed to be a major source of nutrients for marine water column primary productivity.[121,123] Nutrient cycles are maintained by a complex combination of benthic and pelagic biogeochemical processes. The requisite destruction of one half of this interaction by severe sediment contamination would likely result in adverse effects in the water column. This type of work, when performed, will provide a valuable ecological dimension to our understanding of the water column impacts of contaminated sediments.

Laboratory

The design of acute and sublethal toxicity tests on a water column that has been contaminated by sediments is critical to the extrapolation of sediment/water column effects. For instance, in traditional water-only exposures, the experimental design is such that a section of the water column is modeled in an aquarium. In a sediment toxicity test the experimental design models a piece of the benthic environment. Generally, scaling for these studies is a function of the test organism's size. Simplistic designs that are typically employed ignore many of the previously described processes. An accurate water column/sediment exposure should model processes occurring at the water column/sediment interface. Determining the construct of these models raises a myriad of questions that ultimately can affect the experimental design and interpretability of the toxicity tests. Several chemical processes occur in the sediments, which affect the forms of chemicals that are transported into the water column. These processes raise the following questions:

- Do standard sediment sampling and holding procedures significantly alter critical sediment parameters, i.e., stratigraphy, porosity and chemical reactivity?
- Should only intact sediment cores be used in tests without homogenization?

The dominant physical transport processes, advection and diffusion, occur between the water column and sediment. Considering these processes:

- Should toxicity tests rely on molecular diffusion or should turbulent diffusion events be modeled as well?
- Should advection also be included?

Finally, biological activity also has a significant effect on the rate of material transport from the sediments and the form in which the contaminants will be transported. Appreciation of the role of organisms in the transport of pollutants between the water column and sediments causes us to consider the following:

- Should sediments be manipulated in terms of sampling and storage if those actions result in the mortality of bioturbating organisms and chemically active bacteria?

Clearly, practice of the traditional sediment-handling and -testing methodologies for the types of exposures discussed in this chapter *can* be hypothesized to result in severe alteration of the chemical, physical, and biological processes critical to designing a realistic study. This issue is one that is also of considerable current concern relative to sediment solid-phase toxicity tests. So it appears that, while we are trying to expand our understanding of contaminant behavior relative to sediment/water column interactions, we have created artifacts that reduce the interpretability of every toxicologic observation (also see Chapters 3 and 8 for additional discussion).

Practically speaking, one problem that reduces interpretability is a design where water column and sediment phases are combined into one exposure. Using such a design, questions can be raised as to whether organisms are exposed to contaminants in the water column or in the sediment phase or a combination of both.[124-126] If the objective of the experimental design is one of water column/sediment interface exposure, then this design is sufficient. But when the design is concerned with water-column-only exposures, a design with both phases, accessible to the organisms, is not appropriate. A conclusion based on this discussion is that, when performing these types of water column/sediment exposures, the physical separation of the sediment and water column during the exposure is critical.

Few standard acute and sublethal laboratory methods exist for assessing the impact of pollutants released from contaminated sediments, on water column organisms. Some of the best-known methods are the standard battery of toxicity tests, including the elutriate test[127,128] commonly used to assess the environmental impact of dredge material identified for ocean disposal. Because of the dramatic alterations that the dredging process imparts to the physical, chemical, and biological nature of dredged materials, resulting toxicity to water column organisms is beyond the scope of this chapter, but is reviewed elsewhere.[129-133]

Studies that have been conducted in which the water column and sediments have been separated prior to the organism's introduction to the exposure, have demonstrated significant acute and sublethal water column toxicity (Table 4). The demersal estuarine spot, *Leiostomus xanthurus*, was exposed to contaminated sediments from the Elizabeth River, Virginia which contains high levels of PAHs.[134] Initially, the study was designed as a sediment exposure with the use of flow-through water conditions. However, mortality was observed in the first few days of the exposure, and a second exposure aquarium was added to the design. In the second aquarium, the spot was exposed to the effluent that flowed out of the primary exposure system. During the 24-day exposure, acute and sublethal effects were observed. Mortality in the effluent exposure was 16%, with several of the remaining fish displaying lesions, eye damage, and fin erosion (Table 4).

Two studies investigated the significance of contaminated sediments on water column toxicity from three contaminated sites in the New England region.[54] The sea urchin *(Arbacia punctulata)* fertilization test was used to measure the release of pollutants from contaminated sediments, from Providence River, Rhode Island, Black Rock Harbor, Connecticut, and New Bedford Harbor, Massachusetts, to the water column. Exposure conditions were designed to model processes such as molecular diffusion (bedded conditions) and turbulent diffusion (suspended conditions). The toxicity of interstitial water was also measured for each sediment. Water column toxicity was observed for the most contaminated sediments, Black Rock Harbor and New Bedford Harbor, and was proportional to the physical processes each exposure modeled (for example, Figure 8). Treatments modeling molecular diffusive processes showed the least toxicity, while treatments designed to model turbulent diffusion showed the most water column toxicity. The specific classes of toxicants identified as causing toxicity in the water column were compared to those of the interstitial water. This comparison indicated that water column toxicity was associated with metals and nonionic organics, while interstitial water toxicity was not associated with either, but was probably related to the presence of ammonia and hydrogen sulfide. Chemical processes likely resulted in differences in the

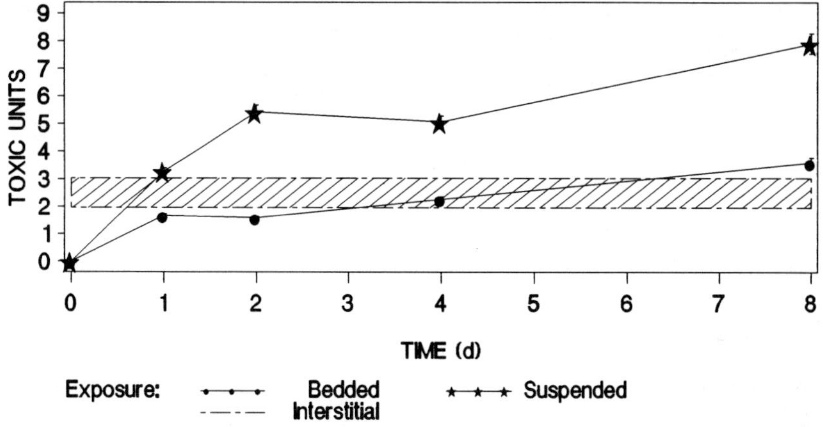

Figure 8. Toxicity of water column samples exposed to contaminated sediments from New Bedford Harbor (shaded area represents interstitial water toxicity). Toxic units are equivalent to $100/EC_{50}$.[54]

availability of contaminants among the three exposures regimes. The transition metals and nonionic organics in the interstitial water appear to have been bound to colloidal material, causing them to be unavailable. Conversely, the concentration of hydrogen sulfide and probably ammonia, which are both formed under anoxic conditions, were elevated relative to water column concentrations. However, as the bound contaminants and naturally occurring toxicants diffuse into the oxic water column, the metals and organics are released in an available form as the colloidal material is oxidized; similarly, both ammonia and hydrogen sulfide are oxidized into nontoxic forms or diluted (Table 4).[54]

In a second study[54] involving the same three sediments, three marine species, several acute and sublethal endpoints were employed to measure toxicity. The macroalga, *Champia parvula*, the mysid, *Mysidopsis bahia*, and the sea urchin, *Arbacia punctulata*, were exposed to samples of the contaminated water column. All species were affected adversely, with *C. parvula* being the most sensitive species (Table 4).

Other methods of contaminating water column samples to model water column/ sediment interactions have been developed, but have not been used very often for marine applications (also see Chapter 8). These systems include a chamber that recycles water across the sediment surface,[135,136] and microcosm-like chambers used for monitoring sediment and water interactions.[137,138]

The acute and sublethal studies performed to date have demonstrated that contaminated sediments can release bioavailable pollutants in sufficient concentrations to cause toxic effects on a variety of marine organisms, including algae, echinoderms, arthropods, and fish. However, these results were reached using

Table 6. Comparison of the Results of Toxicity Tests Performed on Water Column Samples from Laboratory and Field Exposures to New Bedford Harbor Sediments

Species	Toxicity (% sample)[a]	
	Laboratory	Field
Macroalga		
Champia parvula	<3.1	<100
Sea Urchin		
Arbacia punctulata	12.5	No effect
Mysid		
Mysidopsis bahia		
Survival	15	No effect
Fecundity	15	No effect
Growth	3.8	No effect

Source: Torello et al.[139]

[a] No observed effect concentration (NOEC).

experimental designs that represented "worst-case" scenarios, i.e., generally static exposures. Studies have not been performed in the laboratory that model the hydrological conditions of the New England sites that demonstrated significant water column toxicity. If these studies were performed, would natural turnover rates result in the dilution of contaminants that are released from the sediment to no observed effect concentrations (NOEC)? One direct approach to answering this hypothetical question is to compare the laboratory-derived, static exposure results to field-collected water column samples from the same field locations. A problem with this approach is that it must be assumed that any water column toxicity observed in the field-collected sample is due to sediment sources. Generally, this cannot be done due to existing terrestrial inputs to sites like the Providence River and Black Rock Harbor by anthropogenic runoff and effluents. However, New Bedford Harbor may be a site meeting the assumptions that the majority of the contaminants are originating from the sediments.[14] A comparison of the laboratory and the field results[139] for the three marine species (Table 6) tested indicates that, unsurprisingly, the toxicity under real-world conditions is reduced relative to static exposures. The nature of these "real-world" effects on contaminant bioavailability are open for discussion and probably range from simple dilution of released pollutants to complexation of pollutants by organic and inorganic ligands.[140]

In Situ Testing of Organisms

The in situ testing of organisms, particularily filter-feeding bivalves, represents a promising approach to determining the magnitude of the release of bioavailable pollutants from in-place contaminated sediments. A primary value of deployed organisms is their ability to integrate, under realistic conditions, over long periods of pollutant exposure relative to short laboratory studies. Unfortunately, it is necessary to make the assumption that any adverse effects observed on deployed organisms are due exclusively to sediment-released pollutants.

The most commonly used deployed organism in the marine environment is the mussel *Mytilus edulis*.[141-143] Several studies have illustrated the utility of this species in demonstrating contaminant bioaccumulation[144,145] and adverse impacts on cellular, metabolic, organismal and population parameters[141,143] in field exposures (also see Chapter 12). Work performed in New Bedford Harbor indicated that several bioavailable pollutants had probably originated from the sediments (Table 4).[48,49]

Recently, several other biomonitoring techniques have been developed that could be utilized to address the role of sediments on water column contamination. These include a simple, cost-effective exposure cage for field deployment of mysids, *Mysidopsia bahia*,[146] which would be analogous to the standard laboratory 7-day sublethal toxicity test performed as part of other biological monitoring programs.[147] Other methods include a variety of elaborate devices for the exposure of fish (also see Chapter 13).[148-150]

Microcosms/Mesocosms

Due to the complicated nature of water column/sediment exposure designs, the microcosm and/or mesocosm approach may provide the most relevant information. Relevance is defined here as the ability of the exposure methods to accurately reproduce many of the processes that occur between the water column and sediments. The use of these types of designs to study the effects of contaminated sediments on the water column have been performed, though indirectly, through recovery studies.[151,152] These studies consider how long is required for a contaminated site to return to the ecological and chemical state of an uncontaminated site. Generally, studies of this kind deliver "clean" seawater into the experimental system that contains the contaminated sediment. This design enhances the release of contaminants from the sediments into the water column by creating a large concentration gradient between the "clean" water column and the "dirty" sediment, thus accelerating the diffusive processes. Also, the deposition of particulate material from the water column may bury the contaminated sediment. A consequence of this design is that the water column experiences an increase in the concentration of many pollutants that originated from the sediments.

Two studies have looked at the recovery of sites along a pollution gradient in Narragansett Bay, Rhode Island (Table 4). The Marine Ecosystem Research Laboratory (MERL) mesocosms, 13-m^3 tanks with both sediment and water column components, performed a gradient study with sediments from Narragansett Bay.[151] Sediments from three stations in the bay were examined: Providence River (contaminated), midbay (moderate), and Rhode Island Sound (clean). No long-term adverse effects were observed in the water column (an initial reduction in phytoplankton diversity was observed) despite the measurable release of several metals and PAHs from the oxidizing and bioturbated contaminated sediments.[52,71] A second microcosm study using sediments from the same sites in Narragansett Bay observed similiar findings, again despite the release of several pollutants by bioturbation and diffusional processes.[152]

When using the microcosms or mesocosms with clean seawater, it can be assumed that any adverse effects observed originated from the sediments alone. Further, the results of the microcosm scenario can be compared to the field scenario. Comparison of the water column samples from the MERL treatments with their in situ counterparts, i.e., water samples from the same stations where the sediments were collected originally, indicated that there were many significant differences in several population and community parameters between stations. A conclusion drawn from this apparent contradiction between the mesocosm and in situ response is that water column impacts caused by Providence River sediments are temporary and minor relative to the impact of contemporary inputs (e.g., industrial and municipal effluents).[151]

At this point it is tempting to conclude that contaminated sediments have no lasting adverse effects on the water column, based on the microcosm/mesocosm approach. However, to conclude that contaminated sediments do not have the potential to cause significant water column effects, based only on these two studies, would be premature. These studies demonstrated that moderately contaminated sediments[153,154] in a relatively dynamic estuary did not have an observable effect on the water column. This conclusion may not hold in severely contaminated areas.

CONCLUSIONS

This chapter has described the dominant chemical, physical, and biological processes that result in the transport of bioavailable contaminants from the sediments into the water column. Furthermore, some of the observed impacts of these processes have been described. At this time the eventual and full spectrum of effects that these contaminants will have on the water column environment is not known. We cannot definitively answer the third question posed earlier: *Does a problem exist?* Too little data have been collected across all types of exposures (Table 4) and between different locations with varying levels of contamination. Obviously, the demonstration of adverse effects in the field, or the lack of adverse effects on the water column environment, would supply a definitive answer, but this work, like much of the work in this area, has yet to be performed. Likewise, meso- and microcosm studies would provide invaluable information, but to date, these studies have been performed only in Narragansett Bay, where sediments are only moderately contaminated and thus cannot be used to predict effects in severely contaminated sites. Therefore, the performance of field surveys at heavily contaminated locations and meso/microcosm studies with sediments that are extremely contaminated are required.

Laboratory acute and sublethal toxicity tests that have been performed, strongly suggest that extremely contaminated sediments may cause detrimental effects in the water column. Although some of these studies were performed under worst-case conditions, more accurate follow-up work, possibly using models of the physical, chemical, and hydrological parameters of the sites, is necessary before definitive conclusions can be drawn.

In situ organism exposures have been used in some instances, but many more applications are needed. Unfortunately, as with almost all field studies, a problem with deployments is the ability to discriminate between causes of adverse effects. In the field, caged organisms could be impacted by the contaminated sediments or terrestrial point and nonpoint sources of contaminants. Despite these limitations, more deployments specifically designed to address water column effects resulting from contaminated sediments are required.

As discussed, oceanographic studies have illustrated the significance of the interactions between the pelagic and benthic environments, in terms of nutrient cycling and water column productivity. Much of the primary productivity that occurs in the water column in coastal areas is supported by the remineralization of nutrients by the benthos. The concept that contaminants in the benthos may have detrimental impacts on the water column via the disruption of nutrient cycling and pelagic/benthic ecological dynamics is worth further consideration.

What is the significance of in-place contaminated marine sediment on the water column? Clearly, much more work is required in all areas of this question before a decisive answer can be reached. This work should include a consideration of the tools we are using to measure toxicity and other adverse impacts in the water column. For example, the sensitivity of methods for measuring water column toxicity are in question when we are concerned with detecting the effects of very low concentrations of chronically released contaminants. Grossly contaminated sites with low water turnover rates may produce significant water column toxicity, using our current tests, because of the lack of dilution of sediment-released contaminants. However, the same site with high water turnover rates, which result in high dilution, may not produce short-term and observable toxic effects while the chronic release of toxic contaminants goes undetected. The application of appropriate biomarkers and sublethal endpoints to assess the impact of contaminated sediments on the water column is crucially needed. Along with toxicological tools, we must consider chemical and physical approaches. These tools have yet to be exploited to their fullest so that we may better understand the release and transport of sediment-associated contaminants.

The versatility and efficacy of our suite of "tools" will need to increase for us to begin to address the significance of in-place contaminated marine sediments, particularly as the terrestrial sources of water column contamination are reduced by environmental regulations. Despite these reductions, marine sediments undoubtedly will continue to persistently release their unsavory contaminants into the water column for many years into the future.

ACKNOWLEDGMENTS

The authors thank the reviewers of this chapter for their comments; additionally, the comments of Edward Dettman (U.S. EPA) relative to the physical processes are much appreciated. This chapter was partially supported by U.S. Environmental Protection Agency contracts to Science Applications International Corporation #68-03-3529 and #68-C1-0005. This report has been reviewed by the U.S. EPA Environmental Research Laboratory, Narragansett, R.I. and approved for publication (Contribution No. X187). Approval does not signify that the contents necessarily reflect the views and policies of the Agency.

REFERENCES

1. Rand, G. M., and S. R. Petrocelli. *Fundamentals of Aquatic Toxicology* (Washington, D.C.: Hemisphere, 1985), p. 666.
2. Baker, R. A. *Contaminants and Sediments: Volume 1 Fate and Transport, Case Studies, Modeling, Toxicity* (Ann Arbor, MI: Ann Arbor Science, 1980), p. 558.
3. Baker, R. A. *Contaminants and Sediments: Volume 2 Analysis, Chemistry, Biology* (Ann Arbor, MI: Ann Arbor Science, 1980), p. 627.
4. Dickson, K. L., A. W. Maki, and W. A. Brungs. *Fate and Effects of Sediment-Bound Chemicals in Aquatic Systems* (New York: Pergamon Press, 1987), p. 449.
5. National Research Council. *Contaminated Marine Sediments: Assessments and Remediation* (Washington, D.C.: National Academy Press, 1989), p. 493.
6. Baudo, R., J. P. Giesy, and H. Muntau. *Sediments: Chemistry and Toxicity of In-Place Pollutants* (Chelsea, MI: Lewis Publishers, 1990), p. 405.
7. Lee, G. F. "Factors Affecting the Transfer of Materials Between Water and Sediments," University of Wisconsin, Water Resources Center, Eutrophication Information Center, p. 50 (1970).
8. Forstner, U., and G. T. W. Wittman. *Metal Pollution in the Aquatic Environment* (New York: Springer-Verlag, 1979) p. 486.
9. Brannon, J. M., R. H. Plumb, Jr., and I. Smith, Jr. "Long-Term Release of Heavy Metals from Sediments," in *Contaminants and Sediments: Volume 2 Analysis, Chemistry, Biology,* R.A. Baker, Ed. (Ann Arbor, MI: Ann Arbor Science, 1980), p. 221–266.
10. Horzempa, L. M., and D. M. Di Toro. "The Extent of Reversibility of Polychlorinated Biphenyl Adsorption," *Water Res.* 17:851–859 (1983).
11. Thoma, G. J., A. C. Koulermos, K. T. Valsaraj, D. D. Reible, and L. J. Thibodeaux. "The Effects of Pore Water Colloids on the Transport of Hydrophobic Organic Compounds from Bed Sediments," in *Organic Substances and Sediments in Water: Volume I Humic and Other Substances*, R.A. Baker, Ed. (Boca Raton, FL: Lewis Publishers, 1991).
12. Larsson, P. "Contaminated Sediments of Lakes and Oceans Act as Sources of Chlorinated Hydrocarbons for Release to Water and Atmosphere," *Nature* 317:347–349 (1985).
13. Salomons, W., N. M. de Rooij, H. Kerdijk, and J. Bril. "Sediments as a Source for Contaminants?" *Hydrobiologia* 149:13–30 (1987).

14. Thibodeaux, L. J., and D. D. Reible. "A Theoretical Evaluation of the Effectivenss of Capping PCB Contaminated New Bedford Harbor Sediment," Louisiana State University Hazardous Waste Research Center: Final Report, p. 92 (1990).
15. Baudo, R., and H. Muntau. "Lesser Known In-Place Pollutants and Diffuse Source Problems," in *Sediments: Chemistry and Toxicity of In-Place Pollutants*, R. Baudo, J.P. Giesy, and H. Muntau, Eds. (Chelsea, MI: Lewis Publishers, 1990), pp. 1–14.
16. Reuber, B., D. MacKay, S. Patterson, and P. Stokes. "A Discussion of Chemical Equilibria and Transport at the Sediment-Water Interface," *Environ. Toxicol. Chem.* 6:731–739 (1987).
17. Huggett, R. J., M. E. Bender, and M. A. Unger. "Polynuclear Aromatic Hydrocarbons in the Elizabeth River, Virginia," in *Fate and Effects of Sediment-Bound Chemicals in Aquatic Systems*, K. L. Dickson, A. W. Maki, and W. A. Brungs, Eds. (New York: Pergamon Press, 1987), pp. 327–341.
18. Sanders, J. E. "PCB Pollution in the Upper Hudson River" in *Contaminated Marine Sediments: Assessments and Remediation*, National Research Council (Washington, D.C.: National Academy Press, 1989), pp. 365–400.
19. Huggett, R. J. "Kepone and the James River" in *Contaminated Marine Sediments: Assessments and Remediation*, National Research Council (Washington, D.C.: National Academy Press, 1989), pp. 417–424.
20. Ginn, T. C. "Assessment of Contaminated Sediments in Commencement Bay (Puget Sound, Washington)," in *Contaminated Marine Sediments: Assessments and Remediation*, National Research Council (Washington, D.C.: National Academy Press, 1989), pp. 425–439.
21. Boehm, P. D., and J. G. Quinn. "The Effects of Dissolved Organic Matter in Sea Water on the Uptake of Mixed Individual Hydrocarbons and Number 2 Fuel Oil by a Marine Filter Feeding Bivalve (*Mercenaria mercenaria*)," *Estuar. Coast. Mar. Sci.* 4:93–105 (1976).
22. Sunda, W. G., and R. R. Guillard. "Relationship Between Cupric Ion Activity and the Toxicity of Copper to Phytoplankton," *J. Mar. Res.* 34:511–529 (1976).
23. Sunda, W. G., D. W. Engel, and R. M. Thoutte. "Effect of Chemical Speciation on Toxicity of Cadmium to Grass Shrimp, *Paleomonetes pugio*: Importance of Free Cadmium Ion," *Environ. Sci. Technol.* 12:409–413 (1978).
24. Louma, S. N. "Bioavailability of Trace Metals to Aquatic Organisms — A Review," *Sci. Tot. Environ.* 28:1–22 (1983).
25. McCarthy, J. F., B. D. Jimenez, and T. Barbee. "Effect of Dissolved Humic Material on Accumulation of Polycyclic Aromatic Hydrocarbons: Structure-Activity Relationships," *Aquat. Toxicol.* 7:15–24 (1985).
26. Landrum, P. F., S. R. Nihart, B. J. Eadie, and L. R. Herche. "Reduction in Bioavailability of Organic Contaminants to the Amphipod *Pontoporeia hoyi* by Dissolved Organic Matter of Sediment Interstitial Waters," *Environ. Toxicol. Chem.* 6:11–20 (1987).
27. Day, K. E. "Effects of Dissolved Organic Carbon on Accumulation and Acute Toxicity of Fenvalerate, Deltamethrin and Cyhalothrin to *Daphnia magna* (Straus)," *Environ. Toxicol. Chem.* 10:91–101 (1991).
28. Di Toro, D. M., J. D. Mahony, D. J. Hansen, K. J. Scott, M. B. Hicks, S. M. Mayr, and M. S. Redmond. "Toxicity of Cadmium in Sediments: the Role of Acid Volatile Sulfide," *Environ. Toxicol. Chem.* 9:1487–1502 (1990).

29. Wu, S., and P. M. Gschwend. "Sorption Kinetics of Hydrophobic Organic Compounds to Natural Sediments and Soils," *Environ. Sci. Technol.* 20:717–725 (1986)
30. U. S. Environmental Protection Agency. "Briefing Report to the EPA Science Advisory Board on the Equilibrium Partitioning Approach to Generating Sediment Quality Criteria," Criteria and Standards Division (1989).
31. Moore, J. W., and S. Ramamoorthy. *Organic Chemicals in Natural Waters* (New York: Springer-Verlag, 1984), p. 289.
32. Thurman, E. M. *Organic Geochemistry of Natural Waters* (Boston: Martinus Nijhoff/ D. W. Junk Publishers, 1985), p. 497.
33. Knezovich, J. P., F. L. Harrison, and R. G. Wilhelm. "The Bioavailability of Sediment-Sorbed Organic Chemicals: A Review," *Water, Air, Soil Pollut.* 32:233–245 (1987).
34. Chiou, C. T., L. J. Peters, and V. H. Freed. "A Physical Concept of Soil-Water Equilibria for Nonionic Organic Compounds," *Science* 206:831–832 (1979).
35. Karickhoff, S. W., D. S. Brown, and T. A. Scott. "Sorption of Hydrophobic Pollutants on Natural Sediments," *Water Res.* 13:241–248 (1979).
36. Gschwend, P. M., and S. Wu. "On the Constancy of Sediment-Water Partition Coefficients of Hydrophobic Organic Pollutants," *Environ. Sci. Technol.* 19:90–96 (1985).
37. Brownawell, B. J., and J. W. Farrington. "Biogeochemistry of PCBs in Interstitial Waters of a Coastal Marine Sediment," *Geochim. Cosmochim. Acta* 50:157–169 (1986).
38. Eadie, B. J., N. R. Morehead, and P. F. Landrum. "Three-Phase Partitioning of Hydrophobic Organic Compounds in Great Lake Waters," *Chemosphere* 20:161–178 (1990).
39. Means, J. C., and R. Wijayaratne. "Role of Natural Colloids in the Transport of Hydrophobic Pollutants," *Science* 215:968–970 (1982).
40. Enfield, C. G., G. Bengtsson, and R. Lindqvist. "Influence of Macromolecules on Chemical Transport," *Environ. Sci. Technol.* 23:1278–1286 (1989).
41. McCarthy, J. F., and J. M. Zachara. "Subsurface Transport of Contaminants," *Environ. Sci. Technol.* 23:496–502 (1989).
42. Sigleo, A. C., and J. C. Means. "Organic and Inorganic Components in Estuarine Colloids: Implications for Sorption and Transport of Pollutants," *Rev. Environ. Contam. Toxicol.* 112:123–147 (1990).
43. Brannon, J. M., T. E. Myers, D. Gunnison, and C. B. Price. "Nonconstant Polychlorinated Biphenyl Partitioning in New Bedford Harbor Sediment during Sequential Batch Leaching," *Environ. Sci. Technol.* 25:1082–1087 (1991).
44. Pavlou, S. P., and R. N. Dexter. "Distribution of Polychlorinated Biphenyls (PCB) in Estuarine Ecosystems; Testing the Concept of Equilibrium Partitioning in the Marine Environment," *Environ. Sci. Technol.* 13:65–71 (1979).
45. Sodergren, A., and P. Larsson. "Transport of PCBs in Aquatic Laboratory Model Ecosystems from Sediment to the Atmosphere Via the Surface Microlayer," *Ambio* 11:41–45 (1982).
46. Larsson, P. "Transport of ^{14}C-Labelled PCB Compounds from Sediment to Water and from Water to Air in Laboratory Model Systems," *Water Res.* 17:1317–1326 (1983).
47. Weaver, G. "PCB Contamination In and Around New Bedford, Mass.," *Environ. Sci. Technol.* 18:22A–27A (1984).

48. Nelson, W. G., and D. J. Hansen. "Development and Use of Site-Specific Chemical and Biological Criteria for Assessing the New Bedford Harbor Pilot Dredging Project," *Environ. Manage.* 15:105–112 (1991).
49. Bergen, B. J. "Bioaccumulation of PCB Congeners in Mussels Deployed in New Bedford Harbor," M.S. Thesis, University of Rhode Island, Narragansett, RI (1991).
50. Pruell, R. J., J. L. Lake, W. R. Davis, and J. G. Quinn. "Uptake and Depuration of Organic Contaminants by Blue Mussels (*Mytilus edulis*) Exposed to Environmentally Contaminated Sediments," *Mar. Biol.* 91:497–507 (1986).
51. Wildish, D. J., C. D. Metcalfe, H. M. Akagi, and D. W. McLeese. "Flux of Aroclor 1254 Between Estuarine Sediments and Water," *Bull. Environ. Contam. Toxicol.* 24:20–26 (1980).
52. Pruell, R. J., and J. G. Quinn. "Polycyclic Aromatic Hydrocarbons in Surface Sediments Held in Experimental Mesocosms," *Toxicol. Environ. Chem.* 10:183–200 (1985).
53. McElroy, A. E., J. W. Farrington, and J. M. Teal. "Influence of Mode of Exposure and the Presence of a Tubiculous Polychaete on the Fate of Benz[a]anthracene in the Benthos," *Environ. Sci. Technol.* 24:1648–1655 (1990).
54. Burgess, R. M. "Water Column Toxicity from Contaminated Sediments," M.S. Thesis, University of Rhode Island, Narragansett, RI (1990).
55. O'Connor, D. J., and J. P. Connolly. "The Effect of Concentration of Adsorbing Solids on the Partition Coefficient," *Water Res.* 14:1517–1523 (1980).
56. Di Toro, D. M. "A Particle Interaction Model of Reversible Organic Chemical Sorption," *Chemosphere* 14:1503–1538 (1985).
57. Di Toro, D. M, J. D. Mahony, P. R. Kirchgraber, A. L. O'Byrne, L. R. Pasquale, and D. C. Piccirilli. "Effects of Nonreversibility, Particle Concentration, and Ionic Strength on Heavy Metal Sorption," *Environ. Sci. Technol.* 20:55–61 (1986).
58. Di Toro, D. M., L. J. Dodge, and V. C. Hand. "A Model for Anionic Surfactant Sorption," *Environ. Sci. Technol.* 24:1013–1020 (1990).
59. Jafvert, C. T. "Sorption of Organic Acid Compounds to Sediments: Initial Model Development," *Environ. Toxicol. Chem.* 9:1259–1268 (1990).
60. Jafvert, C. T., and J. K. Heath. "Sediment- and Saturated-Soil-Associated Reactions Involving an Anionic Surfactant (Dodecylsulfate). 1. Precipitation and Micelle Formation," *Environ. Sci. Technol.* 25:1031–1038 (1991).
61. Jafvert, C. T. "Sediment- and Saturated-Soil-Associated Reactions Involving an Anionic Surfactant (Dodecylsulfate). 2. Partition of PAH Compounds among Phases," *Environ. Sci. Technol.* 25:1039–1045 (1991).
62. Jennett, J. C., S. W. Effler, and B. G. Wixson. "Mobilization and Toxicological Aspects of Sedimentary Contaminants," in *Contaminants and Sediments: Volume 1 Fate and Transport, Case Studies, Modeling, Toxicity,* R.A. Baker, Ed. (Ann Arbor, MI: Ann Arbor Science, 1980), pp. 429–444.
63. Jenne, E. A., and J. M. Zachara. "Factors Influencing the Sorption of Metals," in *Fate and Effects of Sediment-Bound Chemicals in Aquatic Systems*, K.L. Dickson, A. W. Maki, and W. A. Brungs, Eds. (New York: Pergamon Press, 1987).
64. Louma, S. N. "Can We Determine the Biological Availability of Sediment-Bound Trace Elements ?" *Hydrobiologia* 176/177:379–396 (1989).
65. Presley, B. J., Y. Kolodny, A. Nissenbaum, and I. R. Kaplan. "Early Diagenesis in a Reducing Fjord, Saanich Inlet, British Columbia. II. Trace Element Distribution in Interstitial Water and Sediment," *Geochim. Cosmochim. Acta* 36:1073–1099 (1972).

66. Duinker, J. C., G. T. M. Van Eck, and R. F. Nolting. "On the Behavior of Copper, Zinc, Iron and Manganese, and Evidence for Mobilization Processes in the Dutch Wadden Sea," *Neth. J. Sea Res.* 8:214–239 (1974).
67. Bothner, M. H., R. A. Jahnke, M. L. Peterson, and R. Carpenter. "Rate of Mercury Loss from Contaminated Estuarine Sediments," *Geochim. Cosmochim. Acta* 44:273–285 (1980).
68. Muller, G., and U. Forstner. "Heavy Metals in Sediments of the Rhine and Elbe Estuaries: Mobilization or Mixing Effects," *Environ. Geol.* 1:33–39 (1975).
69. Troup, B. N., and O. P. Bricker. "Processes Affecting the Transport of Materials from Continents to Oceans," in *Marine Chemistry in the Coastal Environment* (Washington, D.C.: American Chemical Society, 1975), pp. 133–151.
70. Lu, J. C. S., and K. Y. Chen. "Migration of Trace Metals in Interfaces of Seawater and Polluted Surficial Sediments," *Environ. Sci. Technol.* 11:174–182 (1977).
71. Hunt, C. D., and D. L. Smith. "Remobilization of Metals from a Polluted Marine Sediment," *Can. J. Fish. Aquat. Sci.* 40(Suppl.2):132–142 (1982).
72. Baeyens, W., G. Gillain, M. Hoenig, and F. Dehairs. "Mobilization of Major and Trace Elements at the Water-Sediment Interface in the Belgian Coastal Area and the Scheldt Estuary," in *Marine Interfaces Ecohydrodynamics* (New York: Elsevier Oceanography Series 42, 1986), pp. 453–466.
73. Kester, D. R. "Equilibrium Models in Seawater: Applications and Limitations," in *The Importance of Chemical "Speciation" in Environmental Processes* (New York: Springer-Verlag, 1986), pp. 337–363.
74. Morrison, G. M. P. "Approaches to Metal Speciation Analysis in Natural Waters," in *Lecture Notes in Earth Sciences:Speciation of Metals in Water, Sediment and Soil Systems* (New York: Springer-Verlag, 1986), pp. 55–73.
75. Lund, W. "Electrochemical Methods and Their Limitations for the Determination of Metal Species in Natural Waters," in *The Importance of Chemical "Speciation" in Environmental Processes* (New York: Springer-Verlag, 1986), pp. 533–561.
76. Lerman, A. "Migrational Processes and Chemical Reactions in Interstitial Waters," in *The Sea,* Vol. 6 (New York: John Wiley & Sons, 1976), pp. 695–700.
77. Lerman, A. *Geochemical Processes: Water and Sediment Environments* (New York: John Wiley & Sons, 1979), p. 481.
78. Thibodeaux, L. J. *Chemodynamics: Environmental Movement of Chemicals in Air, Water, and Soil* (New York: John Wiley & Sons, 1979), p. 501.
79. Hillel, D. *Introduction to Soil Physics* (New York: Academic Press, 1982), p. 365.
80. Thibodeaux, L. J., and J. D. Boyle. "Bed-Form Generated Convective Transport in Bottom Sediment," *Nature* 325:341–343 (1987).
81. Kennett, J. P. *Marine Geology* (Englewood Cliffs, NJ: Prentice-Hall, 1982) p. 813.
82. Turekian, K. K. "The Fate of Metals in the Oceans," *Geochim. Cosmochim. Acta* 41:1139–1144 (1977).
83. Olsen, C. R., N. H. Cutshall, and I. L. Larsen. "Pollutant-Particle Associations and Dynamics in Coastal Marine Environments: A Review," *Mar. Chem.* 11:501–533 (1982).
84. Voice, T. C., and W. J. Weber. "Sorption of Hydrophobic Compounds by Sediments, Soils and Suspended Solids-I Theory and Background," *Water Res.* 17:1433–1441 (1983).
85. Voice, T. C., C. P. Rice, and W. J. Weber. "Effect of Solids Concentration on the Sorptive Partitioning of Hydrophobic Pollutants in Aquatic Systems," *Environ. Sci. Technol.* 17:513–518 (1983).

86. Weber, W. J., T. C. Voice, M. Pirbazari, G. E. Hunt, and D. M. Ulanoff. "Sorption of Hydrophobic Compounds by Sediments, Soils and Suspended Solids-II Sorbent Evaluation Studies," *Water Res.* 17:1443–1452 (1983).
87. Karickhoff, S. W., and K. R. Morris. "Sorption Dynamics of Hydrophobic Pollutant in Sediment Suspensions," *Environ. Toxicol. Chem.* 4:469–479 (1985).
88. Berner, R. A. "The Benthic Boundry Layer from the Viewpoint of a Geochemist," in *The Benthic Boundary Layer* (New York: Plenum Press, 1976), pp. 33–55.
89. Lick, W. "The Transport of Sediments in Aquatic Systems," in *Fate and Effects of Sediment-Bound Chemicals in Aquatic Systems,* K. L. Dickson, A. W. Maki, and W. A. Brungs, Eds. (New York: Pergamon Press, 1987), pp. 61–74.
90. Sheng, Y. P. "Predicting the Dispersion and Fate of Contaminated Marine Sediments," in *Contaminated Marine Sediments: Assessments and Remediation* (Washington, D.C.: National Academy Press, 1989), pp. 166–177.
91. Duursma, E. K. "Migration in the Seabed: Some Concepts," in *Biogeochemistry of Estuarine Sediments.* Proc. UNESCO/SCOR Workshop, Melreux, Belgium, pp. 179–188 (1976).
92. Duursma, E. K., and M. Smies. "Sediments and Transfer at and in the Bottom Interfacial Layer," in *Pollutant Transfer and Transport in the Sea Volume 2* (Boca Raton, FL: CRC Press, 1982), pp. 101–139.
93. Balzer, W., H. Erlenkeuser, M. Hartmenn, P. J. Moller, and F. Pollehne. "Diagenesis and Exchange Processes at the Benthic Boundary," in *Seawater-Sediment Interactions in Coastal Waters* (New York: Springer-Verlag, 1987), pp. 111–161.
94. Davis, W. R., and J. C. Means. "A Developing Model of Benthic-Water Contaminant Transport in Bioturbated Sediment," in *Proc. 21st EMBS* (Gdańsk, Poland: Polish Academy of Sciences-Institute of Oceanology, 1989), pp. 215–226.
95. Rhoads, D. C., and L. F. Boyer. "The Effects of Marine Benthos on Physical Properties of Sediments: A Successional Perspective," in *Animal-Sediment Relations* (New York: Plenum Press, 1982), pp. 3–52.
96. Aller, R. C. "The Effects of Macrobenthos on Chemical Properties of Marine Sediment and Overlying Water," in *Animal-Sediment Relations* (New York: Plenum Press, 1982), pp. 53–102.
97. Zehnder, A. J. B., and W. Stumm. *Biology of Anaerobic Microorganisms* (New York: Wiley-Interscience, 1990), pp. 1–36.
98. Pritchard, P. H. "Assessing the Biodegradation of Sediment Associated Chemicals," in *Fate and Effects of Sediment-Bound Chemicals in Aquatic Systems*, K. L. Dickson, A. W. Maki, and W. A. Brungs, Eds. (New York: Pergamon Press, 1987), pp. 109–135.
99. Wood, J. M. "Biological Cycles for Toxic Elements in the Environment," *Science* 183:1049–1052 (1974).
100. D'Itri, F. M. "The Biomethylation and Cycling of Selected Metals and Metalloids in Aquatic Sediments," in *Sediments: Chemistry and Toxicity of In-Place Pollutants* (Chelsea, MI: Lewis Publishers, 1990), pp. 163–214.
101. Adams, D. D., N. J. Fendinger, and D. E. Glotfelty. "Biogenic Gas Production and Mobilization of In-Place Sediment Contaminants by Gas Ebullition," in *Sediments: Chemistry and Toxicity of In-Place Pollutants* (Chelsea, MI: Lewis Publishers, 1990), pp. 215–236.
102. Rhoads, D. C. "Organism-Sediment Relations on the Muddy Sea Floor," *Oceanogr. Mar. Biol. Ann. Rev.* 12:263–300 (1974).

103. Petr, T. "Bioturbation and Exchange of Chemicals in the Mud-Water Interface," in *Interactions Between Sediments and Freshwater* (The Hague, Netherlands: D. W. Junk Publishers, 1977), pp. 216–226.
104. Lee, H., II, and R. C. Swartz. "Biological Processes Affecting the Distribution of Pollutants in Marine Sediments. Part II. Biodeposition and Bioturbation," in *Contaminants and Sediments:Volume 2 Analysis, Chemistry, Biology,* R. A. Baker, Ed. (Ann Arbor, MI: Ann Arbor Science, 1980), pp. 555–606.
105. Swartz, R. C., and H. Lee, II. "Biological Processes Affecting the Distribution of Pollutants in Marine Sediments. Part I. Accumulation, Trophic Transfer, Biodegradation and Migration," in *Contaminants and Sediments:Volume 2 Analysis, Chemistry, Biology*, R. A. Baker, Ed. (Ann Arbor, MI: Ann Arbor Science, 1980), pp. 533–553.
106. Krantzberg, G. "The Influence of Bioturbation on Physical, Chemical and Biological Parameters in Aquatic Environments: A Review," *Environ. Pollut.* 39 (Series A): 99–122 (1985).
107. Matisoff, G. "Mathematical Models of Bioturbation," in *Animal-Sediment Relations* (New York: Plenum Press, 1982), pp. 289–330.
108. Fowler, S. W. "Biological Transfer and Transport Processes," in *Pollutant Transfer and Transport in the Sea: Volume II* (Boca Raton, FL: CRC Press, 1982), pp. 1–65.
109. Aller, R. C. "The Effects of Animal-Sediment Interactions on Geochemical Processes Near the Sediment-Water Interface," in *Estuarine Interactions* (New York: Academic Press, 1978), pp. 157–172.
110. Van der Loeff, M. M. R., L. G. Anderson, P. O. J. Hall, A. Iverfeldt, A. B. Josefson, B. Sundby, and S. F. G. Westerlund. "The Asphyxiation Technique: An Approach to Distinguishing Between Molecular Diffusion and Biological Mediated Transport at the Sediment-Water Interface," *Limnol. Oceanogr.* 29:675–686 (1984).
111. Rhoads, D. C., and J. D. Germano. "Interpreting Long-Term Changes in Benthic Community Structure: A New Protocol," *Hydrobiologia* 142:291–308 (1986).
112. Yingst, J. Y., and D. C. Rhoads. "The Role of Bioturbation in the Enhancement of Bacterial Growth Rates in Marine Sediments," in *Marine Benthic Dynamics* (Columbia, SC: University of South Carolina Press, 1980), pp. 195–212.
113. Pearson, T. H., and R. Rosenberg. "Macrobenthic Succession in Relation to Organic Enrichment and Pollution of the Marine Environment," *Oceanogr. Mar. Biol. Ann. Rev.* 16:229–311 (1978).
114. Swartz, R. C., W. A. Deben, K. A. Sercu, and J. O. Lamberson. "Sediment Toxicity and the Distribution of Amphipods in Commencement Bay, Washington, USA," *Mar. Pollut. Bull.* 13:359–364 (1982).
115. Scott, K. J. "Effects of Contaminated Sediments on Marine Benthic Biota and Communities," in *Contaminated Marine Sediments: Assessments and Remediation* (Washington, D.C.: National Academy Press, 1989), pp. 132–154.
116. Valiela, I. *Marine Ecological Processes* (New York: Springer-Verlag, 1984), pp. 546.
117. Swartz, R. C., F. A. Cole, D. W. Schults, and W. A. DeBen. "Ecological Changes in the Southern California Bight Near a Large Sewage Outfall: Benthic Conditions in 1980 and 1983," *Mar. Ecol. Prog. Ser.* 31:1–13 (1986).
118. Chapman, P. M., R. N. Dexter, and E. R. Long. "Synoptic Measures of Sediment Contamination, Toxicity and Infaunal Community Composition (the Sediment Quality Triad) in San Francisco Bay," *Mar. Ecol. Prog. Ser.* 37:75–96 (1987).

119. Becker, D. S., G. R. Bilyard, and T. C. Ginn. "Comparisons Between Sediment Bioassays and Alterations of Benthic Macroinvertebrate Assemblages at a Marine Superfund Site: Commencement Bay, Washington," *Environ. Toxicol. Chem.* 9:669–685 (1990).
120. Hargraves, B. T. "Coupling Carbon Flow Through Some Pelagic and Benthic Communities," *J. Fish. Res. Board Can.* 30:1317–1326 (1973).
121. Zeitzschel, B. F. "Sediment-Water Interactions in Nutrient Dynamics," in *Marine Benthic Dynamics* (Columbia, SC: University of South Carolina Press, 1980), pp. 195–212.
122. Nixon, S. W. "Remineralization and Nutrient Recycling in Coastal Marine Ecosystems," in *Estuaries and Nutrients* (Clifton, NJ: Humana Press, 1981), pp. 111–138.
123. Boynton, W. R., W. M. Kemp, and C. G. Osborne. "Nutrient Flux Across the Sediment-Water Interface in the Turbid Zone of a Coastal Plain Estuary," in *Estuarine Perspectives* (New York: Academic Press, 1980), pp. 93–109.
124. McLeese, D. W., and C. D. Metcalfe. "Toxicities of Eight Organochlorine Compounds in Sediment and Seawater to *Crangon septemspinosa*," *Bull. Environ. Contam.* 25:921–928 (1980).
125. Rogerson, P. F., S. C. Schimmel, and G. Hoffman. "Chemical and Biological Characterization of Black Rock Harbor Dredged Material," United States Army Corps of Engineers Technical Report D-85-9, p. 123 (1985).
126. Clark, J. R., J. M. Patrick, Jr., J. C. Moore, and E. M. Lores. "Waterborne and Sediment-Source Toxicities of Six Organic Chemicals to Grass Shrimp (*Palaemonetes pugio*) and Amphioxus (*Branchiostoma caribaeum*)," *Arch. Environ. Contam. Toxicol.* 16:401–407 (1987).
127. "Ecological Evaluation of Proposed Discharge of Dredged Material into Ocean Waters," Environmental Effects Laboratory, U.S. Army Engineer Waterways Experiment Station (1977).
128. "Ecological Evaluation of Proposed Discharge of Dredged Material into Ocean Waters," Office of Marine and Estuarine Protection, U.S. Environmental Protection Agency (1991).
129. Lee, G. F., and R. H. Plumb. "Literature Review on Research Study for Development of Dredged Material Disposal Criteria," U.S. Army Corps of Engineers, Dredged Material Research Program (1974).
130. Windom, H. L. "Environmental Aspects of Dredging in the Coastal Zone," in *CRC Critical Reviews in Environmental Control* (Boca Raton, FL: CRC Press, 1976), pp. 91–109.
131. Murden, W. R. "The National Dredging Program in Relationship to the Excavation of Suspended and Settled Solids," in *Fate and Effects of Sediment-Bound Chemicals in Aquatic Systems*, K. L. Dickson, A. W. Maki, and W. A. Brungs, Eds. (New York: Pergamon Press, 1987), pp. 40–47.
132. Lee, G. F., and R. A. Jones. "Water Quality Significance of Contaminants Associated with Sediments: An Overview," in *Fate and Effects of Sediment-Bound Chemicals in Aquatic Systems*, K. L Dickson, A. W. Maki, and W. A. Brungs, Eds. (New York: Pergamon Press, 1987), pp. 1–34.

133. Melzian, B. D. "Toxicity Assessment of Dredged Materials: Acute and Chronic Toxicity as Determined by Bioassays and Bioaccumulation Tests," in *Proceedings of the International Seminar on the Environmental Aspects of Dredging Activities* (Nantes, France: Service Etudes Des Eaux, 1990), pp. 49–64.
134. Hargis, W. J., Jr., M. H. Roberts, Jr., and D. E. Zwerner. "Effects of Contaminated Sediments and Sediment-Exposed Effluent Water on an Estuarine Fish: Acute Toxicity," *Mar. Environ. Res.* 14:337–354 (1984).
135. Prater, B. L., and M. A. Anderson. "A 96-Hour Bioassay of Otter Creek, Ohio," *J. Water Pollut. Contr. Fed.* 10:2099–2106 (1977).
136. LeBlanc, G. A., and D. C. Suprenant. "A Method of Assessing the Toxicity of Contaminated Freshwater Sediments," in *Aquatic Toxicology and Hazard Assessment: Seventh Symposium,* STP 854 (Philadelphia: American Society for Testing and Materials, 1985), pp. 269–283.
137. Gunnison, D., J. M. Brannon, I. Smith, Jr., G. A. Burton, Jr., and K. M. Preston. "A Reaction Chamber for Study of Interactions Between Sediments and Water under Conditions of Static or Continuous Flow," *Water Res.* 14:1529–1532 (1980).
138. Brannon, J. M., R. E. Hoeppel, and D. Gunnison. "Capping Contaminated Dredged Materials," *Mar. Pollut. Bull.* 18: 175–179 (1987).
139. Torello, E. "Report on the Toxicity Tests Conducted on Receiving Water from Sites in New Bedford Harbor, New Bedford, Massachusetts: Pilot Project July 9–16, 1987," U.S. Environmental Protection Agency, Environmental Research Laboratory Technical Report, p. 22 (1987).
140. Craig, P. J. "Chemical Species in Industrial Discharges and Effluents," in *The Importance of Chemical "Speciation" in Environmental Processes* (New York: Springer-Verlag, 1986), pp. 443–464.
141. Phelps, D. K., C. H. Katz, K. J. Scott, and B. H. Reynolds. "Coastal Monitoring: Evaluation of Monitoring Methods in Narragansett Bay, Long Island Sound and New York Bight, and A General Monitoring Strategy," in *New Approaches to Monitoring Aquatic Ecosystems,* STP 940 (Philadelphia: American Society for Testing and Materials, 1987), pp. 107–124.
142. DiBona, P., W. Heyman, and H. Schultz. "Biomonitoring for Control of Toxicity in Effluent Discharges to the Marine Environment," Center for Environmental Research Information, Office of Research and Development, U.S. EPA/625/8-89/015 (1989).
143. Nelson, W. G. "Use of the Blue Mussel, *Mytilus edulis*, in Water Quality Toxicity Testing and *in situ* Marine Biological Monitoring," in *Aquatic Toxicology and Risk Assessment: Thirteenth Volume,* STP 1096 (Philadelphia: American Society for Testing and Materials, 1990), pp. 167–175.
144. Lake, J. L., J. L. Hoffman, and S. C. Schimmel. "Bioaccumulation of Contaminants from Black Rock Harbor Dredged Material by Mussels and Polychaetes," U.S. Army Corps of Engineers Technical Report D-85-2, p. 150 (1985).
145. Lake, J. L., W. Galloway, G. L. Hoffman, W. Nelson, and K. J Scott. "Comparison of Field and Laboratory Bioaccumulation of Organic and Inorganic Contaminants from Black Rock Harbor Dredged Material," U.S. Army Corps of Engineers Technical Report D-87-6, p. 195 (1987).
146. Comeleo, R., R. Hastings, and D. Bailey. "A Cage for *In Situ* Deployment of Mysids for Use in Biomonitoring and Field Validation Studies," poster presentation at the 1990 Society of Toxicology and Chemistry Annual Meeting; Arlington, VA.

147. "Short-Term Methods for Estimating the Chronic Toxicity of Effluents and Receiving Waters to Marine and Estuarine Organisms," Environmental Monitoring and Support Laboratory, U.S. Environmental Protection Agency Report EPA 600/4-87/028 (1988).
148. Hall, L. W., Jr., W. S. Hall, S. J. Bushong, and R. L. Herman. "*In Situ* Striped Bass (*Morone saxatilis*) Contaminant and Water Quality Studies in the Potomac River," *Aquat. Toxicol.* 10:73–99 (1987).
149. Hall, L. W., Jr., S. J. Bushong, M. Ziegenfuss, and W. S. Hall. "Concurrent Mobile On-site and *In Situ* Striped Bass Contaminant and Water Quality Studies in the Choptank River and Upper Chesapeake Bay," *Environ. Toxicol. Chem.* 7:815–830 (1988).
150. Ziegenfuss, M. C., L. W. Hall, Jr., S. J. Bushong, J. A. Sullivan, and M. A. Unger "A Remote *In Situ* Apparatus for Ambient Toxicity Testing of Larval and Yearling Fish in River or Estuarine Systems," *Environ. Toxicol. Chem.* 9:1311–1315 (1990).
151. Oviatt, C. A., M. E. Q. Pilson, S. W. Nixon, J. B. Frithsen, R. T. Rudnick, J. R. Kelly, J. F. Grassle, and J. P. Grassle. "Recovery of a Polluted Estuarine System: a Mesocosm Experiment," *Mar. Ecol. Prog. Ser.* 16:203–217 (1982).
152. Perez, K. T., E. W. Davey, J. Heltshe, J. A. Cardin, N. F. Lackie, R. L. Johnson, R. J. Blasco, A. E. Soper, and E. Read. "Recovery of Narragansett Bay, RI: A Feasibility Study," Office of Pesticides and Toxic Substances, United States Environmental Protection Agency, p. 56 (1990).
153. Redmond, M. S., K. J. Scott, K. M. McKenna, and D. Robson. "Evaluation of the Relative Toxicity of Near Coastal Sediments in Narragansett Bay: Interim Report," Science Applications International Corporation Report to Narragansett Bay Program Office, p. 24 (1989).
154. Hinga, K. R. "Contaminated Sediments in Narragansett Bay: Severity and Extent," Report to Narragansett Bay Project, p. 107 (1990).

CHAPTER 8

Plankton, Macrophyte, Fish, and Amphibian Toxicity Testing of Freshwater Sediments

G. Allen Burton, Jr.

INTRODUCTION

History

Nonbenthic aquatic species, that is, species that do not reside in the sediment for an extended portion of their life cycle, have been useful in assessments of sediment contamination, since testing began in the late 1970s. The first reported sediment toxicity testing using nonbenthic species was by Prater and Anderson.[1,2] They used four aquatic species, including *Daphnia magna* (cladoceran), *Asellus communis* (isopod), *Pimephales promelas* (Fathead minnow), and *Hexagenia limbata* (a burrowing mayfly), in a recirculating system with a sediment-to-water ratio of 1:9.5. Adult *D. magna, A. communis,* and *P. promelas* were exposed for 2 to 4 days. This system was used repeatedly to evaluate sediments to be dredged.[1-5] Some relationships were observed between *D. magna* and *H. limbata* responses and bulk sediment contaminant concentrations; however, poor replication and correlations were also noted. Also in 1977, Birge et al.[6] investigated embryopathic effects of sediment-associated cadmium, mercury, and zinc on several fish and amphibians. Exposure of embryos through 4 to 10 days post hatching showed that embryonic mortality and teratogenesis paralleled contaminant levels.

Applications and Justifications

Nonbenthic organisms that have been *successfully* used in sediment evaluations include bacteria (see Chapter 10), protozoa, phytoplankton, macrophytes, cladocera, fish, and amphibians. Some have questioned the utility of using nonbenthic species to evaluate the benthic environment. In some cases this may be a justifiable concern: for example, if the study objectives are to only study effects in benthic communities, only one species is to be used in the sediment assessment, interstitial water toxicity or bioaccumulation is the primary concern, or validation studies are comparing laboratory and indigenous benthic community responses. However, there are numerous justifications and advantages of nonbenthic testing in sediment evaluations, which have been demonstrated in previous studies.

An aquatic ecosystem is a complex milieu comprised of interacting physicochemical and biological components whose dynamics are often integrated, and their respective spatial and temporal scales may vary by several orders of magnitude. The availability of sediment-associated contaminants to aquatic organisms is dependent on numerous sediment and water characteristics/processes, toxicant characteristics, and the organisms exposure via contact and feeding. Since the aquatic ecosystem is a holistic system, it is naive to assume its components can be separated into individual, noninteracting compartments, using the more traditional reductionist research approach (see Chapter 14). This approach has been essential, however, in defining processes and understanding how components interact. But, when one considers the ever-present interactions of a more difficult experimental design, such as the effects of multiple toxicants, intercompartmental transfer rates, temporal patterns, spatial heterogeneity, patch dynamics, and multitrophic level responses, it is easy to understand why the reductionist approach has been popular with both researchers and regulators. Aquatic scientists have been successful in defining some basic processes, toxicant effects, and interactions that predominate in many aquatic environments. A good example is the U.S. Environmental Protection Agency's (EPA) water quality criteria for metals, many of which are linked to water hardness (as $CaCO_3$ mg/L). These criteria and the toxicity-based control approach have been shown in numerous studies[7] to be protective of water quality. The same example, however, can be used to support the weakness of the reductionist approach. Studies have shown the EPA criteria may not be protective of resident species,[8] and toxicant interactions are not always additive in nature.[9-11] Though the sediment compartment is known to dominate many aquatic ecosystem processes, such as the cycling of C, N, P, S, and to strongly influence overlying water quality and food chains,[12-14] it is still poorly understood and has not been incorporated, to any significant degree, into water pollution control programs (e.g., the EPA National Pollutant Discharge Elimination System).

Given the holistic nature of toxicant perturbations on aquatic ecosystems and the strong interaction between the water and sediment compartments, it is essential that nonbenthic community impacts be considered. Sediment associated contaminants enter the nonbenthic community environment through numerous pathways: storm, dredging, current, or bioturbation-related resuspension; desorption; incorporation into benthic biota that are the prey of nonbenthic species; ingestion of sediments by nonbenthic species thatoccasionally act as epibenthic feeders (bottom-feeding fish and cladocerans); and adsorption to, or uptake through, membranes during sediment contact (such as fish embryos).

Bioturbation by benthic invertebrates pumps pore water constituents out of the sediment into overlying waters, pumping particulates to the sediment-water interface, depositing fecal pellets on the sediment surface, and disrupting horizontal and vertical layering.[15] There are numerous examples of the influence of bioturbation on overlying water quality and biotic communities. Karickhoff and Morris[16] reported that 90% of an organic chemical spiked into the biotic zone of sediment was transported by tubificids to the sediment surface in 30 to 50 days, water concentrations were increased 4 to 6 times, and fecal pellets released less than 20% of the incorporated organic. The mayfly larvae *Hexagenia limbata* increased sediment toxicity to *Daphnia magna* when tested in beakers together,[17] and *H. limbata* burrowing had significant effects on organic sorption partitioning in sediments.[18] Chironomid and trichopterid larvae consolidated sand grains in their tubes and affected ammonia and phosphorus concentrations. The tubes were covered by diatoms that elevated chlorophyll concentrations. The larvae thus influenced algal community dynamics by altering nutrient levels. When the larvae emerged, the nutrient dynamics changed.[19] Bioturbation increases sediment aeration and thus can affect Eh and pH gradients.[20-23]

Comparisons of benthic and nonbenthic species sensitivity to a wide variety of toxicants have shown similar sensitivity.[24] In addition, toxicity responses of benthic and nonbenthic test species to sediments contaminated with synthetic organics and metals were significantly correlated, and the nonbenthic species responses were usually better overall sediment contamination assessment tools, based on their sensitivity, response range, and discriminatory ability.[25,26] This does not suggest, however, that one should routinely use a nonbenthic test species as an indicator of benthic community responses, because few such generalities exist in sediment toxicology, and exceptions are likely.

Laboratory Testing: Critical Issues

Test conditions in sediment toxicity studies have varied widely,[27] making data comparisons infeasible. Extrapolations to in situ conditions are also somewhat tenuous at this point because of our inability to separate laboratory manipulative effects from sample-induced toxicity responses. In other words, have the sample collection, storage, and testing conditions created an artificial assay response (either increased or reduced toxicity), thereby nullifying any comparisons to real-world conditions or relative comparisons between samples? A limited amount of research has been published that focused on methodological effects in freshwater sediment toxicity testing. The effects of sediment contact time (e.g., static, recirculation, static-renewal, flow-through, exposure period), test phase (e.g., whole sediments, interstitial waters, elutriates), and feeding in chronic exposures are discussed in Chapter 10 and also apply to nonbenthic species testing. Whole-sediment and interstitial water sampling, sample storage, toxicant dosing (spiking), sediment dilution, and elutriate method effects are discussed in Chapter 3.

Multitrophic Level Testing in the Laboratory

A wide variety of single and multispecies assays using bacteria and protozoan systems have been used in benthic and water column studies of sediment toxicity. These communities exist in all aquatic compartments and are discussed in Chapter 10.

Phytoplankton

The critical role of algae in ecosystem functioning has been demonstrated for many years through limiting nutrient, eutrophication, and primary productivity studies.[28] They are also a basic fisheries resource via zooplankton grazing and primary consumption by fish,[29] provide a major carbon source for the sediment microbial food web,[30] and cycle nutrients and toxicants.[28,31] As with microbial metabolic endpoints, photosynthetic organisms frequently show stimulatory activity in response to nutrients or perhaps alterations of feedback mechanisms,[32] which may be as useful a response in the assessment process as is inhibition. Though regulatory and research interests have shifted towards toxicants in recent years, nutrients still cause extensive water quality degradation across North America, Europe, and many developing countries. As with toxicants, nutrients accumulate in sediments and may be slowly released to overlying waters.

Algae have been proposed as surrogates for plants;[33] however, this generalization is unrealistic, as shown with herbicide compounds.[34] An evaluation of the sensitivity of 16 microalgal strains to 19 compounds revealed that no one species was consistently the most sensitive to the toxicants. Therefore, they recommended testing a wide range of taxonomic types of algae.[35] Assay responses are difficult to compare between laboratories, due to the significant effect that experimental conditions have.[36] A review of the freshwater algal literature found results can vary by three orders of magnitude due to physical and chemical methodological parameters. Biomass and gas exchange differences, due to shaking or continuous aeration effects on CO_2 limitation and pH, were the apparent cause of this variation.[36] This sensitivity to the laboratory test environment reflects the rapid uptake and metabolic characteristics of algae, as also shown in assays with microbial systems. Nonlinear dose responses may occur, and response patterns may also change with differing test conditions, as shown with varying incubation times.[26] Temporal effects are likely due to changing water quality, toxicant availability, and adaptation.[36,37]

As with the protozoan and rotifer assays, most sediment quality testing with planktonic algae has used elutriates, interstitial water, or overlying waters.[38-40] The principal test species has been *Selenastrum capricornutum* and the test consisted of the standard growth assay for 96 hr.[41] Other assays with this species have used 48-hr incubation periods[26] or measured photosynthesis based on $^{14}CO_2$ assimilation during a 24-hr period.[40] Test battery studies of pure chemicals, effluents, and contaminated soils have shown *S. capricornutum* to be a sensitive test species.[38,42-44]

Other phytoplankton assays involving the use of sediment elutriates or overlying waters include (1) the algal fractionation bioassay (short [4 hr] and long term [24–96 hr]) using natural assemblages, in laboratory or in situ exposures, or precultures of micro- and ultraplankton, where $^{14}CO_2$ uptake and chlorophyll are measured;[39,40,45-47] (2) microcomputer-based video analysis of chlorophyll fluorescence (4-hr incubation);[39,46] (3) microplate ATP analyses;[48] and (4) flow cytometry measures of cell size or biochemical integrity.[49]

Whole-sediment exposure to *Chlorella vulgaris* was also reported, whereby sediments were carefully added to the bottom of the test vessel via tubing, and ^{14}C assimilation measured.[50]

Macrophytes

Duckweed, *Lemna* sp., a nonrooted, floating, vascular aquatic macrophyte, has recently been used with sediment elutriates and whole-sediment assays.[26] *L. minor* and *Lemna* sp. have a wide geographic distribution and are common in many lentic environments. It has recently been described for use as an effluent test method. Frond number and chlorophyll production were the most sensitive endpoints when compared to *C. dubia* and *P. promelas* response, for some effluents.[51] Laboratory responses of *D. magna* and *L. minor* were correlated with pond mesocosm responses.[52] Other endpoints recommended have included root length and ^{14}C uptake.[51]

Klaine et al.[53] recently used a rooted aquatic macrophyte, *Hydrilla verticillata* to measure toxicity in whole-sediment assays with stream and lake sediments contaminated with a variety of metals and synthetic organics. Endpoints included root and shoot length, peroxidase, dehydrogenase, and chlorophyll. Some sediments showed plant growth (root and shoots) to be sensitive to contamination, while others showed peroxidase activity to be most sensitive. As with all other photosynthetic system, studies of both inhibitory and stimulatory effects were noted.

Cladocerans

The importance of cladocerans (e.g., *Ceriodaphnia dubia*) in aquatic systems has been well documented since 1883.[54] They play a significant role in the food web and in phytoplankton and protozoan dynamics. Fish stomach contents have been reported from 1 to 95% cladocera by volume.[28,54] Some species are selective filter feeders, while others, e.g., *D. magna*, are nonselective. An extensive data base exists for pure-compound toxicity testing with *Daphnia* sp., and their relative sensitivity compared to other organisms.[55] These organisms are well recognized as useful toxicity test species[33,56] due to their sensitivity to toxicants and ease of culture. In addition, standard methods exist for effluent and pure-compound testing,[41,57,58] while draft ASTM methods have been proposed for sediment assays. These factors, and the significant role they have had in aquatic toxicology and criteria development, make them obvious candidates for routine sediment toxicity assessments.[56,59]

D. magna and *Ceriodaphnia* are planktonic; however, in sediment assays they spend an extensive amount of time feeding on the sediment surface.[56] *D. magna*, a nonselective filter feeder, ingests sediment particles[60] down to 0.5 μm,[61] whether suspended or settled; thus, in sediment assays they function as epibenthic species. *D. magna's* relative sensitivity to a wide variety of contaminants in whole-sediment interstitial water, elutriate, and suspended-sediment assays is well established. Fewer assays have been reported with *C. dubia*,[25,26,62] which is commonly used in determining the "chronic" toxicity of effluents by using the 3-brood (7-day) survival and reproduction test.[41] In most comparative studies, *C. dubia* acute and chronic toxicity sensitivity appears to be slightly greater than *D. magna*.[26,63] Some laboratories have reported occasional culture problems with *C. dubia*, which can be traced to their high water and food quality requirements.[64] A principal advantage of *C. dubia* over *D. magna* is their rapid reproductive rate after birth. In addition, smaller test volumes are needed for *C. dubia* than with *D. magna*. Reproductive effects may be studied in *D. magna* three-brood (7-day) assays by initiating the test with 5-day old organisms.[56,63,65]

Most sediment assays with the cladocerans have measured acute toxicity[17,56] in 48-hr static (1:4 sediment-to-water ratio) systems. The original studies and some more recent ones used recirculating systems to study whole-sediment, suspended-sediment, or slurry effects.[1,2,60,66] In suspended-sediment assays some stress was observed due to recirculation and high turbidity. The recirculating systems used lower sediment-to-water ratios, e.g., 1:9.5. Recently, subchronic toxicity studies of whole sediments and elutriates have been conducted with *D. magna* and *C. dubia*. These assays were static renewal (only overlying water replaced in whole-sediment assays) and measured survival, reproduction, growth, and pigmentation as endpoints.[67] These comparative studies showed the cladocerans to be useful assays: sensitive and discriminated between contaminated sediments. It is of interest that responses with cladocerans were often similar to benthic species, both in laboratory assays and with in situ communities.[26,68-70] Responses with cladocerans and benthic species were similar in pure-compound and ambient-water studies.[24,55] Many comparative studies have recommended *D. magna* as an optimal screening species and have recommended it as a routine sediment toxicity screening tool.[56,59] *D. magna* was observed (in pure-compound studies) to be a good predictor of fish response ($r > 88\%$);[71] was poorly related to rainbow trout, but similar to *P. promelas* responses;

or was compound specific in its similarity with fish.[72] *D. magna* responses in sediment assays have also been related effectively to the concentration of contaminants in whole sediments, pore water, or those dissolved from the sediment to overlying waters.[3,70,73]

Fish

Toxicity testing with fish in sediment systems has been limited primarily to the fathead minnow *(Pimephales promelas);* however, other species have been used, such as the rainbow trout, goldfish, largemouth bass, and bluegill bream.[6,73,74] Since *P. promelas* has a widespread geographic distribution, it is easily cultured; it has been widely used in the development and validation of water quality criteria, and in pure-compound, and effluent testing, and standard methods exist for testing.[41,57,58] It is an obvious choice for sediment testing. A significant amount of fish testing has focused on the bioaccumulation of sediment-associated toxicants.[75] These investigations have involved laboratory and in situ exposures usually ranging from 10 to 28 days (see Chapter 12). In addition, fish are the principal focus of "biomarker" studies using a wide range of genotoxicity, biochemical, histopathological methods, as well as other endpoints indicative of sublethal exposures to sediment contaminants (see Chapter 11 and McCarthy and Shugart[76]).

Most sediment testing has been as acute exposures (96 hr) to the adult, which is relatively insensitive as compared to early life stage and full life cycle endpoints.[77,78] Norberg and Mount[78] developed a 7-day subchronic larval survival and growth assay for effluent testing, which has been adapted for sediment extracts,[79] elutriates,[80] interstitial waters,[81] and whole-sediment assays.[9,26,82] Another 7-day early life stage assay developed by Birge et al.[6] begins with the embryo stage and continues through 3 days of larval development, with endpoints of survival, teratogenicity, and growth (length). This assay was used in whole-sediment testing[83,84] and, as the larval growth assay, has been found to be a useful and sensitive sediment assay.[26,79] Embryos and larvae are exposed to overlying water, interstitial water, and ingested sediment.[85] Because the embryo stage may be susceptible to fungal infections, it may be necessary to gently aerate overlying waters.[83] As in the cladoceran whole-sediment assays, larvae tend to feed extensively on the sediment surface during the test, thereby potentially increasing their exposure to sediment-related contaminants. Increased uptake of PCB was attributed to fish "mouthing" sediments and desorption occurring in their buccal cavity.[86] In addition, sediments serve as a spawning substrate for many pelagic and epibenthic organisms, so effects such as reproductive behavior, hatchability, development (terata) and growth are critical endpoints to monitor.[85]

Correlations between endpoints and sediment contaminant levels have been reported in some studies and not in others.[3,84,85] Pure chemical toxicity data evaluations comparing *P. promelas,* rainbow trout, and bluegill bream, showed fish surrogates were good predictors of fish response,[87] with increasing similarity being related to the degree of taxonomic similarity.[88] Endpoints in zinc-contaminated sediment exposures ranked from terata, growth inhibition, to mortality, in order of sensitivity. Terata EC_{50} in *P. promelas* were four to six times lower than frog embryo terata effect levels.[79] A sublethal endpoint, cough response, was used in interstitial water exposures to bluegills, but was difficult to interpret.[73]

Amphibians

A limited number of sediment toxicity studies have been reported that used amphibians.[74,79,83] Dawson et al.[79] used the frog embryo teratogenesis assay (FETAX), using *Xenopus laevis,* to measure sediment extract effects from zinc-contaminated sediments. Sediments were extracted for 24 hr in reconstituted water at various pH levels. EC_{50} levels (96 hr) for terata were from 2.5 to 3.6 mg/L Zn at 100 mg/L hardness, 2.0 to 4.2 mg/L for growth, and 34.5 mg/L for survival. Peddicord and McFarland[74] exposed *Bufo boreas* to suspended sediments (2 to 20 g/L) for 21 d. Whole-sediment embryo studies were conducted on the leopard frog *(Rana utricularia)* and narrow-mouthed toad *(Gastrophoryne carolinesis).*[85] Tissue concentrations were related to metal exposure, but not mortality. The duration of embryo contact with the sediment appeared to be an important factor mediating exposure.

IN SITU ASSAYS

A newer approach in studies of sediment contamination or verification of laboratory results is through use of in situ assays. These may involve enclosures such as lake limno-corrals or in situ mesocosms[89] that partition a column of water to the sediment surface; artificially constructed streams allowed to colonize with biota indigenous to a nearby stream;[90,91] experimental ponds that are seeded and allowed to colonize with indigenous species;[89] the placement of caged species in situ (plankton, mussels, zooplankton, leeches, fish, amphibians);[39,91-95] and the placement of litter bags or leaf packs in situ.[96,97]

Of these approaches, few have focused directly on sediment toxicity effects. The limno-corrals and experimental ponds have primarily been used to study the environmental fate of pure compounds, such as pesticides, and polychlorinated dibenzo-compounds and have effectively demonstrated the influence of sediment partitioning and food-chain effects.[89,91] In situ experimental pond studies showed filamentous algae was the most sensitive species in pentachlorophenol-dosed systems.[98] Ponds dosed with trichloroethylene showed decreased phytoplankton diversity and increased abundance.[99] As would be expected in test systems more closely mimicking the "real world," variability is typically greater through time and between replicates than observed in more controlled laboratory environments. Artificial substrates (including leafbags and litterpacks) allow one to study multiple trophic levels, from a community structure and function perspective, using a range

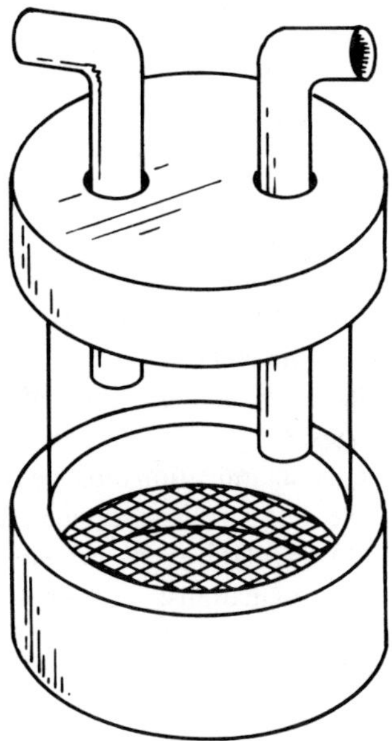

Figure 1. In situ sediment exposure chamber for cladoceran effects testing. (From Sasson-Brickson, G. and G. A. Burton, Jr. "In Situ and Laboratory Sediment Toxicity Testing with *Ceriodaphnia dubia*," *Environ. Toxicol. Chem.* 10:201–207 (1991). With permission.)

of endpoints ranging from simple to complex. The uniform test system reduces habitat (substrate) effects, historical/temporal effects, and sample collection-laboratory-related error. If one is interested primarily in sediment toxicity, however, the substrates create an artificial barrier and thus likely underestimate sediment effects on species that may contact the sediment during their life cycle.

An effective way to study single-species effects of contaminated sediments is possible with in situ sediment test chambers.[92] This approach was originally recognized by Nebeker et al.,[56] however, no studies were published using the in situ chamber approach until 1991. The approach is relatively simple (Figure 1). The chamber removed sampling- and laboratory-induced error from the assessment process while maintaining in situ conditions whose importance in determining sediment toxicity may not be known, such as sunlight, diurnal effects of temperature and oxygen, sediment integrity, spatial and temporal variability effects, flow-through conditions with site water, resident meio-microfaunal interactions, and turbidity. Significant differences were observed between in situ and laboratory responses, with greater sediment toxicity and less overlying water toxicity occurring in the lab.[92] Site toxicity changed seasonally. Elutriate toxicity was generally less than that for whole sediment or interstitial waters, and filtration reduced toxicity significantly. Recent in situ studies with *P. promelas* larvae[94] measured larval weight change in 7-day exposures. The larvae were more susceptible than *C. dubia* to turbidity and storm events, but effectively demonstrated sediment toxicity and laboratory differences.

CONCLUSIONS

For routine screening of sediment contamination, it appears that whole-sediment testing using a benthic invertebrate (e.g., *H. azteca, C. tentans, C. riparius,* or *H. limbata;* see Chapter 10), plankton (*D. magna* or *C. dubia*), and fish *(P. promelas)* should be used, since acute and chronic toxicity assays exist that have been shown to be sensitive and useful in discriminating degrees of contamination. Tests should use 1:4 sediment-to-water ratios with static renewal conditions when exposure periods exceed 48 hr. Studies concerned with criteria development or dredging should also test the interstitial water and elutriate phases, respectively, to aid in extrapolations to field conditions.

Definitive studies of sediment contamination must be more in depth than screening studies, must include multiple test species, replicate samples to define spatial variance, and include a field validation component (such as in situ testing, community surveys, and habitat evaluations; see Chapter 14).

REFERENCES

1. Prater, B. L., and M. A. Anderson. "A 96-Hour Bioassay of Otter Creek, Ohio," *J. Water Pollut. Contr. Fed.* 49:2099–2106 (1977a).
2. Prater, B. L., and M. A. Anderson. "A 96-h Bioassay of Duluth and Superior Harbor Basins (Minnesota) Using *Hexagenia limbata, Asellus communis, Daphnia magna,* and *Pimephales promelas* as Test Organisms," *Bull. Environ. Contam. Toxicol.* 18:159–169 (1977b).
3. Prater, B., and R. A. Hoke. "A Method for the Biological and Chemical Evaluation of Sediment Toxicity," in *Contaminants and Sediments,* Vol. 1, R. A. Baker, Ed. (Ann Arbor, MI: Ann Arbor Science, 1980), pp. 483–499.
4. Hoke, R. A., and B. L. Prater. "Relationship of Percent Mortality of Four Species of Aquatic Biota from 96-Hour Sediment Bioassays of Five Lake Michigan Harbors and Elutriate Chemistry of the Sediments," *Bull. Environ. Contam. Toxicol.* 25:394–399 (1980).
5. Laskowski-Hoke, R. A., and B. L. Prater. "Dredged Material Evaluation: Correlations Between Chemical and Biological Procedures," *J. Water Pollut. Contr. Fed.* 53:1260–1262 (1981).
6. Birge, W. J., J. A. Black, A. G. Westerman, P. C. Francis, and J. E. Hudson. "Embryopathic Effects of Waterborne and Sediment-Accumulated Cadmium, Mercury, and Zinc on Reproduction and Survival of Fish and Amphibian Populations in Kentucky," Research Report No. 100, U.S. Dept. Interior, Washington, D.C. (1977).
7. U.S. Environmental Protection Agency. "Biomonitoring to Achieve Control of Toxic Effluents," Office of Water, Washington D.C., EPA/625/8-87/013 (1987).
8. Pontasch, K.W., B. R. Kiederlehner, and J. Cairns, Jr. "Comparisons of Single-Species, Microcosm and Field Responses to a Complex Effluent," *Environ. Toxicol. Chem.* 8:521–532 (1989).

9. Winks, K. L. "Effects of Metal Mixtures on *Pimephales promelas* Larval Growth in Water and Sediment Exposures," M.S. Thesis, Wright State University, Dayton, OH (1990).
10. Spehar, R. L., and J. T. Fiandt. "Acute and Chronic Effects of Water Quality Criteria-Based Metal Mixtures on Three Aquatic Species," *Environ. Toxicol. Chem.* 5:917–931 (1986).
11. Birge, W. J., J. A. Black, A. G. Westerman, and J. E. Hudson. "The Effects of Mercury on Reproduction of Fish and Amphibians," in *The Biogeochemistry of Mercury in the Environment*, J. O. Nriagu, Ed. (New York: Elsevier North-Holland, 1979).
12. Salomons, W., N. M. de Rooij, H. Keedijk, and J. Bril. "Sediments as a Source for Contaminants?" *Hydrobiologia* 149:13–30 (1987).
13. Carpenter, S. R. *Complex Interactions in Lake Communities.* (New York: Springer-Verlag, 1988).
14. Baker, R.., Ed. *Contaminants and Sediments,* Vol. 1. (Ann Arbor, MI: Ann Arbor Science, 1980).
15. Petr, T. "Bioturbation and Exchange of Chemicals in the Mud-Water Interface," in *Interactions Between Sediments and Freshwater*, H. L. Golterman, Ed. (The Hague, The Netherlands: D. W. Junk Publishers, 1977), pp. 216–266.
16. Karickhoff, S. W., and K. R. Morris. "Impact of Tubificid Oligochaetes on Pollutant Transport in Bottom Sediments," *Environ. Sci. Technol.* 19:51–56 (1985).
17. Malueg, K. W., G. S. Schuytema, J. H. Gakstatter, and D. F. Krawczyk. "Effect of *Hexagenia* on *Daphnia* Responses in Sediment Toxicity Tests," *Environ. Toxicol. Chem.* 2:73–82 (1983).
18. Gerould, S., and S. P. Gloss. "Mayfly-Mediated Sorption of Toxicants into Sediments," *Environ. Toxicol. Chem.* 5:667–673 (1980).
19. Pringle, C. M., R. J. Naiman, G. Brelschko, J. R. Karr, M. W. Oswood, J. R. Webster, R. L. Welcomme, and M. J. Winterbourn. "Patch Dynamics in Lotic Systems: The Stream as a Mosaic," *J. N. Am. Benthol. Soc.* 7:503–524 (1988).
20. Fisher, J. B., and G. Matisoff. "High Resolution Profiles of pH in Recent Sediments," *Hydrobiologia* 79:277–284 (1981).
21. Graneli, W. "Influence of *Chironomus plumosus* Larvae on the Oxygen Uptake of the Sediment," *Arch. Hydrobiol.* 87:385–403 (1979).
22. Hargave, B. T. "Stability in Structure and Function of the Mud-Water Interface," *Verh. Int. Ver. Limnol.* 19:1073–1079 (1975).
23. Lawrence, G. B., M. J. Mitchell, and D. H. Landers. "The Effects of the Burrowing Mayfly on Nitrogen and Sulphur Fractions in Lake Sediment Microcosms," *Hydrobiologia* 87:273–283 (1982).
24. Zarba, C. S. "Equilibrium Partitioning Approach," in U.S. Environmental Protection Agency's Sediment Classification Methods Compendium Watershed Protection Division, Washington D.C., pp. 5.1–5.19 (1989).
25. Burton, G. A., Jr., B. L. Stemmer, and K. L. Winks. "A Multitrophic Level Evaluation of Sediment Toxicity in Waukegan and Indiana Harbors," *Environ. Toxicol. Chem.* 8:1057–1066 (1989).
26. Burton, G. A., Jr., L. Burnett, M. Henry, S. Klaine, P. Landrum, and M. Swift. "A Multi-Assay Comparison of Sediment Toxicity at Three 'Areas of Concern'," Abstr. Annu. Meet Soc. Environ. Toxicol. Chem., Arlington, VA, No. 213, p. 53 (1990).

27. Burton, G. A., Jr. "Assessment of Freshwater Sediment Toxicity," *Environ. Toxicol. Chem.* 10: 1585–1627 (1991).
28. Wetzel, R. G. *Limnology*. (Philadelphia: W. B. Saunders, 1975).
29. Ross, P. E., and M. Munawar. "Zooplankton Feeding Rates at Offshore Stations in the North American Great Lakes," in Proc. Internat. Symp. on Phycology of Large Lakes of the World, M. Munawar, Ed. *Arch. Hydrobiol. Beih. Ergebn. Limnol.* 25:157–164 (1987).
30. Meyer-Reil, L.-A. "Seasonal and Spatial Distribution of Extracellular Enzymatic Activities and Microbial Information of Dissolved Organism Substrates in Marine Sediments," *Appl. Environ. Microbiol.* 53:1748–1755 (1987).
31. Andreae, M. A. "Arsenic Speciation in Seawater and Interstitial Waters: The Influence of Biological-Chemical Interactions on the Chemistry of a Trace Element," *Limnol. Oceanogr.* 24:440–452 (1979).
32. Pratt, J. R., N. J. Bowers, B. R. Niederlehner, and J. Cairns, Jr. "Effects of Atrazine on Freshwater Microbial Communities," *Arch. Environ. Contam. Toxicol.* 17:449–457 (1988).
33. Giesy, J. P., and R. A. Hoke. "Freshwater Sediment Toxicity Bioassessment: Rationale for Species Selection and Test Design," *J. Great Lakes Res.* 15:539–569 (1989).
34. Wang, W., W. Lower, and J. Gorsuch. "Use of Plants for Toxicity Assessment," *ASTM Stand. News* Apr:31–33 (1989).
35. Wangberg, S., and H. Blanck. "Multivariate Patterns of Algal Sensitivity to Chemicals in Relation to Phylogeny," *Ecotoxicol. Environ. Saf.* 16:72–82 (1988).
36. Nyholm, N., and T. Källqvist. "Methods for Growth Inhibition Toxicity Tests with Freshwater Algae," *Environ. Toxicol. Chem.* 8:689–703 (1989).
37. Kuwabara, J. S., and H. V. Leland. "Adaptation of *Selenastrum capricornutum* (Chlorophyceae) to Copper," *Environ. Toxicol. Chem.* 5:197–203 (1986).
38. Miller, W. E., S. A. Peterson, J. C. Greene, and C. A. Callahan. "Comparative Toxicology of Laboratory Organisms for Assessment Hazardous Waste Sites," *J. Environ. Qual.* 14:569–574 (1985).
39. Munawar, J., I. F. Munawar, and G. G. Leppard. "Early Warning Assays: An Overview of Toxicity Testing with Phytoplankton in the North American Great Lakes," *Hydrobiologia* 188/189:237–246 (1989).
40. Ross, P., V. Jarry, and H. Sloterdijk. "A Rapid Bioassay Using the Green Alga *Selenastrum capricornutum* to Screen for Toxicity in St. Lawrence River Sediment Elutriates," in *Function Testing of Aquatic Biota for Estimating Hazards of Chemicals,* J. Cairns, Jr., and J. R. Pratt, Eds., STP 988 (Philadelphia: American Society for Testing and Materials, 1988), pp. 68–73.
41. U.S. Environmental Protection Agency. "Short-Term Methods for Estimating the Chronic Toxicity of Effluents and Receiving Waters to Freshwater Organisms," Environmental Monitoring Systems Laboratory, Cincinnati, OH, EPA/600/4-89/001 (1989).
42. Blank, H., and S. A. Wängberg. "Validity of an Ecotoxicological Test System: Short-Term and Long-Term Effects of Arsenate on Marine Periphyton Communities in Laboratory Systems," *Can. J. Fish. Aquat. Sci.* 45:1807–1815 (1988).
43. Thomas, J. M., J. R. Skalski, J. F. Cline, M. C. McShane, J. C. Simpson, W. E. Miller, S. A. Peterson, C. A. Callahan, and J. C. Greene. "Characterization of Chemical Waste Site Contamination and Determination of its Extent Using Bioassays," *Environ. Toxicol. Chem.* 5:487–501 (1986).

44. Greene, J. C., W. E. Miller, M. Debacon, M. A. Long, and C. L. Bartels. "Use of *Selenastrum capricornutum* to Assess the Toxicity Potential of Surface and Groundwater Contamination Caused by Chromium Waste," *Environ. Toxicol. Chem.* 7:35–39 (1988).
45. Gannon, J. E., and A. M. Beeton. "Procedures for Determining the Effects of Dredged Sediments on Biota-Benthos Viability and Sediment Selectivity Tests," *J. Water Pollut. Contr. Fed.* 43:392–398 (1971).
46. Munawar, M., R. L. Thomas, W. Norwood, and A. Mudroch. "Toxicity of Detroit River Sediment-Bound Contaminants to Ultraplankton," *J. Great Lakes Res.* 11:264–274 (1985).
47. Flint, R. W., and G. J. Lorefice. "Elutriate-Primary Productivity Bioassays of Dredge Spoil Disposal in Lake Erie," *Water Resource Res.* 14:1159–1163 (1978).
48. Blaise, C., R. Legault, N. Bermingham, R. van Coillie, and P. Vasseur. "A Simple Microplate Alga Assay Technique for Aquatic Toxicity Assessment," *Tox. Assess.* 1:261–281 (1986).
49. Berglund, D. L., and S. Eversman. "Flow Cytometric Measurement of Pollutant Stress on Algal Cells," *Cytometry* 9:150–155 (1988).
50. Munawar, M., and I. F. Munawar. "Phytoplankton Bioassays for Evaluating Toxicity of *In Situ* Sediment Contaminants," *Hydrobiologia* 149:87–105 (1987).
51. Taraldsen, J. E., and T. J. Norberg-King. "New Method for Determining Effluent Toxicity Using Duckweed *(Lemna minor)*," *Environ. Toxicol. Chem.* 9:761–767 (1990).
52. Stephenson, R. R., and D. F. Kane. "Persistence and Effects of Chemicals in Small Enclosures in Ponds," *Arch. Environ. Contam. Toxicol.* 13:313–326 (1984).
53. Klaine, S. J., K. Brown, T. Byl, and M. L. Hinman. "Phytotoxicity of Contaminated Sediments," Abstr. Annu. Meet. Soc. Environ. Toxicol. Chem. No. P092, Arlington, VA, p. 140 (1990).
54. Pennak, R. W. *Freshwater Invertebrates of the United States: Protozoa to Mollusca*, 3rd ed. (New York: John Wiley & Sons, 1989).
55. U.S. Environmental Protection Agency. "Recalculation of State Toxic Criteria," Office of Water Regulation and Standards. Washington, D.C. (1982).
56. Nebeker, A. V., M. A. Cairns, J. H. Gakstatter, K. W. Malueg, G. S. Schuytema, and D. F. Krawczyk. "Biological Methods for Determining Toxicity of Contaminated Freshwater Sediments to Invertebrates," *Environ. Toxicol. Chem.* 3:617–630 (1984).
57. U.S. Environmental Protection Agency. "Methods for Measuring the Acute Toxicity of Effluents to Freshwater and Marine Organisms," EPA 600/4-85/013, Cincinnati, OH (1985).
58. American Society for Testing and Materials. *Standard Practice for Conducting Acute Toxicity Test with Fish, Macroinvertebrates, and Amphibians.* E729-80 (Philadelphia: American Society for Testing and Materials, 1980).
59. Great Lakes Water Quality Board. "Procedures for the Assessment of Contaminated Sediment Problems in the Great Lakes," International Joint Commission, Windsor, Ontario (1988).
60. Schuytema, G. S., P. O. Nelson, K. W. Malueg, A. G. Nebeker, D. F. Krawczyk, A. K. Ratcliff, and J. H. Gakstatter. "Toxicity of Cadmium in Water and Sediment Slurries to *Daphnia magna*," *Environ. Toxicol. Chem.* 3:293–308 (1984).
61. Gerritsen, J., and K. G. Porter. "The Role of Surface Chemistry in Filter Feeding by Zooplankton," *Science* 216:1225–1227 (1982).

62. Stemmer, B. L., G. A. Burton, Jr., and G. Sasson-Brickson. "Effect of Sediment Spatial Variance and Collection Method on Cladoceran Toxicity and Indigenous Microbial Activity Determinations," *Environ. Toxicol. Chem.* 9:1035–1044 (1990).
63. Winner, R. W. "Evaluation of the Relative Sensitivities of 7-d *Daphnia magna* and *Ceriodaphnia dubia* Toxicity Tests for Cadmium and Sodium Pentachlorophenate," *Environ. Toxicol. Chem.* 7:153–159 (1988).
64. DeGraeve, G. M., and J. D. Cooney. "*Ceriodaphnia:* An Update on Effluent Toxicity Testing and Research Needs," *Environ. Toxicol. Chem.* 6:331–333 (1987).
65. Lewis, P. A., and W. B. Horning. "A Seven-Day Mini-Chronic Toxicity Test Using *Daphnia magna,*" in *Aquatic Toxicology and Hazard Assessment*, Vol. 10, STP 971, W. J. Adams, G. A. Chapman and W. G. Landis, Eds. (Philadelphia: American Society for Testing and Materials, 1988), pp. 548–555.
66. Hall, W. S., K. L. Dickson, F. Y. Saleh, and J. H. Rogers, Jr. "Effects of Suspended Solids on the Bioavailability of Chlordane to *Daphnia magna,*" *Arch. Environ. Contam. Toxicol.* 15:509–534 (1986).
67. Leibfritz-Frederick, S. "Toxicity of Metals to *Daphnia magna* and *Hyalella azteca* in Sediment Assays and Methodological Variables Within the Test. M.S. Thesis. Wright State University, Dayton, OH (1990).
68. Giesy, J. P., R. L. Graney, J. L. Newsted, C. J. Rosiu, A. Benda, R. G. Kreis, Jr., and F. J. Horvath. "Comparison of Three Sediment Bioassay Methods Using Detroit River Sediments," *Environ. Toxicol. Chem.* 7:483–498 (1988).
69. Giesy, J. P., C. J. Rosieu, R. L. Graney, and M. G. Henry. "Benthic Invertebrate Bioassays with Toxic Sediment and Pore Water," *Environ. Toxicol. Chem.* 9:233–248 (1990).
70. Malueg, K. W., G. S. Schuytema, D. F. Krawczyk, and J. H. Gakstatter. "Laboratory Sediment Toxicity Tests, Sediment Chemistry and Distribution of Benthic Macroinvertebrates in Sediments from the Keweenaw Waterway, Michigan," *Environ. Toxicol. Chem.* 3:233–242 (1984).
71. Doherby, F. G. "Interspecies Correlations of Acute Aquatic Median Lethal Concentrations for Four Standard Testing Species," *Environ. Sci. Technol.* 17:661–665 (1983).
72. Holcombe, G. W., C. L. Phipps, A. H. Sulaimon, and A. D. Huffman. "Simultaneous Multiple Species Testing: Acute Toxicity of 13 Chemicals to 12 Diverse Freshwater Amphibian, Fish and Invertebrate Families," *Arch. Environ. Contam. Toxicol.* 16:697–710 (1987).
73. U.S. Environmental Protection Agency. "Development of Bioassay Procedures for Defining Pollution of Harbor Sediments," Environmental Research Laboratory, Duluth, MN (1981).
74. Peddicord, R. K., and V. A. McFarland. "Effects of Suspended Dredged Material on Aquatic Animals," U.S. Army Eng. Waterways Exper. Sta., Dept. No. WES-TR-D-78-29, NTIS #ADA058 489/GST (1978).
75. Mac, M., and C. Schmidt. "Bioavailability of Sediment Contaminants to Fish," in *Contaminated Sediment Toxicity Assessment*, G. A. Burton, Jr., Ed. (Boca Raton, FL: Lewis Publishers, 1991) (in press).
76. McCarthy, J. F., and L. R. Shugart. *Biomarkers of Environmental Contamination.* (Boca Raton, FL: Lewis Publishers, 1990).
77. Suter, G. W. "Seven-Day Tests and Chronic Tests," *Environ. Toxicol. Chem.* 9:1435–1436 (1990).

78. Norberg-King, T. J., and D. Mount. "An Evaluation of the Fathead Minnow Seven-Day Subchronic Test for Estimating Chronic Toxicity," *Environ. Toxicol. Chem.* 8:1075–1089 (1989).
79. Dawson, D. A., E. F. Stebler, S. L. Burks, and J. A. Bantle. "Evaluation of the Developmental Toxicity of Metal-Contaminated Sediments Using Short-Term Fathead Minnow and Frog Embryo-Larval Assays," *Environ. Toxicol. Chem.* 7:27–34 (1988).
80. Hoke, R. A., J. P. Giesy, G. T. Ankley, J. L. Newsted, and J. R. Adams. "Toxicity of Sediments from Western Lake Erie and the Maumee River at Toledo, Ohio, 1987: Implications for Current Dredged Material Disposal Practices," *J. Great Lakes Res.* 16:457–470 (1990).
81. Ankley, G. T., A. Katko, and J. W. Arthur. "Identification of Ammonia as an Important Sediment Associated Toxicant in the Lower Fox River and Green Bay, Wisconsin," *Environ. Toxicol. Chem.* 9:313–322 (1990).
82. U.S. Environmental Protection Agency. "Assessment and Remediation of Contaminated Sediments (ARCS) Work Plan," Great Lakes National Program Office, Chicago, IL (1990).
83. Francis, P. C., W. J. Birge, and J. A. Black. "Effects of Cadmium-Enriched Sediment on Fish and Amphibian Embryo-Larval Stages," *Ecotoxicol. Environ. Saf.* 8:378–387 (1984).
84. Westerman, A. G. 1989. "Yearly Chronic Toxicity Comparisons of Stream Sediments and Waters," in *Aquatic Toxicology and Hazard Asessment,* STP 1007, G. W. Suter and M. A. Lewis, Eds. (Philadelphia: American Society for Testing and Materials, 1989), pp. 204–214.
85. Birge, W. J., J. A. Black, A. G. Westerman, and P. C. Francis. "Toxicity of Sediment-Associated Metals to Freshwater Organisms: Biomonitoring Procedures," in *Fate and Effects of Sediment-Bound Chemicals in Aquatic Systems*, K. L. Dickson, A. W. Maki, and W. A. Brungs, Eds. (New York: Pergamon Press, 1984), pp. 199–218.
86. Hatter, M. T., and H. E. Johnson. "A Model System to Study to Desorption and Biological Availability of PCB in Hydrosoils," in *Aquatic Toxicology and Hazard Evaluation*, STP 634, F. E. Mayer, and J. L. Hamelink, Eds. (Philadelphia: American Society for Testing and Materials, 1977), pp. 178–195.
87. Mayer, F. L., Jr., K. S. Mayer, and M. R. Ellersieck. "Relation of Survival to Other Endpoints in Chronic Toxicity Tests with Fish," *Environ. Toxicol. Chem.* 5:737–748 (1986).
88. Suter, G. W., II, and D. S. Vaughan. "Extrapolation of Ecotoxicity Data: Choosing Tests to Suit the Assessment," Synthetic Fossil Fuel Technologies (Ann Arbor, MI: Ann Arbor Sci., 1984), pp. 387–399).
89. Ravera, O. "The 'Enclosure' Method: Concepts, Technology, and Some Examples of Experiments with Trace Metals," in *Aquatic Ecotoxicology,* Vol. 1, pp. 249–272.
90. Hartwell, S. I., D. S. Cherry, and J. Cairns, Jr. "Field Validation of Avoidance of Elevated Metals by Fathead Minnows *(Pimephales promelas)* Following *In Situ* Acclimation," *Environ. Toxicol. Chem.* 6:189–200 (1987).
91. Cairns, J. *Multispecies Toxicity Testing.* (Elmsford, NY: Pergamon Press, 1985).
92. Sasson-Brickson, G., and G. A. Burton, Jr. "*In Situ* and Laboratory Sediment Toxicity Testing with *Ceriodaphnia dubia,*" *Environ. Toxicol. Chem.* 10:201–207 (1991).
93. Metcalfe, J. L., and A. Hayton. "Comparison of Leeches and Mussels as Biomonitors for Chlorophenol Pollution," *J. Great Lakes Res.* 15:654–668 (1989).

94. Skalski, C., R. Fisher, and G. A. Burton, Jr. "An *In Situ* Interstitial Water Toxicity Test Chamber," Abstr. Annu. Meet. Soc. Environ. Toxicol. Chem. P058, Arlington, VA, p. 132 (1990).
95. Linder, G. "Laboratory and *In Situ* Toxicity Testing with Amphibians," Abstr. Annu. Meet. Soc. Environ. Toxicol. Chem., P220, Arlington, VA, p. 169 (1990).
96. Allred, P. M., and J. P. Giesy. "Use of *In Situ* Microcosms to Study Mass Loss and Chemical Composition of Leaf Litter Being Processed in a Blackwater Stream," *Arch. Hydrobiol.* 114:231–250 (1988).
97. Swift, M. C., R. A. Smucker, and K. W. Cummins. "Effects of Dimilin® on Freshwater Litter Decomposition," *Environ. Toxicol. Chem.* 7:161–166 (1988).
98. Crossland, N. O., and C. J. M. Wolff. "Fate and Biological Effects of Pentachlorophenol in Outdoor Ponds," *Environ. Toxicol. Chem.* 4:73–86 (1985).
99. Lay, J. P., W. Schauerte, W. Klein, and F. Korte. "Influence of Tetrachloroethylene on the Biota of Aquatic Systems: Toxicity to Phyto- and Zooplankton Species in Compartments of a Natural Pond," *Arch. Environ. Contam. Toxicol.* 13:135–142 (1984).

CHAPTER 9

Assessment of Sediment Toxicity to Marine Benthos*

Janet O. Lamberson, Theodore H. DeWitt, and Richard C. Swartz

INTRODUCTION

Most chemical contaminants entering the marine environment eventually accumulate in sediments and, thereby, potentially render the sediments toxic to benthic and demersal organisms. Through deposition, adsorption, diffusion, resuspension, and emigration, sediments serve as both a sink and source for toxic contaminants in the marine environment.[1,2] The relationship between the concentrations of chemicals in sediments and in the tissues of benthic biota is well established (see Chapters 11 and 12). Although the linkage between bioaccumulation and toxicological responses is poorly documented, logic indicates a strong association. Chemical contaminants in sediments have been implicated as the cause of the abnormal pathology observed in benthic and demersal organisms and the alterations in the structure of benthic invertebrate populations and communities. In Puget Sound and Chesapeake Bay, aromatic hydrocarbons, metals, and other contaminants in sediments have been correlated with observations of liver tumors, fin erosion, and mortality in bottom-dwelling fish and invertebrates, as well as with the degradation of benthic community structure.[3-7] Chemical analyses revealed high contaminant concentrations in Puget Sound sediments that were acutely toxic and had reduced benthic species diversity.[8-10] Mortality and reductions in somatic growth

* Contribution No. N-175, U.S. Environmental Protection Agency Environmental Research Laboratory, Narragansett, RI/Newport, OR.

rate and gonad production were observed in sea urchins exposed to contaminated sediments from Southern California.[11] Synoptic measurements of chemical concentrations in sediments, sediment toxicity, and benthic community structure analysis revealed a pattern of toxicity related to chemical contamination in sediments in San Francisco Bay.[12,13] Reductions in the chemical discharge from a large sewage outfall in Los Angeles County apparently have resulted in a reduction in sediment toxicity, and improvements in benthic community structure.[14,15] Oysters exposed to chemically contaminated sediment from Black Rock Harbor, Connecticut, in the laboratory and the field, accumulated high concentrations of polychlorinated biphenyls (PCBs), polyaromatic hydrocarbons (PAHs), and chlorinated pesticides, and developed neoplastic disorders, as did winter flounder fed with mussels exposed to the contaminated sediment, demonstrating trophic transfer of sediment-derived carcinogens.[16]

The assessment of the toxicological effects of sediment-associated chemicals can be made with a variety of relatively simple bioassays. These tests directly measure the toxicological effects of the bioavailable fractions of the contaminants under controlled conditions. Sediment toxicity tests provide a rapid, integrated measure of the effects of a substrate and the chemicals found in it on a toxicologically sensitive representative of the benthic fauna. Sediment toxicity tests may be used alone or in conjunction with sediment chemistry and benthic community structure data, thus forming the "sediment quality triad" (see Chapter 15). Sediment toxicity tests have a distinct logistical advantage over the other two legs of the triad in that they usually are less expensive, provide data more rapidly, and require less technical expertise than do chemical analyses or evaluation of benthic community structure. These toxicity tests provide information that is unique in several ways: (1) they integrate the additive and interactive effects of the complex mixtures of chemicals found in contaminated sediments, including chemicals that are not measured; (2) cause and effect relationships can be demonstrated by either concentration-response experiments with single compounds or dilution series experiments with contaminated field sediments;[17-20] (3) a variety of endpoints may be examined over different exposure periods to predict potential impact at several levels of biological organization, including physiological (luminescence and growth), ontogenetic (development), behavioral (burrowing and choice), and population (reproduction and mortality); and (4) toxicological parameters can be incorporated into regulatory strategies either directly, as a test of the toxicity of field-collected sediment, or indirectly, as part of several methods to develop sediment quality criteria.

This chapter focuses on sediment toxicity testing methodology for determining the effects of chemical contaminants on marine benthos. Several reviews addressing these methods have been published previously.[21-28] The objectives of this chapter are to review the factors that affect sediment toxicity, compare current sediment toxicity testing methods and their applications, promote methodological standardization and quality control, discuss problems and limitations of current methods, and speculate on directions for new research.

FACTORS AFFECTING SEDIMENT TOXICITY

Abiotic Factors: Sediment Geochemistry

The toxicity of contaminated sediments may be modified by abiotic factors, in addition to the absolute concentration of specific chemicals. Chemical factors that may influence the apparent toxicity of a sediment-associated chemical include sorption to particulate or dissolved organic matter, the physical chemical form of the compound (including its ionic state), the presence of other ions (e.g., salinity, acid volatile sulfides [AVS], hardness), interstitial redox potential, and concentrations of limiting compounds (e.g., dissolved oxygen, ammonia, hydrogen sulfide). Interactions among these factors and between geochemical and physical variables (especially temperature and disturbance) may further modify sediment toxicity, although few studies have examined these issues (see Chapters 3, 8, and 9).

The amount of organic carbon in sediments regulates the partitioning of nonionic organic compounds between particulate and interstitial water phases,[29,30] and organic carbon content has been shown to have a substantial effect on the toxicity of these compounds to benthic organisms.[31-34] Swartz et al.[19] found that the LC_{50} of total sediment fluoranthene increased with the concentration of sedimentary organic carbon, indicating that the contaminant was less bioavailable in more organically enriched sediments. DeWitt et al.[35] found that the source of organic matter in sediment had relatively little effect on the toxicity of fluoranthene, which suggests that the bioavailability of nonionic organic chemicals in sediment may be predicted from knowledge of the whole-sediment concentration of the contaminant and the sediment organic content.

The bioavailability of metals in sediment can be affected by the binding of the metal ions to sediment constituents. Metal ions can be free in solution, complexed to dissolved or colloidal organic materials in sediment interstitial water, or bound to sediment particles. Kemp and Swartz[36] demonstrated that the interstitial water concentration, rather than the total bulk concentration of cadmium in sediment, determined the toxicity of cadmium to the amphipod *Rhepoxynius abronius*. In sediment, metal ions that are bound to sediment constituents may be unavailable to sediment-dwelling organisms; thus, sediments with relatively high concentrations of metals might have unexpectedly low toxicity.[37] Di Toro et al.[38,39] demonstrated that cadmium binds to AVS in sediment and that the molar concentration of AVS could be used to predict the toxicity of unbound cadmium and, presumably, other metals in reduced sediments. Metal ions also have an affinity to iron and manganese oxides and organic carbon in sediments;[38,39] and in oxidized sediments, the presence of these substances might determine bioavailability.[40,41] Swartz et al.[37] showed that the toxicity of cadmium in a sandy sediment was inversely correlated with the concentration of organic carbon in the form of sewage sludge.

Abiotic Factors: Sediment Geology

The geological properties of sediments can also affect the apparent toxicity of a sediment by modifying the bioavailability of the contaminants or by exerting a direct stress on the test organism. Organic contaminants and metals bind more readily to the finer particles in sediment (silts and clays) than to coarser particles (sand and gravel).[38] This may happen simply because the finer particles have more surface area per volume and, thus, more binding sites, or because the finer particles are more likely to be ionized. Fine particles include clays, metal oxides, sulfides, and organic matter aggregates, all of which may bind organic compounds and metals.[38]

Sediment geological properties may also affect apparent toxicity by having a direct effect on the health of the test species. Densely packed sediments with low water content may impede burrowing, tube building, or feeding, thus modifying sediment toxicity by either reducing exposure (i.e., the test organism remains above the substrate) or enhancing stress (i.e., the test organism is fatigued by its attempt to work the sediment). Sediment grain size can alter sediment toxicity if the particle-size distribution of test sediments is unlike that of the test organism's native habitat. For example, organisms that utilize fine particles to build tubes might not be able to build normal tubes in sandy sediment, while organisms that burrow freely in sand might have problems with digging into fine-grained muds or might experience fouling of their gills.[42] DeWitt et al.[43,44] developed regression-based statistical models to separate the effects of fine particles and chemical contaminants in test sediments, on the acute mortality of the free-burrowing amphipods *R. abronius* (Figure 1) and *Eohaustorius estuarius,* respectively. The models are based on the regression of amphipod survival vs. the percent fines (i.e., % silt + % clay) in uncontaminated sediments. The lower 95% prediction limit (95% LPL) equation of that regression is used to predict the lowest amphipod survival one would expect to see (with 95% confidence) in an uncontaminated sediment of a given percent fines, assuming a sample size of five replicates. Thus, if *R. abronius* survival in a test sediment was 12 out of 20 and percent fines equals 90%, one would conclude from the model in Figure 1 that mortality may have been caused by grain-size effects alone, since 12 is above the 95% LPL line. However, if percent fines equals 50%, one would conclude that mortality was caused by other factors (such as chemical contaminants) in addition to fine-grained sediment, since 12 is below the 95% LPL line (Figure 1).

The 95% LPL will vary with (1) the species of interest; (2) the number of data points in the underlying regression, using uncontaminated sediment (i.e., the line approaches the mean as regression sample size increases, and becomes more hyperbolic in shape as sample size decreases), and (3) the number of replicates in the test (i.e., the line approaches the mean as the number of replicates increases). This same approach could be used to adjust for other potential sources of background error, such as sediment organic content, sediment water content, interstitial dissolved oxygen concentration, or sediment pH.

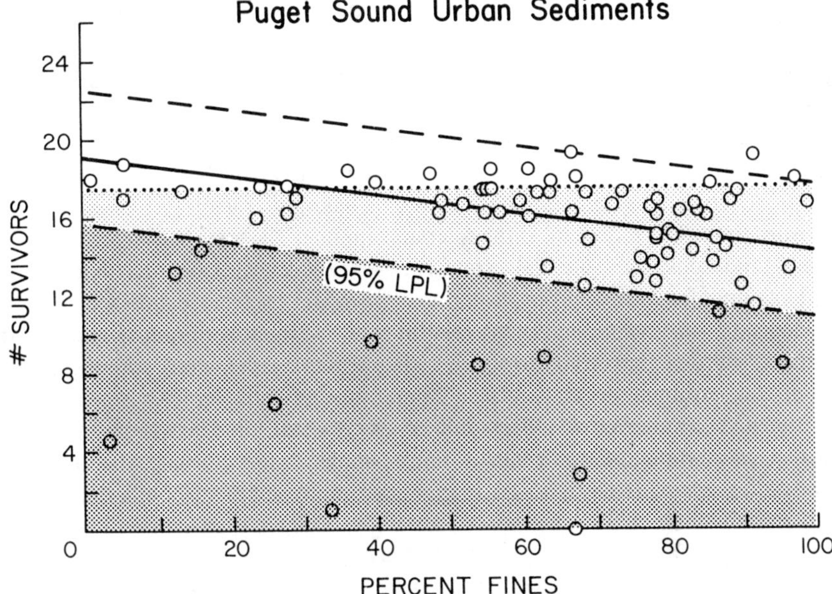

Figure 1. Statistical model to discriminate the "toxic" effects of fine particles from those of chemical contaminants in test sediments for the free burrowing amphipod, *Rhepoxynius abronius*.[43] If survival in sediment with a particular silt-clay content falls in the clear area at the top of the graph, the sediment is nontoxic (i.e., not different from the control). If it falls in the dark shaded area, below the 95% LPL, the sediment is toxic, since mortality is caused by factors in addition to particle size. If it falls within the lightly shaded area, mortality may be due to particle-size "toxicity" or some other factor, such as chemical contamination, but the distinction between the factors cannot be made statistically.

Biological Factors

The response of the test organisms to the toxicant or test sediment may also be affected by its life stage, the route of contaminant uptake, the general state of health of the test organism, the acclimation of the test organisms to test conditions, the duration of exposure, and the mode of toxicant action. In general, larval or juvenile life stages are more sensitive than adults. Animals that burrow freely in sediment are more directly exposed to chemicals dissolved in interstitial water and have greater direct contact with contaminated particulates than do tube-building or epibenthic species or animals that emerge from the substrate to swim or crawl on the sediment surface. Animals that ingest sediment particles may be more directly exposed to sediment-associated contaminants than species that filter overlying water for suspended particulate food. Predators in the test substrate may also have an undesirable impact by consuming the toxicity test species.

Ideally, conditions within the test chamber should be well within the tolerance range of the test organisms so that mortality can be attributed solely to chemical contamination of the test sediment. Factors to consider include interstitial water salinity, temperature, dissolved oxygen in interstitial or overlying water, sediment oxidation state (i.e., redox potential), metabolic byproducts such as ammonia and hydrogen sulfide, and grain size, as discussed previously. The test species should be able to survive in uncontaminated sediment that has similar geological and chemical properties to the test sediment. If the tolerance limits of the test species are exceeded by test conditions, the potential effects of the excedence(s) on the toxicological endpoints must be quantified by either an appropriate control treatment or *a priori* knowledge of the effects, such as the DeWitt et al.[43,44] model of grain-size effects on amphipod survival. Sensitivity to specific chemical toxicants varies with the species used in the sediment toxicity test.[13] Some species show greater sensitivity to contaminated sediments than closely related species, and some taxa seem to be more sensitive, as a whole, than other taxa. For example, in tests to date, the amphipod *Rhepoxynius abronius* tends to be more sensitive to sediment contaminants than other amphipods,[44] while amphipods, as a group, tend to be more sensitive to sediment contaminants than polychaetes and bivalves, in side-by-side comparisons.[45] On the other hand, some taxa with low general sensitivity may have high sensitivity to specific contaminants; for example, bivalves and gastropods are very sensitive to tributyl tin (TBT), but much less so to polynuclear aromatic hydrocarbons (PAHs) and metals. These generalizations may not hold up over time, as they are biased by the small number of species compared. In addition, sensitivity varies with age and life stage of the test organism, and different species vary in which life stage is most sensitive.[46] Therefore, the toxicity of sediments with unknown chemical content should be evaluated with more than one test species if decisions on toxicity are to be made using toxicity tests alone, without chemical and community structure analyses.[23,27]

OVERVIEW OF REPRESENTATIVE SEDIMENT TOXICITY TESTS WITH MARINE AND ESTUARINE BENTHIC ORGANISMS

Sediment toxicity tests were first developed in the 1970s[47,48] and were recommended in the U.S. Environmental Protection Agency/U.S. Army Corps of Engineers Implementation Manual for dredged material evaluation (commonly referred to as the "Green Book").[49] Methodology proliferated in the 1980s, and sediment toxicity tests were recommended, along with chemical and community structure analysis, to provide information on the ecological impact of sediment contamination.[50] A wide variety of studies utilizing the biological assessment of sediment toxicity in marine sediments have previously been reviewed and tabulated.[22,23,25-27,50,51] Responses have ranged from sublethal physiological effects, such as changes in respiration, to alterations in community structure and function.

The duration of the tests ranges from a few minutes to more than a year, and the quantities of sediment tested ranges from a few grams to over a metric ton.[27] Only a few of these methods have been standardized (i.e., published as ASTM Standard Guides,[52] Puget Sound Guidelines,[53] or recommended in the "Green Book"[49,54]) and are currently in common use as sediment toxicity tests (Table 1). Commonly used sediment toxicity tests may be classified as "acute" or "chronic," and test whole sediment (e.g., solid phase), suspended sediment, sediment liquid phases (e.g., pore water, interstitial water), or sediment extracts (e.g., elutriates, solvent extracts). The exact definitions of the terms "acute" and "chronic" have been the subject of long discussions among sediment toxicologists, and agreement has yet to be established. The American Society of Testing and Materials (ASTM) currently defines an *acute* toxicity test as "a comparative study in which organisms that are subjected to different treatments are observed for a short period, usually not constituting a substantial portion of their life span." A *chronic* test is defined as "a comparative study in which organisms that are subjected to different treatments are observed for a long period or a substantial portion of their life span."[55] A chronic test should include a substantial portion of the life cycle or number of life stages and allow sufficient time so that the contaminant steady state is approached in the tissues of the exposed organisms. There is no specific test duration for any species that defines a boundary between acute and chronic tests. Acute tests often utilize mortality as the only measure of effect, while chronic tests usually include measures of growth, reproduction, behavioral effects, or other sublethal endpoints.

Tests of whole or solid-phase sediment differ from tests with other sediment phases in that the whole, intact sediment is used to test the exposed organisms.[52] Suspended-sediment tests utilize a slurry of sediment and water to expose the organisms where sediment particles are held in suspension by stirring or agitation of the sediment/water mixture. Sediment elutriate tests examine the toxicity of a liquid supernatant withdrawn after suspended-sediment particles settle. Many of the liquid-phase test methods (suspended sediment and elutriate tests) are adaptations of test procedures developed for effluent or water toxicity testing. Sediment extract tests examine the toxicity of a liquid extract acquired by aqueous or organic solvent extraction of the whole sediment. Organic extract tests are usually conducted on a small scale, using microorganisms or cellular responses. Pore water tests expose the test organism to the interstitial water phase. Table 1 focuses on species used in standardized sediment toxicity tests appropriate for each of these sediment phases (see Chapter 3 for additional discussion of the strengths and weaknesses of the various test phases).

Whole Sediment

Acute lethality tests have been developed for amphipods, cumaceans, copepods, shrimps, isopods, bivalves, polychaetes, and fish.[22,23,25-27,50,51] Sublethal tests have utilized various physiological, pathological, behavioral, growth, and life-cycle

Table 1. Marine and Estuarine Sediment Toxicity Test Species for Which Standardized Protocols Have Been Developed

Whole Sediments
- Amphipods: mortality, reburial, 10 day
  - *Rhepoxynius abronius*[52-54]
  - *Eohaustorius estuarius*[52-54]
  - *Ampelisca abdita*[52,54]
  - *Grandidierella japonica*[52,56]
  - *Corophium insidiosum*[56]
- Polychaetes: mortality, juvenile growth, 96 hr and 20 day
  - *Neanthes arenaceodentata*[53,54,56]
  - *Hediste (Nereis)* sp.[54]
  - *Capitella capitata*[56]
- Shrimp: mortality, 96 hr, bioaccumulation
  - Ridgeback prawn, *Sicyonia ingentis*[56]
- Pelecypods: mortality, 96 hr, bioaccumulation
  - Littleneck clam, *Protothaca staminea*[56]
  - Bentnose clam, *Macoma nasuta*[56]
- Fish: mortality, 96 hr, bioaccumulation
  - Arrow goby, *Clevelanda ios*[56]

Sediment Liquid-Phases
- Bivalve larvae: mortality, abnormal growth, 48 hr
  - Blue mussel, *Mytilus edulis*[53,54,78]
  - Eastern oyster, *Crassostrea virginica*[54,78]
  - Pacific oyster, *Crassostrea gigas*[53,78]
  - Oyster, *Ostrea* sp.[54]
- Echinoderm embryos: mortality, abnormal growth, 48–96 hr
  - Purple sea urchin, Strongylocentrotus purpuratus[53,56]
  - Green sea urchin, *Strongylocentrotus droebachiensis*[53]
  - Sea urchin, *Strongylocentrotus franciscanus*[53]
  - Sea urchin, *Arbacia punctulata*[53]
  - Sea urchin, *Lytechinus pictus*[56]
  - Sand dollar, *Dendraster excentricus*[53]
- Zooplankton: mortality, 48 hr
  - Copepods, *Acartia* sp.[54]
  - Grunion (embryos), *Leuresthes tenuis*[56]
- Mysids: mortality, 96 hr
  - *Holmesimysis costata*[54,56]
  - *Mysidopsis* sp.[54]
  - *Neomysis* sp.[54]
- Fish: mortality, 96 hr
  - Silversides, *Menidia* sp.[54]
  - Shiner perch, *Cymatogaster aggregata*[54]
  - Top smelt, *Atherinops affinis*[56]
  - Speckled sanddab, *Citharichthys stigmaeus*[56]

Sediment Extracts
- Organic Solvent Extract
  - Anaphase aberration: 48 hr
    - Rainbow trout, *Salmo gairdneri*[53]
  - Microtox®: luminescence, 5, 15, 30 min
    - *Photobacterium phosphorum*[53]
- Saline Extract
  - Microtox®: luminescence, 15 min
    - *photobacterium phosphorum*[53]

responses in crustaceans, polychaetes, bivalves, fish, and plants, as well as community structure based on larval settlement. Whole-sediment tests that have been more or less standardized have been developed for amphipods and polychaetes (Table 1); however, pelecypods, shrimp, and fish have also been used.[56]

Amphipods

Amphipods are among the most sensitive of benthic species. They are among the first to disappear from benthic communities in sediments impacted by pollution[8,57] and were the most sensitive of several taxa tested in a multispecies whole-sediment test.[45] Amphipods are available in large numbers in uncontaminated marine and estuarine waters worldwide, and are easily collected and handled in the laboratory. Culture procedures are available, or under development, for some species.

Most whole-sediment tests with amphipods are short-term static tests conducted for 10 days,[52,53,58] with the primary endpoints being mortality and the ability of test survivors to rebury in clean sediment at the end of the exposure period. The latter endpoint is a sublethal measure of the test organism's ability to survive under real-world conditions, because a benthic amphipod that is unable to bury will most likely be swept away from a suitable habitat by water currents or be consumed by predators. The test is initiated when amphipods (usually 20 individuals) are introduced into the test chamber (typically a standard 1-L glass beaker or a 1-qt canning jar) containing test sediment and overlying water. Most tests are static (i.e., no change of overlying water during the test) with aeration, though some test designs may call for a static renewal (water changed at intervals) or continuous-flow procedure. Test animals may be from a laboratory culture or field collected and are usually not fed during the 10-day sediment exposure because food is available in the test sediment.[52] Salinity, temperature, and light regime are specific to the test species. At the end of the exposure period, the amphipods are sieved from the sediment, and the survivors counted. Survivors also may be tested for sublethal behavioral responses such as reburial. The species selected depends upon local availability, salinity, and temperature at the collection site, and grain size distribution of the test sediment. Standard guidelines[52] have been published for 10-day tests with *Ampelisca abdita, Eohaustorius estuarius, Grandidierella japonica,* and *Rhepoxynius abronius.* Test methods have also been developed for *Leptocheirus plumulosus*[59] and are being developed for other species, such as *Eogammarus confervicolus,*[60] generally following the ASTM[52] guidelines. Reish and LeMay[56] recommend using *G. japonica* or *Corophium insidiosum* in short-term exposures to test dredged material in southern California. Procedures for longer-term tests have been used for *G. japonica* (28 days),[61] *G. lutosa* and *G. lignorum* (39 to 90 days),[62,63] and *A. abdita* (56 days).[64] In longer-term tests, amphipods are fed, and endpoints include mortality, growth, and reproduction. Tests with *G. lignorum/lutosa*[63] spanned four generations and followed population changes. Chronic test procedures are under development at several laboratories but, unfortunately, are not yet standardized for any species of amphipods.

Polychaetes

The species *Neanthes arenaceodentata* was selected for sediment toxicity testing because of its availability from laboratory cultures and the existence of a large data base of its responses to a variety of chemicals and sediment types.[56,65,66] This species has been used successfully in acute, chronic, and life-cycle testing of water and sediments since the 1970s.[56,65-67] Other species of polychaetes that have been used to some extent in whole marine sediment toxicity testing include *Capitella capitata, Ophrotrocha diadema,* and *Ctenodrilus serratus*.[56,68] The "Green Book"[54] recommends *Neanthes* sp. or *Hediste (Nereis)* sp. and also lists several other genera of polychaetes for sediment toxicity testing.

The *Neanthes arenaceodentata* test procedure[53,66-68] characterizes the toxicity of marine sediments, based on juvenile polychaete survival in a 96-hr to 10-day test, or growth and survival over 20 d. Test animals have been available only from laboratory cultures. Typically, five juvenile worms, 2 to 3 weeks postemergence, are exposed to sediment in 1-L jars, with salinity and temperature at 28‰ and 20°C, respectively. During the exposure the worms are fed every other day during the test, with a commercial fish food.[70] At the end of the exposure period, worms are sieved from the sediment, counted, and weighed to obtain total biomass and average individual biomass.

Other Taxa

Reish and LeMay[56] also recommend the ridgeback prawn, *Sicyonia ingentis,* the clams, *Protothaca staminea* and *Macoma nasuta,* and the arrow goby, *Clevelanda ios,* for whole-sediment toxicity testing, because they live within, or in contact with, the sediment and are large enough to analyze for bioaccumulation of toxicants. Sediment toxicity tests have also been developed using clam burrowing behavior[12,71-73] and oyster *(Crassostrea gigas)* larval mortality and metamorphosis.[74] Other species of shrimps, crabs, and pelecypods are listed in the "Green Book".[54] The sea urchin, *Lytechinus pictus,* has been used in southern California to test chronic effects of contaminated sediments on growth, gonad production, and bioaccumulation over 60-day periods in continuous-flow exposure systems.[11]

Pore Water

Pore water concentrations of contaminants are often correlated with toxicity.[31,37,38] A Microtox® procedure using pore water in place of the saline extract has been described by True and Heyward.[75] This procedure proved to be more sensitive than the saline-extract test, as is discussed below.

A short-term life-cycle test for pore water toxicity has been developed using the polychaete *Dinophilus gyrociliatus*.[76] Four or more 1- to 2-day-old postemergence female cultured worms are introduced into 20 mL of pore water test solution adjusted to a salinity of 25‰. Females are fertilized by dwarf males prior to emergence from egg capsules. Animals in the test chambers are examined after 1-, 4-, and 7-day exposure, to determine mortality and reproductive effects (egg production and numbers of eggs and juveniles produced per female).

Sediment Elutriates

Toxicity tests for other sediment liquid phases have utilized methods developed for water or effluent toxicity testing, involving bacteria, phytoplankton, zooplankton, oligochaetes, polychaetes, crustaceans, bivalves, echinoderms, tunicates, and fish. Endpoints range from acute lethality to sublethal effects on osmoregulation, respiration, fertilization, larval development, and life history parameters.[22,23,25-27,50,51]

The "Green Book"[54] recommends testing the effects of dredged material on aquatic organisms, using sediment elutriates. Dredged-material discharge regulations require testing multiple species representing several phyla, including zooplankton, crustaceans, and fish. In this test, a subsample of homogenized dredged material is mixed into seawater at a 1:4 ratio, and the slurry vigorously mixed for 30 min. The mixture is allowed to settle for 1 hr, and the supernatant withdrawn and transferred to test chambers. Test organisms are exposed to several dilutions of the supernatant in glass exposure chambers for 48 to 96 hr at controlled salinity, temperature, dissolved oxygen, and pH levels appropriate to the test species. At the end of the exposure period, the numbers of surviving test animals in each test chamber are determined. Test species recommended by the EPA/COE[54] are listed in Table 1.

Bivalve Larvae

The bivalve larvae test procedure[53,77,78] is a well-established and reliable indicator of water quality. Two species recommended for sediment toxicity testing on the U.S. west coast[53] are the Pacific oyster *Crassostrea gigas* and the bay mussel *Mytilus edulis*. Endpoints are mortality and abnormal development in survivors, after a 48- to 60-hr liquid-phase exposure. In this test, 20 g of sediment are shaken in 1 L of sea water (salinity is 28‰) in a 1-L test chamber and allowed to settle for 4 hr. Bivalve larvae are obtained from laboratory-cultured adult brood stock, which are induced to spawn. Within 2 hr of fertilization, 20,000 to 40,000 developing embryos are introduced into each test chamber and held at 20°C for 48 to 60 hr. At the termination of the test, subsamples of at least 20 larvae per sample are examined under 100× magnification, and the percentages of mortality and abnormal survivors determined.

Echinoderm Embryos

The echinoderm embryo test[46,53,79] is similar to the bivalve larvae test in that newly fertilized embryos are exposed to 20 g of sediment per liter of 28‰ sea water. For both methods, sediments with interstitial water salinity levels below 28‰ can be used, since salinity may be adjusted to 28‰. Echinoderm species recommended include purple sea urchins *(Strongylocentrotus purpuratus),* green sea urchins *(S. droebachiensis),* sand dollars *(Dendraster excentricus),* and the Atlantic urchin, *Arbacia punctata.* Other species of echinoderms may be appropriate for other areas. Since these species naturally spawn at different times of the year, the echinoderm embryo test can be run at any time during the year, but not always with the same species. The echinoderm embryo test is run in 1-L beakers, using about 25,000 embryos per test chamber and a temperature of 12°C. The test is terminated when embryos reach the pluteus stage (e.g., 72 to 96 hr), and a minimum of 20 embryos per sample are evaluated for mortality and percentage of abnormal development.

Sediment Extracts

Toxicity tests utilizing aqueous (including saline) or organic extracts of the sediment are typically small in scale, using microorganisms or cellular responses to test small quantities of extracts. Procedures have been developed for two assays using an organic solvent extract (anaphase aberration and Microtox® tests) and a saline water extract (Microtox® test).

Anaphase Aberration

This procedure is used to test the genotoxicity of nonionic organic compounds, such as aromatic and chlorinated hydrocarbons, in sediments. This toxicity test has been applied in the Puget Sound area, using cultured rainbow trout gonad cells.[53,60,80-82] Other fish cell lines, as well as amphibian, plant, and human cells, also may be used. Extraction follows the procedures of Chapman et al.[60] and MacLeod et al.[83] using dimethylsulfoxide (DMSO), which results in a dried quantity of extracted organic material. Some contaminants, such as metals and highly acidic or basic organic compounds, would not be efficiently extracted by this procedure. Trout cells are grown on microscope slides and exposed for 48 hr to the extract added to culture medium in a dilution series typically ranging from 50 to 1 μg/mL. The cells are fixed and examined to determine the inhibition of mitosis and the percentage of anaphase aberration effects.[60,80-82]

Microtox®

The Microtox® test for marine sediments[84,85] is based on the inhibition of light emission by the luminescent bacterium *Photobacterium phosphoreum* in the presence of toxicants. The toxicant may be extracted by either an organic solvent procedure,[83] which removes nonionic aromatic and chlorinated hydrocarbons, or a saline water extraction,[86] which removes water-soluble contaminants in pore water or adsorbed to sediment particles.[53] In the saline-extract procedure the sediment sample is centrifuged, and the supernatant is withdrawn for testing. Freeze-dried luminescent bacteria are reconstituted in control and test solutions and incubated, and luminescence is measured on serial dilutions after 5 to 15 min exposure. The percent inhibition of light emission is converted to an EC_{50} value.

Selection of Species for Sediment Toxicity Tests

Criteria for choosing the best species for sediment toxicity tests have been published previously[44,56,87] and are summarized as follows:

1. *Availability*. Organisms should be readily available in sufficient numbers, supplied from field-collected or cultured populations. If the test animals are from cultures, the possible modification of toxicant sensitivity (relative to field-collected organisms) should be documented.[88]
2. *Sensitivity*. There should be temporal consistency in the sensitivity to contaminants. The sensitivity of different life-stage or age/size classes used also should be documented. Reference toxicants can be used to determine whether ontogenetic, age/size, or seasonal variations in sensitivity exist, in addition to other potential sensitivity-modifying factors such as temperature, salinity, light intensity, photoperiod, duration of holding organisms prior to exposure,[88] sources of test organisms, or culturing conditions.
3. *Tolerance to handling*. Cultures should be tolerant of handling (e.g., sieving, transferring) and large enough to recover and accurately count following exposure.
4. *Tolerance to abiotic factors*. Species should be tolerant to the range of natural abiotic factors found in the study habitats, including grain size, organic matter, salinity, and temperature. If there is doubt, appropriate controls should be incorporated into the test design, or models should be used to measure the potential background interference that may be caused by these factors.
5. *Route of exposure*. Sediment toxicity test organisms should be directly exposed to sediment contaminants either dissolved in pore water or a sediment extract or adsorbed to sediment particles.
6. *Ecological or economic relevance*. The ecological or economic value of test species should be known and appropriate to the environments studied or monitored. Ecological considerations include the representation of biota from reference habitats or that have a similar functional role in the community, such as a predator or bioturbator. Economic considerations include direct commercial or recreational value (i.e., fishery species), or indirect value, such as dominant prey item of fisheries species.

7. *Geographic distribution.* Species with broad geographic distributions are preferred because they have direct ecological relevance to many sites.
8. *Standardized methodology.* A species that has been used extensively in sediment toxicity tests offers clear advantages to species for which methodologies have not been developed for this purpose. Some methodologies are sufficiently general that they may be adapted to accommodate new species fairly easily.[44,52] If a standard procedure does not exist for the selected test species, its sensitivity should be compared to that of a well-established test species, using reference toxicants or by comparing responses of both species to the same test sediment.
9. *Field validation.* The observed sensitivity of a test species in the laboratory should relate to the distribution of sediment toxicity, benthic community structure, and chemical contamination in the field.[8,14,15]
10. *Regulatory requirements.* Sediment toxicity tests conducted to comply with certain requirements, such as dredged-material discharge permit regulations, may be restricted to specific test procedures and species recommended by the regulations.
11. *Life history.* Chronic tests should be able to test various stages of the life history of the test organism and to detect sublethal endpoints related to the life cycle of the species.

QUALITY ASSURANCE AND QUALITY CONTROL CONSIDERATIONS

A laboratory-based sediment toxicity test procedure is not designed or intended to exactly simulate the exposure of toxicity test organisms to chemical contaminants under "natural" conditions in the field, but rather to provide a rapid and consistent test procedure that yields a reasonably sensitive indication of the degree of toxicity of materials in sediments.[52] To this end, quality assurance/quality control (QA/QC) guidelines and standardized procedures should be consistently followed so that test results can be compared between laboratories and within a laboratory at different times and with different personnel. The test procedure and test organisms function comparably to an analytical instrument that quantifies toxicity, and QA/QC procedures are analogous to calibrating the instrument.

Reference Toxicants

Reference toxicants are used to "calibrate" the test organism population, i.e., to ascertain whether the test population is consistently sensitive to the reference toxicant, compared to previously tested populations. The assumption is that comparable sensitivity to the reference toxicant indicates comparable sensitivity to all toxicants. For marine sediment toxicity tests, the recommended procedure for using reference toxicants is to expose the test organisms to a concentration series of the reference toxicant in seawater for 96 hr to obtain an LC_{50} value (or an EC_{50} value for some assays) for comparison to previously derived values.[53,54] Exposing the test organisms in water without sediment, under standard conditions specific for the toxicity test procedure, eliminates the problems of variations in sediment parameters, which can affect chemical partitioning and the test organism's sensitivity. The stress on the organism in being out of sediment is assumed to be a constant. The LC_{50} or EC_{50} should fall within two standard deviations of previously derived values.[53,54] Appropriate reference toxicants may include cadmium chloride, silver chloride, phenol, and sodium pentachlorophenol and will vary with the test species.[53] The reference toxicant treatment serves as a positive control for the sediment toxicity test.

Negative and Solvent Controls

A negative control of uncontaminated substrate must also accompany each run of a toxicity test procedure. This serves to indicate whether sediment handling, laboratory procedures, and the viability of the test organisms were adequate to yield acceptable survival (or no effect) in uncontaminated substrates, and to test the hypothesis that observed biological responses were due to the toxicity of the sample. In whole-sediment tests, the negative control is an uncontaminated sediment, preferably one which supports a healthy population of the test species (i.e., the native habitat sediment or culture substrate for field-collected and laboratory-reared organisms, respectively). In a liquid-phase test the negative control is uncontaminated water. In both cases the negative control is handled in the same way as test samples, and observed mortality or other biological effects should occur in 10% or less of the control test population, unless the test guidelines allow for a higher control effect level. A solvent control includes the same level of a solvent used to dissolve a test chemical incorporated into spiked sediment, but does not contain any of the test chemical. It is handled in the same way as the spiked sediment treatments, and results are combined with the negative control if there is no difference. If there is a difference in results between the negative and solvent controls, the solvent control results should be used to calculate the statistical differences between treatments.[52]

Reference Sediment

An uncontaminated reference sediment control may be included in whole-sediment toxicity tests to test for background interference effects of geochemical and geophysical properties of the sediment. The reference sediment should come from an area near the test sediment and should resemble the test sediment in characteristics such as grain size, water content, Eh, pH, interstitial water salinity, and total organic carbon, but should not be contaminated. This sediment provides a site-specific basis for comparing the results of test sediment treatments with nontoxic sediment.[52] Alternatively, statistical models might be used to mathematically remove the interfering effects of specific parameters, such as sediment grain size.[43]

Water Quality

Water quality parameters for overlying or dilution water should be constant during all sediment toxicity tests; this includes dissolved oxygen, salinity, temperature, pH, hydrogen sulfide, and ammonia. Deterioration in water quality within toxicity-test treatments may cause an increase in apparent toxicity and/or high variance among replicates.

Test Organisms

If laboratory-cultured populations of test organisms are used, their sensitivity to reference toxicants should be compared with populations of the same species either freshly collected from the field or from previous generations in culture. Prolonged holding of field-collected organisms can modify their toxicological sensitivity,[88] but field-collected animals also need time to acclimate to laboratory conditions (especially temperature and salinity). The holding time should thus be standardized and minimized to balance between these competing concerns.

Experimental Design and Test Monitoring

Statistical comparisons and experimental hypotheses should be explicitly considered in the design of research or in monitoring toxicity tests. Treatments should be replicated sufficiently to statistically distinguish "hits" from negative controls and reference sediments, given the amount of variation seen in previous toxicity tests with the test species. Statistical power analysis is useful to determine the sample size.[89,90] Treatments should be randomly assigned to test chambers, and chambers or samples should be coded so that personnel monitoring the test or gathering response data have no knowledge of the sample identities. Proper laboratory procedures and maintenance of consistent test conditions should be documented, and any deviations should be taken into consideration in the evaluation of the results.

APPLICATIONS OF SEDIMENT TOXICITY TESTS

Sediment toxicity tests are applicable to a wide variety of regulatory and environmental management considerations, including the disposal of dredged material, monitoring cleanup of severely contaminated sites, monitoring of downstream effects of point- and nonpoint-source effluent discharges, and mapping of the temporal and spatial distribution of sediment contamination. Additionally, sediment quality criteria may incorporate sediment toxicity endpoints in a manner similar to water quality criteria that are based on water column toxicity tests.[38,91]

Vertical profiles of the toxicity in sediments have been used to document historical changes in sediment contamination.[92]

Sediment toxicity tests have been used in research to compare the relative toxicological sensitivities of indigenous benthic organisms and to explore the basic mechanisms underlying biological responses to contaminants in sediments.[93,94] Contaminant-spiked sediments have been used in the laboratory to explore the integrated toxicological effects of multiple chemical contaminants (i.e., synergism, interactions, etc.)[17] and to determine the effects of sediment geochemical properties on contaminant bioavailability and chemical partitioning.[19,35-38,95] Sediment toxicity tests, combined with sediment fractionation procedures, have been used to determine effective toxic agents in sediments contaminated with a complex mixture of chemicals.[96,97]

Sediment toxicity test data eventually should be combined in a common data base that includes information on chemical concentrations, geographic locality of samples, species and response endpoints, exposure protocols, and QA/QC. This data base could be used to support the development of sediment quality criteria, to monitor geographic and temporal trends in sediment toxicity, and for comparisons of the toxicological sensitivities of test species, endpoints, and toxicity test procedures.

PROBLEMS FACING SEDIMENT TOXICITY TESTS: CHALLENGES FOR RESEARCH

Sediment toxicity tests have their problems, limitations, and underlying assumptions, most of which are well known by their advocates and critics. These problems and limitations fall into four classes: (1) disruption of sediment geochemistry and the kinetic activity of bedded contaminants through sampling and handling (a generic problem in sediment contaminant work), (2) sensitivity to natural sedimentary features and laboratory conditions, (3) toxicological uncertainties, and (4) poorly understood ecological interactions and relevance. It is anticipated that much of the near-term research in sediment toxicity will address the following topics.

The Assumption of Equilibrium: Disruption of Sediment Geochemistry Through Sampling, Handling, and Bioturbation

Sediment contaminants are expected to reach an equilibrium in their partitioning between "particulate/colloid-bound" and "freely dissolved" states if the sediment is undisturbed. Since the toxicity of several organic and metallic chemicals is proportional to their interstitial dissolved concentrations, the proximity of the contaminant-substrate system to kinetic equilibrium can have a dramatic influence on the toxicity of the sediment, due to the variation in the interstitial concentration of the contaminant(s). Disturbance of sediment through handling (i.e., sampling, mixing, distribution to exposure chambers, etc.) can disrupt this equilibrium, leading to increased interstitial concentrations of the contaminant. The subsequent storage of field-collected or laboratory-spiked sediment may permit equilibrium to return, but the rate at which this occurs will vary with the kinetic properties of each chemical and the geochemical and structural properties of the sediment. Bioturbation of test sediments caused by the burrowing and feeding by test organisms in the exposure chamber may upset the substrate-contaminant equilibrium by facilitating the flux of overlying water into the interstitial *milieu*, resuspending particulate matter, or creating vertical gradients of contaminants through feeding activities (i.e., biogenic grading). Finally, contaminants may degrade between sampling (or spiking) and the completion of toxicity testing. Degradation rates may be affected by changes in the sediment geochemistry or microbial microenvironment, which are sensitive to the handling of the sediment. All of these factors have the same net effect, which is to complicate the estimation of exposure concentrations.

Few studies have explicitly examined the effects of sampling, handling, or storage on the toxicity (or geochemistry) of contaminated sediments (see Chapter 3). Often these issues are considered of secondary or tertiary importance to the development of toxicity tests or to studies of contaminant geochemistry. However, despite all of the potential complications that sampling, handling, and storage may introduce, toxicity tests have produced repeatable and predictable results under conservative experimental conditions.

Background Noise: Sensitivity of Test Organisms to Natural Sedimentary Features and Laboratory Conditions

All organisms live within physical and chemical limits that define the habitat portion of their niche. Existence at (or beyond) the margins of these boundaries is, by definition, limited or impossible, and typically these boundaries are not sharply defined. Thus, a mud-dwelling worm may experience higher stress when placed in coarse sand than when placed in a finer-grained silty sand. The tolerance limits of a sediment toxicity test organism to microenvironmental variables of sedimentary habitats will determine the types of sediments that may be tested by that species. Unfortunately, these limits are poorly known for most species currently used for sediment toxicity tests. Without adequate controls, this will likely result in stress responses caused by natural features being erroneously attributed to chemical contaminants.

The sensitivities of sediment toxicity test species to sediment-associated characteristics, such as particle size, organic content, penetrability, water content, salinity, hydrogen sulfide, oxygen and ammonia content, temperature, or photoperiod, may bias their responses to chemical contaminants. Sensitivities to some of these natural variables are relatively easy to determine by experimental manipulation (e.g., particle size, water content, salinity, temperature, and photoperiod), while sensitivities to the other variables (e.g., hydrogen sulfide, oxygen, ammonia, and organic content) may have to be determined by observation and correlation. Factoring these sensitivities into the interpretation of routine toxicity tests may be accomplished by controlling exposure conditions (e.g., temperature and photoperiod), additional controls (e.g., particle size and/or organic content, salinity, and water content), or through use of statistical models.[43,44]

Toxicological Uncertainties

Two classes of toxicological problems facing sediment toxicity tests are the narrow range of chemical compounds as yet tested, and difficulties in measuring exposure concentrations. These issues are of concern when trying to identify the cause of toxicity in contaminated sediments, but not to investigations of the overall toxicity of the substrate.

The first problem will improve as more compounds are tested, though quantitative structure/activity relationship (QSAR) approaches will likely be necessary to estimate the toxicity of most of the thousands of chemicals present in coastal sediments. Determination of actual exposure concentrations is critical in order to relate cause to effect in any toxicity test, and this problem is particularly difficult in sediment toxicity tests.

Quantification of the routes of exposure affects the identification of the bioavailable fraction(s) of the sediment and, thus, the selection of the relevant phase of the sediment in which to measure contaminant concentrations (i.e., the whole substrate, the silt-clay fraction, the total interstitial water, the freely dissolved interstitial water phase). Comparisons of routes of exposure rarely have been conducted in benthic organisms, due to the difficulty in controlling exposure. Boese et al.[99] identified ten possible routes of exposure for the deposit-feeding bivalve *Macoma nasuta,* but narrowed this to three routes that accounted for 80 to 100% of the uptake of hexachlorobenzene.

What Does It All Mean? Ecological Relevance of Sediment Toxicity Tests

If the purpose of testing the toxicity of sediments is to protect marine benthic populations, communities, and ecosystems from degradation from anthropogenic chemical pollution, then sediment toxicity tests need to have a demonstrable connection to these levels of ecological organization. The ecological relevance of sediment toxicity tests can be approached from three vantages: (1) the choice of test species; (2) the choice of response endpoint; and (3) comparison to field-measured effects at the population, community, or ecosystem level (i.e., field validation). These problems are interrelated and should be foremost in the consideration of the development and implementation of all sediment toxicity tests.

Test Species

Moderately few species are presently used for marine or estuarine sediment toxicity tests, and the toxicological sensitivities of many phyla (let alone species) are unknown. Several U.S.-related geographic regions are not represented by the species used routinely in sediment toxicity tests (Table 1), particularly the mid- and southern Atlantic Ocean, the Gulf of Mexico, the Caribbean Sea, and Alaskan and Hawaiian waters. Questions repeatedly arise concerning the relevance of using species that are not represented in the benthic habitat(s) of concern. Demonstration that these tests are predictive of ecological changes in a given habitat needs to be made. Organisms that are being considered for new sediment toxicity tests should be selected from species assemblages known to be sensitive to sediment contamination, as would be found at the clean end of a pollution gradient.

Response Endpoints

The endpoints of current sediment toxicity tests are often difficult to interpret ecologically; this is especially true for sublethal endpoints. Most current tests allow only acute exposure over a small portion of the life span of the organism, though considerable effort is underway to develop chronic toxicity tests that at least incorporate exposures over critical periods of the life cycle. High acute mortality in a bioassay is a sign of likely extinction of the test species in a contaminated sediment. But what is the ecosystem-level impact of developmental abnormalities, changes in behavior, and impaired somatic growth? Each is an element in the population biology of a species, but none alone may be sufficient to predict the risk of extinction of a population that would be exposed to a tested sediment. Toxicological endpoints must be explicitly tied to predictions of population growth and risk of extinction, if the tests are to be used to protect populations of benthic organisms. One powerful approach to achieve this linkage is to determine the sensitivity of population growth rate to changes in potential test endpoints, using established theoretical ecological demographic models, such as the Leslie age-based or Lefkovitch stage-based matrix population models.[100]

Ultimately, sediment toxicity test endpoints must be shown to be predictive of community- and ecosystem-level responses. Relating currently measured endpoints to higher levels of ecological organization can be presently accomplished only by correlation with observed changes in spatial or temporal distributions of benthic species. Theoretical ecological models have yet to be developed to allow *a priori* predictions of community or ecosystem change from laboratory-based measurements. Toxicity tests that effectively measure community or ecosystem responses are in their infancy, though early work on the structure of the benthic community colonizing trays of contaminated sediments[101,102] is a promising start in this direction. Clearly, these are areas needing much theoretical and empirical research.

Field Validation

Since theoretical ecological models cannot, as yet, be used to predict ecological responses to sediment contamination above the population level, verification of the ecological protectiveness of sediment toxicity tests depends on demonstrating correlations between toxicological responses and changes in the distribution of species. Difficulties arise in the interpretation of cause and effect in both the toxicity tests and changes in community structure, both of which respond to natural environmental features as well as pollution. Factoring out natural environmental variation in community structure data is a difficult process that relies upon the appropriate choice of unpolluted reference sites against which contaminated sites are compared.[103] Guidelines for the optimal design of field validation studies are nascent,[104,105] but are clearly needed.

Despite all these realized and potential problems, sediment toxicity tests have been enormously successful as both research and management tools. Enunciation of their shortcomings should not be interpreted as a criticism of the utility of the technique, but rather a recognition of their limitations and a guide for their refinement.[106,107]

REFERENCES

1. Lee, H., II, and R. C. Swartz. "Biological Processes Affecting the Distribution of Pollutants in Marine Sediments. Part II. Biodeposition and Bioturbation," in *Contaminants and Sediments*, Vol. 2, R. A. Baker, Ed. (Ann Arbor, MI: Ann Arbor Scientific 1980), pp. 555–606.
2. Swartz, R. C., and H. Lee, II. "Biological Processes Affecting the Distribution of Pollutants in Marine Sediments. Part I. Accumulation, Trophic Transfer, Biodegradation and Migration," in *Contaminants and Sediments*, Vol. 2, R. A. Baker, Ed. (Ann Arbor, MI: Ann Arbor Scientific 1980), pp. 533–553.
3. Chapman, P. M., R. N. Dexter, R. D. Kathman, and G. A. Erikson. "Survey of Biological Effects of Toxicants Upon Puget Sound Biota. IV. Interrelationships of Infauna, Sediment Bioassay and Sediment Chemistry Data," NOAA Technical Memorandum NOS OMA 9 (Rockville, MD: National Oceanic and Atmospheric Administration, 1984).
4. Hargis, W. J., Jr., M. H. Roberts, and D. E. Zwerner. "Effects of Contaminated Sediments and Sediment-Exposed Effluent Water on an Estuarine Fish: Acute Toxicity," *Mar. Environ. Res.* 14:337–354 (1984).
5. Long, E. R., and P. M. Chapman. "A Sediment Quality Triad: Measures of Sediment Contamination, Toxicity and Infaunal Community Composition in Puget Sound," *Mar. Pollut. Bull.* 16:405–415 (1985).
6. Myers, M. S., J. T. Landahl, M. M. Krahn, L. L. Johnson, and B. B. McCain. "Overviews of Studies on Liver Carcinogenesis in English Sole from Puget Sound; Evidence for a Xenobiotic Chemical Etiology I: Pathology and Epizootiology," *Sci. Tot. Environ.* 94:33–50 (1990).

7. Stein, J. E., W. L. Reichert, M. Nishimoto, and U. Varanasi. "Overview of Studies on Liver Carcinogenesis in English Sole from Puget Sound; Evidence for a Xenobiotic Chemical Etiology II: Biochemical Studies," *Sci. Tot. Environ.* 94:51–69 (1990).
8. Swartz, R. C., W. A. DeBen, K. A. Sercu, and J. O. Lamberson. "Sediment Toxicity and the Distribution of Amphipods in Commencement Bay, Washington, USA," *Mar. Pollut. Bull.* 13:359–364 (1982).
9. Becker, D. S., G. R. Bilyard, and T. C. Ginn. "Comparisons Between Sediment Bioassays and Alterations of Benthic Macroinvertebrate Assemblages at a Marine Superfund Site: Commencement Bay, Washington," *Environ. Toxicol. Chem.* 9:669–685 (1990).
10. Long, E. R., and L. G. Morgan. "The Potential for Biological Effects of Sediment-Sorbed Contaminants Tested in the National Status and Trends Program," NOAA Technical Memorandum NOS OMA 52 (Seattle, WA: National Oceanic and Atmospheric Administration, National Ocean Service, 1990).
11. Thompson, B. E., S. M. Bay, J. W. Anderson, J. D. Laughlin, D. J. Greenstein, and D. T. Tsukada. "Chronic Effects of Contaminated Sediments on the Urchin *Lytechinus pictus*," *Environ. Toxicol. Chem.* 8:629–637 (1989).
12. Chapman, P. M., R. N. Dexter, and E. R. Long. "Synoptic Measures of Sediment Contamination, Toxicity and Infaunal Community Composition (The Sediment Quality Triad) in San Francisco Bay," *Mar. Ecol. Prog. Ser.* 37:75–96 (1987).
13. Long, E. R., M. F. Buchman, S. M. Bay, R. J. Breteler, R. S. Carr, P. M. Chapman, J. E. Hose, A. L. Lissner, J. Scott, and D. A. Wolfe. "Comparative Evaluation of Five Toxicity Tests with Sediments from San Francisco Bay and Tomales Bay, California," *Environ. Toxicol. Chem.* 9:1193–1214 (1990).
14. Swartz, R. C., D. W. Schults, G. R. Ditsworth, W. A. DeBen, and F. A. Cole. "Sediment Toxicity, Contamination and Macrobenthic Communities Near a Large Sewage Outfall," in *Validation and Predictability of Laboratory Methods for Assessing the Fate of Contaminants in Aquatic Ecosystems,* STP 865, T. P. Boyle, Ed. (Philadelphia: American Society for Testing and Materials, 1985), pp. 152–175.
15. Swartz, R. C., F. A. Cole, D. W. Schults, and W. A. DeBen. "Ecological Changes in the Southern California Bight Near a Large Sewage Outfall: Benthic Conditions in 1980 and 1983," *Mar. Ecol. Prog. Ser.* 31:1–13 (1986).
16. Gardner, G. R., P. P. Yevich, J. C. Harshbarger, and A. R. Malcolm. "Carcinogenicity of Black Rock Harbor Sediment to the Eastern Oyster and the Trophic Transfer of Black Rock Harbor Carcinogens from the Blue Mussel to the Winter Flounder," *Environ. Health Perspect.* 90:53–66 (1991).
17. Swartz, R. C., P. F. Kemp, D. W. Schults, and J. O. Lamberson. "Effects of Mixtures of Sediment Contaminants on the Marine Infaunal Amphipod, *Rhepoxynius abronius*," *Environ. Toxicol. Chem.* 7:1013–1020 (1988).
18. Swartz, R. C., P. F. Kemp, D. W. Schults, G. R. Ditsworth, and R. O. Ozretich. "Acute Toxicity of Sediment from Eagle Harbor, Washington, to the Infaunal Amphipod *Rhepoxynius abronius*," *Environ. Toxicol. Chem.* 8:215–222 (1989).
19. Swartz, R. C., D. W. Schults, T. H. DeWitt, G. R. Ditsworth, and J. O. Lamberson. "Toxicity of Fluoranthene in Sediment to Marine Amphipods: A Test of the Equilibrium Partitioning Approach to Sediment Quality Criteria," *Environ. Toxicol. Chem.* 9:1071–1080 (1990).

20. Pastorok, R. A., and D. S. Becker. "Comparative Sensitivity of Sediment Toxicity Bioassays at Three Superfund Sites in Puget Sound," in *Aquatic Toxicology and Risk Assessment: Thirteenth Volume*, STP 1096, W. G. Landis, and W. H. van der Schalie, Eds. (Philadelphia: American Society for Testing and Materials, 1990), pp. 123–139.
21. Anderson, J., W. Birge, J. Gentile, J. Lake, J. Rogers, Jr., and R. Swartz. "Biological Effects, Bioaccumulation, and Ecotoxicology of Sediment-Associated Chemicals," in *Fate and Effects of Sediment-Bound Chemicals in Aquatic Systems,* K. L. Dickson, A. W. Maki, and W. A. Brungs, Eds. (New York: Pergamon Press, 1987), pp. 267–295.
22. Chapman, P. M. "Sediment Bioassay Tests Provide Toxicity Data Necessary for Assessment and Regulation," in *Proceedings of the Eleventh Annual Aquatic Toxicology Workshop (11/13–15, 1984),* G. H. Green, and K. L. Woodward, Eds. Canadian Technical Report of Fisheries and Aquatic Sciences 1480, Vancouver, Canada, pp. 178–197 (1986).
23. Chapman, P. M. "Marine Sediment Toxicity Tests," in *Chemical and Biological Characterization of Sludges, Sediments, Dredge Spoils, and Drilling Muds,* STP 976, J. J. Lichtenberg, F. A. Winter, C. I. Weber, and L. Fradkin, Eds. (Philadelphia: American Society for Testing and Materials, 1988), pp. 391–402.
24. Chapman, P. M., R. N. Dexter, R. M. Kocan, and E. R. Long. "An Overview of Biological Effects Testing in Puget Sound, Washington: Methods, Results and Implications," in *ASTM Spec. Tech. Publ.,* STP 854, (Philadelphia: American Society for Testing and Materials, 1985), pp. 344–363.
25. Lamberson, J. O., and R. C. Swartz. "Use of Bioassays in Determining the Toxicity of Sediment to Benthic Organisms," in *Toxic Contaminants and Ecosystem Health: A Great Lakes Focus,* M. S. Evans, Ed. (New York: John Wiley & Sons, 1988), pp. 257–279.
26. Swartz, R.C. "Toxicological Methods for Determining the Effects of Contaminated Sediment on Marine Organisms," in *Fate and Effects of Sediment-Bound Chemicals in Aquatic Systems,* Dickson, K. L., A. W. Maki, and W. A. Brungs, Eds. (New York: Pergamon Press, 1987), pp. 183–198.
27. Swartz, R. C. "Marine Sediment Toxicity Tests," in *Proc. National Research Council Symposium on Contaminated Marine Sediments,* K. Kamlet, Ed. (Washington, D.C.: National Academy Press, 1989), pp. 115–129.
28. Melzian, B. D. "Toxicity Assessment of Dredged Materials: Acute and Chronic Toxicity as Determined by Bioassays and Bioaccumulation Tests," in *Proc. International Seminar on the Environmental Aspects of Dredging Activities* (Nantes, France: Goubault Imprimeur, S.A., 1989), pp. 49–64.
29. Karickhoff, S. W. "Semi-Empirical Estimation of Sorption of Hydrophobic Pollutants on Natural Sediments and Soils," *Chemosphere* 10:833–846 (1981).
30. Brownawell, B. J., and J. W. Farrington. "Biogeochemistry of PCBs in Interstitial Waters of a Coastal Marine Sediment," *Geochim. Cosmochim. Acta* 50:157–169 (1986).
31. Adams, W. J., R. A. Kimerle, and R. G. Mosher. "Aquatic Safety Assessment of Chemicals Sorbed to Sediments," in *Aquatic Toxicology and Hazard Assessment: Seventh Symposium,* STP 854, R. D. Cardwell, R. Purdy, and R. C. Bahner, Eds. (Philadelphia: American Society for Testing and Materials, 1985), pp. 429–453.

32. Landrum, P. F. "Bioavailability and Toxicokinetics of Polycyclic Aromatic Hydrocarbons Sorbed to Sediments for the Amphipod *Pontoporeia hoyi*," *Environ. Sci. Technol.* 23:588–595 (1985).
33. Nebeker, A. V., G. S. Schuytema, W. L. Griffis, J. A. Barbitta, and L. A. Carey. "Effect of Sediment Organic Carbon on Survival of *Hyalella azteca* Exposed to DDT and Endrin," *Environ. Toxicol. Chem.* 8:705–718 (1989).
34. Ziegenfuss, P. S., W. J. Renaudette, and W. J. Adams. "Methodology for Assessing the Acute Toxicity of Chemicals Solved to Sediments: Testing for the Equilibrium Partitioning Theory," in *Aquatic Toxicology and Environmental Fate: Ninth Volume,* STP 921, T. J. Poston, and R. Purdy, Eds. (Philadelphia: American Society for Testing and Materials, 1986), pp. 479–493.
35. DeWitt, T. H., R. J. Ozretich, R. C. Swartz, J. O. Lamberson, D. W. Schults, G. R. Ditsworth, J. K. P. Jones, L. Hoselton, and L. M. Smith. "The Influence of Organic Matter Quality on the Toxicity and Partitioning of Sediment-Associated Fluoranthene," *Environ. Toxicol. Chem.* 11:(in press) (1992).
36. Kemp, P. F., and R. C. Swartz. "Acute Toxicity of Interstitial and Particle-Bound Cadmium to a Marine Infaunal Amphipod," *Mar. Environ. Res.* 26:135–153 (1988).
37. Swartz, R. C., G. R. Ditsworth, D. W. Schults, and J. O. Lamberson. "Sediment Toxicity to a Marine Infaunal Amphipod: Cadmium and Its Interaction with Sewage Sludge," *Mar. Environ. Res.* 18:133–153 (1986).
38. Di Toro, D. M., J. D. Mahony, D. J. Hansen, K. J. Scott, M. B. Hicks, S. M. Mayr, and M. S. Redmond. "Toxicity of Cadmium in Sediments: The Role of Acid Volatile Sulfide," *Environ. Toxicol. Chem.* 9:1487–1502 (1990).
39. Di Toro, D. M., J. D. Mahony, D. J. Hansen, K. J. Scott, A. R. Carlson, and G. T. Ankley. "Acid Volatile Sulfide Predicts the Acute Toxicity of Cadmium and Nickel in Sediments," *Environ. Sci. Technol.* (in press) (1991).
40. San Francisco Estuary Project. "Status and Trends Report on Dredging and Waterway Modification in the San Francisco Estuary," San Francisco Estuary Project, Oakland, CA (1990).
41. San Francisco Estuary Project. "Status and Trends Report on Pollutants in the San Francisco Bay Estuary," Draft Report, San Francisco Estuary Project, Oakland, CA (1991)
42. Ott, F. S. "Amphipod Sediment Bioassays: Experiments with Naturally-Contaminated and Cadmium-Spiked Sediments to Investigate Effects on Response of Methodology, Grain Size, Grain Size-Toxicant Interactions, and Variations in Animal Sensitivity Over Time," Ph.D. Thesis, University of Washington, Seattle (1986).
43. DeWitt, T. H., G. R. Ditsworth, and R. C. Swartz. "Effects of Natural Sediment Features on the Phoxocephalid Amphipod, *Rhepoxynius abronius:* Implications for Sediment Toxicity Bioassays," *Mar. Environ. Res.* 25:99–124 (1988).
44. DeWitt, T. H., R. C. Swartz, and J. O. Lamberson. "Measuring the Acute Toxicity of Estuarine Sediments," *Environ. Toxicol. Chem.* 8:1035–1048 (1989).
45. Swartz, R. C., W. A. DeBen, F. A. Cole, P. J. Schuba, and H. E. Tatum. "Guidance for Performing Solid Phase Bioassays," in *Ecological Evaluation of Proposed Discharge of Dredged Material into Ocean Waters* (Vicksburg, MS: U. S. Army Engineer Waterways Experiment Station, 1977), Appendix F.

46. Dinnel, P. A., J. M. Link, Q. J. Stober, M. W. Letourneau, and W. E. Roberts. "Comparative Sensitivity of Sea Urchin Sperm Bioassays to Metals and Pesticides," *Arch. Environ. Contam. Toxicol.* 18:748–755 (1989).
47. Gannon, J. E., and A. M. Beeton. "Procedures for Determining the Effects of Dredged Sediments on Biota — Benthos Viability and Sediment Selectivity Tests," *J. Water Pollut. Contr. Fed.* 43:392–398 (1971).
48. Swartz, R. C., W. A. DeBen, and F. A. Cole. "A Bioassay for the Toxicity of Sediment to the Marine Macrobenthos," *J. Water Pollut. Contr. Fed.* 51:944–950 (1979).
49. U.S. Environmental Protection Agency/U.S. Army Corps of Engineers. "Ecological Evaluation of Proposed Discharge of Dredged Material Into Ocean Waters," (Vicksburg, MS: U.S. Army Engineer Waterways Experiment Station, 1977).
50. Chapman, P. M., and E. R. Long. "The Use of Bioassays as Part of a Comprehensive Approach to Marine Pollution Assessment," *Mar. Pollut. Bull.* 14:81–84 (1983).
51. Benedict, A. B., and E. R. Long. "A Catalog of Biological Effects Measurements Along the Pacific Coast," in *NOAA Technical Memorandum NOS OMA 32* (Rockville, MD: National Oceanic and Atmospheric Administration, 1987).
52. American Society for Testing and Materials. "E 1367-90. Guide for Conducting 10-Day Static Sediment Toxicity Tests with Marine and Estuarine Amphipods," in *Annual Book of ASTM Standards, Water and Environmental Technology,* Vol. 11.04 (Philadelphia: American Society for Testing and Materials, 1990).
53. Puget Sound Estuary Program. "Recommended Guidelines for Conducting Laboratory Bioassays on Puget Sound Sediments," Draft Report, U.S. Environmental Protection Agency, Region X, Office of Puget Sound, Seattle, WA (1991).
54. U. S. Environmental Protection Agency/U.S. Army Corps of Engineers. *Evaluation of Dredged Material Proposed for Ocean Disposal—Testing Manual* (Washington, D.C.: U.S. Environmental Protection Agency, 1991), EPA-503-8-91/001.
55. American Society for Testing and Materials. "E 943-91. Definitions of Terms Relating to Biological Effects and Environmental Fate," in *Annual Book of ASTM Standards, Water and Environmental Technology,* Vol. 11.04. (Philadelphia: American Society for Testing and Materials, 1991).
56. Reish, D. J., and J. A. Lemay. "Bioassay Manual for Dredged Materials," U.S. Army Corps of Engineers, Los Angeles District, Los Angeles, CA, Technical Report DACW-09-83R-005 (1988).
57. Mearns, A. J., and J. Q. Word. "Forecasting Effects of Sewage Solids on Marine Benthic Communities," in *Ecological Stress and the New York Blight: Science and Management,* G. F. Mayer, Ed. (Columbia, SC: Estuarine Research Federation, 1982), pp. 495–512.
58. Swartz, R. C., W. A. DeBen, J. K. P. Jones, J. O. Lamberson, and F. A. Cole. "Phoxocephalid Amphipod Bioassay for Marine Sediment Toxicity," in *Aquatic Toxicology and Hazard Assessment: Seventh Symposium,* STP 854, R. D. Cardwell, R. Purdy, and R. C. Bahner, Eds. (Philadelphia: American Society for Testing and Materials, 1985), pp. 284–307.
59. Schlekat, C. E., B. L. McGee, and E. Reinharz. "Testing Sediment Toxicity in Chesapeake Bay Using the Amphipod *Leptocheirus plumulosus:* An Evaluation," *Environ. Toxicol. Chem.* (in press) (1991).

60. Chapman, P. M., G. A. Vigers, M. A. Farrell, R. N. Dexter, E. A. Quinlan, R. M. Kocan, and M. Landolt. "Survey of Biological Effects of Toxicants upon Puget Sound Biota. I. Broad-Scale Toxicity Survey," in *NOAA Technical Memorandum OMPA-25,* (Boulder, CO: National Oceanic and Atmospheric Administration, 1982).
61. Nipper, M. G., D. J. Greenstein, and S. M. Bay. "Short- and Long-Term Sediment Toxicity Test Methods with the Amphipod *Grandidierella japonica,*" *Environ. Toxicol. Chem.* 8:1191–1200 (1989).
62. Connell, A. D., and D. D. Airey. "Life-Cycle Bioassays Using Two Estuarine Amphipods, *Grandidierella lutosa* and *G. lignorum,* to Determine Detrimental Levels of Marine Pollutants," *S. Afr. J. Sci.* 75:313–314 (1979).
63. Connell, A. D., and D. D. Airey. "The Chronic Effects of Fluoride on the Estuarine Amphipods *Grandidierella lutosa* and *G. lignorum,*" *Water Res.* 16:1313–1317 (1982).
64. Scott, K. J., and M. S. Redmond. "The Effects of a Contaminated Dredged Material on Laboratory Populations of the Tubicolous Amphipod, *Ampelisca abdita,*" in *Aquatic Toxicology and Hazard Assessment: Twelfth Volume,* STP 1027, U. M. Cowgill, and L. R. Williams, Eds. (Philadelphia: American Society for Testing and Materials, 1989)
65. Pesch, C. E. "Influence of Three Sediment Types on Copper Toxicity to the Polychaete *Neanthes arenaceodentata,*" *Mar. Biol.* 52:237–245 (1979).
66. Reish, D. J. "The Use of the Polychaetous Annelid *Neanthes arenaceodentata* as a Laboratory Experimental Animal," *Tephtys* 11:335–341 (1985).
67. Pesch, C. E., and G. L. Hoffman. "Interlaboratory Comparison of a 28-Day Toxicity Test with the Polychaete *Neanthes arenaceodentata,*" in *Aquatic Toxicology and Hazard Assessment: Sixth Symposium,* STP 802, W. E. Bishop, R. D. Cardwell, and B. B. Heidolph, Eds. (Philadelphia: American Society for Testing and Materials, 1983), pp. 482–493.
68. Chapman, P. M., and R. Fink. "Effects of Puget Sound Sediments and Their Elutriates on the Life Cycle of *Capitella capitata,*" *Bull. Environ. Contam. Toxicol.* 33:451–459 (1984).
69. Johns, D. M., and T. C. Ginn. "Development of a *Neanthes* Sediment Bioassay for Use in Puget Sound," in *Technical Report 910/9-90-005,* U.S. Environmental Protection Agency, Region X, Seattle, WA (1990).
70. Pesch, C. E., P. S. Schauer, and M. A. Balboni. "Efffect of Diet on Copper Toxicity to *Neanthes arenaceodentata* (Annelida: Polychaeta)," in *Aquatic Toxicology and Environmental Fate: Ninth Volume,* STP 921, T. M. Poston, and R. Purdy, Eds. (Philadelphia: American Society for Testing and Materials, 1986), pp. 369–383.
71. McGreer, E. R. "Sublethal Effects of Heavy Metal Contaminated Sediments on the Bivalve *Macoma balthica* (L.)," *Mar. Pollut. Bull.* 10:259–262 (1979).
72. Phelps, H. L., W. H. Pearson, and J. T. Hardy. "Clam Burrowing Behavior and Mortality Related to Sediment Copper," *Mar. Pollut. Bull.* 16:309–313 (1986).
73. Phelps, H. L. "Clam Burrowing Bioassay for Estuarine Sediment," *Bull. Environ. Contam. Toxicol.* 43:838–845 (1989).
74. Phelps, H. L., and K. A. Warner. "Estuarine Sediment Bioassay With Oyster Pediveliger Larvae *(Crassostrea gigas),*" *Bull. Environ. Contam. Toxicol.* 44:197–204 (1990).
75. True, C. J., and A. A. Heyward. "Relationships Between Microtox® Test Results, Extraction Methods, and Physical and Chemical Compositions of Marine Sediment Samples," *Tox. Assess.* (in press) (1992).

76. Carr, R. S., J. W. Williams, and C. T. B. Fragata. "Development and Evaluation of a Novel Marine Sediment Pore Water Toxicity Test with the Polychaete *Dinophilus gyrociliatus*," *Environ. Toxicol. Chem.* 8:533–543 (1989).
77. American Public Health Association. *Standard Methods for the Examination of Water and Wastewater*, 16th ed., p. 1268 (1985).
78. American Society for Testing and Materials. "E 724-89. Standard Guide for Conducting Static Acute Toxicity Tests Starting with Embryos of Four Species of Saltwater Bivalve Molluscs," in *Annual Book of ASTM Standards, Water and Environmental Technology*, Vol. 11.04. (Philadelphia: American Society for Testing and Materials, 1989).
79. Dinnel, P. A., and Q. J. Stober. "Methodology and Analysis of Sea Urchin Embryo Bioassays," in *Fisheries Research Institute Circular No. 85-3*, (Seattle, WA: School of Fisheries, University of Washington, 1985).
80. Kocan, R. M., M. L. Landolt, and K. M. Sabo. "Anaphase Aberrations: A Measure of Genotoxicity in Mutagen-Treated Fish Cells," *Environ. Mutagen.* 4:181–189 (1982).
81. Kocan, R. M., and D. B. Powell. "Anaphase Aberrations: An *In Vitro* Test for Assessing the Genotoxicity of Individual Chemicals and Complex Mixtures," in *Short-Term Bioassays in the Analysis of Complex Environmental Mixtures, IV*, M. D. Waters, S. S. Sandhu, J. Lewtas, L. Claxton, G. Strauss, and S. Nesnow, Eds. (New York: Plenum Press, 1985).
82. Kocan, R. M., K. N. Sabo, and M. L. Landolt. "Cytotoxicity/Genotoxicity, the Application of Cell Culture Techniques to the Measurement of Marine Sediment Pollution," *Aquat. Toxicol.* 6:165–177 (1985).
83. MacLeod, W. D., Jr., D. W. Brown, A. J. Friedman, O. Maynes, and R. W. Pearce. "Standard Analytical Procedures of the NOAA National Analytical Facility, 1984–1985: Extractable Toxic Organic Compounds," in *NOAA Technical Memorandum NMFS-F/NWC-64* (Washington, DC: National Oceanic and Atmospheric Administration, 1985).
84. Bulich, A. A., M. W. Greene, and D. L. Isenberg. "Reliability of the Bacterial Luminescence Assay for Determination of the Toxicity of Pure Compounds and Complex Effluent," in *Aquatic Toxicology and Hazard Assessment: Proceedings of the Fourth Annual Symposium*, D. R. Branson, and K. L. Dickson, Eds. (Philadelphia: STP 737, American Society for Testing and Materials, 1981), pp. 338–347.
85. Schiewe, M. H., E. G. Hawk, D. I. Actor, and M. M. Krahn. "Use of Bacterial Bioluminescence Assay to Assess Toxicity of Contaminated Marine Sediments," *Can. J. Fish. Aquat. Sci.* 42:1244–1248 (1985).
86. Williams, L. G., P. M. Chapman, and T. C. Ginn. "A Comparative Evaluation of Sediment Toxicity Using Bacterial Luminescence, Oyster Embryo, and Amphipod Sediment Bioassays," *Mar. Environ. Res.* 19:225–249 (1986).
87. Shuba, P. J., S. R. Petrocelli, and R. E. Bentley. "Considerations in Selecting Bioassay Organisms for Determining the Potential Environmental Impact of Dredged Material," U.S. Army Engineer Waterways Experimental Station, Vicksburg, MS, Technical Report EL-81-8 (1981).
88. Robinson, A. M., J. O. Lamberson, F. A. Cole, and R. C. Swartz. "Effects of Culture Conditions on the Sensitivity of a Phoxocephalid Amphipod, *Rhepoxynius abronius*, to Cadmium in Sediment," *Environ. Toxicol. Chem.* 7:953–959 (1988).
89. Zar, J. H. *Biostatistical Analysis*, 2nd Ed. (Englewood, NJ: Prentice-Hall, 1984).

90. Cohen, J. *Statistical Power Analysis for the Behavioral Sciences* (Orlando, FL: Academic Press, 1977).
91. Chapman, P. M. "Current Approaches to Developing Sediment Quality Criteria," *Environ. Toxicol. Chem.* 8:589–599 (1989).
92. Swartz, R. C., D. W. Schults, J. O. Lamberson, R. J. Ozretich, and J. K. Stull. "Vertical Profiles of Toxicity, Organic Carbon, and Chemical Contaminants in Sediment Cores from the Palos Verdes Shelf and Santa Monica Bay, California," *Mar. Environ. Res.* (31:215–225) (1991).
93. Reichert, W. L., B.-T. L. Eberhart, and U. Varanasi. "Exposure of Two Species of Deposit-Feeding Amphipods to Sediment-Associated [^{3}H]Benzo[a]pyrene: Uptake, Metabolism and Covalent Binding to Tissue Macromolecules," *Aquat. Toxicol.* 6:45–56 (1985).
94. Varanasi, U., W. L. Reichert, J. E. Stein, D. W. Brown, and H. R. Sanborn. "Bioavailability and Biotransformation of Aromatic Hydrocarbons in Benthic Organisms Exposed to Sediment from an Urban Estuary," *Environ. Sci. Technol.* 19:836–841 (1985).
95. Ankley, G. T., G. L. Phipps, P. A. Kosian, D. J. Hansen, J. D. Mahoney, A. M. Cotter, E. N. Leonard, J. R. Dierkes, D. A. Benoit, and V. R. Mattson. "Acid Volatile Sulfide as a Factor Mediating Cadmium and Nickel Bioavailability in Contaminated Sediments," *Environ. Toxicol. Chem.* (submitted) (1991).
96. Burkhard, L. P., and G. T. Ankley. "Identifying Toxicants: NETAC's Toxicity-Based Approach," *Environ. Sci. Technol.* 23:1438–1443 (1989).
97. Ankley, G. T., A. Katko, and J. W. Arthur. "Identification of Ammonia as an Important Sediment-Associated Toxicant on the Lower Fox River and Green Bay, Wisconsin," *Environ. Toxicol. Chem.* 9:313–322 (1990).
98. Samoiloff, M. R., J. Bell, D. A. Birkholz, G. R. B. Webster, E. G. Arnott, R. Pulak, and A. Madrid. "Combined Bioassay-Chemical Fractionation Scheme for the Determination and Ranking of Toxic Chemicals in Sediments," *Environ. Sci. Technol.* 17:329–334 (1983).
99. Boese, B. L., H. Lee, II, D. T. Specht, R. C. Randall, and M. H. Winsor. "Comparison of Aqueous and Solid-Phase Uptake for Hexachlorobenzene in the Tellinid Clam *Macoma nasuta* (Conrad): A Mass Balance Approach," *Environ. Toxicol. Chem.* 9:221–231 (1990).
100. Caswell, H. *Matrix Population Models* (Sunderland, MA: Sinauer Associates, 1989).
101. Tagatz, M. E., and M. Tobia. "Effects of Barite ($BaSO_4$) on Development of Estuarine Communities," *Estuar. Coast. Mar. Sci.* 7:401–407 (1978).
102. Hansen, D. J., and M. E. Tagatz. "A Laboratory Test for Assessing Impacts of Substances on Developing Communities of Benthic Estuarine Organisms," in *Aquatic Toxicology,* STP 707, J. G. Eaton, P. R. Parrish, and A. C. Hendricks, Eds. (Philadelphia: American Society for Testing and Materials, 1980), pp. 40–57.
103. Green, R. H. *Sample Design and Statistical Methods for Environmental Biologists* (New York, NY: John Wiley & Sons, 1979).
104. Ferraro, S.P., F. A. Cole, W. A. DeBen, and R. A. Swartz. "Power-Cost Efficiency of Eight Macrobenthic Schemes in Puget Sound, Washington, USA," *Can. J. Fish. Aquat. Sci.* 46:2157–2165 (1989).
105. Ferraro, S. P., and F. A. Cole. "Taxonomic Level and Sample Size Sufficient for Assessing Pollution Impacts on the Southern California Bight Macrobenthos," *Mar. Ecol. Prog. Ser.* 67:251–262 (1990).

106. Spies, R. B. "Sediment Bioassays, Chemical Contaminants and Benthic Ecology: New Insights or Just Muddy Water?" *Mar. Environ. Res.* 27:73–75 (1989).

107. Chapman, P. M., E. R. Long, R. C. Swartz, T. H. DeWitt, and R. Pastorok. "Sediment Toxicity Tests, Sediment Chemistry and Benthic Ecology *Do* Provide New Insights into the Significance and Management of Contaminated Sediments — A Reply to Robert Spies," *Environ. Toxicol. Chem.* 10:1–4 (1991).

CHAPTER 10

Freshwater Benthic Toxicity Tests

G. A. Burton, Jr., M. K. Nelson, and C. G. Ingersoll

INTRODUCTION

Benthic macroinvertebrates, as a group, are often the optimal assessment tool in determinations of sediment toxicity. As discussed in Chapters 4 and 5, macroinvertebrate community structure indices have been used for many years as effective and sensitive indicators of ecosystem pollution.[1-3] A substantial data base exists on macroinvertebrate responses to xenobiotics, nutrients, and other physicochemical perturbations. In addition, life cycles and habitat and culturing requirements are known for a number of species that play a major role in the function of many aquatic ecosystems, such as *Chironomus* (midges), *Tubifex* (aquatic earthworms), *Hyalella* (scuds), *Gammarus* (scuds), and *Hexagenia* (mayfly nymphs).[4-6] Their intimate contact with bottom sediments and interstitial and overlying waters for extended periods of their life cycle increases the likelihood for adverse effects occurring in the presence of contaminated sediments. Benthic macroinvertebrates fill a multitude of ecological niches: functioning as prey, predators, herbivores, omnivores, collectors, gatherers, shredders, filter feeders, and thus interacting with multiple trophic levels, controlling energy/nutrient/organic matter cycling dynamics in many ecosystems.[7-9]

HISTORY

Sediment toxicity testing began with freshwater benthic macroinvertebrates, namely the mayfly, *Hexagenia limbata*,[10] and the midge, *Chironomus tentans*,[11,12] in

1977. These studies using acute (96 hr) and chronic exposures (11 to 28 days) indicated survival, growth, and emergence were related to bulk sediment contaminant concentrations.[10-15] Most sediment testing in the late 1970s and early 1980s[10,13,16-19] was focused on concerns of dredging of contaminated sediments and the potential impact of dredge material (acute effects) on water quality and biota.[14,16-21] The U.S. Army Corps of Engineers (COE) and the U.S. Environmental Protection Agency (EPA) developed guidance for the testing of dredge and fill materials in marine systems, using three appropriate species and conducting acute exposures in whole sediments, suspended sediments, and elutriate (water-extractable) fractions.[20] Exposure periods and test phases used in dredging evaluations were designed to mimic dredging conditions: that is, short-term perturbations primarily involving suspended solids and water-extractable toxicants. Some have questioned the realism of the recommended test conditions and the sensitivity of many of the test species.[22]

During the past 10 years the research and literature concerned with assessing sediment contamination has expanded substantially.[23] Laboratory studies involving benthic invertebrate species have varied widely in their experimental design, species selection, endpoints of toxicity, and manipulation of sediments. These approaches, and their associated strengths and weaknesses, will be discussed in the following sections. Standardization of methods is advantageous for some study objectives and regulatory usage and has recently begun within the American Society for Testing and Materials (ASTM), Subcommittee E47.03 on Sediment Toxicology.[24] Standard guides for whole-sediment toxicity testing with the midges, *Chironomus tentans, C. riparius,* and the amphipod, *Hyalella azteca,* were approved in 1990. Draft methods also exist for *Daphnia* and *Ceriodaphnia* sp., *Hexagenia,* and oligochaete acute- and short-term chronic toxicity testing.

ASTM Subcommittee E47.03 on Sediment Toxicology was established in May of 1987 and is one of 11 subcommittees in the ASTM Committee E47 on Biological Effects and Environmental Fate. The goal of the sediment subcommittee is to develop guides for assessing the bioavailability of contaminants associated with sediments. These guides are used to evaluate the toxicological hazard of contaminated sediment, soil, sludge, drilling fluids, and similar materials. The subcommittee initially decided to develop guides, not test methods, because most testing procedures for sediment have been recently developed. Eventually, test methods can be developed from the guides when definitive procedures for a particular test are established. Over the past four years the subcommittee has gained ASTM approval for three sediment toxicity testing guides.

TEST CONDITIONS

As discussed in Chapters 1 and 3, the sediment environment is very complex, consisting of a quasi-stable physical system in which numerous physicochemical and microbiological gradients exist and interact. Inorganic and organic substances, both of natural (e.g., carbonates, oxyhydroxides, humics, low-molecular-weight fatty acids) and anthropogenic (e.g., arsenic, cadmium, copper, lead, polycyclic aromatic hydrocarbons, biphenyls, dibenzodioxins, pesticides) origin, partition between sediments, interstitial water, overlying water, and resident biota. Many sampling and laboratory manipulations can have a dramatic impact on partitioning (Chapter 3) and, thereby, affect toxicity responses in the test species. Some obvious conditions that may affect results include overlying water quality, sediment/water contact time, exposure period, and exposure phase.

Overlying Water Quality

Sediment/water contact time in sediment toxicity assays may exert substantial effects on overlying water quality and therefore organism response.[25] Sediment oxygen demand (biochemical and chemical) can be significant in some sediments rich in nutrients and reduced substances,[26] requiring aeration of the overlying water.[27] The dissolution of sediment components, such as carbonates, may elevate hardness,[28] which would affect the availability of some metals such as Cd, Cu, and Pb. Disturbance of redox gradients and increased oxygenation may result in reduced levels of acid volatile sulfides (AVS), and thus the possible release of available metal.[29] Exposure conditions have consisted of static,[28] recirculating,[10,17,30] static-renewal,[31] and flow-through systems,[17,28] and system comparisons have shown significant differences in toxicity response in some studies.[28,32] Static and flow-through tests of sediments from Waukegan Harbor were equally toxic to amphipods, but when other sites were included in the statistical analyses, flow-through exposures provided greater survival. This was likely a result of the flushing of contaminants from overlying water that had desorbed from the sediment.[28] Ingersoll and Nelson[28] recommend whole-sediment tests be conducted with low turnover rates (<4 chamber volume per day) to maintain more consistent overlying water quality and to reduce the flushing of contaminants from the exposure system.

In whole-sediment assays, a 1:4 ratio of sediment to water has been common;[31,32] however, the Prater-Anderson recirculating system has been used at a ratio of 1:9.5.[10,18] The 1:4 ratio probably originated from the COE elutriate preparation procedure. The interaction between the sediment and the overlying water in the test chambers, and the ratio of sediment to overlying water in the test chambers, may influence the availability of the contaminants. Tests may need to be conducted with the range of environmental conditions expected in the overlying water of sediment. For example, water hardness or pH of the overlying water may alter sediment toxicity.[22] Stemmer et al.[33] investigated the influence of sediment volume and surface area on the toxicity of selenium-spiked sediment to *Daphnia magna*. Varying surface area within a constant 1:4 ratio of sediment to water did not alter daphnid survival; however, a decrease in the sediment-to-water ratio (1:8) and an increased surface area decreased survival of the test organisms. These results indicate that test conditions that deviate substantially from more conventional test methods using a 1:4

sediment-to-water ratio may affect contaminant availability, and standardizing sediment-to-water ratios may be necessary in order to make comparisons between species.

Test Phase

Perhaps the most important issue in sediment toxicity testing is the appropriate sediment phase to test. Sediment phase can be categorized as follows: extractable (solute other than water) phase, elutriate (water-extractable) phase, interstitial water phase, whole sediment, and in situ assays. Each has associated strengths and weaknesses that prevent the recommendation of any one phase to meet all study objectives. The issues discussed previously regarding sediment integrity and contaminant sorption and desorption are particularly pertinent when attempting to interpret assay responses between different sediment phases. These considerations are summarized in Table 1.

Few studies have compared test phases as treatments.[34,35] Some studies have compared phases, but using different assays,[34,36] which does not allow a true comparison of phase effects on toxicity. The elutriate phase has been shown to be more toxic[37] and less toxic[34,35,38] than other phases. In studies of four areas in the Great Lakes[34,38] and one stream in Ohio,[35] the elutriate fraction was always less toxic than whole-sediment assays using the same endpoints. Some sediment toxicity effects are only associated with the whole phase.[39] Interstitial waters, however, were more toxic or of equal toxicity to whole sediment.[35] The greater toxicity may be due to elevated ammonia concentrations[40] that are diluted in overlying waters in whole-sediment assays. This, however, may be an artifact of pH shifts, which may increase when interstitial waters are isolated, thereby increasing ammonia toxicity. Higher metal concentrations have been observed in interstitial waters compared to elutriates.[19,41] The sediment interstitial water toxicity test was developed for evaluating the potential in situ effects of contaminated sediment on aquatic organisms. Once the interstitial water or elutriate samples are isolated from the whole sediment, the toxicity testing procedures are similar to effluent toxicity testing with nonbenthic species, described in Chapter 8. If benthic species are used as test organisms, they may be stressed by the absence of sediment.[42] Methods for sampling interstitial water have not been standardized. Isolating sediment interstitial water has been accomplished using several methods, including centrifugation, squeezing, suction, and dialysis.[43,44]

Knezovich et al.[45] stated that organism morphology, ecological niche, feeding mechanism, and physiology will determine toxicant, uptake, pathway, and, thus, hazard. For example, oligochaetes are sediment ingesters, while many benthic and epibenthic species are filter feeders,[46] and thus are exposed to interstitial and overlying waters to varying degrees.[46,47] It is likely that no consistent relationship between relative toxicity of all interstitial, elutriate, and whole-sediment assays will ever exist, due to the multitude of physicochemical and biological process variables.

Some acute toxicity assays using benthic invertebrates have been conducted in sediment-free systems such as interstitial water, elutriate phase, or spiked waters.[34-40,48] Most benthic test organisms (such as *Hyalella azteca, Chironomus* sp., or *Hexagenia limbata*) require substrate contact or burrowing capabilities during their life cycles;[4,5] the absence of sediment may induce artificial and perhaps stressful conditions.[42,49] Stress has been observed in exposures greater than 48 h by decreased control survival or cannibalism. The relationship of this unnatural stress factor on acute-effect-level determinations is unknown, but should be considered.

EXPOSURE CONDITIONS

Given the sensitive and tenuous nature of sediment integrity, exposure conditions are particularly crucial in determining contaminant behavior and organism or community response. Parameters of concern include time of exposure, feeding, and both the chemical and physical environment (e.g., light, temperature, dissolved oxygen).

Time

Most marine and freshwater sediment toxicity testing has been limited to acute testing where exposure periods typically were 15 min for Microtox®, 48 hr for cladocerans, 96 hr for fish, and 4 to 10 days for amphipods, oligochaetes, chironomids, and Ephemeroptera. Greater sensitivity to toxicants occurs with extended exposure.[17,28,50-53] Concentrations of contaminants in sediments may not be acutely lethal, but may interfere with the ability of an organism to develop, grow, or reproduce. However, only a limited amount of freshwater subchronic and chronic toxicity testing has been conducted and has usually consisted of amphipod reproduction and growth (28 day), oligochaeta growth (10 day), and chironomid growth and emergence (10 to 15 day). Debate continues in aquatic toxicology over the definitions, adequacy, or relationships between acute, subchronic, and chronic toxicity testing.[54-56] Early life stage tests that monitor fecundity and growth are often more sensitive than survival studies using adults.[57] However, long-term survival in chronic toxicity tests may be more sensitive than other endpoints.[58] The sensitivity of molecular and cellular endpoints is greater than community structure and ecosystem function endpoints; however, determining their ecosystem relevance is much less clear (Figure 1). While there are obvious advantages in conducting subchronic tests (e.g., shorter testing period, thus less resource intensive, allowing more testing), chronic tests do not require extrapolation from shorter test exposure periods.[55,56] In addition, different lethal or sublethal biochemical endpoints can be studied, so both types of assays are useful and necessary.

Table 1. Sediment Toxicity Exposure Phases

Phase	Strengths	Weaknesses	Routine Uses
Extractable Phase (XP) (solutes vary)	Use with all sediment types Sequentially extract different degrees of bioavailable fractions Greater variety of assay endpoints available Determine dose-response	Ecosystem realism: bioavailability unknown, chemical alternation	Rapid screen Unique endpoints, so component of test battery
Elutriate Phase (EP) (water extractable)	Use with all sediment types Readily available fraction Mimics oxic toxic environmental process Large variety of assay endpoints available Methods more standard Determine dose-response	Ecosystem realism: only one oxidizing condition used; only one solid: water ratio; exposure for extended period of one phase condition which never occurs in situ or never occurs in equilibrium in situ Extract conditions vary with investigator Filtration affects response, sometimes used	Rapid screen Endpoints not possible with WS Dredging evaluations
Interstitial water (IW)	Direct route of uptake for some species Indirect exposure phase for some species Large variety of assay endpoints available Methods of exposure more standard Determine dose-response Sediment criteria may be determined	Can't collect IW from some sediments Limited volumes can be collected efficiently Optimal sampling method unknown, constituents altered by all methods Exposure phase altered chemically and physically when isolated from WS Flux between overlying water and sediment unknown Relationship to and between some organisms, uncertain: burrowers, epibenthic, water column species, filter feeders, selective filtering, life cycle vs. pore water exposure	Rapid screen Endpoints not possible with WS Initial surveys Sediment criteria

Whole sediment (WS)	Use with all sediment Relative realism high Determine dose-response Holistic (whole) vs. reductionist toxicity approach (water, IW, EP, and XP) Sediment quality criteria may be determined Use site or reconstituted water to isolate WS toxicity	Some physical/chemical/microbiological alteration from field collection Dose-response methods tentative Testing more difficult with some species and some sediments Few standard methods Indigenous biota may be present in sample	Rapid screen Chronic studies Initial surveys Sediment criteria
In situ[a] (IS)	Real measure integrating all key components, eliminating extraneous influences Criteria may be determined Resuspension/suspended solids effects assessed	Few methods and endpoints Not as rapid as some assay systems Mesocosms variable Predation by indigenous biota	Resuspension effects Intensive system monitoring Sediment criteria

[a] Organisms exposed in situ in natural systems, pond/stream mesocosms, or lake limnocorrals.

Source: Burton[152]

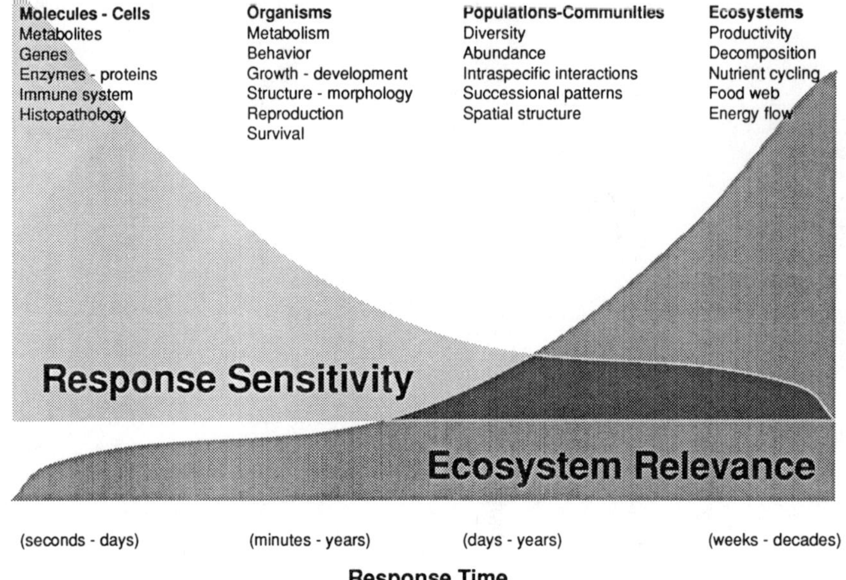

Figure 1. Levels of biological organization. (From Burton, G. A., Jr. "Assessing Toxicity of Freshwater Sediments," *Environ. Toxicol. Chem.* 10:1585–1627 (1991). With permission.)

Feeding

The same testing factors (feeding, physical and chemical conditions) that have been recognized as important in controlling toxicity responses in effluent pure-chemical and water column assays,[59] are also important in assays of sediment toxicity. Feeding may alter toxicant exposure and elimination rates and must be considered (see Chapters 12 and 13). However, some different considerations and parameters do exist, such as water quality changing through time.

In longer exposures, supplemental food is often added to the test chambers. Without the addition of food, the test organisms may starve during longer exposures, but the addition of the food may alter the availability of the contaminants in the sediment.[60] Furthermore, if too much food is added to the test chamber or if mortality of test organisms is high, mold-bacterial growth will develop on the sediment surface. If test organisms are fed in whole-sediment tests, the amount of food should be kept to a minimum. If food accumulates on the sediment or if a mold or bacterial growth is observed on the surface of the sediment, feeding should be suspended for 1 or more days. Detailed records of feeding rates and the appearance of the sediment should be made daily.

Species Selection

The species tested should be selected based on (1) their behavior in sediment (e.g., habitat, feeding habits), (2) their sensitivity to test material(s), (3) their ecological relevance, (4) their geographical distribution, (5) their taxonomic relation to indigenous animals, (6) their acceptability for use in toxicity assessment (e.g., a replicable and standardized method, ease of test method), (7) their availability, and (8) their tolerance to natural geochemical sediment characteristics such as grain size. Many species that might be appropriate for sediment testing do not meet these criteria because, historically, an emphasis has been placed on developing testing procedures for water column exposures. Test species should not be collected at or near a disposal site, since these populations may have developed an enhanced resistance to contaminant perturbations. Unfortunately, culturing methods and testing procedures have not been developed for many benthic animals.[61]

Sensitivity is related to the degree of contact between the sediment and the organism. Feeding habits, including the type of food and feeding rate, will control the dose of contaminant from sediment.[46] Infaunal deposit-feeding organisms can receive a dose of sediment contaminants from three sources: interstitial water, whole sediment, and overlying water. Benthic invertebrates may selectively consume particles with higher organic carbon concentration and higher contaminant concentrations. Organisms in direct contact with sediment may also accumulate contaminants by direct adsorption to the body wall or exoskeleton or by absorption through the integument.[45] Thus, estimates of bioavailability will be more complex for epibenthic animals that inhabit both the sediment and the overlying water. Tests with elutriate samples measure the water-soluble constituents potentially released from sediment to the water column during dredge disposal operations.

Geochemical Characteristics

Natural geochemical properties, such as sediment texture, may influence the response of infaunal animals in sediment tests. It is important to select test organisms that have a wide range in tolerance to natural sediment properties. The natural geochemical properties of test sediment collected from the field need to be within the tolerance limits of the test species. The limits for the test species should be determined experimentally. Controls for such factors as particle size and organic carbon should be run if the limits of the test animal are exceeded in the sediments. The effects of sediment characteristics such as grain size and organic carbon concentration can either be addressed experimentally using toxicity tests or be addressed using normalizing equations (see Chapter 9). Studies of the influence of additional "noncontaminant" factors, such as sediment moisture, organic content, and water quality (e.g., hardness, pH, Eh, ammonia), on the response of test animals are required to differentiate between effects resulting from the influence of natural sediment characteristics and effects caused by contaminants.

The route of exposure may be uncertain, and data generated in sediment toxicity tests may be difficult to interpret, if normalizing factors for bioavailability are unknown. Bulk sediment chemical concentrations need to be normalized to factors other than dry weight. For example, concentrations of nonpolar organic compounds might be normalized to sediment organic carbon content, and metals normalized to acid volatile sulfides.

Indigenous Animals

Indigenous animals may be present in field-collected sediments.[28] An abundance of the same species, or of species taxonomically similar to the test species in the sediment sample, may make interpretation of treatment effects difficult. Previous investigators have inhibited the biological activity of sediment with heat, mercuric chloride, antibiotics, or gamma irradiation. Gamma irradiation is probably the most desirable method because it causes the least alteration in either the physical or chemical characteristics of the sediment.[44] Further research is needed to determine the effects on the bioavailability of contaminants from treating sediment to destroy indigenous organisms before testing.

TESTS OF SEDIMENT TOXICITY

The benthic community is comprised of several taxonomic levels of organisms, and many have been used in toxicity assessments, including bacteria, protozoans, nematodes, bryozoans, oligochaetes, amphipods, gastropods, pelecypods, insects, and periphyton.[23] The following section will focus those assays that have been reported several times to be useful in sediment toxicity assessments.

Bacteria

Microbial assays can be divided into testing groups of either indigenous communities or laboratory cultured strains and assay endpoints that are biochemical (such as enzyme activity, bioluminescence, lipopolysaccharides, muramic acid, and ATP content) or other metabolic processes (such as growth, uptake, respiration, substrate transformation, viability, and microcalorimetry). These endpoints have been measured by a multitude of methods, primarily in studies of water, wastewater, and soil systems, and have been applied to sediment systems, to some degree.

Biomonitoring by EPA has not routinely included testing of the microbial community. The Toxic Substances Control Act recommends premanufacturing testing of chemical effects on several microbial processes. Microbial testing is also a component of new-product testing under the Federal Insecticide, Fungicide, and

Rodenticide Act. Microtox® has been listed by the EPA as a supplemental test to use in Tier 1 screening tests in the *Technical Support Document for the Water Quality-Based Toxics Control Approach.*[62] However, limited use is actually made of microbial toxicity tests in any of the EPA program activities.[63,64]

Microbial responses have been recommended as early warning indicators of ecosystem stress[65] and as a means of establishing toxicant criteria for terrestrial and aquatic ecosystems.[64] The resulting changes at the species level should be accompanied by changes in respiration and/or decomposition rates.[65] The usefulness of monitoring the microbial community is due, in part, to its ability to respond so quickly to environmental conditions (e.g., toxicant exposure) and the major role they play in ecosystem biogeochemical cycling processes and the food web.[66,67]

When investigating chronic toxicity and other early warning indicators of toxicant stress, stimulatory effects are often noted at low toxicant concentrations in fish, cladoceran, algal, macrophyte, and microbial indicator assay responses.[38,68,69] This phenomenon, known as hormesis, is common when using microbial and photosynthetic organisms as indicators. Stimulatory effects can be attributed to nutrients, adapted microbial communities, the Arndt-Schultz phenomenon, or feedback mechanism disruption.[70,71] Elevated structure and function responses were initial stress indicators, probably reflecting a disruption of normal feedback mechanisms controlling microbial nutrient dynamics and species interactions.[71] Stimulation or inhibition of activity may also result when carbon or nutrient substrates are altered, so that one enzyme system, e.g., alkaline phosphatase activity, is stimulated while another, such as galactosidase activity, is inhibited.[72] When comparing test samples with reference samples, inhibitory *and* stimulatory effects should be regarded as possible perturbations.

Pure microbial culture systems used in assessments of sediment extracts include Microtox® *(Photobacterium phosphoreum)*[73] and *Spirillum volutans,*[74] *Escherichia coli,*[75] *Nitrobacter* sp.,[76] *Azotobacter vinelandii,*[77] *Aeromonas hydrophila,*[78] *Pseudomonas fluorescens,*[79] and *P. putida.*[80] In most comparative surveys *Spirullum* was the least sensitive to toxicity, but not in other studies.[81] Pure-culture studies with bacteria and fungi have demonstrated that sensitivity to metals is equal to or less than plant or animal systems.[82]

Microtox® testing has recently been incorporated into sediment toxicity test strategies (Chapter 15) and was originally used in marine sediment assessments.[73] Some toxicity has been attributed to the extraction of natural organics.[83] The insensitivity of Microtox® to some elements and compounds may result from an inappropriate diluent (ionic strength adjustor).[40,73,84,85] Interstitial water in large-grain contaminated sediments was more toxic than fine-grain sediments; however, the opposite was observed for solvent-extracted sediments.[85] Numerous comparisons of Microtox® sensitivity to pure compounds and effluents with *Daphnia* sp. and fish (primarily *P. promelas*) indicate similar effect levels,[86,87] and Microtox® was generally more sensitive than other microbial tests.[88] In three of the four sediment comparison studies,[34,89,90] Microtox® was very sensitive and discriminatory of sediments contaminated with a wide variety of synthetic organic and metal compounds. In a fourth study where the primary toxicant was ammonia, interstitial water effects were observed on *P. promelas, C. dubia,* and *S. capricornutum,* but not Microtox®.[40] Recently, a whole-sediment exposure method using *Photobacterium phosphoreum* was presented and appeared to be more sensitive to hydrophobic chemicals than the elutriate Microtox® assay.[91]

Metabolic processes such as methanogenesis, sulfate reduction, denitrification, and carbon dioxide evolution, or enzymes involved in key metabolic systems such as dehydrogenases, alkaline phosphatase, and glucosidase, have been measured in contaminated sediments.[68,72,92-99] Some of these processes were depressed, whereas other processes were stimulated.[66] Discrimination of contaminated and noncontaminated sediments requires several response endpoints.[68,97,98] Burton and Stemmer[68] evaluated five stream profiles across the U.S. in which several indigenous oxidoreductase and hydrolase enzyme activities in waters and sediments were compared to in situ chemical concentrations, biological community structure, and laboratory test animals. At all five sites significant relationships were observed between indigenous enzyme activities and in situ conditions, indicating toxicant effects, natural spatial variation, and food web interactions. Activity of β-galactosidase indicated significant relationships in 80% of the studies to 37.5% of the biological and chemical stream parameters measured. β-glucosidase and dehydrogenase activity were significant indicators of stream conditions in 60% of the studies, whereas *C. dubia* reproduction (water only) was related in 50% of the studies to 22.5% of the stream parameters. Hydrolases were also effectively used to define sediment spatial variance in creosote-contaminated sediments.[33]

In summary, studies to date have demonstrated that indigenous microbial enzyme activity and bioluminescence (e.g., Microtox®) are generally as sensitive as fish and invertebrate toxicity assays to metals, some organics, and contaminated sediments. Microbial assays are effective at discriminating degrees of sediment contamination[34,89] and are related to in situ conditions at most sites. Assays of microbial processes and natural assemblages of microorganisms tend to be superior to pure-culture test systems. Bacteria should be considered in perturbation studies, since microbes play such an important role in energy flow and ecosystem functioning. However, since bacteria reproduce and adapt relatively quickly, pollutant effects must be greater than macrofaunal responses to be ecologically relevant.

Protozoa

Protozoan tests of sediment contamination have primarily evaluated overlying waters or the elutriate phase. Pratt and Cairns[100] grouped freshwater protozoa into six functional groups based on food requirements: dissolved mineral nutrients; bacteria and detritus; algae, bacteria, and detritus; diatoms; dissolved organics;

and, rotifers and protozoans. Protozoans feeding on the "bacteria and detritus" component comprise the majority of genera. Protozoan communities are a complex structure of herbivores, carnivores, omnivores, and detritus feeders.[101] The majority of species are cosmopolitan and tolerate a wide range of freshwater quality.[4] Protozoa play an important role in food web dynamics and the "microbial loop".[1,4] Holophytic and saprozoic protozoans are producers that use dissolved nutrients and are food for meiofauna, as are holozoic species that consume particulate living and dead material.

Few studies of sediment contamination have been conducted with protozoans. Acute toxicity assays using the ciliate protozoa *Tetrahymena* sp. have only involved water exposures. Growth of *Colpidium campylum* was used to evaluate the toxicity of elutriates and sediment slurries.[1,102] Protozoan colonization of artificial substrates (polyurethane foam) were used in laboratory tests with elutriates and in situ tests with substrates suspended over contaminated sediments.[103] Community structure and functional endpoints reported by Henebry and Ross[103] include decolonization, protozoan abundance, taxa number, phototroph and heterotroph abundance, respiration, and island-epicenter colonization rates. Functional endpoints and phototrophs were the most sensitive endpoints. Stimulatory and inhibitory results were observed, and careful interpretation of effects was required.[103]

Periphyton

Benthic-associated algae (periphyton) dominate primary production in many streams[104] and shallow-lake regions.[105] Attached algal (periphyton) communities are useful indicators of aquatic pollution,[106,107] but are infrequently included in sediment studies. Shifts in community structure from pollution-sensitive groups to tolerant groups occurred in streams receiving waterborne metal[106,107] and organic pollution.[108] A continuous-flow in situ periphyton bioassay was described that measured nutrient limitation using chlorophyll and $^{14}CO_2$ uptake.[109] Outdoor experimental stream periphyton communities were sensitive to μg/L levels of pentachlorophenol, based on periphyton biomass and pigment production.[110]

Nematodes

As with the preceding biological groups, most nematode testing has been conducted on water, water-extract, or elutriate phases. Little is known about free-living freshwater nematodes. Many species are cosmopolitan in nature, can survive in a wide variety of conditions, and are primarily in the meiobenthos. They may reach densities of 100,000/m^2 and up to a depth of 2 cm in soft sediments. Nematodes can survive anoxic conditions for several weeks and have highly resistant eggs.[4] Sediment extracts and elutriate toxicity were evaluated using *Panagrellus redivivus* in 4-day exposures.[111,112] Survival, growth, and molting frequency were evaluated in tests started with the second embryonic stage. A free-living nematode, *Caenorhabditis elegans,* was recently proposed as a promising test organism, based on ease of culture and sensitivity to metals.[113]

Oligochaetes

Oligochaetes have primarily been tested in whole-sediment exposures. Oligochaetes are a major component of benthic systems in many aquatic systems[114] and transport deeper sediments to the surface as fecal pellets. The tubificids are common in polluted areas and effectively mix sediment surface layers and play a major role in the cycling of metal and organic contaminants out of the sediments.[4,115-117] The "aquatic earthworms" used for freshwater sediment toxicity assessments are limited primarily to *Tubifex* sp. *Tubifex tubifex* is considered an indicator of organic pollution, particularly in waters with low dissolved-oxygen saturation. *Lumbriculus* has been used in whole-sediment tests to a limited extent,[118] as have some other species in Sweden.[60] However, the usefulness of oligochaetes as sediment toxicity indicators has received mixed reviews.[60,119] Taxonomy, variable sensitivity, and fragility make oligochaetes difficult to use.[60] Growth and reproduction of five oligochaete species were followed during 0.5 to 1.5-year exposures in contaminated oligotrophic sediments, and reproduction was the most sensitive endpoint.[60] *Limnodrilus* and *Stylodrilus* burrowing avoidance was a sensitive indicator of sediment contamination.[120] *T. tubifex* and *L. hoffmeisteri* avoidance behavior was observed in copper- and zinc-spiked sediments.[121] The oligotrophic *Stylodrilus heringianus* will acclimate to sediment perturbations such as mixing. Sediment reworking rates, survival, and weight were sensitive indicators of a variety of sediment contaminants[122] and Endrin-spiked sediment.[120]

Amphipods

About 150 North American freshwater species of "scuds" or "sideswimmers" have been identified.[4] The dominant species include *Hyalella azteca, Gammarus pseudolimnaens, Gammarus fasciatus, Crangonyx gracillus,* and, in the Great Lakes, *Pontoporeia hoyi* (now *Diporeia* sp.). Amphipods are widely distributed and common in unpolluted lotic and lentic systems; however, they are less common in large rivers. *Hyalella azteca* is a common and widely distributed Talitrid amphipod inhabiting permanent lakes, ponds, and streams throughout the Nearctic and Neotropical biogeographical realms.[4,123-124] In addition, *H. azteca* is euryhaline and occurs in waters of varying salinities, from 5 g/L in the estuary Barataria Bay, Louisiana[125] and athalassic Pyramid Lake, Nevada[126] to 22 g/L in saline lakes of Canada.[127] If slowly acclimated, *H. azteca* will reportedly survive at 30 g/L.[124] They are a primary food source for fish and voracious feeders of animal, plant, and detrital material.[4] The life cycle of *H. azteca* can be divided into three stages according to Cooper[128] and Pennak:[4] (1) immature (includes instars 1 to 5), (2) juvenile (includes instars 6 and 7), and (3) adult (includes the 8th instar and beyond). The potential number of adult instars is large, and growth is indeterminate.[129] The number of molts that may occur during the adult period is variable, but may be as high as 15 or 20.[4] DeMarch[130] reported that juvenile *Hyalella azteca* completes a life cycle in 27 days or longer, depending on temperature. *H. azteca* exhibits sexual dimorphism; the adult male is larger than females and has an enlarged propodus on the second gnathopod. Eggs in the female are visible in both the ovaries and brood pouch. The epibenthic species *H. azteca,* has been used frequently in sediment toxicity testing.[28] *Hyalella azteca* has many desirable characteristics of a toxicity test organism: short generation time; easily collected from natural sources or cultured in the laboratory;[28,32,128,130,131] and data on survival, growth, and development can be obtained in toxicity tests.[28,132,133] *H. azteca* is successfully used in sediment toxicity testing and is a sensitive indicator of the presence of contaminants associated with sediments.[28,32,39,131,134-136] The amphipod juvenile stage is more sensitive to sediment contamination than the adult. *Hyalella* are easily cultured,[6,28,31,32] and standard sediment toxicity test methods were recently developed for *H. azteca* whole-sediment testing.[24,31] *Hyalella azteca* and *Gammarus* sp. have been used frequently in acute toxicity studies of pure compounds or ambient waters and found to be relatively sensitive in comparative studies.[28,34] *Pontoporeia* sp. have been used in Great Lakes studies, since it is a primary benthic species there;[16,17] unfortunately, culturing methods have not been developed; thus, deepwater collections must be made for testing.

Sediment testing with *H. azteca* has consisted primarily of whole-sediment exposures (1:4 ratios of sediment to water) in static renewal systems for 7-, 10-, 14-, 28-, or 29-day periods.[3,28,31,32,38,136] Survival is most frequently used as the endpoint in studies; however, in a 28-day chronic exposure, growth and reproductive maturation are measured.[28] *Hyalella azteca* large juveniles-young adults and *Gammarus lacustris* adults were less sensitive than *D. magna* or *C. tentans* to Cu in sediment-spiked sediment 10-day exposures.[39] *Hyalella azteca* was more sensitive than *D. magna* to Cd in static-spiked sediment tests, and only free Cd contributed to toxicity. No toxicity was observed in flow-through tests;[135] however, the flow rate was 60 volume additions per day. *Hyalella azteca* were one of the most sensitive and discriminatory of 20 different sediment toxicity assays in studies of three contaminated Great Lakes areas during 7 to 14-day whole-sediment exposures[34] and has been recommended as a tool to measure acute[32] and chronic sediment toxicity.[3,28]

Pelecypods

Mussels have been used, to a limited extent, for environmental assessments[137] and only recently in studies of sediment toxicity.[138] Bivalve mollusks are common in large rivers and vary in size from 2 to 250 mm in length. Their primary food is fine organic detritus that has been resuspended;[4] the significance of plankton as a food varies with the species and habitat. Particles as small as 1 μm can be removed by mollusks from water. Some species burrow during their life cycle, well below the sediment surface (up to 25 cm) and have an interstitial water suspension-feeding mechanism. The life cycle of mollusks ranges from 1 month to 3 years, and they are a common food of fish, reptiles, amphibians, and mammals. The filtration capacity of mollusks is massive. An estimated 7 billion clams inhabit Lake St. Clair and theoretically filter the entire lake every 13 days, assuming each filters 4 L/day.[115] This has profound implications on their role in ecosystem dynamics. A drastic decline in species and population numbers of this ecologically and economically important group has been observed over the past three decades.[4] Recently, mollusks have been used in surveys of aquatic toxicity, both in the laboratory and in situ.[139,140] Preference-avoidance tests with *Acroneuria* indicated increased drift and locomotor activity with exposure to insecticides.[141] Celluolytic activity was sensitive to effluent toxicity in laboratory and field experiments with caged individuals.[142] Longer exposures were necessary in the laboratory to elicit response levels observed in situ. Mollusks have also been useful in long-term field monitoring studies,[32] and growth in situ seems to be a sensitive endpoint.[140] Keller and Zam[143] reported a simplified method for in vitro culturing of *Anodonta, Lampsilis,* and *Villosa* sp.

Insects

The mayfly (Ephemeroptera), *Hexagenia limbata,* and midges (Diptera), including *Chironomus tentans* and *C. riparius,* are the aquatic insects primarily used in sediment toxicity testing. For *H. limbata,* the sediment-dwelling nymph life stage may last from 1 to 2 years.[5] Mayflies are collectors-gatherers, with possibly some filtering at the mouth of their burrow, and have a wide distribution.[4] Midge larvae often inhabit eutrophic lakes and streams. In lotic and lentic habitats with soft sediments, about 95% of the chironomid larvae occur in the upper 10 cm of substrate.[31] The life cycle of *C. riparius* and *C. tentans* consists of three distinct stages: (1) a larval stage consisting of four instars, (2) a pupal stage, and (3) an adult stage.[31]

Hexagenia limbata has been used since the late 1970s in sediment toxicity evaluations and is sensitive to the presence of toxicants, both in laboratory and field surveys.[36,52,53] Most toxicity testing has used field-collected organisms because mayflies are difficult to culture and may only reproduce once a year. Toxicity tests have been conducted with water, interstitial water, elutriate, artificial burrows, and whole-sediment systems, using both static, static renewal, and recirculating 10-day exposures.[36,52,146,147] The measured endpoints have included mortality, molting, growth, and avoidance. Mayflies are reportedly more sensitive than other simultaneously tested species (such as *C. tentans, P. promelas, Asellus*),[10] and their response has been correlated to other species.[36] Responses have also been representative of contaminant concentrations in the sediment extracts, whole sediments, and with in situ community profiles.[18,36,145] A failure of acute responses in the laboratory to correlate with effects of benthic communities in contaminated areas was attributed to comparing acute (10-day exposures) with in situ chronic effects.[52,53] *Hexagenia* may be less sensitive than *D. magna,*[146] but sensitivity was increased with increased exposure time (5 to 10 days).[52,53] The burrowing behavior of mayflies alters Eh, pH, organic carbon, and contaminant profiles[147] and may affect overlying water toxicant concentrations and toxicity to zooplankton.[146] The International Joint Commission (IJC) recommends their use in sediment evaluations with 14-day exposures at 20°C.[3]

Whole-sediment testing with *Chironomus* sp. was first reported by Wentsel et al.[11,12] Unlike *H. limbata,* many midge species can be easily cultured in the laboratory.[6,28,32] *Chironomus* sp. have been widely used in assays with water, interstitial water, elutriate, and whole sediment, ranging from 48-hr to 29-day exposures. Midges were generally perceived to be relatively insensitive organisms in toxicity testing. This conclusion was based on the practice of conducting short-term tests with fourth instar larvae, a procedure that may underestimate midge sensitivity to toxicants.[32,149] Midge exposures started with older larvae may underestimate midge sensitivity to toxicants. For instance, first instar *C. tentans* larvae were 6 to 27 times more sensitive than fourth instar larvae to acute copper exposure,[32,148,149] and first instar *C. riparius* larvae were 127 times more sensitive than second instar larvae to acute cadmium exposure.[148]

Chironomus sp. has been recommended as a routine whole-sediment[32] and interstitial water[49] toxicity test species. Standard sediment toxicity test methods are available for *C. tentans* and *C. riparius*.[31] The most common endpoints include mortality and growth (dry weight).[32,36] The IJC recommends growth and emergence of *C. tentans* beginning with a 13-day-old organism and continuing for 10 days or emergence.[3] Nebeker et al.[32] recommends beginning with 10-day-old organisms and continuing the assay for 15 days. Growth and survival are typical endpoints measured in chronic midge toxicity tests.[28] Wentsel measured *C. tentans* growth (length) with early instars in 17-day tests and found responses were related to bulk metal concentrations.[11,12] Emergence of mature larvae was also related to metal contamination.[15] Emergence of adult midges is a sensitive indicator of contaminant stress,[150] but is seldom monitored in toxicity tests and may be related to the choice of test species. For instance, the survival of *Chironomus tentans* in chronic toxicity tests typically exceeds 80% when exposures continue to the fourth instar; however, many of these larvae fail to pupate and emerge as adults.[150] In contrast, adult emergence by both *C. riparius* and *C. plumosus* typically exceeds 80% in chronic toxicity tests.[151]

Ingersoll and Nelson[28] observed a growth of mold and bacteria on the surface of sediment in adult emergence studies with *C. riparius.* When feeding levels were reduced enough to eliminate visible mold-bacterial growth on the surface of the sediment, larval survival was not affected, but the emergence of adults was delayed beyond 30 days, because of the dependence of adult emergence on feeding and the problem of mold-bacterial growth at higher feeding levels. Ingersoll and Nelson[28] recommend conducting *C. riparius* whole-sediment tests for 14 days, using flow-through exposures. In this period, the first instar larvae develop to the fourth instar at 20°C, and larval survival, growth, and development can be monitored as toxicity endpoints.

In chronic toxicity tests started with first instar animals, midges were often as sensitive as daphnids to inorganic and organic compounds.[28] Sublethal response (growth) of *C. tentans* correlated with the response of Microtox®, *H. limbata, D. magna,* benthic community structure, and discriminated areas of contamination.[36,89] In recent comparative testing, *C. riparius* was more sensitive than *C. tentans,* in several contaminated whole-sediment assays.[34]

CONCLUSIONS

Benthic organisms, as a group, are the best overall indicators of toxic sediments due to (1) their direct contact with sediment solids and interstitial waters; (2) our knowledge of the relative pollution sensitivity and life histories of many species; and (3) the proven effectiveness of amphipod, midge, and mayfly larvae assays in detecting sediment toxicity in a wide range of studies. Unfortunately, the most commonly used assays (e.g., *H. azteca* 10-day survival, *C. tentans* 10-day survival and growth, *H. limbata* 10-day survival) are not ideal toxicity indicators. Problems such as laboratory culturing and recovery of early instar organisms can and will be relatively difficult to overcome. Reliable, efficient, and sensitive chronic toxicity assays have not been widely reported, but represent an active area of current research. In the interim period it seems prudent to include more conventional chronic toxicity indicators, such as *Pimephales promelas* early life stage and *Ceriodaphnia dubia* three-brood reproduction and survival assays, in sediment toxicity test batteries that also include benthic indicators. In the near future our understanding of sublethal indicators (biomarkers) and of the relationship between acute and chronic effects may allow relatively short-term exposures, e.g., hours to several days, to be reliably used in estimates of chronic effects. Presently, they represent a key component of integrated ecosystem health assessments.

REFERENCES

1. Munawar, M., I. F. Munawar, C. I. Mayfield, and L. H. McCarthy. "Probing Ecosystem Health: A Multi-Disciplinary and Multi-Trophic Assay Strategy," *Hydrobiologia* 188/189:93–116 (1989).
2. Great Lakes Science Advisory Board. "Literature Review of the Effects of Persistent Toxic Substances on Great Lakes Biota," International Joint Commission, Windsor, Ontario (1986).
3. Great Lakes Water Quality Board. "Procedures for the Assessment of Contaminated Sediment Problems in the Great Lakes," International Joint Commission, Windsor, Ontario (1988).

4. Pennak, R. W. *Freshwater Invertebrates of the United States: Protozoa to Mollusca*, 3rd Ed. (New York: John Wiley & Sons, 1989).
5. Merritt, R. W., and K. W. Cummins. *An Introduction to the Aquatic Insects of North America* (Dubuque, IA: Kendall/Hunt, 1984).
6. Lawrence, S. G., Ed. *Manual for the Culture of Selected Freshwater Invertebrates*. Can. Spec. Publ. Fish. Aquat. Sci. 54:169 (1981).
7. Cummins, K. W. "Structure and Function of Stream Ecosystems," *Bioscience* 24:631–641 (1974).
8. Carpenter, S. R. *Complex Interactions in Lake Communities* (New York: Springer-Verlag, 1988).
9. Minshall, G. W. "Stream Ecosystem Theory: A Global Perspective," *J. N. Am. Benthol. Soc.* 7:263–288 (1988).
10. Prater, B. L., and M. A. Anderson. "A 96-h Bioassay of Duluth and Superior Harbor Basins (Minnesota) Using *Hexagenia limbata, Asellus communis, Daphnia magna,* and *Pimphales promelas* as Test Organisms," *Bull. Environ. Contam. Toxicol.* 18:159–169 (1977).
11. Wentsel, R., A. McIntosh, W. P. McCafferty, G. Atchison, and V. Anderson. "Avoidance Response of Midge Larvae *(Chironomus tentans)* to Sediments Containing Heavy Metals," *Hydrobiologia* 55:171–175 (1977).
12. Wentsel, R., A. McIntosh, and G. Atchison. "Sublethal Effects of Heavy Metal Contaminated Sediment on Midge Larvae *(Chironomus tentans),*" *Hydrobiologia* 56:153–156 (1977).
13. Prater, B., and R. A. Hoke. "A Method for the Biological and Chemical Evaluation of Sediment Toxicity," in *Contaminants and Sediments,* Vol. 1, R. A. Baker, Ed. (Ann Arbor, MI: Ann Arbor Science, 1980), pp. 483–499.
14. Laskowski-Hoke, R. A., and B. L. Prater. "Dredged Material Evaluation: Correlations Between Chemical and Biological Procedures," *J. Water Pollut. Contr. Fed.* 53:1260–1262 (1981).
15. Wentsel, R., A. McIntosh, and W. P. McCafferty. "Emergence of the Midge *Chironomus tentans* When Exposed to Heavy Metal Contaminated Sediment," *Hydrobiologia* 57:195–196 (1978).
16. Gannon, J. E., and A. M. Beeton. "Procedures for Determining the Effects of Dredged Sediments on Biota-Benthos Viability and Sediment Selectivity Tests. *J. Water Pollut. Contr. Fed.* 43:392–398 (1971).
17. U.S. Environmental Protection Agency. *Development of Bioassay Procedures for Defining Pollution of Harbor Sediments.* U.S. EPA Environmental Research Laboratory, Duluth, MN (1981).
18. Prater, B. L., and M. A. Anderson. "A 96-Hour Bioassay of Otter Creek, Ohio," *J. Water Pollut. Contr. Fed.* 49:2099–2106 (1977).
19. Peddicord, R. K. "Direct Effects of Suspended Sediments on Aquatic Organisms," in *Contaminants and Sediments,* Vol. 1, R. A. Baker, Ed. (Ann Arbor, MI: Ann Arbor Science, 1980), pp. 501–536.
20. U. S. Environmental Protection Agency, U.S. Army Corps of Engineers. *Ecological Evaluation of Proposed Discharge of Dredged Material into Ocean Waters.* Environmental Effects Laboratory, Experimental Waterways Experiment Station, Vicksburg, MI (1977).

21. Lee, G. F., and R. A. Jones. "Water Quality Significance of Contaminants Associated With Sediments: An Overview," in *Fate and Effects of Sediment-Bound Chemicals in Aquatic Systems*, K. L. Dickson, A. W. Maki, and W. A. Brungs, Eds. (New York: Pergamon Press, 1984), pp. 1–34.
22. Anderson, J., W. Birge, J. Gentile, J. Lake, J. Rodgers, Jr., and R. Swartz. "Biological Effects, Bioaccumulation, and Ecotoxicology of Sediment-Associated Chemicals," in *Fate and Effects of Sediment-Bound Chemicals in Aquatic Systems*, K. L. Dickson, A. W. Maki, and W. A. Brungs, Eds. (New York: Pergamon Press, 1984) pp. 267–296.
23. Burton, G. A., Jr., Ed. *Sediment Toxicity Assessment*. (Boca Raton, FL: Lewis Publishers, 1992).
24. Ingersoll, C. "Sediment Toxicity and Bioaccumulation Testing," American Society for Testing and Materials, *ASTM Standardization News*, (pp. 29–33, April 1991).
25. Burton, G. A., Jr., J. M. Lazorchak, W. T. Waller, and G. R. Lanza. "Arsenic Toxicity Changes in the Presence of Sediment," *Bull. Environ. Contam. Toxicol.* 38:491–499 (1987).
26. David, W. S., L. A. Fay, and C. E. Herdendorf. "Overview of USEPA/Clear Lake Erie Sediment Oxygen Demand Investigations During 1979," *J. Great Lakes Res.* 13:731–737 (1987).
27. Francis, P. C., W. J. Birge, and J. A. Black. "Effects of Cadmium-Enriched Sediment on Fish and Amphibian Embryo-Larval Stages. *Ecotoxicol. Environ. Saf.* 8:378–387 (1987).
28. Ingersoll, C., and M. K. Nelson. "Testing Sediment Toxicity With *Hyalella azteca* (Amphipoda) and *Chironomus riparius* (Diptera)," in *Aquatic Toxicology and Risk Assessment: Thirteenth Volume*, STP 1096, W. G. Landis, and W. H Vander Schalie, Eds., (Philadelphia: American Society of Testing and Materials, 1990), pp. 93–109.
29. Di Toro, D. M., J. D. Mahony, D. J. Hansen, K. J. Scott, M. B. Hicks, S. M. Mayr, and M. S. Redmond. "Toxicity of Cadmium in Sediments: The Role of Acid Volatile Sulfide," *Environ. Toxicol. Chem.* 9:1487–1502 (1990).
30. Schuytema, G. S., P. O. Nelson, K. W. Malueg, A. G. Nebeker, D. F. Krawczyk, A. K. Ratcliff, and J. H. Gakstatter. "Toxicity of Cadmium in Water and Sediment Slurries to *Daphnia magna*. *Environ. Toxicol. Chem.* 3:293–308 (1984).
31. American Society for Testing and Materials. *Standard Guide for Conducting Toxicity Tests with Freshwater Invertebrates,* (Philadelphia: American Society for Testing and Materials, 1990), Standard E (in press).
32. Nebeker, A. V., M. A. Cairns, J. H. Gakstatter, K. W. Malueg, G. S. Schuytema, and D. F. Krawczyk. "Biological Methods for Determining Toxicity of Contaminated Freshwater Sediments to Invertebrates," *Environ. Toxicol. Chem.* 3:617–630 (1984).
33. Stemmer, B. L., G. A. Burton, Jr., and G. Sasson-Brickson. "Effect of Sediment Spatial Variance and Collection Method on Cladoceran Toxicity and Indigenous Microbial Activity Determinations," *Environ. Toxicol. Chem.* 9:1035–1044 (1990).
34. Burton, G. A., Jr., L. Burnett, M. Henry, S. Klaine, P. Landrum, and M. Swift. "A Multi-Assay Comparison of Sediment Toxicity at Three 'Areas of Concern'," Abstr. Annu. Meet Soc. Environ. Toxicol. Chem., Arlington, VA No. 213, p. 53, (1990).

35. Sasson-Brickson, G., and G. A. Burton, Jr. "*In Situ* and Laboratory Sediment Toxicity Testing With *Ceriodaphnia dubia*," *Environ. Toxicol. Chem.* 10:201–207 (1991).
36. Giesy, J. P., C. J. Rosieu, R. L. Graney, and M. G. Henry. "Benthic Invertebrate Bioassays With Toxic Sediment and Pore Water," *Environ. Toxicol. Chem.* 9:233–248 (1990).
37. Hoke, R. A., J. P. Giesy, G. T. Ankley, J. L. Newsted, and J. R. Adams. "Toxicity of Sediments from Western Lake Erie and the Maumee River at Toledo, Ohio, 1987: Implications for Current Dredged Material Disposal Practices," *J. Great Lakes Res.* 16:457–470 (1990).
38. Burton, G. A., Jr., B. L. Stemmer, and K. L. Winks. "A Multitrophic Level Evaluation of Sediment Toxicity in Waukegan and Indiana Harbors," *Environ. Toxicol. Chem.* 8:1057–1066 (1989).
39. Cairns, M. A., A. V. Nebeker, J. N. Gakstatter, and W. L. Griffis. "Toxicity of Copper-Spiked Sediments to Freshwater Invertebrates," *Environ. Toxicol. Chem.* 3:435–445 (1984).
40. Ankley, G. T., A. Katko, and J. W. Arthur. "Identification of Ammonia as in Important Sediment-Associated Toxicant in the Lower Fox River and Green Bay, Wisconsin," *Environ. Toxicol. Chem.* 9:313–322 (1990).
41. Brannon, J. M., R. H. Plumb, Jr., and I. Smith, Jr. "Long-Term Release of Heavy Metals for Sediments, in *Contaminants and Sediments,* Vol. 1, R.A. Baker, Ed., (Ann Arbor, MI: Ann Arbor Science, 1980) pp. 221–266.
42. Lamberson, J., and R. Swartz. "Use of Bioassays in Determining the Toxicity of Sediment to Benthic Organisms," in *Toxic Contaminants and Ecosystem Health: A Great Lakes Focus,* M. S. Evans, Ed. (New York: John Wiley & Sons, 1988), pp. 257–279.
43. Pittinger, C. A., V. C. Hand, J. A. Masters, and L. F. Davidson. "Interstitial Water Sampling in Ecotoxicological Testing: Partitioning of a Cationic Surfactant," in *Aquatic Toxicology and Hazard Assessment: 10th Volume*, STP 971, W. J. Adams, G. A. Chapman, and W. G. Lundis, Eds. (Philadelphia: American Society for Testing and Materials, 1988), pp. 138–148.
44. American Society for Testing and Materials. *Standard Guide for Collection, Storage, Characterization, and Manipulation of Sediments for Toxicological Testing*, ASTM Standard No. E1391 (Philadelphia: American Society for Testing and Materials, 1991).
45. Knezovich, J. P., F. L. Harrison, and R. G. Wilhelm. "The Bioavailability of Sediment-Sorbed Organic Chemicals: A Review," *Water, Air, Soil Pollut.* 32:233–245 (1987).
46. Adams, W. J., R. A. Kimerle, and R. G. Mosher. "Aquatic Safety Assessment of Chemicals Sorbed to Sediments," in *Aquatic Toxicology and Hazard Assessment,* Seventh Symposium, STP 854 (Philadelphia: American Society for Testing and Materials, 1985), pp. 429–453.
47. Eadie, B. J., P. F. Landrum, and W. Faust. "Polycyclic Aromatic Hydrocarbons in Sediments, Pore Water and the Amphipod *Pontoporeia hoyi* from Lake Michigan," *Chemosphere* 11:847–858 (1982).
48. Spehar, R. L., R. L. Anderson, and J. T. Fiandt. "Toxicity and Bioaccumulation of Cadmium and Lead in Aquatic Invertebrates," *Environ. Pollut.* 15:195–208 (1978).

49. Giesy, J. P., and R. A. Hoke. "Freshwater Sediment Toxicity Bioassessment: Rationale for Species Selection and Test Design," *J. Great Lakes Res.* 15:539–569 (1989).
50. Birge, W. J., J. A. Black, A. G. Westerman, and P. C. Francis. "Toxicity of Sediment-Associated Metals to Freshwater Organisms: Biomonitoring Procedures," in *Fate and Effects of Sediment-Bound Chemicals in Aquatic Systems*, K. L. Dickson, A. W. Maki, and W. A. Brungs, Eds. (New York: Pergamon Press, 1984) pp. 199–218.
51. LeBlanc, G. A., and D. C. Surprenant. "A Method of Assessment the Toxicity of Contaminated Freshwater Sediments," in *Aquatic Toxicology and Hazard Assessment*, Seventh Symposium, STP 854, R. D. Cardwell, R. Purdy, and R. C. Bahner, Eds. (Philadelphia: American Society of Testing and Materials, 1985), pp. 269–283.
52. Malueg, K. W., G. S. Schuytema, D. F. Krawczyk, and J. H. Gakstatter. "Laboratory Sediment Toxicity Tests, Sediment Chemistry and Distribution of Benthic Macroinvertebrates in Sediments from the Keweenaw Waterway, Michigan," *Environ. Toxicol. Chem.* 3:233–242 (1984).
53. Malueg, K. W., G. S. Schuytema, J. H. Gakstatter, and D. F. Krawczyk. "Toxicity of Sediments from Three Metal-Contaminated Areas," *Environ. Toxicol. Chem.* 3:279–291 (1984).
54. Chapman, P. M. "A Bioassay by Another Name Might Not Smell the Same," *Environ. Toxicol. Chem.* 8:551 (1989).
55. Suter, G. W. "Seven-Day Tests and Chronic Tests," *Environ. Toxicol. Chem.* 9:1435–1436 (1990).
56. Norberg-King, T. J. "An Evaluation of the Fathead Minnow Seven-Day Subchronic Test for Estimating Chronic Toxicity," *Environ. Toxicol. Chem.* 8:1075–1089 (1984).
57. Giesy, J. P., and R. L. Grancy. "Recent Developments in and Intercomparisons of Acute and Chronic Bioassays and Bioindicators," *Hydrobiologia* 188/189:21–60 (1989).
58. Winner, R. "Multigeneration Life-Span Tests of the Nutritional Adequacy of Several Diets and Culture Waters for *Ceriodaphnia dubia*. *Environ. Toxicol. Chem.* 8:513–520 (1988).
59. Rand, G. M., and S. R. Petrocelli. *Fundamentals of Aquatic Toxicology*. (New York: Hemisphere, 1985).
60. Wiederholm, T., A. Wiederholm, and G. Milbrink. "Bulk Sediment Bioassays with Five Species of Fresh-Water Oligochaetes," *Water, Air Soil Pollut.* 36:131–154 (1987).
61. Swartz, R., W. A. DeBen, J. K. Jones, J. O. Lamberson, and F. A. Cole. "Phoxocephalid Amphipod Bioassay for Hazard Assessment," in *Aquatic Toxicology and Hazard Assessment: Proceedings of the Seventh Annual Symposium*, STP 854, R. D. Cardwell, R. Purdy, and R. C. Bahner, Eds. (Philadelphia: American Society for Testing and Materials, 1985), pp. 284–307.
62. U.S. Environmental Protection Agency. *Technical Support Document for Water Quality-Based Toxics Control*, Office of Water, Washington, D.C. (1985).
63. Walker, J. D. "An U.S. EPA Perspective on Ecotoxicity Testing Using Microorganisms," in *Toxicity Testing Using Microorganisms*, Vol. II, B. J. Dutka, and G. Bitton, Eds. (Boca Raton, FL, CRC Press, 1986) pp. 175–186.
64. Babich, H., and G. Stotzky. "Developing Standards for Environmental Toxicants: The Need to Consider Abiotic Environmental Factors and Microbe-mediated Ecologic Processes," *Environ. Health Perspect.* 49:247–260 (1983).
65. Odum, E. P. "Trends Expected in Stressed Ecosystems," *Bioscience* 35:419–422 (1985).

66. Griffiths, R. P. "The Importance of Measuring Microbial Enzymatic Functions While Assessing and Predicting Long-Term Anthropogenic Perturbations," *Mar. Pollut. Bull.* 14:162–165 (1983).
67. Porter, K. G., H. Paerl, R. Hodson, M. Pace, J. Priscu, B. Riemann, D. Scavia, and J. Stockner. "Microbial Interactions in Lake Food Webs," in *Complex Interactions in Lake Communities*, S. R. Carpenter, Ed. (New York: Springer-Verlag, 1987), pp. 209–227.
68. Burton, G. A., Jr., and B. L. Stemmer. "Evaluation of Surrogate Tests in Toxicant Impact Assessments," *Tox. Assess.* 3:255–269 (1988).
69. Stebbing, A. R. D. "Hormesis — The Stimulation of Growth by Low Levels of Inhibitors," *Sci. Tot. Environ.* 22:213–234 (1982).
70. Calabrese, E. J., and M. McCarthy. "Hormesis: A New Challenge to Current Approaches for Estimating Cancer Risks Associated with Low Doses," *Water Res. Q.* 4:12–15 (1986).
71. Pratt, J. R., N. J. Bowers, B. R. Niederlehner, and J. Cairns, Jr. "Effects of Atrazine on Freshwater Microbial Communities," *Arch. Environ. Contam. Toxicol.* 17:449–457 (1988).
72. Griffiths, R. P., B. A. Caldwell, W. A. Broich, and R. Y. Morita. "Long-Term Effects of Crude Oil on Microbial Processes in Subarctic Marine Sediments," *Mar. Pollut. Bull.* 13:273–278 (1982).
73. Schiewe, M. H., E. G. Hawk, D. I. Actor, and M. M. Krahn. "Use of Bacterial Bioluminescence Assay to Assess Toxicity of Contaminated Marine Sediments," *Can. J. Fish. Aquat. Sci.* 42:1244–1248 (1985).
74. Dutka, B. J. "Method for Determining Acute Toxicant Activity in Water, Effluents and Leachates Using *Spirillum volutans*," *Tox. Assess.* 1:139–145 (1986).
75. Palmateer, G. A., D. E. McLean, M. J. Walsh, W. L. Kutus, E. M. Janzen, and D. E. Hocking. "A Study of Contamination of Suspended Stream Sediments With *Escherichia coli*," *Tox. Assess.* 4:377–397 (1989).
76. Williamson, K. S., and D. G. Johnson. "A Bacterial Bioassay for Assessment of Wastewater Toxicity," *Water Res.* 15:383 (1981).
77. Tam, T.-Y., and J. T. Trevors. "Toxicity of pentachlorophenol to *Azobacter vinelandii*," *Bull. Environ. Contam. Toxicol.* 27:230 (1981).
78. Flemming, C. A., and J. T. Trevors. "Copper Toxicity in Freshwater Sediment and *Aeromonas hydrophila* Cell Suspensions Measured Using an O_2 Electrode," *Tox. Assess.* 4:473–485 (1989).
79. Trevors, J. T., C. I. Mayfield, and W. E. Inniss. "A Rapid Toxicity Test Using *Pseudomonas fluorescens*," *Bull. Enviorn. Contam. Toxicol.* 26:433 (1981).
80. De Zwart, D., and W. Sloof. "The Microtox as an Alternative Assay in the Acute Toxicity Assessment of Water Pollutants," *Aquat. Toxicol.* 4:129 (1983).
81. Dutka, B. J., K. Jones, K. K. Kwan, H. Bailey, and R. McInnis. "Use of Microbial and Toxicant Screening Tests for Priority Size Selection of Degraded Areas in Water Bodies," *Water Res.* 22:503–510 (1988).
82. Babich, H., and G. Stotzky. "Heavy Metal Toxicity to Microbe-Mediated Ecologic Processes: A Review and Potential Application to Regulatory Policies," *Environ. Res.* 36:111–137 (1985).
83. Bihari, N., M. Najdek, R. Floris, R. Batel, and R. K. Zahn. "Sediment Toxicity Assessment Using Bacterial Bioluminescence: Effect of an Unusual Phytoplankton Bloom," *Mar. Ecol. Prog. Ser.* 57:307–310 (1989).

84. Hinwood, A. L., and M. J. McCormick. "The Effect of Ionic Strength of Solutes on EC_{50} Values Measured During the Microtox Test," *Tox. Assess.* 2:449–461 (1987).
85. True, C. J., and A. A. Heyward. "Relationships Between Microtox Test Results, Extraction Methods, and Physical and Chemical Compositions of Marine Sediment Samples," *Tox. Assess.* 5:29–45 (1990).
86. Bulich, A. A. "Bioluminescence Assays," in *Toxicity Testing Using Microorganisms,* Vol. 1, G. Bitton, and B. J. Dutka, Eds. (Boca Raton, FL, CRC Press, 1986) pp. 57–74.
87. Greene, J. C., W. E. Miller, M. K. Debacon, M. A. Long, and C. L. Bartels. "A Comparison of Three Microbial Assay Procedures for Measuring Toxicity to Chemical Residues," *Arch. Environ. Contam. Toxicol.* 14:657–667 (1985).
88. Dutka, B. J., and K. K. Kwan. "Studies on a Synthetic Activated Sludge Toxicity Screening Procedure with Comparison to Three Microbial Toxicity Tests," in *Toxicity Screening Procedures Using Bacterial Systems*, B. J. Dutka, and D. Liu, Eds. (New York: Marcel Dekker, 1984), pp. 125–138.
89. Giesy, J. P., R. L. Graney, J. L. Newsted, C. J. Rosiu, A. Benda, R. G. Kreis, Jr., and F. J. Horvath. "Comparison of Three Sediment Bioassay Methods Using Detroit River Sediments," *Environ. Toxicol. Chem.* 7:483–498 (1988).
90. Athey, L. A., J. M. Thomas, W. E. Miller, and J. Q. Word. "Evaluation of Bioassays for Designing Sediment Cleanup Strategies at a Wood Treatment Site," *Environ. Toxicol. Chem.* 8:223–230 (1989).
91. Brouwer, H., T. Murphy, and L. McArdle. "A Sediment-Contact Bioassay With *Photobacterium phosphoreum,*" *Environ. Toxicol. Chem.* 9:1353–1358 (1990).
92. Capone, D. G., D. D. Reese, and R. P. Kline. "Effects of Metals on Methanogenesis, Sulfate Reduction, Carbon Dioxide Evolution, and Microbial Biomass in Anoxic Salt Marsh Sediments," *Appl. Environ. Microbiol.* 45:1586–1591 (1983).
93. Furutani, A., and J. W. M. Rudd. "A Method for Measuring the Response of Sediment Microbial Communities to Environmental Perturbations," *Can. J. Microbiol.* 30:1408–1414 (1984).
94. Baker, J. H., and R. Y. Morita. "A Note on the Effects of Crude Oil on Microbial Activities in Stream Sediment," *Environ. Pollut.* 31:149–157 (1983).
95. Buikema, A. L., C. L. Rutherford, and J. Cairns, Jr. "Screening Sediments for Potential Toxicity by In Vitro Enzyme Inhibition," in *Contaminants and Sediments,* Vol. 1, R. A. Baker, Ed. (Ann Arbor, MI: Ann Arbor Science, 1980), pp. 463–475.
96. Sayler, G. S., M. Puziss, and M. Silver. "Alkaline Phosphatase Assay for Freshwater Sediments: Application to Perturbed Sediment Systems," *Appl. Environ. Microbiol.* 38:922–927 (1979).
97. Sayler, G. S., T. W. Sherrill, R. E. Perkins, L. M. Mallory, M. P. Shiaris, and D. Pedersens. "Impact of Coal-Coking Effluent on Sediment Microbial Communities: A Multivariate Approach," *Appl. Environ. Microbiol.* 44:1118–1129 (1982).
98. Sayler, G. S., R. E. Perkins, T. W. Sherrill, B. K. Perskins, M. C. Reid, M. S. Shields, H. L. Kong, and J. W. Davis. "Microcosm and Experimental Pond Evaluation of Microbial Community Response to Synthetic Oil Contamination in Freshwater Sediments," *Appl. Environ. Microbiol.* 46:211–219 (1983).

99. Heitkamp, M. A., and B. T. Johnson. "Impact of an Oil Field Effluent on Microbial Activities in a Wyoming River," *Can. J. Microbiol.* 30:786–792 (1984).
100. Pratt, J. R., and J. Cairns. "Functional Groups in the Protozoa: Roles in Differing Ecosystems," *J. Protozool.* 32:415–423 (1985).
101. Picken, L. E. R. "The Structure of Some Protozoan Communities," *J. Ecol.* 25:368–384 (1937).
102. Dive, D., S. Robert, E. Angrand, C. Bel, H. Bonnemain, L. Brun, Y. Demarque, A. Le Du, R. El Bouhouti, M. N. Fourmaux, L. Guery, O. Hanssens, and M. Murat. "A Bioassay Using the Measurement of the Growth Inhibition of a Ciliate Protozoan: *Colpidium campylum* Stokes," in *Environmental Bioassay Techniques and Their Application*, M. Munawar, G. Dixon, C. I. Mayfield, T. Reynoldson, and H. Sadar, Eds., *Hydrobiologia* 188/189: 181–188 (1989).
103. Henebry, M. S., and P. E. Ross. "Use of Protozoan Communities to Assess the Ecotoxicological Hazard of Contaminated Sediments," *Tox. Assess.* 4:209–227 (1989).
104. Bott, T. L., and L. A. Kaplan. "Bacterial Biomass, Metabolic State, and Activity in Stream Sediments: Relation to Environmental Variables and Multiple assay Comparisons," *Appl. Environ. Microbiol.* 50:508–522 (1985).
105. Carlton, R. G., and M. J. Klug. "Spatial and Temporal Variations in Microbial Processes in Aquatic Sediments: Implications for the Nutrient Status of Lakes," in *Sediments: Chemistry and Toxicity of In-Place Pollutants*, R. Baudo, J. Giesy,, and H. Muntau, Eds. (Boca Raton, FL: Lewis Publishers, 1990), pp. 107–130.
106. Burton, G. A., Jr., A. Drotar, J. M. Lazorchak, and L. L. Bahls. "Relationship of Microbial Activity and *Ceriodaphnia* Responses to Mining Impacts on the Clark Fork River, Montana," *Arch. Environ. Contam. Toxicol.* 16:523–530 (1987).
107. Genter, R. B., D. S. Cherry, E. P. Smith, and J. C. Cairns, Jr. "Attached-Algal Abundance Altered by Individual and Combined Treatments of Zinc and pH," *Environ. Toxicol. Chem.* 7:723–733 (1988).
108. Lange-Bertalot, H. "Pollution Tolerance of Diatoms as a Criterion for Water Quality Estimation," *Nova Hedwigia, Beiheft* 64:285–304 (1979).
109. Peterson, B. J., J. E. Hobbie, T. L. Corliss, and K. Kriet. "A Continuous-Flow Periphyton Bioassay: Tests of Nutrient Limitation in a Tundra Stream," *Limnol. Oceanogr.* 28:583–591 (1983).
110. Yount, J. D., and J. E. Richter. "Effects of Pentachlorophenol on Periphyton Communities in Outdoor Experimental Streams," *Arch. Environ. Contam. Toxicol.* 15:51–60 (1986).
111. Samoiloff, M. R., J. Bell, D. A. Birkholz, G. R. Webster, E. G. Arnott, R. Pulak, and A. Madrid. Combined Bioassay-Chemical Fraction Scheme for the Determination and Ranking of Toxic Chemicals in Sediments," *Environ. Sci. Technol.* 17:329–334 (1983).
112. Ross, P. E., L. C. Burnett, and M. S. Henebry. "Chemical and Toxicological Analyses of Lake Calumet (Cook County, Illinois) Sediments," Report No. HWRIC RR-036, Illinois Hazardous Waste Research and Information Center, Champaign, IL (1989).
113. Williams, P. L., and D. B. Dusenbery. "Aquatic Toxicity Testing Using the Nematode, *Caenorhabditis elegans*," *Environ. Toxicol. Chem.* 9:1285–1290 (1990).
114. Brinkhurst, R. O. *The Benthos of Lakes* (London: Macmillan Press, 1974).

115. Reynoldson, T. B. "Interactions Between Sediment Contaminants and Benthic Organisms," *Hydrobiologia* 149:53–66 (1987).
116. Karickhoff, S. W., and K. R. Morris. "Impact of Tubificid Oligochaetes on Pollutant Transport in Bottom Sediments," *Environ. Sci. Technol.* 19:51–56 (1985).
117. Jernelöv, A. "Release of Methyl Mercury from Sediments with Layers Containing Inorganic Mercury at Different Depths," *Limnol. Oceanogr.* 15:958–960 (1970).
118. Bailey, H. C., and D. H. Liu. "*Lumbriculus variegatus,* a Benthic Oligochaete, as a Bioassay Organism," in *Aquatic Toxicology,* STP 707, (Philadelphia: American Society of Testing and Materials, 1980), pp. 205–215.
119. Chapman, P. M., and R. O. Brinkhurst. "Lethal and Sublethal Tolerances of Aquatic Oligochaetes with Reference to Their Use as a Biotic Index of Pollution," *Hydrobiologia* 115:139–144 (1984).
120. Keilty, T. J., D. S. White, and P. F. Landrum. "Short-Term Lethality and Sediment Avoidance Assays with Endrin-Contaminated Sediment and Two Oligochaetes from Lake Michigan," *Arch. Environ. Contam. Toxicol.* 17:95–101 (1988).
121. McMurthy, M. J. "Avoidance of Sublethal Doses of Copper and Zinc by Tubificid Oligochaetes," *J. Great Lakes Res.* 10:267–272 (1984).
122. Keilty, T. J., and P. F. Landrum. "Population-Specific Toxicity Responses by the Freshwater Oligochaete, *Stylodrilus heringianus,* in Natural Lake Michigan Sediments," *Environ. Toxicol. Chem.* 9:1147–1154 (1990).
123. Bousfield, E. L. "Fresh-Water Amphipod Crustaceans of Glaciated North America," *Can. Field Nat.* 72:55–113 (1958).
124. deMarch, B. G. E. "*Hyalella azteca* (Saussure)," in *A Manual for the Culture of Selected Freshwater Invertebrates*, S. G. Lawrence, Ed., *Can. Spec. Publ. Fish. Aquat. Sci.* 54:61–78 (1981).
125. Thomas, J. D. "A Survey of Gammarid Amphipods of the Barataria Bay, Louisiana Region," *Cont. Mar. Sci.* 20:87–100 (1976).
126. Galat, D. L., M. Coleman, and R. Robinson. "Experimental Effects of Elevated Salinity on Three Benthic Invertebrates in Pyramid Lake, Nevada," *Hydrobiologia* 158:133–144 (1988).
127. Timms, B. V., U. T. Hammer, and J. W. Sheard. "A Study of Benthic Communities in Some Saline Lakes in Saskatchewan and Alberta, Canada," *Int. Rev. ges. Hydrobiol.* 71:759–777 (1986).
128. Cooper, W. E. "Dynamics and Production of a Natural Population of a Fresh-Water Amphipod, *Hyalella azteca,*" *Ecol. Monogr.* 35:377–394 (1965).
129. Strong, D. R. "Life History Variation Among Populations of an Amphipod (*Hyalella azteca*)," *Ecology* 53:1103–1111 (1972).
130. deMarch, B. G. E. "The Effects of Constant and Variable Temperatures on the Size, Growth, and Reproduction of the Freshwater Amphipod *Hyalella azteca* (Saussure)," *Can. J. Zool.* 56:1801–1806 (1978).
131. Borgmann, U., and M. Munawar. "A New Standardized Sediment Bioassay Protocol Using the Amphipod *Hyalella azteca* (Saussure)," *Hydrobiologia* 188/189:425–531 (1989).
132. Maciorowski, H. D. "Comparison of the Lethality of Selected Industrial Effluents Using Various Aquatic Invertebrates Under Laboratory Conditions," Technical Report No. CEN/T-75-3, Department of the Environment, Fisheries and Marine Service, Central Region, Winnipeg, Canada (1975).

133. Borgmann, U., K. M. Ralph, and W. P. Norwood. "Toxicity Test Procedures for *Hyalella azteca*, and Chronic Toxicity of Cadmium and Pentachlorophenol to *H. azteca, Gammarus fasciatus,* and *Daphnia magna,*" *Arch. Environ. Contam. Toxicol.* 18 (1989).
134. Landrum, P. F., and D. Scavia. "Influence of Sediment on Anthracene Uptake, Depuration, and Biotransformation by the Amphipod *Hyalella azteca,*" *Can. J. Fish. Aquat. Sci.* 40:298–305 (1983).
135. Nebeker, A. V., S. T. Onjukka, M. A. Cairns, and D. F. Krawczyk. "Survival of *Daphnia magna* and *Hyalella azteca* in Cadmium-Spiked Water and Sediment," *Environ. Toxicol. Chem.* 5:933–938 (1986).
136. Nebeker, A. V., and C. E. Miller. "Use of the Amphipod Crustacean *Hyalella azteca* in Freshwater and Estuarine Sediment Toxicity Tests," *Environ. Toxicol. Chem.* 7:1027–1033 (1988).
137. Imlay, M. J. "Use of Shells of Freshwater Mussels in Monitoring Heavy Metals and Environmental Stresses: A Review," *Mamacol. Rev.* 15:1–14 (1982).
138. Wade, D. C., and R. G. Hudson. "The Use of Juvenile Freshwater Mussels as a Laboratory Test Species for Evaluating Environmental Toxicity," Abstr. Annu. Meet. Soc. Environ. Toxicol. & Chem., Toronto, Ontario, No. P232, p. 247 (1989).
139. Gentile, J. H., and K. J. Scott. "The Application of a Hazard Assessment Strategy to Sediment Testing: Issues and Case Study," in *Fate and Effects of Sediment-Bound Chemicals in Aquatic Systems*, K. L. Dickson, A. W. Maki, and W. A. Brungs, Eds. (New York: Pergamon Press, 1984) pp. 167–182.
140. Dickson, K. L., W. T. Waller, J. H. Kennedy, W. R. Arnold, W. P. Desmond, S. D. Dyer, J. F. Hall, J. T. Knight, Jr., D. Malas, M. L. Martinez, and S. L. Matzner. "A Water Quality and Ecological Survey of the Trinity River. Vol. 1," City of Dallas Water Utilities, Dallas, TX (1989).
141. Scherer, E., and R. E. McNicol. "Behavioral Responses of Stream-Dwelling *Acroneuria lycorias* (Ins., Plecopt.) Larvae to Methoxychlor and Fenitrothion," *Aquat. Toxicol.* 8:251–263 (1986).
142. Farris, J. L., J. H. van Hassel, S. E. Belanger, D. S. Cherry, and J. Cairns, Jr. "Application of Cellulolytic Activity of Asiatic Clams (*Corbicula* sp.) to In-Stream Monitoring of Power Plant Effluents," *Envrion. Toxicol. Chem.* 7: 701–713 (1988).
143. Keller, A. E., and S. G. Zam. "Simplification of In Vitro Culture Techniques for Freshwater Mussels," *Environ. Toxicol. Chem.* 9:1291–1296 (1990).
144. Fremling, C. D., and W. L. Mauck. "Methods for Using Nymphs of Burrowing Mayflies (Ephemeroptera, Hexagenia) as Toxicity Test Organisms," in *Aquatic Invertebrate Bioassays*, STP 715, A. L. Buikema, Jr., and J. Cairns, Jr., Eds. (Philadelphia: American Society of Testing and Materials, 1980), pp. 81–97.
145. Hoke, R. A., and B. L. Prater. "Relationship of Percent Mortality of Four Species of Aquatic Biota from 96-Hour Sediment Bioassays of Five Lake Michigan Harbors and Elutriate Chemistry of the Sediments," *Bull. Environ. Contam. Toxicol.* 25:394–399 (1980).
146. Malueg, K. W., G. S. Schuytema, J. H. Gakstatter, and D. F. Krawczyk. "Effect of *Hexagenia* on *Daphnia* Responses in Sediment Toxicity Tests," *Environ. Toxicol. Chem.* 2:73–82 (1983).
147. Gerould, S., and S. P. Gloss. "Mayfly-Mediated Sorption of Toxicants into Sediments," *Environ. Toxicol. Chem.* 5:667–673 (1980).

148. Gauss, J. D., P. E. Woods, R. W. Winner, and J. H. Skillings. "Acute Toxicity of Copper to Three Life Stages of *Chironomus tentans* as Affected by Water Hardness-Alkalinity," *Environ. Bull. (Ser. A)* 37:149–157 (1985).
149. Williams, K. A., D. W. J. Green, D. Pascoe, and D. E. Gower. "The Acute Toxicity of Cadmium to Different Larval Stages of *Chironomus riparius* (Diptera: Chironomidae) and Its Ecological Significance for Pollution Regulation," *Oecologia* 70:362–366 (1986).
150. Nebeker, A. V., M. A. Cairns, and C. M. Wise. "Relative Sensitivity *Chironomus tentans* Life Stages to Copper," *Environ. Toxicol. Chem.* 3:151–158 (1984).
151. Ingersoll, C. G., F. J. Dwyer, and T. W. May. "Toxicity of Inorganic and Organic Selenium to *Daphnia magna* (Cladocera) and *Chironomus riparius* (Diptera)," *Environ. Toxicol. Chem.* 9:1171–1181 (1990).
152. Burton, G. A., Jr. "Assessing Toxicity of Freshwater Sediments," *Environ. Toxicol. Chem.* 10: 1585–1627 (1991).

CHAPTER 11

Biomarkers in Hazard Assessments of Contaminated Sediments

William H. Benson and Richard T. Di Giulio

INTRODUCTION

Biomarkers can be defined as biochemical, physiological, or pathological responses measured in individual organisms, which provide information concerning exposures to environmental contaminants and/or sublethal effects arising from such exposures. The focus of ecotoxicology, however, is the elucidation of effects at levels of organization higher than that of the individual — that is, upon populations, communities and ecosystems. Furthermore, the uncertainties associated with attempting to predict effects at higher levels of organization, based upon responses at the organismal level, likely will remain insurmountable for the foreseeable future. Nonetheless, there exists considerable interest in the development of biomarkers, due to the recognition of the significant roles they can play in ecological assessments, including assessments of contaminated sediments.

Table 1 presents several underlying considerations for employing biomarkers in hazard assessments of contaminated sediments. First, biomarkers, particularly those associated with mechanisms of action, can provide the most sensitive measures of exposure and response achievable for many contaminants. This feature of sensitivity suggests the utility of biomarkers as early warning signals of ecological degradation and for delineating zones of contaminant impact. Additionally, many contaminants, at concentrations typically encountered in natural systems, exert relatively subtle

Table 1. Rationale for Use of Biomarkers in Contaminated Sediment Toxicity Assessment

Sensitive measures of exposure and response
Early warning signals of ecological degradation
Delineation of zones of contaminant impact
Detection of subtle detrimental effects
Short response time
Bridge between hazard assessments addressing ecological and human health

detrimental effects that are difficult to discern with direct measurements at higher levels of organization in the time frames provided by hazard assessments. In addition to sensitivity, most biomarkers exhibit relatively short response times (hours, days). More fundamentally, the biomarker approach can provide a bridge between hazard assessments addressing ecological and human health. That is, concomitant with their role in ecological assessments, biomarkers can play a role analogous to that of the canary in the coal mine. This role is particularly significant in the context of environmental contamination by carcinogens. These chemicals can produce profound and quantifiable biochemical and pathological responses in organisms (e.g., benthic fish, bivalves) without affecting higher levels of organization. In its various roles, the biomarker approach may prove highly appropriate for assessments of sediments that often serve as the major reservoir for persistent contaminants in aquatic systems. The fundamental biology underlying a number of biomarkers and their potential utility in ecological hazard assessments have been the subject of a recent symposium[1] as well as workshops.[2,3]

This chapter describes selected biomarkers that may be of particular potential utility in hazard assessments of contaminated sediments. In addition, an overview of considerations required in the actual application of biomarkers is provided. For more detailed descriptions of most of the biomarkers described in this chapter, including methodologies, the reader is referred to Huggett et al.[3]

SELECTED BIOMARKERS

This section describes a number of organismal-level responses that appear to have significant potential as biomarkers. Clearly, these biomarkers do not represent a unique set of biological responses, but simply an aggregate of responses elucidated by scientists concerned with fundamental toxicological phenomena that have potential in the context of hazard assessment, based upon general consensus. Criteria underlying this consensus, not all of which are necessarily applicable to every biomarker described, include their sensitivity, reliability, relationship to mechanisms of action, importance to organismal health, technical difficulty, and practicality for routine use in typical hazard assessments (Table 2). The biomarkers described here include responses associated with (1) various biochemical aspects of contaminant

Table 2. Criteria for Selection of Biomarkers

Sensitivity
Reliability
Relationship to mechanisms of action
Relative ease of measurement
Practicality

biotransformation, mode of action, and adaptation; (2) specific indices of DNA damage (genotoxicity); (3) responses of the immune system to toxicants; (4) physiological and nonspecific responses; and (5) structural changes in tissues due to contaminants (histopathology).

Biochemical Indices

Biochemical responses to contaminants have received the greatest amount of attention as potential biomarkers for environmental quality. This interest is based upon a number of factors, including the proximity of some biochemical responses to mechanisms of action, and the short response times generally required for these effects. Collectively, these factors suggest the potential for biochemical responses to provide very sensitive markers for environmental degradation, an attribute that has motivated considerable interest in the biomarker concept. Following are described a number of biochemical systems of potential utility in assessments of contaminated sediments. For more thorough descriptions of these systems in the context of biomarkers, see Stegeman et al.[4]

Biotransformation Enzymes

A complex array of enzyme systems functions in many organisms to transform lipophilic organic compounds to more hydrophilic products (for reviews, see Sipes and Gandolfi[5] and Buhler and Williams[6]). These systems have been primarily examined in mammalian liver, but components occur in many other tissues of vertebrates (such as kidney, intestinal, gonadal, lung, and gill tissue) and invertebrates. An overall strategy of these systems is to enhance water solubility, and hence excretability, of lipophilic compounds, including both endogenous substrates (e.g., steroid hormones) and foreign compounds (such as plant secondary products and anthropogenic xenobiotics). In the case of lipophilic toxicants, these systems generally result in less toxic products; however, enhanced toxicities can occur. In fact, these systems are sufficiently complex to allow for both detoxifications and enhanced toxicities for a given compound in an individual organism. The interest in these systems as biomarkers stems from the ability of many xenobiotics to markedly enhance components of this system. In addition to implications for biomonitoring, inductions of these systems (i.e., increases in the proteins underlying catalytic activities) have important bearings upon more basic issues, such as adaptations to polluted environments and increased risks from activation-requiring toxicants such as procarcinogens.

Biotransformation systems are generally treated as two subsystems: the phase I and phase II systems. Phase I systems essentially serve to add or expose reactive functional groups on lipophilic substrates. Hydroxylations of aliphatic or aromatic structures are classic examples; dealkylations of N-, O-, or S-based substituents of organic compounds provide other important examples of phase I activities. Phase I products, due to the reactive functional groups, are generally somewhat more water soluble than are starting substrates. However, in the overall scheme of metabolism and excretion, the key role of phase I metabolism is to provide suitable substrates for phase II metabolism. In phase II, various enzymes catalyze the conjugation of phase I products (and other suitable substrates) with highly polar endogenous compounds, such as sulfate, glutathione, and glucuronic acid. The resulting phase II product is generally far more water soluble than the original compound and is readily excreted.

Central to phase I activities are reactions catalyzed by a family of enzymes referred to as cytochrome P-450s (or P450); these reactions are called mixed-function oxidase (MFO) or monooxygenase reactions.[7] The bulk of P450 is associated with microsomes, which are artifacts of smooth endoplasmic reticula obtained in the course of differential centrifugation. The majority of interest in biotransformation enzymes as biomarkers emanates from the remarkable inducibility of P450. Various isozymes of P450 are induced in species-specific and chemical-specific manners. Of particular interest in the context of aquatic contamination is the pronounced inducibility in fish of a particular gene family of P450, referred to as P450 IA.[4] This family is induced by polycyclic aromatic hydrocarbons (PAHs) and a variety of halogenated hydrocarbons (notably certain chlorinated biphenyls, dibenzofurans, and dibenzodioxins). These inducers clearly include important sediment-associated contaminants.

A large and growing number of field studies have indicated the utility of P450 as a biomarker for aquatic contamination (see review by Payne et al.[8]). Recent examples include benthic fish inhabiting portions of the Puget Sound exhibiting elevated sediment concentrations of PAHs,[9] and coastal fish inhabiting a bleached kraft mill effluent-receiving system.[10] However, when considering the use of P450 as a biomarker, it is important to consider the nature of contamination anticipated (i.e., the presence/absence of known inducers), potential environmental influences such as temperature, potential physiological influences such as sex and reproductive status, and potential inhibitory effects of some contaminants such as hepatotoxicants.[11]

A number of techniques have been employed for examining P450 inductions. The most promising appear to be measures of catalytic activities, quantification of specific isozyme proteins immunochemically, and quantification of messenger RNAs (mRNA) for these proteins with complementary DNA (cDNA) probes.[4] In fish, the most widely employed and readily performed techniques are measurements of enzyme activities, particularly aryl hydrocarbon hydroxylase (AHH) and ethoxyresorufin O-deethylase (EROD). These activities are highly associated with the P450 IA proteins. More recently, quantifications of IA isozymes and their mRNAs in fish have been performed with the use of specific antibodies and cDNA probes, respectively. While enzyme activities appear to provide generally greater sensitivity and are easily measured, they are more sensitive to sample handling and storage and inhibitory factors that may mask inductions. Ideally, P450 responses are best examined with a combination of enzyme activity and isozyme determinations.

In addition to P450, there has been interest in the utility of phase II biotransformation enzymes as biomarkers in aquatic animals. However, results to date have been less encouraging than those obtained for P450, and the use of phase II activities as biomarkers of aquatic contamination is likely to be more limited.[4] Nevertheless, the critical role these enzymes play in metabolism and detoxification motivates continued basic research on these systems; such research may reveal greater utility for biomonitoring than indicated currently.

Key phase II enzymes include glutathione S-transferases (GST), uridine diphosphate glucuronosyltransferases (UDPGT), and sulfotransferases. These catalyze the conjugations of substrates with glutathione, glucuronic acid, and sulfate, respectively.[6,12-14] While not a conjugating enzyme, epoxide hydrolase is often treated with phase II enzymes. This enzyme catalyzes the insertion of a water molecule into an epoxide, giving rise to a diol. As with the conjugating reactions, this reaction can serve to detoxify reactive intermediates, including some P450 metabolites.

Metallothionein

Metallothioneins (MT) comprise the major metal-binding proteins in animals and occur in plants and prokaryotes as well. Mammalian MT is characterized by features of low molecular weight (about 60 amino acids), high cysteine content (about one third of amino acids), and a lack of aromatic amino acids. The precise function of MT is unclear, but it appears to play a key role in Cu and Zn homeostasis, the nutrient metals with which it is typically associated under normal conditions. However, under in vitro conditions, most metals can bind to the protein. Of particular interest is the ability of a number of metals to induce MT synthesis in vivo. In addition to the essential metals Cu and Zn, other metals of importance, mainly as contaminants, such as Cd, Hg and Ag, can readily induce MT synthesis in a variety of animals. Other factors, such as stress response, cold, and hypoxia, can also induce MT.[15] For detailed information on the chemistry and biology of MT, see reviews by Hamer[16] and Engel and Brouwer.[17]

The ability of relatively low exposures to trace metal contaminants (e.g., Cd) to induce MT has generated interest in the use of MT as a biomarker for metal pollution. Also, while the function of MT in the metabolism of nonnutrient metals, such as Cd, is not fully understood, it appears that MT often plays a protective function by sequestering these metals and inhibiting interactions with sensitive cellular components, such as enzymes. Cellular toxicities may ensue after the metal-binding capacity of MT has been exceeded. Therefore, it has been proposed that MT measures may provide considerably more information about potential health hazards of metals in exposed animals than tissue metal residues alone. However, more information concerning the physiological functions of MT, with respect to both essential metal homeostasis and nonessential metal metabolism, is required before the potential utility of MT as a biomarker can be realized.

Stress Proteins

Stress proteins include a diverse array of proteins that are inducible by a variety of stressors. As with MT, the functions of these proteins are not fully understood, but appear to include aspects of ameliorating cellular injury following environmental perturbation, such as the resolubilization and transport of denatured proteins. A key group of these proteins are referred to as heat-shock proteins because of their association with heat shock and thermal acclimation in early studies. Other natural stressors, such as anoxia and salinity stress, can also induce heat-shock proteins. Another group is referred to as glucose-regulated proteins; these are induced by glucose or oxygen deprivation. Reviews of the basic biology of stress proteins are provided by Subjeck and Shyy[18] and Welch.[19]

A number of studies have indicated that many stress proteins, including representatives of both groups, are readily induced by a variety of environmental contaminants, hence their potential role as biomarkers. This role, and techniques for stress protein quantification, are provided by Sanders[20] and Stegeman et al.[4] The utility of stress proteins as biomarkers is presently a very active area of inquiry. It appears that several proteins are induced by environmentally realistic concentrations of a variety of metals and organic contaminants in aquatic vertebrates and invertebrates. However, considerable research is still required to determine the chemical specificities of particular proteins, the relationships between inductions and organismal health, and aspects of these responses, such as length of exposure, that will determine protocols for field monitoring.

Oxidative Stress

Oxidative stress, or oxygen toxicity, refers to injurious cellular effects due to activated oxygen species ("oxyradicals"). Important oxyradicals include the superoxide anion radical (O_2^-), hydrogen peroxide (H_2O_2), the hydroxyl radical ($\cdot OH$), and singlet oxygen ($^1O_2$). These first three compounds are, respectively, the one-, two-, and three-electron reduction products of molecular oxygen (O_2); the four-electron product is water. Oxyradicals are produced in the course of aerobic metabolism, and the cells of all aerobic organisms examined are equipped with antioxidants that serve to detoxify these compounds. Oxidative stress ensues when the production of oxyradicals outstrips the capacities of antioxidant defenses.

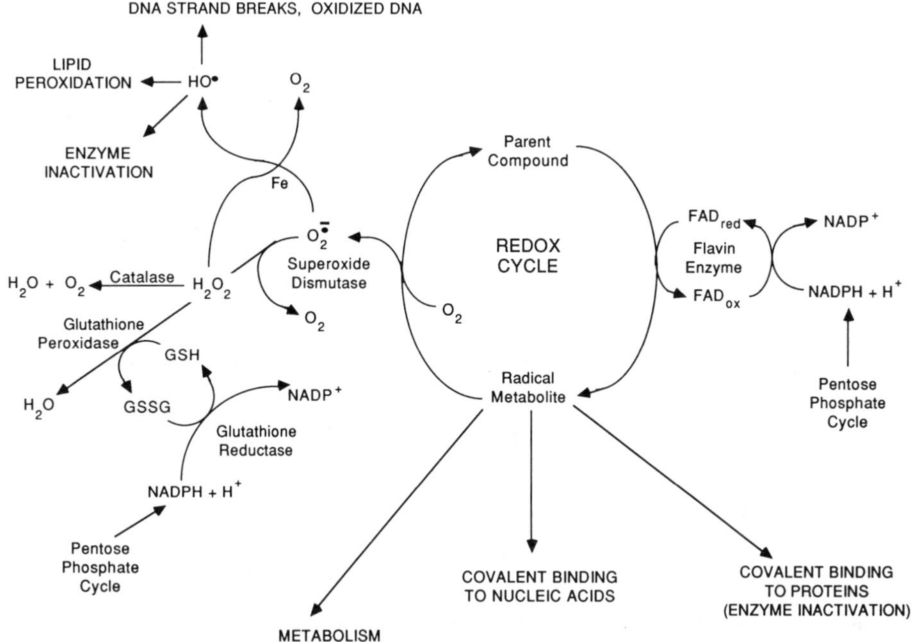

Figure 1. Overview of redox cycling, antioxidant defenses, and toxicological sequalae. (Adapted from Kappus (1987).[22])

In recent years a large literature has developed indicating the ability of a variety of compounds, including many environmental contaminants, to enhance the production of oxyradicals and exert oxidative stress. A key pathway by which chemicals can enhance the flux of oxyradicals is referred to as redox cycling (Figure 1). In this cycle, a redox-active compound first accepts an electron from an intracellular reductant (e.g., NADPH), typically in an enzymatically catalyzed reaction (e.g., by microsomal NADPH-P450 reductase). This reaction produces a radical metabolite of the parent compound (i.e., containing an unshared electron). This radical can donate the unshared electron to O_2, producing O_2– and the parent compound, which can undergo another cycle. At each turn of the cycle, two potentially deleterious events occur — a reductant has been oxidized and an oxyradical has been generated (Figure 1). Redox active compounds include aromatic diols and quinones, nitroaromatics, aromatic hydroxylamines, bipyridyls (e.g., paraquat and diquat), and various transition metal chelates. For overviews of free-radical biology and oxidative stress, see Halliwell and Gutteridge[21] and Kappus.[22] Reviews focusing upon these phenomena in aquatic systems are provided by Di Giulio et al.[23] and Winston and Di Giulio.[24]

Increased productions of oxyradicals have been shown to produce a variety of responses, both adaptive and toxic, and a number of these have received attention as potential biomarkers. Adaptive responses are largely comprised by inductions of antioxidant enzymes, such as superoxide dismutase, peroxidases, and catalase (Figure 1). The tripeptide glutathione (GSH), alluded to under *Biotransformation Enzymes,* also serves an important antioxidant function, and its production can be enhanced in response to oxyradicals. Inductions of antioxidants can provide sensitive early warning signals of incipient oxidative stress. Biochemical manifestations of oxygen toxicity may also provide useful biomarkers. Important responses here include lipid peroxidation, DNA oxidations (described under *Indices of Genotoxicity* below), and perturbed cellular redox status (e.g., lowered NADPH/$NADP^+$ or GSH/GSSG ratios).

The use of indices of oxidative stress as biomarkers in aquatic systems is in its infancy. However, results from mammalian studies and studies of air pollutant effects on plants suggest a strong potential for this approach. For more detailed discussions of these biomarkers, including techniques and examples, see Di Giulio.[25]

Indices of Genotoxicity

In the context of hazard assessments of contaminated sediments, the elucidation of genotoxic effects provides a particularly compelling role for the biomarker approach. Genotoxins, such as PAHs, aromatic amines, and carbazoles, comprise very important, ubiquitous, sediment-associated contaminants and produce toxicities, such as genotoxicity, that are often relatively subtle and therefore difficult to discern with traditional approaches. However, highly sensitive probes for detecting damage to DNA have been developed and, in some cases, applied to aquatic systems. In addition to the information these biomarkers can provide concerning aquatic ecosystem health, they also can provide a link to human health assessments, which in the case of carcinogens can be very important.

A number of approaches are available for assessing genotoxicity. Broadly, these include direct structural alterations to DNA, perturbations in DNA repair processes, and genomic mutations.[26] In aquatic studies for determining structural DNA damage, the techniques that have received the most attention are the quantification of bulky adducts by phosphorous-32 postlabeling (PPL) and the measurement of DNA strand breaks by alkaline elution unwinding. The formation of DNA adducts is believed to be a critical event underlying the mutagenicity of many chemicals. PPL provides a very sensitive assay for measuring bulky adducts (e.g., PAHs) arising from complex media such as contaminated sediments, and this approach has been employed successfully to assess DNA damage in benthic fish inhabiting predominantly PAH-contaminated systems.[27,28] The alkaline elution technique provides an index of DNA strand breaks by determining the rate of unwinding. The rate of unwinding, which is facilitated by alkaline conditions, is enhanced by the presence of strand breaks. These strand breaks may arise as a direct result of adduct formation or in the course of DNA repair. This technique, and its application in fish, is described by Shugart.[29]

Other techniques for assessing DNA damage, that merit consideration include quantifications of minor nucleosides (such as 5-methyldeoxycytodine, 5-MC) and measures of oxidized bases (such as 8-hydroxydeoxyguanosine, 8-OHdG). 5-MC appears to play an important regulatory role in gene expression, and some compounds have been shown to result in reduced levels of 5-MC (hypomethylations).[30] This assay has been adapted recently for aquatic animals by Shugart.[31] As mentioned earlier, oxyradicals can damage DNA and result in oxidized bases, such as 8-OHdG, for which very sensitive assays have been developed.[32] This base was recently measured in benthic fish from the Puget Sound, and higher 8-OHdG concentrations were observed in tumorous hepatic tissue than in nontumorous tissue.[33]

In addition to the quantification of structural DNA alterations, a number of techniques exist for assessing chemical effects at the level of the genome, many of which have been employed on feral animals. These include (with referenced examples of applications to biomonitoring) standard chromosomal analysis for aberrations,[34] sister-chromatid exchange,[35] and micronuclei formation.[36] A particularly exciting area of inquiry, due to its relationship to current theories of chemical carcinogenesis, is oncogene activation. Recently, c-Ki-*ras* oncogenes have been reported in benthic fish from polluted sites in the Boston Harbor[37] and the Hudson River.[38]

Approaches other than those described here are also available. For a thorough discussion of biomarkers for genotoxicity, see Shugart et al.[26] This is a very active area of research presently, and new approaches are likely to emerge in the near future. And again, this topic comprises a biomarker application of great relevance to the concern for contaminated sediments.

Immunological Responses

The evaluation of immunological responses has potential for assessing the toxicological impact of contaminated sediments. Since there is considerable information concerning the cellular, humoral, and molecular components of the immune system, immune responses may be well suited for comparative analyses that emphasize mechanisms of action. A summary of selected immune parameters used in toxicity assessment is presented. For a more detailed description on the organization of the immune system as well as a thorough review of the current and potential use of immunological responses, the reader is referred to Stolen et al.[39] and Weeks et al.[40]

Phagocytosis is an important parameter of nonspecific immunity, as it is well conserved throughout phylogeny and important in immune surveillance. Therefore, assays such as fish macrophage function have potential as immunological biomarkers useful in assessing the impact of contaminants in aquatic ecosystems.[40] Macrophages are an important part of the cell-mediated immune system of fish and function to protect the host by phagocytizing foreign material, including disease-causing agents.[41] Since phagocytosis is an indicator of macrophage function, it is of interest to ascertain the effects of contaminants on the normal phagocytic activity of macrophages. Weeks et al.[42-44] and Warinner et al.[45] have demonstrated that fish exposed to PAHs, both in the field and in the laboratory, show significant alterations of the kidney macrophage functions, such as phagocytosis of bacteria and yeast, chemotaxis, pinocytosis, chemiluminescence, and the accumulation of melanin.

Humoral antibody assays also have been used to determine altered immune capacity caused by environmental contaminants. The humoral-mediated immune response is characterized by the production of antibody molecules that react specifically with antigen. Assays measuring humoral or circulating antibody concentrations may be conducted by taking blood samples without killing the organism. Robohm[46] demonstrated that cadmium stimulated antibody production in cunner (*Tautogalabrus adspersus*), but suppressed it in striped bass (*Morone saxatilis*). In field investigations, confounding results were sometimes obtained due to interference by background antibody levels. O'Neill[47] also demonstrated immunosuppression of humoral antibody titers in brown trout (*Salmo trutta*) exposed to heavy metals. A reduction in antibody-producing cells following bath immunization was observed in rainbow trout that were exposed to phenol, formalin, or detergent solutions.[48]

Due to the complexity of the immune system and the number of assays available, a tiered system for assessing the effects of contaminants on the immune system has been proposed.[49-52] The battery of immune-function tests presented in Table 3 have been used to determine the mechanism of action of various individual chemicals in the laboratory. As such, these tests have been used primarily in controlled laboratory exposures. In addition, several of the assays require laboratory facilities that are difficult to provide at a field site, and capture and handling stress may cause immunosuppression. Therefore, caution must be taken when considering these tests for use in assessing the toxicity of contaminated sediments.

Tier I

Tier I provides a general screening of immune function and comprises tests (Table 3) that are relatively easy as well as inexpensive and require little in the way of specialized equipment. Complete and differential blood cell counts, in addition to the weights and histological integrity of lymphoid organs, provide a general overview of the adequacy of the structural parts of the immune system. Natural killer-cell activity (as measured by a chromium release assay), macrophage phagocytosis of fluorescent beads or stained yeast cells, the killing of bacteria, and the measurement of lysozyme activity by agar gel immunodiffusion can be performed with almost any organism and provide information concerning nonspecific immune responses.[40]

Tier II

Tier II tests comprise a comprehensive evaluation of all the components of the immune response, such as immune cell quantitation, native immunoglobin

Table 3. Tier Approach to Assess the Effects of Contaminants on the Immune System

Tier I — General Screening
Complete and differential blood cell counts
Organ weights
Histological integrity
NK cell activity
Macrophage phagocytosis and killing
Lysozyme activity
Tier II — Comprehensive Evaluation
Immune cell quantitation
Native immunoglobin quantitation
Plaque-forming cell assay
Lymphocyte blastogenesis
Mixed leukocyte response
Cytotoxic T cell activity
Macrophage responses
Tier III — Host Resistance Challenge
Mortality
Bacteremia/viremia/parasitemia/tumor quantitation and duration
Specific antibody quantitation

quantitation, plaque-forming cell assay, lymphocyte blastogenesis, mixed leukocyte response, cytotoxic T-cell (leukocyte) activity, and macrophage responses (melanin accumulation, chemotaxis, pinocytosis) (Table 3). There is some question, however, as to whether some Tier II tests that have been developed in mammalian species are applicable to aquatic organisms. For example, B- and T-cell surface markers, used to quantitate lymphocytes and differentiate them into subgroups, are commercially available for mice and have been produced for chickens,[53,54] but are not yet available for use with other organisms.

Tier III

The final tier of testing requires the use of a host resistance challenge study. In such studies, organisms are exposed to appropriate bacterial, viral, or parasitic pathogens, or to tumor cells. Subsequent survivorship, the amount and duration of pathogen replication, and specific protective immune responses are measured.[40]

Physiological Indices

To maintain homeostasis, organisms must compensate for the biochemical and physiological alterations incurred by exposure to contaminated sediments. The measure of physiological and nonspecific indices may serve as a means to assess sublethal effects of contaminated sediments. In addition, such measures may be used to predict the impact of contaminants on growth and reproduction. The following briefly describes the current and potential use of selected physiological indices as biomarkers. For more detailed descriptions of the biomarkers presented, the reader is referred to Mayer et al.[55]

Corticosteroids

The corticosteroid hormones are the most extensively studied group of stress-related hormones in fish. This group of hormones is comprised of cortisol, cortisone and 11-deoxycortisol, with cortisol, the predominant form. Release of these hormones from the interrenal tissue is mediated through the hypothalamus (corticotropic releasing factor, CRF) and the rostral pars distalis of the pituitary (adrenocorticotropic hormone, ACTH). The hypothalamic-pituitary-interrenal axis (HPI) is activated by a wide variety of stressors. In fish, corticosteroids induce metabolic activity in various organs and osmoregulatory changes at the gill and kidney. Cortisol produces hyperglycemia due to reduced peripheral glucose utilization and gluconeogenesis. Cortisol also decreases total white blood cell counts and alters differential white cell counts.[56] Corticosteroids may be measured in plasma by either radioimmunoassay, high-pressure liquid chromatography, or enzyme-linked immunoabsorbent assays.

The HPI axis is rapidly stimulated by a variety of physical and chemical stressors, resulting in the release of cortisol into the plasma. The elevation of plasma cortisol depends on the intensity and duration of the stressor. After a short-term stressor is removed, plasma cortisol levels will rapidly return to normal. However, with long-term stress the return of plasma cortisol to normal levels is highly variable and is dependent upon the duration, intensity, and nature of the stressor. Data interpretation is somewhat difficult because of the rapid release, metabolism, and cellular uptake of cortisol. In assessing toxicity, handling stress may mask effects associated with contaminant exposure. Also, many biotic factors may affect plasma cortisol concentrations, including diet,[57] photoperiod,[58] temperature,[59] dissolved oxygen,[60] social stress,[61] and sex.[62] Regarding contaminant exposures, there does not appear to be any uniform response to toxicants. For example, cadmium exposure produced no change in plasma cortisol,[63] and naphthalene and polychlorinated biphenyls (PCB) caused decreases in plasma cortisol.[64,65]

Certainly, corticosteroids provide information on general stress. However, care must be taken in sampling to ensure that alterations in the plasma concentrations of these hormones are due to environmental contamination, not handling stress. In organisms collected from a contaminated site, such measures may provide information about the general health of the organism.

Glycogen

The glycogen content of fish tissue is influenced by exposure to inorganic[66-68] as well as organic contaminants.[69-71] In most cases, the increased energy demand associated with stress results in a depletion of glycogen reserves.[72,73] Glycogen storage and mobilization is restricted to certain tissues; therefore, observed alterations in glycogen concentrations are often tissue specific. In mussels, for example, the hepatopancreas and mantle are the primary storage organs for glycogen and generally the first to be utilized.[74,75] In vertebrates the liver is the primary glycogen storage location.[76] Glycogen content can be influenced by diet,[74,75] reproductive condition, and season.[77] Therefore, the influence of accessory factors present in the environment often hinders the interpretation of glycogen levels relative to contaminant-induced stress.

Lipids

Lipids provide an essential, readily available energy source for a large number of aquatic organisms. Their importance as a primary energy source vs. a secondary one varies with species and season.[78,79] In addition, the tissue distribution of lipids may vary. For example, in fish the lipid content of dark muscle is much greater than that of white muscle. Therefore, the effects of contaminants on lipid may be more meaningful to the analysis of specific tissues rather than whole body content. In both invertebrates and vertebrates, the total and tissue-specific lipid content may be influenced by a variety of factors, including diet,[80,81] reproductive condition,[82,83] salinity,[84] and temperature.[85] In many cases lipid content of fish decreases during toxicant exposure.[69,86] These decreases, however, may be accompanied by increases in specific components of lipid constituents (i.e., triglycerides, fatty acids, and phospholipids). Increases in total lipid content also have been noted in fish exposed to PCB-contaminated sediments.[87] As with glycogen, influencing accessory factors must be qualified and understood before lipids can be used as a biomarker of contaminated sediment.

Blood Chemistry

Hematological measures have historically been used to assess the physiological and nutritional status of fish populations. More recently, blood chemistry profiles have been used in hazard assessments of contaminant exposure. Such investigations have demonstrated the potential utility of hematological and blood-chemistry parameters as indicators of sublethal stress in organisms exposed to contaminants.[88-91] It may prove that clinical tests in aquatic organisms will serve as valuable diagnostic tools in the evaluation of deleterious effects due to exposures to contaminated sediments.

The specific mechanism by which serum enzyme activities increase is not known; however, it is generally agreed that it can be due to, and diagnostic of, tissue damage.[92,93] Particular tissues often display characteristic enzyme profiles; therefore, damage to a given tissue can give rise to increased activities of certain enzymes in serum. For example, serum transaminases, specifically aspartate aminotransferase (ASAT) and alanine aminotransferase (ALAT), have been widely utilized as biomarkers of specific organ dysfunction. ASAT is a nonspecific cytosolic and mitochondrial enzyme found in a variety of tissues, including liver, skeletal muscle, cardiac muscle, and kidney. ALAT also is a cytosolic enzyme, but is more tissue-specific and is normally associated with liver. Both of these enzymes have been measured in invertebrates and fish under stress. Increased serum transaminase activity is usually associated with hepatocyte dysfunction, since many compounds are metabolized in the liver where transaminase activities are high.[94] However, increased activities of the transaminases also occur in other organs, such as the heart.[95,96] Increased serum activities of the transaminases are good biomarkers of contaminant effects at the cellular level; however, transaminase activities in serum do not necessarily correlate with mortality or other adverse effects.[97]

Metals have been shown to influence the activities of transaminases in fish. For example, fish exposed to Cd, Hg, or Cu displayed increased transaminase activites.[98-100] Increases in serum transaminases also have been associated with exposure to organic contaminants. Serum ASAT:ALAT ratios in the bluegill sunfish increased following exposure to carbon tetrachloride.[96] In addition to serum enzyme activites, ion levels (e.g., sodium, chloride, calcium, and magnesium) as well as serum glucose and lipids have been used as diagnostic tools in the evaluation of the toxicological effects of contaminants.

RNA:DNA

The use of biomarkers in ecological assessments can be enhanced by examining endpoints indicative of detrimental effects on growth rate. RNA:DNA ratios of whole fish and of various tissues have been used for this purpose. RNA content per individual, as well as tissue-specific RNA concentrations, also have been utilized as indicators of the rate of protein synthesis, the level of metabolic activity, and the relative condition.[101]

RNA:DNA ratios were initially used as predictors of growth rates of zooplankton[102] and fish.[103] More recently, research with RNA:DNA ratios as well as RNA concentration has focused on their potential use as biomarkers. Many of these investigations have led to the identification of variations that may limit the use of these biomarkers with in situ investigations. Major sources of variation have been attributed to the age of the organism and changes in environmental conditions, such as food supply, temperature, and pH.[101]

Variation due to environmental conditions cannot be eliminated with in situ investigations. However, it may be reduced by the selection of appropriate control sites. In a study using yellow perch exposed to Cd and Zn contamination, Kearns and Atchison[104] demonstrated both RNA:DNA ratios and growth rates to be negatively correlated to the extent of contamination. However, in a study using salmonids exposed to carbaryl in situ, Wilder and Stanely[105] identified an increase in RNA:DNA ratios, suspected to be a result of increased consumption of dead insects by the fish. These contrasting studies suggest that RNA concentration may be useful in ecological assessments of growth rates, but indicate the need for identifying factors that can influence the activity of nucleic acids.

Histopathological Indices

Biochemical and physiological alterations, if severe enough or protracted, can lead to structural alterations in organelles, cells, tissue, and organs. Detection of specific alterations using anatomical and cytological endpoints may indicate prior and/or current exposure to chemical contaminants. For such reasons, histopathological indices have been useful in assessing the toxicological impact of contaminated sediments. The documentation of neoplasms in aquatic organism was perhaps the first use of histopathological indices in ecotoxicology.

The documentation of neoplasms in aquatic organisms began in the 1800s with the description of tumors in reptiles, amphibians, fish, and mollusks. By 1963 several epizootic neoplasms had been reported in feral fish. Yet, until 1964, the connection between epizootic neoplasms in feral fish and chemical contamination of aquatic ecosystems was not suggested. Dawe et al.[106] first postulated that liver neoplasms observed in white sucker and brown bullhead from Deep Creek Lake, Maryland might have resulted from chemical contamination. Since then, several dozen feral fish tumor epizootics have been identified in North America, dominated by liver tumors in bottom feeders in the vicinity where chemical contaminants were concentrated. For example, Pierce et al.[107] found hepatic tumors in English sole at Puget Sound, Washington. Malins et al.[108] also have studied sediment-associated contaminants from near-coastal areas. These investigators made the assumption that PAHs were related to the induction of marine fish tumors. In fact, there was a good correlation between the indices of hepatocellular carcinoma or papilloma of the fish examined and PAH sediment concentration. Black[109] reported that a high prevalence of epidermal papilloma, epidermal carcinoma, and hepatocellular carcinoma was observed in brown bullhead inhabiting the Buffalo River in which the sediment was contaminated with PAHs. Many other investigators have indicated a relationship between cancer in aquatic organisms and chemical contamination.[110-115] In addition, unexpected prevalences of neoplasms in certain feral fish from localized sites in North America,[115,116] Japan,[117] and Europe[118] have been reported.

There are a number of characteristics of anatomical and cytological alterations that favor the use of histological indices as biomarkers. These include sensitivity and ease of recognition of various biomarker alterations. However, as with the other biomarkers, histological indices may be influenced by factors other than chemical contaminants. Given age, diet, environmental factors, seasonal variation, and reproductive cycle, a number of structural states may represent normality and could be potentially confounding issues in an attempt to use histological criteria as biomarkers.[119] Regarding specific alterations and detail on the use of histological indices, the reader is referred to Hinton et al.[120]

BIOMARKER APPLICATIONS

Integration With Other Approaches

Biological monitoring has been a critical component in assessing the biological impact of chemical contaminants on aquatic ecosystems. Because of the importance of accurately assessing the biological impact of chemicals, attention has been given more recently to the use of biomarkers in hazard assessments. There is, therefore, a need to develop a more comprehensive risk-assessment approach. Such an approach would, when possible, include the concurrent use of traditional approaches to ecological assessments and examination of a suite of biochemical, physiological, and pathological responses. An obvious advantage to an integrated approach is that if toxicity is demonstrated, the mechanisms of action may be discernible. Furthermore, biomarkers can provide a sensitive measure of exposure as well as response and therefore may serve as early warning signals of environmental contamination.

Other chapters in this book (see Chapters 14, 16, and 17) describe techniques for assessment of sediments that integrate chemical analyses, toxicity assays, and aquatic community data. Biomarkers could be effectively incorporated into both the bioassay and field study components of these approaches. Biomarkers have been employed in laboratory studies of both field-derived and artificially amended sediments, as well as in field studies employing both feral and caged organisms. The addition of biomarkers would likely strengthen integrated sediment assessments and also provide much needed information concerning the relationships among biomarker responses, standard toxicity test results, and ecological effects.

Design Considerations

When implementing the use of biomarkers in ecological assessments, investigators must be aware of problems inherent in such testing programs. The selection of sample stations includes a number of important considerations (see Chapter 2). Choosing a station, using map sitings alone, is usually unsatisfactory, as many limitations of access may not be discernible unless on-site reconnaissance is performed. The proximity of stations to impact sources also is very important in attempting to determine if there is any spatial variability in effect. The establishment of a reference site can be difficult. For instance, in freshwater ecosystems, an upstream location may be contaminated to the extent that it is unsuitable for use as the reference site. Thus, one may be restricted to using laboratory controls. The sites should be evaluated as to their importance as spawning areas or recreational areas and for their potential for producing human health effects. For field validation studies that involve evaluation of benthic invertebrates, fish populations, periphyton, etc., sites for toxicity testing should be selected near the ecological sampling sites. Further, it is important to understand that handling as well as the creation of manipulated exposure conditions (e.g., fish cages) potentially produce noncontaminant-related stresses that may obscure observations of biochemical, physiological, and/or pathological responses. In addition, as previously discussed, age, diet, environmental factors (e.g., temperature, salinity, dissolved oxygen), seasonal variation, and reproductive cycle may influence the response of organisms, and such confounding factors must be considered.

CONCLUSIONS

As new biomarkers are developed for use in ecological assessments, reliable and reproducible methodologies must be established. Several novel biomarker methods for evaluating the deleterious effects of contaminants are currently under development. For example, chemically induced alterations in biotransformation processes of aquatic organisms may be used as potential diagnostic tools for in situ biological monitoring. Such an example, in the case of selected organic contaminants, is the use of detoxification processes such as mixed-function oxygenase enzymes.[10,121,122] As with most of the biomarkers discussed, a variety of biotic (i.e., age, reproductive cycle, etc.) and abiotic (i.e., seasonal variation, temperature, salinity, etc.) factors influence biotransformation processes. As such, these confounding factors must be considered when validating biomarkers for use in ecological hazard assessments.

Despite limitations, biomarkers provide a unique opportunity for evaluating the mechanisms by which contaminants exert deleterious biological effects. It is expected that further research in this area will result in more-refined methodologies that, in time, will assist in the development of improved ecological assessment strategies.

REFERENCES

1. McCarthy, J.F., and L.R. Shugart, Eds. *Biomarkers of Environmental Contamination.* (Boca Raton, FL: Lewis Publishers, 1990), p. 457.
2. Reilly, F.J., Jr., V.A. McFarland, J.U. Clarke, A.S. Jarvis, R.B. Spies, and R.F. Lee. "Evaluation of Sediment Genotoxicity: Workshop Summary and Conclusions," Environmental Effects of Dredging Technical Note EEDP-01-231990. U.S. Army Engineer Waterways Experiment Station, Vicksburg, MS (1990).
3. Huggett, R.J., R.A. Kimerle, P.M. Mehrle, and H.L. Bergmen. *Biomarkers: Biochemical, Physiological, and Histopathological Markers of Anthropogenic Stress* (Boca Raton, FL: Lewis Publishers, 1992), in press.
4. Stegeman, J.J., M. Brouwer, R.T. Di Giulio, B.A. Fowler, L. Forlin, B.M. Sanders, and P.A. Van Veld. "Molecular Responses to Environmental Contamination: Enzyme and Protein Systems as Indicators of Contaminant Exposure and Effect," in *Biomarkers: Biochemical, Physiological, and Histopathological Markers of Anthropogenic Stress*, R.J. Huggett, R.A. Kimerle, P.M. Mehrle, and H.L. Bergmen, Eds. (Boca Raton, FL: Lewis Publishers, 1992), in press.

5. Sipes, I.G., and A.J. Gandolfi. "Biotransformation of Toxicants," in *Casarett and Doull's Toxicology*, 3rd ed., C.D. Klaassen, M.O. Amdur, and J. Doull, Eds. (New York: Macmillan, 1986), pp. 64–98.
6. Buhler, D.R., and D.E. Williams. "The Role of Biotransformation in the Toxicity of Chemicals," *Aquat. Toxicol.* 11:19–28 (1988).
7. Ortiz de Montellano, P.R. *Cytochrome P-450. Structure, Mechanism, and Biochemistry* (New York: Plenum Press, 1986).
8. Payne, J.F., L.L. Fancey, A.D. Rahimtula, and E.L. Porter. "Review and Perspective on the Use of Mixed-Function Oxygenase Enzymes in Biological Monitoring," *Comp. Biochem. Physiol.* C 86:233–245 (1987).
9. Collier, T.K., B.L. Eberhart, J.E. Stien, and U. Varanasi. "Aryl Hydrocarbon Hydroxylase — A 'New' Monitoring Tool in the Status and Trends Program," in *Proceedings Oceans '89*, IEEE Publ. No. 89CH2780-5, pp. 608–610, (1989).
10. Andersson, T., L. Forlin, J. Hardig, and A. Larsson. "Physiological Disturbances in Fish Living in Coastal Water Polluted with Bleached Kraft Pulp Mill Effluent," *Can. J. Fish. Aquat. Sci.* 45:1525–1536 (1988).
11. Jimenez, B.D., A. Oikari, S.M. Adams, D.E. Hinton, and J.F. McCarthy. "Hepatic Enzymes as Biomarkers: Interpreting the Effects of Environmental Physiological and Toxicological Variables," in *Biomarkers of Environmental Contamination*, McCarthy, J.F., and L.R. Shugart, Eds. (Boca Raton, FL: Lewis Publishers, 1990), pp. 123–142.
12. Armstrong, R.N. "Enzyme-Catalyzed Detoxication Reactions: Mechanisms and Stereochemistry," *CRC Crit. Rev. Biochem.* 22:39–88 (1987).
13. James, M.O. "Conjugation of Organic Pollutants in Aquatic Species," *Environ. Health Perspec.* 71:97–103 (1987).
14. Foureman, G.L. "Enzymes Involved in Metabolism of PAH by Fishes and Other Aquatic Animals: Hydrolysis and Conjugation Enzymes (or Phase II Enzymes)," in *Metabolism of Polycyclic Aromatic Hydrocarbons in the Aquatic Environment,* U. Varanasi, Ed. (Boca Raton, FL: CRC Press, 1989), pp. 185–202.
15. Benson, W.H., C.F. Watson, K.N. Baer, and R.A. Stackhouse. "Response of Hematological and Biochemical Parameters to Heavy Metal Exposure: Implications in Environmental Monitoring," *Mar. Environ. Res.* 24:219–222 (1988).
16. Hamer, D.H. "Metallothionein," *Ann. Rev. Biochem.* 55:913–951 (1986).
17. Engel, D.W., and M. Brouwer. "Metallothionein and Metallothionein-Like Proteins: Physiological Importance," *Adv. Comp. Environ. Physiol.* 4:53–75 (1989).
18. Subjeck, J.R., and T.-T. Shyy. "Stress Protein Systems of Mammalian Cells," *Cell. Physiol.* 19:C1–C17 (1986).
19. Welch, W.J. "The Mammalian Stress Response: Cell Physiology and Biochemistry of Stress Proteins," in *The Role of the Stress Response in Biology and Disease,* R. Moromoto and A. Tissieres, Eds. (Cold Spring Harbor, NY: Cold Spring Harbor Laboratory, 1990), pp. 223–278.
20. Sanders, B.M. "Stress Proteins: Potential as Multitiered Biomarkers," in *Biomarkers of Environmental Contamination*, L. Shugart and J. McCarthy, Eds. (Boca Raton, FL: Lewis Publishers, 1990), pp. 165–191.
21. Halliwell, B., and J.M.C. Gutteridge. *Free Radicals in Biology and Medicine* (Oxford: Clarendon Press, 1985).

22. Kappus, H. "Oxidative Stress in Chemical Toxicity," *Arch. Toxicol.* 60:144–149 (1987).
23. Di Giulio, R.T., P.C. Washburn, R.J. Wenning, G.W. Winston, and C.S. Jewell. "Biochemical Responses in Aquatic Animals: A Review of Determinants of Oxidative Stress," *Environ. Toxicol. Chem.* 8:1103–1123 (1989).
24. Winston, G.W., and R.T. Di Giulio. "Prooxidant and Antioxidant Mechanisms in Aquatic Organisms," *Aquat. Toxicol.* 19:137–161 (1991).
25. Di Giulio, R.T. "Indices of Oxidative Stress as Biomarkers for Environmental Contamination," in *Aquatic Toxicology and Risk Assessment: Fourteenth Volume,* STP 1124, M.A. Mayes and M.G. Barron, Eds. (Philadelphia: American Society of Testing and Materials, 1991), pp. 15–31.
26. Shugart, L., J. Bickham, G. Jackim, G. McMahon, W. Ridley, J. Stein, and S. Steinert. "DNA Alterations," in *Biomarkers: Biochemical, Physiological, and Histopathological Markers of Anthropogenic Stress,* R.J. Huggett, R.A. Kimerle, P.M. Mehrle, and H.L. Bergmen, Eds. (Boca Raton, FL: Lewis Publishers, 1992), in press.
27. Dunn, B., J. Black, and A. Maccubbin. "^{32}P-Postlabeling Analysis of Aromatic DNA Adducts in Fish from Polluted Areas," *Cancer Res.* 47:6543–6548 (1987).
28. Varanasi, U., W.L. Reichert, B.-T. Eberhart, and J. Stein. "Formation and Persistence of Benzo[a]pyrene-Diolepoxide-DNA Adducts in Liver of English Sole (*Parophrys vetulus*)," *Chem. Biol. Interact.* 69:203–216 (1989).
29. Shugart, L.R. "Quantitation of Chemically Induced Damage to DNA of Aquatic Organisms by Alkaline Unwinding Assay," *Aquat. Toxicol.* 13:43–52 (1988).
30. Pfeifer, G.P., D. Grungerger, and D. Drahovsky. "Impaired Enzymatic Methylation of BPDE-Modified DNA," *Carcinogenesis* 5:931–935 (1984).
31. Shugart, L.R. "5-methyl Deoxycytidine Content of DNA from Bluegill Sunfish (*Lepomis macrochirus*) Exposed to Benzo[a]pyrene," *Environ. Toxicol. Chem.* (9: 205–208) (1990).
32. Floyd, R.A., J.J. Watson, P.K. Wong, D.H. Altmiller, and R.C. Rickard. "Hydroxyl Free Radical Adduct of Deoxyguanosine: Sensitive Detection and Mechanisms of Formation," *Free Rad. Res. Comm.* 1:163–172 (1986).
33. Malins, D.C., G.K. Ostrander, R. Haimanot, and P. Williams. "A Novel DNA Lesion in Neoplastic Livers of Feral Fish: 1, 6-Diamino-4-Hydroxy-5-Formamidopyrimidine," *Carcinogenesis* 6:1045–1047 (1990).
34. McBee, K., J.W. Bickham, K.C. Donnely, and K.W.Brown. "Chromosomal Aberrations in Native Small Mammals (*Peromyscus leucopus* and *Sigmodon hispidus*) at a Petrochemical Waste Disposal Site: I. Standard Karyology," *Arch. Environ. Contam. Toxicol.* 16:681–688 (1987).
35. Pesch, G.G., and C.E. Pesch. "Neanthes Arenaceodentata (*Polychaeta annelida*): A Proposed Cytogenetic Model for Marine Genetic Toxicology," *Can. J. Fish. Aquat. Mar. Genet. Toxicol.* 37:1225–1228 (1980).
36. Hose, J.E., J.N. Cross, S.C. Smith, and D. Diehl. "Elevated Circulating Erythrocyte Micronuclei in Fishes from Contaminated Sites off Southern California," *Mar. Environ. Res.* 22:167–176 (1987).
37. McMahon, G., L.J. Huber, M.J. Moore, J.J. Stegeman, and G.N. Wogan. "c-K-*ras* Oncogenes: Prevalence in Livers of Winter Flounder from Boston Harbor," in *Biological Markers of Environmental Contaminants,* J.F. McCarthy, and L.R. Shugart, Eds. (Boca Raton, FL: Lewis Publishers, 1990), pp. 229–235.

38. Wirgin, I.I., D. Currie, C. Gorunwald, and S.Y. Garte. "Molecular Mechanisms of Carcinogenesis in a Natural Population of Hudson River Fish," *Proc. AACR Mtg.* 30:194 (1989).
39. Stolen, J.S., T.C. Fletcher, D.P. Anderson, B.S. Robertson, and W.B. van Muiswinkel, Eds. *Techniques in Fish Immunology* (Fair Haven, NJ: SOS Publications, 1990).
40. Weeks, B.A., D.P. Anderson, A.J. Goven, G. Peters, A. Fairbrother, A.P. DuFour, G.P. Lahvis. "Immunological Biomarkers to Assess Environmental Stress," in *Biomarkers: Biochemical, Physiological, and Histopathological Markers of Anthropogenic Stress*, R.J. Huggett, R.A. Kimerle, P.M. Mehrle, and H.L. Bergmen, Eds. (Boca Raton, FL: Lewis Publishers, 1992), in press.
41. Ellis, A.E., A.L.S. Munroe, and R.J. Roberts. "Defense Mechanisms in Fish. 1. A Study of the Phagocytic System and the Fate Intraperitoneally Injected Particulate Material in the Plaice (*Pleuronectes platessa* L.)," *J. Fish Biol.* 8:67–78 (1976).
42. Weeks, B.A., J.E. Warinner, P.L. Mason, and D.S. McGinnis. "Influence of Toxic Chemicals on the Chemotactic Response of Fish Macrophages," *J. Fish Biol.* 28:653–658 (1986).
43. Weeks, B.A., A.S. Keisler, Q.N. Myrvik, and J.E. Warinner. "Differential Uptake of Neutral Red by Macrophages from Three Species of Estuarine Fish," *Dev. Comp. Immunol.* 11:117–124 (1987).
44. Weeks, B.A., A.S. Keisler, J.E. Warinner, and E.S. Mathews. "Preliminary Evaluation of Macrophage Pinocytosis as a Fish Health Monitor," *Mar. Environ. Res.* 22:205–213 (1987).
45. Warinner, J.E., E.S. Mathews, and B.A. Weeks. "Preliminary Investigations of the Chemiluminescent Response in Normal and Pollutant-Exposed Fish," *Mar. Environ. Res.* 24:281–284 (1988).
46. Robohm, R.A. "Paradoxical Effects of Cadmium Exposure on Antibacterial Antibody Responses in Two Fish Species: Inhibition in Cunners (*Tautogolaburs adspersus*) and Enhancement in Stripped Bass (*Morone saxatilis*)," *Vet. Immunol. Immunopathol.* 12:251–262 (1986).
47. O'Neill, J.G. "Effects of Intraperitoneal Lead and Cadmium on the Humoral Immune Response of *Salmo trutta*," *Bull. Environ. Contam. Toxicol.* 27:42–48 (1981).
48. Anderson, D.P., O.W. Dixon, and F.W. van Ginkel. "Suppression of Bath Immunization in Rainbow Trout by Contaminant Bath Pretreatments," in *Chemical Regulation of Immunity in Veterinary Medicine*, M. Kende, J. Gainer, and M. Chirigos, Eds. (New York: Alan R. Liss, 1986), pp. 289–293.
49. Vos, J.G. "Immunotoxicity Assessment: Screening and Function Studies," *Arch. Appl. Toxicol.* 4(Suppl.):95–108 (1980).
50. Kohler, L.D., and J.H. Exon. "The Rat as a Model for Immunotoxicity Assessment," in *Immunotoxicology and Immunopharmacology*, J.H. Dean, M.I. Luster, A.E. Munson, and H. Amos, Eds. (New York: Raven Press, 1985), pp. 99–112.
51. Kerkvliet, N. "Measurements of Immunity and Modifications by Toxicants," in *Safety Evaluation of Drugs and Chemicals*, W.E. Lloyd, Ed. (Washington, D.C.: Hemisphere, 1986), pp. 235–256.
52. Luster, M.I., A.E. Munson, P.T. Thomas, M.P. Holsapple, J.D. Fenters, K.L. White, L.D. Lauer, D.R. Germolec, G.J. Rosenthal, and J.H. Dean. "Development of a Testing Battery to Assess Chemical-Induced Immunotoxicity: National Toxicology Program's Guidelines for Immunotoxicity Evaluation in Mice," *Fund. Appl. Toxicol.* 10:2–19 (1988).

53. Bucy, R.P., C.L. Chen, J. Cihak, U. Losch, and M.D. Cooper. "Avian T Cells Expressing Gamma Delta Receptors Localize in the Splenic Sinusoids and the Intestinal Epithelium," *J. Immunol.* 141:2200–2205 (1988).
54. van de Water, J., L. Haapanen, R. Boyd, H. Abplanalp, and M.E. Gershwin. "Identification of T cells in Early Dermal Lymphocytic Infiltrates in Avian Scleroderma," *Arthritis Rheum.* 32:1031–1040 (1989).
55. Mayer, F.L., D.J. Versteeg, M.J. McKee, L.C. Folmar, R.L. Graney, D.C. McCume, and B.A. Rattner. "Physiological and Nonspecific Biomarkers," in *Biomarkers: Biochemical, Physiological, and Histopathological Markers of Anthropogenic Stress*, R.J. Huggett, R.A. Kimerle, P.M. Mehrle, and H.L. Bergmen, Eds. (Boca Raton, FL: Lewis Publishers, 1992), in press.
56. Johansson-Sjobeck, M.-L., G. Dave, A. Larsson, K. Lewander, and U. Lidman. "Hematological Effects of Cortisol in the European Eel, *Anguilla anguilla* L.," *Comp. Biochem. Physiol.* 60A:165–168 (1978).
57. Bry, C. "Daily Variations in Plasma Cortisol Levels of Individual Female Rainbow Trout *Salmo gairdneri*: Evidence for a Post-Feeding Peak in Well-Adapted Fish," *Gen. Comp. Endocrinol.* 48:462–468 (1982).
58. Peter, R.E., A. Hontela, A.F. Cook, and C.R. Pulenau. "Daily Cycles in Serum Cortisol Levels in the Goldfish: Effects of Photoperiod, Temperature and Sexual Condition," *Can. J. Zool.* 56:2443–2448 (1978).
59. Strange, R.J. "Acclimation Temperature Influences Cortisol and Glucose Concentrations in Stressed Channel Catfish," *Trans. Am. Fish. Soc.* 109:298–303 (1980).
60. Swift, D.J. "Changes in Selected Blood Component Concentrations of Rainbow Trout *Salmo gairdneri* Richardson, Exposed to Hypoxia or Sublethal Concentrations of Phenol or Ammonia," *J. Fish Biol.* 19:45–61 (1981).
61. Ejike, C., and C.B. Schreck. "Stress and Social Hierarchy Rank in Coho Salmon," *Trans. Am. Fish. Soc.* 109:423–426 (1980).
62. Donaldson, E.M., and H.M. Dye. "Corticosteroid Concentrations in Sockeye Salmon (*Oncorhynchus nerka*) Exposed to Low Concentrations of Copper," *J. Fish. Res. Board Can.* 32:533–539 (1975).
63. Thomas, P. "Effect of Cadmium Exposure on Plasma Cortisol Levels and Carbohydrate Metabolism in Mullet (*Mugil cephalus*)," *J. Endrocrinol.* 94(Suppl.):35 (1982).
64. DiMichelle, L., and M.H. Taylor. "Histopathological and Physiological Responses of *Fundulus heteroclitus* to Naphthalene Exposure," *J. Fish. Res. Board Can.* 35:1060–1066 (1978).
65. Sivarajah, K., C.S. Franklin, and W.P. Williams. "The Effects of Polychlorinated Biphenyls of Plasma Steroid Levels and Hepatic Microsomal Enzymes in Fish," *J. Fish. Biol.* 13:401–409 (1978).
66. Gill, T.S., and J.C. Pant. "Cadmium toxicity: Inducement of Changes in Blood and Tissue Metabolites in Fish," *Toxicol. Lett.* 18:195–200 (1983).
67. Sastry, K.V., and K. Sunita. "Enzymological and Biochemical Changes Produced by Chronic Chromium Exposure in a Teleost Fish, *Channa punctatus*," *Toxicol. Lett.* 16:9–15 (1983).
68. Chaudhry, H.S. "Nickel Toxicity on Carbohydrate Metabolism of a Freshwater Fish, *Colisa Fasciatus,*" *Toxicol. Appl. Pharmacol.* 50:241–252 (1984).

69. Murty, A.S., and A.P. Devi. "The Effect of Endosulfan and Its Isomers on Tissue Protein, Glycogen, and Lipids in the Fish (*Channa punctatus*)," *Pest. Biochem. Physiol.* 17:280–286 (1982).
70. Pant, J.C., and T. Singh. "Inducement of Metabolic Dysfunction by Carbamate and Organophosphorus Compounds in a Fish, *Puntius conchonius*," *Pest. Biochem. Physiol.* 20:294–298 (1983).
71. Verma, S.R., S. Rani, I.P. Tonk, and R.C. Dalela. "Pesticide-Induced Dysfunction in Carbohydrate Metabolism in Three Freshwater Fishes," *Environ. Res.* 32:127–133 (1983).
72. Bhagyalakshmi, A., R.S. Reddy, and R. Ramamurthi. "Subacute Stress Induced by Sumithion on Certain Biochemical Parameters in *Oziotelphusa senex*, the Freshwater Ricefield Crab," *Toxicol. Lett.* 21:127–134 (1984).
73. Thomas, P., H.W. Wofford, and J.M. Neff. "Biochemical Stress Responses of Striped Mullet (*Mugil cephalus* L.) to Fluorene Analogs," *Aquat. Toxicol.* 1:329–342 (1981).
74. Bayne, B.L. "Physiological Changes in *Mytilis edulis* L. Induced by Temperature and Nutritive Stress," *J. Mar. Biol. Assoc. U.K.* 53:39–58 (1973).
75. Bayne, B.L. "Aspects of the Metabolism of *Mytilus edulis* During Starvation," *Nether. J. Sea Res.* 7:399–410 (1973).
76. Grizzle, J.A., and W.A. Rogers. "Anatomy and Histology of the Channel Catfish," Auburn University Agric. Expt. Station, Auburn, AL (1976).
77. Newell, R.I.E., and B.L. Bayne. "Seasonal Changes in the Physiology, Reproductive Condition and Carbohydrate Content of the Cockle *Cardium* (Cerastoderma) *edule* (Bivalvia: Cardiidae)," *Mar. Biol.* 56:11–19 (1980).
78. Morris, R.J. "Seasonal and Environmental Effects in the Lipid Composition of *Neomyses integer*," *J. Mar. Biol. Assoc. U.K.* 51:21–31 (1971).
79. Gardner, D., and J.P. Riley. "The Component Fatty Acids of the Lipids of Some Species of Marine and Freshwater Molluscs," *J. Mar. Biol. Assoc. U.K.* 52:827–838 (1972).
80. Phillips, A.M., F.E. Lovelace, H.A. Podoliak, D.R. Brockway, and G.C. Balzer. "The Nutrition of Trout," *Fish. Res. Bull.* 19:56 (1956).
81. Fletcher, D.J. "Plasma Glucose and Plasma Fatty Acid Levels of *Limanda limanda* (L.) in Relation to Season, Stress, Glucose Loads and Nutritional State," *J. Fish Biol.* 25:629–648 (1984).
82. Love, R.M. *The Chemical Biology of Fishes* (London: Academic Press, 1970), p. 547.
83. Frank, J.R., S.D. Sulkin, and R.P. Morgan. "Biochemical Changed During Larval Development of the Xanthid Crab *Rhethropanopeus harrissii* I. Protein, Total Lipid, Alkaline Phosphatase and Glutamic Oxaloacetic Transaminase," *Mar. Biol.* 32:105–111 (1975).
84. Daikoku, T., I. Yano, and M. Musui. "Lipid and Fatty Acid Compositions and Their Changes in the Different Organs and Tissues of Guppy, *Poecilia reticulata* on Sea Water Adaptation," *Comp. Biochem. Physiol.* 73A:167–174 (1982).
85. Hansen, H.J.M., and S. Abraham. "Influence of Temperature, Environmental Salinity and Fasting on the Patterns of Fatty Acids Synthesized by Gills and Liver of the European Eel (*Anguilla anguilla*)," *Comp. Biochem. Physiol.* 75B:581–587 (1983).
86. Dey, A.C., J.W. Kiceniuk, U.P. Williams, R.A. Khan, and J.F. Payne. "Long Term Exposure of Marine Fish to Crude Petroleum I. Studies on Liver Lipids and Fatty Acids in Cod (*Gadus morhua*) and Winter Flounder (*Pseudopleuronectes americanus*," *Comp. Biochem. Physiol.* 75C:93–101 (1983).

87. Dillon, T.M., and W.H. Benson. "Effects of PCB-Contaminated Sediments on Reproductive Success of Fathead Minnows: Relationship Between Tissue Residues and Biological Effects," paper presented at the Eighth Annual Meeting of the Society of Environmental Toxicology and Chemistry, Pensacola, FL, November 9–12, 1987.
88. Wedemeyer, G.A., and W.T. Yasutake. "Clinical Methods for the Assessment of the Effects of Environmental Stress in Fish Health," U.S. Fish and Wildlife Service Technical Paper No. 89. (1977).
89. Lockhart, W.L., and D.A. Metner. "Fish Serum Chemistry as a Pathological Tool," in *Contaminant Effects on Fishes*, Vol. 16, V.W. Cairns, P.V. Hodson, and J.D. Nriagu, Eds. (New York: John Wiley & Sons, 1984), pp. 73–85.
90. Benson, W.H., K.N. Baer, and C.F. Watson. "Metallothionein as a Biomarker of Environmental Metal Contamination: Species-Dependent Effects," in *Biomarkers of Environmental Contamination*, J.F. McCarthy and L.R. Shugart, Eds. (Boca Raton, FL: Lewis Publishers, 1990), pp. 255–265.
91. Watson, C.F., K.N. Baer, and W.H. Benson. "Dorsal Gill Incision: A Simple Method for Obtaining Blood Samples in Small Fish," *Environ. Toxicol. Chem.* 8:457–461 (1989).
92. Galen, R.S. "Multiphasic Screening and Biochemical Profiles: State of the Art," in *Progress in Clinical Pathology*, Vol. 6, M. Stefanini, and H.D. Isenberg, Eds. (New York: Grune and Stratton, 1975), pp. 83–110.
93. Chenery, R., M. George, and G. Krishna. "The Effect of Ionophore A23187 and Calcium on Carbon Tetrachloride-Induced Toxicity in Cultured Rat Hepatocytes," *Toxicol. Appl. Pharmacol.* 50:241–252 (1981).
94. Gingerich, W.H. "Hepatic Toxicology of Fishes," in *Aquatic Toxicology*, Vol. 1., L. J. Weber, Ed. (New York: Raven Press, 1982), pp. 55–105.
95. Gaudet, M., J.G. Racicot, and C. Leray. "Enzyme Activities of Plasma and Selected Tissues in Rainbow Trout *Salmo gairdneri* Richardson," *J. Fish Biol.* 7:505–512 (1975).
96. Versteeg, D.J. "Lysosomal Membrane Stability, Histopathology, and Serum Enzyme Activities as Sublethal Bioindicators of Xenobiotic Exposure in the Bluegill Sunfish (*Lepomis macrochirus* Rafinesque)," Ph.D. Dissertation, Michigan State University, East Lansing (1985).
97. Lane, C.E., and E.D. Scura. "Effects of Dieldrin on Glutamic Oxaloacetic Transaminase in *Poecilia latipinna*," *J. Fish. Res. Board. Can.* 27:1869–1871 (1970).
98. Williams, H.A., and R. Wooten. "Some Effects of Therapeutic Levels of Formalin and Copper Sulphate on Blood Parameters in Rainbow Trout," *Aquaculture* 24:341–353 (1981).
99. Verma, S.R., M. Saxena, and T.P. Tonk. "The Influence of Idet 20 on the Biochemical Composition and Enzymes in the Liver of *Clarias batrachus*," *Environ. Pollut.* 33:245–255 (1984).
100. Versteeg, D.J., and J.P. Giesy. "The Histological and Biochemical Effects of Cadmium Exposure in the Bluegill Sunfish (*Lepomis macrochirus*)," *Ecotoxicol. Environ. Saf.* 11:31–43 (1986).
101. Bulow, F.J. "RNA-DNA Ratios as Indicators of Growth in Fish: A Review," in *Age and Growth of Fish*, R.C. Summerfelt, and G.E. Hall, Eds. (Ames, IA: Iowa State University Press, 1987), pp. 45–64.

102. Sutcliffe, W.H., Jr. "Relationship Between Growth Rate and Ribonucleic Acid Concentration in Some Invertebrates," *J. Fish. Res. Board Can.* 27:606–609 (1970).
103. Bulow, F.J. "RNA-DNA Ratios as Indicators of Recent Growth Rates of a Fish," *J. Fish. Res. Board Can.* 27:2343–2349 (1970).
104. Kearns, P.K., and G.J. Atchison. "Effects of Trace Metals on Growth of Yellow Perch (*Perca flavescens*) as Measured by RNA-DNA Ratios," *Environ. Biol. Fish.* 4:383–387 (1979).
105. Wilder, I.B., and J.G. Stanley. "RNA-DNA Ratio as an Index to Growth in Salmonid Fishes in the Laboratory and in Streams Contaminated by Carbaryl," *J. Fish Biol.* 22:165–172 (1983).
106. Dawe, C.J., M.F. Stanton, and F.J. Schwartz. "Hepatic Neoplasms in Native Bottom-Feeding Fish of Deep Creek Lake, Maryland," *Cancer Res.* 24:1194–1201 (1964).
107. Pierce, K.V., B.B. McCain, and S.R. Willings. "Pathology of Hepatoma and Other Liver Abnormalities in English Sole (*Parophrys vetulis*) from the Duwamish River Estuary, Seattle, Washington," *J. Natl. Cancer Inst.* 50:1445–1449 (1978).
108. Malins, D.C., B.B. McCain, D.W. Brown, S.-L. Chain, M.S. Myers, J.T. Landahl, P.G. Prohaska, A.J. Friedman, L.D. Rhodes, D.G. Burrows, W.D. Gronlund, and H.O. Hodgins. "Chemical Pollutants in Sediments and Disease of Bottom-Dwelling Fish in Puget Sound, Washington," *Environ. Sci. Technol.* 18:705–713 (1984).
109. Black, J. "Field and Laboratory Studies Environmental Carcinogenesis in Niagara River Fish," *J. Great Lake Res.* 9:326–334 (1983).
110. Hendricks, J.D. "Chemical Carcinogens in Fish," in *Aquatic Toxicology,* L.J. Weber, Ed. (New York: Raven Press, 1982), pp. 149–211.
111. Meyers, T.R., and J.D. Hendricks. "A Summary of Tissue Lesions in Aquatic Animals Induced by Controlled Exposures to Environmental Contaminants, Chemotherapeutic Agents, and Potential Carcinogens," *Mar. Fish. Rev.* 44:1–17 (1982).
112. Black, J.J. "Aquatic Animal Neoplasia as an Indicator for Carcinogenic Hazards to Man," in *Hazard Assessment of Chemicals: Current Developments*, Vol. 3, J. Saxena, Ed. (New York: Academic Press, 1984), pp. 181–232.
113. Mix, M.D. "Cancerous Diseases in Aquatic Animals and Their Association with Environmental Pollutants: A Critical Review of the Literature," *Mar. Fish. Rev.* 20:1–141 (1986).
114. Overstreet, R.M. "Aquatic Pollution Problems, Southeastern U.S. Coasts: Histopathological Indicators," *Aquat. Toxicol.* 11:213–239 (1987).
115. Harshbarger, J.C., and J.B. Clark. "Epizootiology of Neoplasms in Bony Fish of North America," *Sci. Tot. Environ.* 94:1–32 (1990).
116. Couch, J.A., and J.C. Harshbarger. "Effects of Carcinogenic Agents on Aquatic Animals: An Environmental and Experimental Overview," *Environ. Carcinogen. Rev.* 3:63–105 (1985).
117. Kimura, I. "Aquatic Pollution Problem in Japan," *Aquat. Toxicol.* 11:287–301 (1988).
118. Wellings, S.R. "Neoplasia and Primitive Vertebrate Phylogeny: Echinoderms, Prevertebrates, and Fishes — A Review," *Natl. Cancer Inst. Monogr.* 31:59–128 (1969).
119. Hinton, D.E., and D.J. Lauren. "Liver Structural Alterations Accompanying Chronic Toxicity in Fishes: Potential Biomarkers of Exposure," in *Biomarkers of Environmental Contamination,* J.F. McCarthy and L.R. Shugart, Eds. (Boca Raton, FL: Lewis Publishers, 1990), pp. 17–57.

120. Hinton, D.E., P.C. Baumann, G.R. Gardner, W.E. Hawkins, J.D. Hendricks, R.A. Murchelano, and M.S. Okihiro. "Histopathological Biomarkers," in *Biomarkers: Biochemical, Physiological, and Histopathological Markers of Anthropogenic Stress*, R.J. Huggett, R.A. Kimerle, P.M. Mehrle, and H.L. Bergmen, Eds. (Boca Raton, FL: Lewis Publishers, 1992), in press.
121. Lindstrom-Seppa, P., and A. Oikari. "Biotransformation and Other Toxicological and Physiological Responses in Rainbow Trout (*Salmo gairdneri* Richardson) Caged in a Lake Receiving Effluent of Pulp and Paper Industry," *Aquat. Toxicol.* 16:187–204 (1990).
122. Mather-Mihaich, E., and R.T. Di Giulio. "Oxidant, Mixed-Function Oxidase, and Peroxisomal Responses in Channel Catfish Exposed to a Bleached Kraft Mill Effluent," *Arch. Environ. Contam. Toxicol.* 20:391–397 (1991).

CHAPTER 12

Models, Muddles, and Mud: Predicting Bioaccumulation of Sediment-Associated Pollutants

Henry Lee II

INTRODUCTION

As is increasingly well documented and frequently stated in introductions, sediments are the ultimate sink for hydrophobic organic pollutants and heavy metals. These pollutants are potentially bioavailable to sediment-dwelling organisms, and acute or chronic health effects can occur if sufficient levels are bioaccumulated.[1] Predation by fishes and epibenthic invertebrates on contaminated benthos introduces sediment contaminants into pelagic food webs that may then biomagnify through the food web.[2] Thus, bioaccumulation by infaunal organisms is the first step in the biological transport of hydrophobic pollutants from the sediment reservoir to higher trophic levels, including shore birds, marine mammals, and human consumers. Accumulation of these pollutants in higher trophic levels can result in impaired reproduction[3,4] and increased cancer risk for human consumers.[5-7] In addition to playing a key role in ecological and human health-risk assessments, bioaccumulation by infaunal organisms generates insights into the bioavailability of sediment contaminants and can be used to test the assumptions of the techniques used to derive sediment quality criteria.[8]

Various approaches have been developed to monitor or predict bioaccumulation by sediment-dwelling organisms. The conceptually simplest measure tissue residues in field-collected or laboratory-exposed organisms, which are direct measures of existing conditions. However, the direct approaches can be costly and have limited ability to predict tissue residues resulting from changes in sediment contamination, such as from a new sewage discharge. Although environmental models must be used cautiously,[9] sediment bioaccumulation models are cost-effective alternatives that allow the prediction of tissue residues when direct measurements are not possible.

Two general types of sediment bioaccumulation models have been developed: equilibrium-based and kinetic approaches. The equilibrium-based approaches assume steady-state conditions between the organism and the environment, whereas the kinetic approaches describe bioaccumulation as the net effect of rate processes. The two equilibrium models are bioaccumulation factors (BAFs) and the equilibrium partitioning bioaccumulation model. The two general types of kinetic approaches are kinetic process models and bioenergetically based toxicokinetic models.

This paper will assess each of the direct and modeling approaches in relation to sediment-dwelling organisms. The evaluation focuses on the ability of each approach to predict rather than to monitor tissue residues. The discussion is weighted towards neutral organics rather than metals or polar compounds, reflecting both the author's expertise (or lesser ignorance) and the more extensive research on neutral organic pollutants. The factors considered in the evaluation include (1) suitability for predicting vs. monitoring tissue residues, (2) appropriate exposure scenarios, (3) applicability to metals and polar organics as well as to neutral organics, (4) cost and requirements for specialized expertise or equipment, (5) accuracy and realism, (6) whether the technique inherently under- or overestimates tissue residues, and (7) whether the approach generates insights into the mechanisms regulating sediment bioavailability.

FIELD MONITORING

Measuring tissue residues in field-collected sediment-dwelling organisms is the most straightforward method of assessing bioaccumulation. The field approach avoids introducing laboratory artifacts, as well as the extra time, expense, and facilities required for laboratory tests. Although more expensive than the bioaccumulation models, this approach avoids the uncertainty inherent in models. However, use of field-collected organisms as a routine monitoring or predictive method has several substantial limitations, the most important of which is collecting sufficient tissue biomass of the appropriate species. Standard analytical techniques require about 1 to 10 g of wet tissue for metals, another 1 to 10 g for organic priority pollutants, and 1 to 5 g for lipids. More sophisticated analytical techniques can reduce the tissue mass required, whereas screening for more compounds or analysis for certain compounds (e.g., dioxins) will increase the tissue required. Sediment-ingesting species must be sampled to accurately assess the bioavailability of sediment contaminants,[10] limiting the number and tissue mass of suitable species, especially as this eliminates large filter-feeding bivalves. Obtaining sufficient biomass can be especially difficult at highly contaminated or disturbed sites, which are characteristically dominated by smaller species and often have reduced total density and/or biomass.[11,12]

Even when sufficient biomass of a particular species can be collected at a station, it often will be difficult to collect the same species from other stations along a pollution gradient, seasonally at a single station, or at an estuarine dredge site and an open ocean disposal site. Tissue residues can be compared among the different species collected, but interspecific ecological and phylogenetic differences may affect both the total residues accumulated and the relative proportions of compounds. For example, some amphipods and polychaetes readily metabolize PAHs, whereas bivalves have a limited ability.[13,14] Such compounding factors could make it unclear whether patterns in tissue residues result from interspecific effects or from differences in bioavailability.

Another problem is the unknown exposure history of field-collected specimens. Many postmetamorphal benthic species, including amphipods, polychaetes, and bivalves, are mobile[15,16] and may have recently migrated into a site. Although pollutant concentrations in sediments are usually considered relatively constant, resuspension and deposition events could obscure sediment-bioaccumulation relationships. Also, field organisms are exposed to suspended sediment, phytoplankton, and pollutants dissolved in the overlying water. Relating tissue residues to sediment concentrations could overestimate sediment bioavailability if these water-column routes are important

Field surveys have limited utility in predicting tissue residues. A survey of existing tissue residues can serve as the "worst-case" prediction if it is unequivocal that a site will get cleaner (e.g., pretreatment reducing pollutant input). Existing tissue residues cannot be used as "worst-case" predictions where it is not clear whether pollutant concentrations will decrease (e.g., pretreatment, but with increase in mass emission rate) or in sites undergoing changes in key environmental parameters (e.g., decrease in total organic carbon [TOC]). Predictions can be made by extrapolating tissue residues from field surveys conducted at surrogate sites, with pollutant concentrations and environmental conditions similar to those projected at the site of concern. However, finding suitable surrogate sites will be difficult.

Because of these limitations, field collections are not as well suited as bioaccumulation tests for predicting tissue residues or for among-site comparisons. Field collections are a powerful regulatory tool if used to periodically monitor existing sites. Field collections also complement laboratory studies as quality assurance checks and by providing data on species difficult to maintain in the laboratory. Both laboratory tests and field assessments may be required for large discharges or dredging operations or with highly contaminated sediments.

LABORATORY BIOACCUMULATION TEST

The basic procedure for conducting laboratory bioaccumulation tests is to place appropriate test organisms into the sediment and allow the organisms to accumulate the sediment-associated pollutants. In most cases, a static-renewal exposure system is sufficient to maintain overlying water quality. Although the test is conceptually simple and has been used widely,[17,18] a variety of experimental procedures have been used. In a step towards standardization, we produced a guideline for bioaccumulation tests with estuarine/marine sediments,[10] and the reader is referred to that document for specific experimental procedures.

A few experimental procedures will be mentioned because of their overriding importance. One key requirement is the need to use sediment-ingesting species as the test organisms. Ingested sediment can be an important, if not dominant, uptake route for compounds with high octanol-water partitioning coefficients (K_{ow}),[19,20] and the use of filter-feeders can substantially underestimate maximum tissue residues.[21,22] Likewise, the addition of a supplemental food source during the test is likely to reduce the exposure to ingested sediment and is not recommended.

Another key factor is the duration of the exposure. Short-term (e.g., 10-days) exposures have been used to qualitatively determine which compounds are bioavailable (i.e., the "bioaccumulation potential"). Because the low tissue residues resulting after only a few days exposure may be statistically indistinguishable from zero, the short-term exposures may underestimate the number of bioavailable compounds. For example, Tracey et al.[23] detected three times as many PCB congeners with "ecologically significant" (≥twofold) increased tissue residues in *Nereis* with 28-d tests than with 10-d tests. The nonsteady-state tissue residues from short-term exposures are also of limited utility. As underestimates of steady-state tissue residues, they should not be used in ecological or human health-risk assessments, compared to any tissue residue guideline (e.g., FDA action limit), or used to derive bioaccumulation or accumulation factors.

Ideally, the duration should be sufficient to obtain steady-state tissue residues with all pollutants. While recognizing that no single duration will be optimal for all compounds, we recommend 28 days as the standard test duration.[10] This is also the recommended duration for dredge-material assessment whenever neutral organics are present.[18] Twenty-eight-day exposures resulted in average tissue residues within about 80% of the steady-state values for 17 PAHs and chlorinated organics.[10] In comparison, 10-day exposures averaged only about 50% of steady-state residues for all 17 compounds and 32% of steady state for PCBs. Durations longer than 28 days may be required when it is important to obtain more than 80% of the steady-state tissue residue or with slowly accumulated compounds, such as 2,3,7,8-TCDD.[24,25]

Laboratory tests offer a number of practical advantages over field surveys, including the opportunity to collect sufficient test organisms in advance of the test. Standardizing on a few test species, compared to the naturally wide variety of species found in field collections, simplifies comparing sediments or time periods.

Laboratory tests also allow a quantification of the exposure regime. Lastly, by taking time series samples, it is possible to document the occurrence of a steady state and determine the kinetics of uptake.

Practically, the main disadvantage of laboratory tests is the additional time and laboratory resources required to conduct the tests. However, analytical chemistry, required in both laboratory and field approaches, will normally constitute the major cost. Another drawback is the possibility of laboratory artifacts. Tissue residues could be overestimated if the seawater or air systems are contaminated, or underestimated if the organisms are in poor health or deplete the bioavailable fraction of the pollutant. These potential artifacts can be minimized with good laboratory practices.

One unavoidable artifact is the alteration of sediment structure and redox conditions during the collection, storage, and distribution of test sediments. Because, uptake of metals depends upon their speciation, the bioavailability of metals should be more affected than the neutral organics. Methods of sampling to minimize sediment disturbance are discussed in Chapter 3 of this volume. Another unavoidable limitation is the impossibility of mimicking the full range of environmental and physiological conditions in laboratory exposures. To the extent these factors affect bioaccumulation, a laboratory test will estimate only a subset of the range of field tissue residues.

The extent of these laboratory limitations is poorly understood, and the bioaccumulation test apparently has never been directly field validated. The similarity of laboratory and field accumulation factors (AFs, discussed below) in the same species from sediments collected from the same general area[22] suggests the tests are reasonably predictive of field tissue residues. More rigorous lab-field comparisons with a broad range of compounds are needed to further validate the laboratory approach. The laboratory test is a powerful tool for assessing existing sediments or dredge materials, when conducted properly. The ability to control the exposure regime and test organism(s) lends the bioaccumulation test to comparing bioavailability among sites or times. The approach is also adaptable to experimental studies using spiked sediment. The laboratory test can be used to predict tissue residues by conducting tests using sediments from a surrogate site. As with the field approach, finding sites with pollutant and environmental conditions similar to the predicted conditions will be difficult.

BIOACCUMULATION FACTOR

The simplest model is the bioaccumulation factor (BAF), which is the ratio between pollutant concentrations in the organism and sediment:

$$\mathrm{BAF} = \mathrm{Ctss/Cs} \qquad (1)$$

where:

Ctss = tissue concentration at steady state (μg/g), and
Cs = sediment concentration (μg/g).

BAFs are analogous to bioconcentration factors (BCF = tissue concentration/water concentration) used to predict tissue residues from water concentrations. The use of "bioconcentration" or "BCF" to refer to uptake from sediment is potentially confusing, and the ASTM practice[26] of limiting "BCF" to water uptake and "BAF" to uptake from all routes is recommended.

BAFs are empirically derived from laboratory-exposed or field-collected organisms. Laboratory tests must be of sufficient duration to estimate the steady-state tissue residue; otherwise the ratio will be underestimated. The value of a BAF depends on whether wet- or dry-weight concentrations were used. In general, a dry/dry ratio is preferable, but in either case, the wet/dry-weight ratios for the tissue and sediment should be reported in order to allow conversions.

BAFs are appealing in their simplicity and facility to predict tissue residues of neutral organics, polar organics, and metals. BAFs, however, vary with sediment type and species.[27] This variation limits the accuracy of any predictions if either the organisms or sediment differs substantially from those used in deriving the BAF. The equilibrium partitioning bioaccumulation model is almost as simple to apply and, to some extent, accounts for among-sediment and -species variations. Unless there is a substantial data base of BAFs, the equilibrium partitioning model is the preferred screening method for neutral organics. There are no equivalent models for metals or polar organics. BAFs could serve as a first-order predictive method for these compounds, while using a kinetic model for more accurate predictions.

EQUILIBRIUM PARTITIONING BIOACCUMULATION MODEL

Description Of Model And Assumptions

The equilibrium partitioning bioaccumulation model assumes that neutral organic pollutants obtain a thermodynamic equilibrium among infaunal organisms, solid phases, and interstitial water (Figure 1). At equilibrium, infaunal organisms and all the environmental phases have equal fugacities, where fugacity is the tendency of a compound to escape from a phase.[28,29] This model is also referred to as the thermodynamic bioaccumulation model.

Assuming that organic carbon is the only sink for neutral organics in the sediment, and lipids are the only sink in the organism, the model becomes (see Lake et al.[30] for derivation)

$$\mathrm{Ctss}/\mathrm{L} = \mathrm{AF} * (\mathrm{Cs}/\mathrm{TOC}) \tag{2}$$

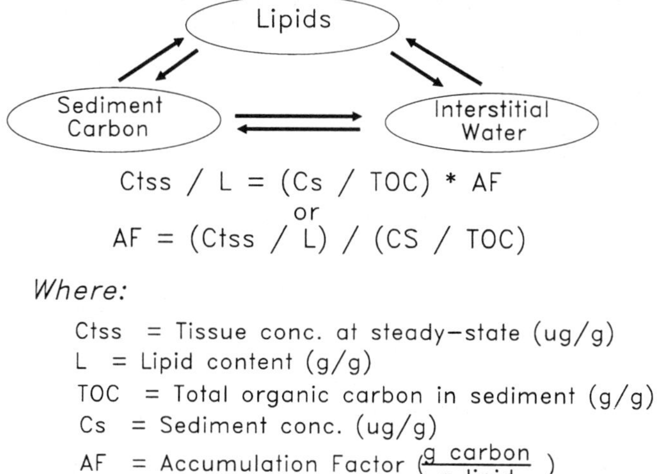

Figure 1. Equilibrium partitioning bioaccumulation model.

or

$$AF = (Ctss/L)/(Cs/TOC) \quad (3)$$

where:

L = concentration of lipids in organism (gL/g tissue) (decimal fraction);
TOC = total organic carbon in sediment (gC/g sediment) (decimal fraction);
AF = accumulation factor (gC/gL).

The accumulation factor accounts for the difference in the fugacity capacity in sediment carbon and lipids. (Note: some previous studies[30,31] reported "preference factors" that are gL/gC, the inverse of AFs.) Multiplying the AF by the carbon-normalized pollutant concentration predicts the lipid-normalized tissue residue. The value of the AF for a compound should not vary among sediments or species, assuming that partitioning is not a function of lipid or carbon type. Because the model assumes equilibrium conditions, the thermodynamic approach can not be used to predict the time-course of uptake when there are varying sediment concentrations or if the organism undergoes substantial growth or other major physiological changes (e.g., spawning). The assumption of equilibrium also precludes the model from predicting tissue residues resulting from exposure to multiple pollutant sources that are not at thermodynamic equilibrium, such as bedded sediment and overlying water.

AFs can be determined empirically for each pollutant from laboratory exposures or field surveys. Alternatively, a theoretical value of 1.7 for all compounds can be used, based on the partitioning of neutral organics between carbon and lipids (i.e., the value for gC/gL).[32] AFs less than 1.7 indicate less partitioning into the lipids than predicted, whereas values greater than 1.7 indicate more uptake than can be explained by simple partitioning. In theory, an AF of 1.7 is an environmentally conservative value, as it represents the thermodynamic limit to tissue residues.

Empirical Results and Interpretation

Because of the regulatory interest in this model,[18] it is worthwhile reviewing the equilibrium approach in detail. Five questions will be addressed: (1) are AFs constant among species; (2) are AFs constant among sediments; (3) are AFs below the theoretical maximum of 1.7; (4) are AFs better predictors than BAFs; and (5) what methodological problems exist? The most data are available for total PCBs (Table 1). Mean AF values for other compounds are given in Table 2. The values in these tables have not been standardized to a single lipid method, which will affect the comparison of AF values to a certain extent.

Rubinstein et al.,[33] McElroy and Means,[31] and Lake et al.[22] exposed different species to the same sediment, allowing an assessment of the species effect. Among-species differences in AF values were observed in all three studies. As expected, filter-feeding bivalves had lower AFs than did either deposit feeders or predators.[22] Feeding type does not explain all the variation, and the largest species difference was between two deposit feeders.[31] This variation indicates that there are biological factors, other than gross feeding type and lipid concentration (at least as presently measured), affecting the bioaccumulation of neutral organics. Based on the available information, species-related variation can result in up to a two- to fourfold variation in AFs.

Several studies have exposed the same species to several sediment types, allowing an estimation of the sediment effect on AFs.[22,31,33-37] The general trend is for an inverse relationship between AF and the sediment pollutant concentration and/or TOC. Lake et al.[22] divided stations into a low TOC (<2.6%) and PCB concentration group (<50 ppb) and a high TOC (>3.5%) and PCB concentration group (>300 ppb). The AF for Aroclor 1254 in the high group was 2.62, vs. 4.94 in the low group. This trend for higher AFs in "cleaner" sediments is consistent with results from laboratory studies using spiked[31] and field[33] sediments. There is no consistent trend among sediments with TOCs less than 3%.[33,37]

This apparent dichotomy in AF values could either be due to TOC/pollutant-concentration-related changes in partitioning or in organism behavior (e.g., feeding rate). Whatever the cause(s), AFs for the same compound in the same species can vary by as much as two- to fourfold among sediments when data from single experiments are analyzed. The effect of TOC on AFs for PCBs is not obvious when all the data are analyzed (Table 1). This may reflect the effects of different methods and species, geographical differences in sediment, or pollutant concentration effects independent of TOC effects.

Table 1. Mean AFs for PCBs for Infaunal Organisms[a]

Organism[h]	Mean AF[b]	TOC%	Reference
N. incisa	10.9	1.83	33[c]
Y. limatula	10.6	4.02	33
N. virens	10.0		30[d]
Y. limatula	8.0	1.83	33
N. incisa	5.9	4.02	33
M. nasuta	5.9	1.85	33
Y. limatula	5.7	7.94	33
N. virens	5.2	4.02	33
N. incisa	5.1	0.67–2.39	22[e]
N. virens	5.0	1.83	33
Y. limatula	4.8	0.67–2.39	22[e]
N. incisa	4.0		30
N. incisa	3.6		30[d]
Y. limatula	3.6	3.57–5.24	22[e]
N. incisa	3.5	7.94	33
M. nasuta	3.4	.5–1.1	37[f]
Y. limatula	3.3		30
N. virens	3.2	7.94	33
N. incisa	3.1	3.57–5.24	22[e]
M. nasuta	3.0	1.1–1.4	37[f]
M. nasuta	2.9	2.1–3.2	37[f]
Glycera sp.	2.8	3.6–5.2	22[e]
M. lateralis	2.0		30
M. nasuta	2.1	2.99	33
N. virens	1.9	1.85	33
M. nasuta	1.8	0.9	34
M. mercenaria	1.3	3.57–5.24	22[e]
M. nasuta	1.3	6.08	33
N. virens	1.0	2.99	33
N. virens	0.9	6.08	33
M. nasuta	0.8	3.7	34
M. nasuta	0.8	5.7	25[g]
P. pugio	0.7	5.7	25[g]
M. nasuta	0.7	4.0	34
M. nasuta	0.5	5.1	34
M. nasuta	0.5	0.8	34
N. virens	0.5	5.7	25[g]
M. nasuta	0.4	7.4	34

[a] PCBs are Aroclor 1254 or total PCBs. Values are the means for each sediment-organism treatment. The AFs are not corrected for lipid method.

[b] Mean AF (SD) = 3.4 (2.82). 0.95 Confidence Interval = 4.2. 5th Percentile = 0.5. 95th Percentile = 10.6.

[c] Values from Rubinstein et al. (1987)[33] calculated from the TOC, lipid, and tissue residues at the end of the exposure.

[d] Value in Lake et al. (1987).[30]

[e] *N. incisa* and *Y. limatula* values from Table 9, while *Glycera* and *M. mercenaria* values from Table 4 of Lake et al. (1990).[22]

[f] Sum of 13 PCB congeners spiked into sediment. Estimated lipid value used.

[g] Tissue residues and sediment concentrations given in Rubinstein et al.[24]

[h] Species: *Y. limatula* = *Yoldia limatula* (bivalve); *M. nasuta* = *Macoma nasuta* (bivalve); *N. incisa* = *Nephtys incisa* (polychaete); *N. virens* = *Nereis virens* (polychaete); *Glycera* sp. = *Glycera* sp. (polychaete); *M. lateralis* = *Mulinea lateralis* (bivalve); *M. mercenaria* = *Mercenaria mercenaria* (bivalve); *P. pugio* = *Palaemonetes pugio* (crustacean – epifaunal, but burrows into sediment).

Table 2. Mean AFs for Compounds Other than PCBs

Compound	Mean AF[a]	Mean Range	N	Reference
Chlordane	4.7	4.0–5.9	4	30[b]
Hexachlorobenzene	3.1	2.1–4.1	3	37[c]
DDD	2.1	0.4–4.8	10	30, 34
DDE	1.3	0.7–2.8	6	34
2,3,7,8-TCDD	0.7	0.5–0.8	3	25[d]
Pyrene	0.4	0.18–0.5	10	34, 36
Benzo[b,(k)] fluoranthene	0.4	0.2–1.0	6	34
Chrysene	0.4	0.2–0.6	6	34
Benz[a]anthracene	0.4	0.2–0.6	6	34
Benzo[a]pyrene	0.2	0.05–0.9	10	34, 36

[a] Mean AF is the overall mean for all sediment–organism treatments. The range is of the mean values for individual treatments.
[b] Value in Lake et al. (1987).[30]
[c] Estimated lipid value used.
[d] Tissue residues and sediment concentrations are given in Rubinstein et al. (1990).[24]

The equilibrium partitioning model could be used as an ecologically conservative screening tool if the theoretical AF predicted maximum tissue residues. However, tissue residues exceeding the theoretical maximum have been reported for several compounds, with the mean AFs for PCBs, DDD, hexachlorobenzene (HCB), and chlordane exceeding the 1.7 value by 25% to more than 250% (Tables 1 and 2). PCB AFs for a single treatment can exceed 10 (Table 1), while values for individual PCB congeners can exceed 30.[33] There is not as much data for PAHs, but it appears that 1.7 may be a maximum value for this group of compounds (Table 2).

The potential sources of error and violations of assumptions of the equilibrium partitioning bioaccumulation model are summarized in Table 3. Though the extent of error introduced by individual processes is not well quantified, it is possible to identify which processes potentially play a key role. Low AFs (less than 1.7) may result from insufficient time to reach steady state or from rapid metabolism of the parent compound. Low AFs may also result from steric hindrances, as with 2,3,7,8-TCDD,[38] some PAHs,[39] and possibly the highly chlorinated PCB congeners. Additionally, slow desorption rates from solid phases may hinder the biological availability of strongly hydrophobic compounds.

High AFs may result from active gut uptake (e.g., phagocytosis) moving pollutants against a thermodynamic gradient.[8] Selective feeding on high TOC particles increases the ingested dose over that predicted from bulk sediment concentration,[40] further contributing to high AFs if uptake is dependent upon the total exposure and not just equilibrium partitioning. Moreover, a decrease in the volume and lipid content of food during digestion increases fugacity in the gut, which may result in pollutants being thermodynamically "driven" into the gut wall.[41] As a result of these internal processes, deposit feeders may equilibrate with the gut rather than with the external environment. AFs calculated using the gut pollutant concentrations, rather than bulk sediment concentrations, may generate more accurate predictions for high-K_{ow} compounds.[42]

Table 3. Potential Violation of the Assumptions or Sources of Variation in the Equilibrium Partitioning Bioaccumulation Model[a]

Processes resulting in tissue residues less than that predicted from the theoretical AF
Rapid metabolism of pollutant by organism
Insufficient time for organism to reach steady state
Steric hindrance at membranes inhibiting uptake
Not all the pollutant associated with sediment carbon is bioavailable
Kinetic limitations of desorption reduce effective bioavailability of pollutant
Processes resulting in tissue residue greater than that predicted from theoretical AF
Active uptake of carbon and associated pollutants in the gut
Increase in gut fugacity as lipids are digested and volume of food is decreased
Multiple uptake pollutant routes are additive
Compartmentalization of pollutants within organism so organs/tissues not at thermodynamic equilibrium
Processes increasing among species or among sediment variability
Tissues other than lipids are sinks for the pollutant
Sediment components other than organic carbon are sinks for sediment-associated pollutant
AFs vary with sediment TOC or pollutant concentration
Uptake is a function of the feeding rate and/or mode of the organism
Potential methodological and application errors
Different lipid methods can result in several-fold difference in AF values
Different TOC methods may result in significant differences in AFs
The tissue values (lipid concentration, tissue concentration) must be expressed in consistent units (i.e., wet or dry weight) as must the sediment values (TOC, sediment concentration)
AFs derived from epifaunal filter-feeders are not valid for infaunal organisms
Accumulation factors (gC/gL) are not to be directly comparable to preference factors (gL/gC)

[a] The theoretical AF (1.7) is the value based on equilibrium partitioning between carbon and lipid.

The calculation of AFs is subject to several potential methodological errors (Table 3). Of these, the greatest source of variation is introduced by the use of different lipid extraction techniques, which can result in a threefold difference in AFs.[43] There is no standard lipid method, but the Bligh-Dyer method has been suggested for use in benthic bioaccumulation studies.[10,18]

Utility of the Model

The existing data demonstrate violations of the assumptions and predictions of the equilibrium partitioning model. AFs are, however, less variable than BAFs calculated from the same data.[22,33,34,37] For example, the coefficient of variation for PCB AFs was less than half that for BAFs.[22] Besides partially accounting for species and sediment effects, the model is simple to apply. Probably the greatest strength of the model is as a screening tool for neutral organics. Using a single screening AF value, such as 1.7 or 4, as in the new dredge implementation manual,[18] simplifies the application of the model, but will underestimate tissue residues in some cases and overestimate them in others.

A more realistic approach would be to use some percentile of an empirically derived distribution of AFs for each compound. This compound-specific screening value would incorporate any variation due to species, sediments, and methodological variations, and the level of protection could be adjusted by varying the percentile chosen. For example, the 95th percentile screening AF value for PCBs is 10.6 (Table 1). The corresponding value for PAHs would be considerably lower. It should be possible to reduce these screening values by deriving AFs for specific sediment and species groupings or by altering the model (e.g., applying AFs based on gut concentrations).

FIRST-ORDER KINETIC MODEL

Description of Model

Bioaccumulation can be modeled as the net result of uptake and elimination kinetics, with the uptake and elimination rates modeled as independent processes. The simplest of these kinetic process models is the first-order model that has been used to describe bioconcentration in fish.[44,45] This discussion focuses on the first-order model rather than on the more complex process models. The first-order equation to predict bioconcentration is

$$dCt/dt = k1 * Cw - k2 * Ct \tag{4}$$

where:

Ct = tissue residue (μg/g tissue);
Cw = pollutant concentration in water (μg/g water);
k1 = uptake rate constant (mL water/(tissue * time));
k2 = elimination rate constant ($time^{-1}$);
t = time.

Units for k1 are usually given as $time^{-1}$ but Stehly et al.[46] suggested that the grams should not be canceled. The elimination rate, k2, includes both depuration of the compound and metabolic degradation of the parent compound. k2 is sometimes referred to as Kd.

The same equation, with sediment pollutant concentration substituted for water concentration, can be used for sediment bioaccumulation. The term k_s, the "sediment uptake rate coefficient," is preferred over k1, as it avoids confusion with uptake from water.[8] k_s has also been referred to as the "sediment uptake clearance."[19] Including the units in k_s illustrates that the value will depend on whether dry- or wet-weight concentrations were used.

Substituting the sediment terms into Equation 4, the first-order equation becomes

$$dCt/dt = k_s * Cs - k2 * Ct \tag{5}$$

where:

Cs = pollutant concentration in sediment (μg/g sediment);
k_s = sediment uptake rate coefficient (g sediment/(g tissue * time)).

Assuming a constant pollutant concentration, Equation 5 can be integrated to predict tissue residues at any time:

$$Ct(t) = Cs * k_s / k2 * \left(1 - e^{-k2*t}\right) \tag{6}$$

where C(t) = tissue residue at time t.

As time approaches infinity, the value of (e^{-k2*t}) approaches zero, so that the maximum or equilibrium tissue concentration (Ct_{max}) becomes

$$Ct_{max} = Cs * k_s / k2 \tag{7}$$

Correspondingly, the bioaccumulation factor for a compound is the ratio between the uptake and elimination rate constants:

$$BAF = Ct_{max} / Cs = k_s / k2 \tag{8}$$

According to this model, BAFs are independent of sediment concentration, but increase with any factor that increases the sediment uptake-rate coefficient or decreases the elimination-rate constant.

The first-order model predicts that equilibrium is obtained only as time becomes infinite. For practical purposes, the apparent steady state can be defined as 95% of the equilibrium tissue residue. Rearranging Equation 6 and setting Ct/Ct_{max} to 0.95, the time to steady state is calculated as

$$S = \ln(1.00 - 0.95) / -k2 = 3.0 / k2. \tag{9}$$

where S = time to steady-state tissue residue (95% of equilibrium).

Equations 4–9 assume no growth. If test organisms grow, the increase in body mass will dilute the pollutant concentrations.[47] An organism that doubled its weight would appear to lose half of its pollutants, even though none had been depurated or metabolized. Growth can be incorporated into the first-order kinetic model as a first-order process:

$$Ct(t) = Cs * k_s / (k2 + k3) * \left[1 - e^{-(k2+k3)*t}\right] \tag{10}$$

where k3 = growth rate constant ($time^{-1}$). The maximum tissue residues becomes:

$$Ct_{max} = Cs * k_s / (k2 + k3) \tag{11}$$

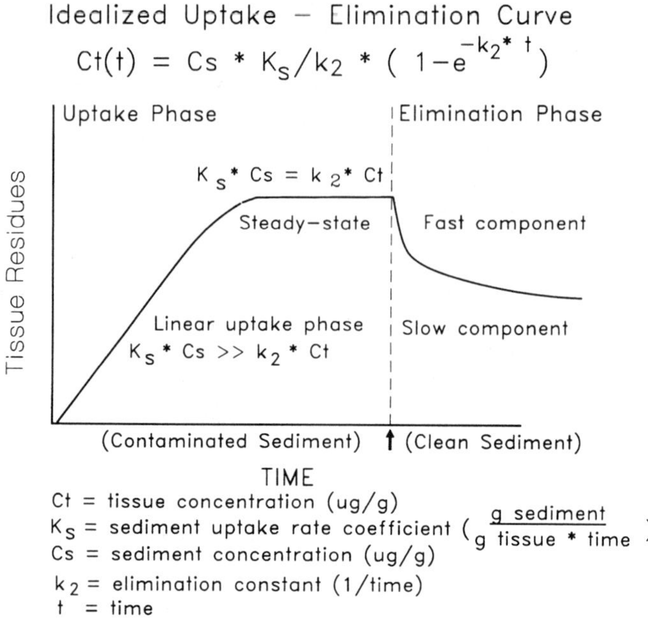

Figure 2. Idealized uptake — elimination curve.

Measurement of k_s and k2

One advantage of the first-order approach is the ability to measure the rate coefficients from relatively short-term, nonequilibrium experiments.[8,26] k_s is estimated from the rate of increase in tissue residues measured during the linear uptake phase (Figure 2). Estimating k_s from tissue residues obtained after the linear uptake phase underestimates the true value. k2 is determined from the elimination rate of the pollutant after the organisms have been placed in a clean environment (Figure 2). The model assumes that the elimination-rate constant is not a function of the tissue residue or the route of exposure. Therefore, the exposure to increase tissue residues in the test organisms before initiating the elimination phase could be from a short-term water exposure.

A substantial number of analytical chemistry samples are required to generate the regressions of tissue residue vs. time. This analytical load can more than negate any cost or time savings from conducting short-term exposures. The analytical load can be reduced by estimating k_s from a single point, as long as the sampling period is within the linear uptake phase.[8] Alternatively, both k_s and k2 can be estimated from the uptake phase alone if the experiment continues past the time when the tissue residues begin to "bend over", indicating that elimination is sufficient to slow net uptake (Figure 2). Since both k_s and k2 are estimated by fitting the data to a nonlinear equation, this method is presumably less reliable than independent measures of k_s and k2. Nonetheless, this approach has utility when time is limited or if a long-term test is terminated before steady state is attained. Branson et al.[48] and Foster et al.[21] discuss the specifics of estimating the values in the nonlinear equations.

Model Assumptions

The first-order model is based on the assumption of reversible uptake and depuration reactions between two compartments: the organism and the environment.[44,45] When applied to sediment pollutants, the entire sediment milieu is considered as a single compartment. Thus, empirically derived k_s's include uptake from both ingested sediment and interstitial water, and the first-order model does not separate uptake routes or model multiple uptake routes. The first-order model also assumes constant uptake and elimination rates. This assumption may not hold if there are major physiological changes (e.g., gonadal maturation) or thresholds in the metabolic processes controlling depuration/metabolism, or if the organism's feeding behavior varies with size or pollutant concentration.

The first-order model, as well as other kinetic process models, is not constrained by assumptions of thermodynamic equilibrium. Accordingly, the model can predict nonsteady-state tissue residues and the time course of uptake. More complex nonlinear kinetic equations can be used if the sediment concentration (or biologically available fraction) varies.[19] These nonlinear equations may be less precise and can generate nonunique solutions. The first-order kinetic model is theoretically applicable to any sediment-associated pollutant, though most of the work has been with neutral organics.[19,21,49,50] The first-order approach may not work as well with heavy metals, as discussed below.

Model Limitations And Verification

When predicting bioconcentration, it is implicitly assumed that an uptake rate constant (k1) derived from one body of water can be applied with reasonable precision to another water body. The bioavailability of sediments, however, can vary dramatically among sediment types, resulting in different k_s's. For example, the k_s's for PCBs declined by more than an order of magnitude as TOC increased from 1.3% to 5.7%,[8] while the k_s's for benzo(a)pyrene and pyrene declined about threefold as TOC increased from 0.46% to 1.03%.[36] The first-order model contains no terms to account for these sediment effects. Consequently, k_s's would have to be determined for each sediment if accurate predictions are required. It may be possible to assign a k_s by sediment type if a sufficient data base can be developed.

The first-order model does not directly account for interspecific differences in uptake rates other than by allowing the use of species-specific k_s's. However, an analysis of Rubinstein's[24,25] uptake data for 2,3,7,8-TCDD by three infaunal species suggests that species effects on k_s are less important than sediment effects.[8] The relative similarity in uptake rates suggests that interspecific differences in elimination rates may be an important source of species variations. Therefore, k2s should be determined for all key species when accurate predictions are required. It may be sufficient to use k2s derived from the same taxonomic group when less accuracy is required, though further data are required to demonstrate that species within the same genus, family, or higher taxonomic grouping have similar elimination rates.

The first-order model has not been extensively field validated with either fish or benthic invertebrates. In laboratory studies with fish exposed to neutral organics, the first-order model generally fits the observed data, though there is a tendency to underestimate BCFs of higher K_{ow} compounds (i.e., DDT) by about half.[45] A kinetic process model that allowed for varying sediment pollutant concentrations fit the observed tissue residues in freshwater benthic organisms exposed to contaminated sediment in the laboratory.[19] However, a related model, which included water and sediment uptake, overestimated tissue residues of higher K_{ow} PAHs in field-collected amphipods by about twofold, and those of lower K_{ow} PAHs (log K_{ow} ≤5) by more than eightfold.[39]

The first-order model may not be as accurate for metals as for neutral organics. The kinetics of uptake of metals may be dependent upon a small fraction of the total sediment load,[51] which in turn may be controlled by sediment characteristics such as acid volatile sulphide concentration[52] or Eh. Thus, metal bioavailability and k_s may vary with sediment characteristics that are only beginning to be understood. Additionally, exposure to high concentrations of metals can lead to the induction of metal-binding proteins, such as metallothionein, which have very low exchange rates.[53] Metal elimination may not be a simple first-order process and may depend upon the exposure history of the organism.

BIOENERGETICIALLY-BASED TOXICOKINETIC MODELS

Bioenergetic-based toxicokinetic bioaccumulation models resemble the first-order kinetic model in describing tissue residues as the balance between uptake and loss rates. The more complex kinetic process models overlap the bioenergetic-based models, with the major difference being the degree to which physiological mechanisms are incorporated. The bioenergetic-based models explicitly model physiological processes affecting uptake or elimination (e.g., feeding rate), whereas the process models aggregate these behaviors into the uptake or elimination terms (e.g., k_s).

The bioenergetic-based models depict increases in tissue residues as the sum of the uptake from each individual phase (e.g., interstitial water, ingested sediment). The general form of the model is

$$dCt/dt = \Sigma(Fx * CPx * EPx) - E - GD \qquad (12)$$

BIOENERGETIC–BASED TOXICOKINETIC MODEL
MAJOR UPTAKE ROUTES FOR DEPOSIT–FEEDERS

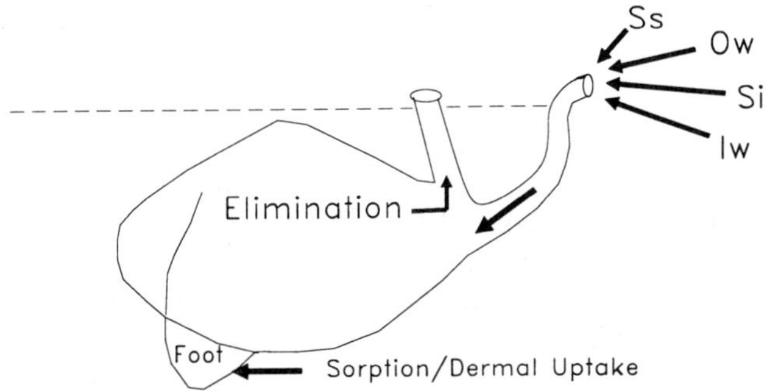

Tissue Residue = Ow + Iw + Si + Ss + Sorption − Elimination

Uptake Sediment = (Flux Sediment) * (Pollutant Conc.) * (Extraction Efficiency)

Flux Sediment = f(Weight, Activity, TOC)

Ss = Suspended Solids Uptake Route
Ow = Overlying Water Uptake Route
Si = Ingested Sediment Uptake Route
Iw = Interstitial Water Uptake Route

Figure 3. Bioenergetics-based toxicokinetic model.

where:

dCt/dt = weight-specific change in tissue residue with time (μg/(g tissue * time));
Fx = weight-specific flux of phase X through organism (g/(g tissue * time));
CPx = concentration of pollutant in phase X (μg/g);
EPx = fraction of pollutant extracted (assimilated) from phase X (decimal fraction);
E = elimination rate of pollutant from depuration of parent compound and metabolic degradation (μg/(g tissue * time));
GD = growth dilution (μg/(g tissue * time));
X = phase (Wo = overlying water, Wi = interstitial water, Si = ingested sediment, F = food)

Uptake from ingested sediment, for example, would be the product of the amount of sediment ingested (FSi) times the pollutant concentration of the ingested sediment (CPSi) times the efficiency with which the sorbed pollutant is extracted in the gut (EPSi) (Figure 3). Modeling sediment uptake requires, at the minimum, values for the fluxes, concentrations, and pollutant extraction efficiencies for ventilated interstitial water and ingested sediment. A complete description would require inclusion of an additional eight uptake routes, though with the exception of sorption or dermal uptake, these other routes make a minor contribution.[20]

Model Assumptions and Validation

The key assumption of the uptake component is that uptake from each route is independent and additive. The model predicts that an organism exposed to both interstitial water and ingested sediment would have a higher tissue residue than an organism exposed to only one uptake route. In contrast, the equilibrium partitioning model predicts no difference in tissue residues in organisms exposed to only interstitial water and organisms exposed to both routes, assuming the water and sediment were at thermodynamic equilibrium. Consequently, the bioenergetic-based model can predict tissue residues greater than the thermodynamic maximum of the equilibrium partitioning model.

Bioenergetic-based toxicokinetic models are highly flexible in their structure, and the other assumptions depend on the specific formulation. In most cases, as in Equation 12, the efficiency of pollutant assimilation terms (e.g., EPW) are assumed to be constant (i.e., do not decline with increasing tissue residues). Equation 12 does not contain an uptake term for the passive sorption of pollutants to the body surface or dermal uptake from nongill surfaces. Passive sorption, which accounted for 5 to 11% of the total uptake of HCB by *Macoma nasuta,*[54] can be included as a simple additive term.[20] Pollutant uptake from nonrespiratory surfaces may be important in certain organisms,[55] in which case a dermal uptake term would have to be added.

Elimination/metabolism is generally not modeled as mechanistically as uptake. Elimination is often modeled as a first-order process (i.e., as in Equations 4–11), which often will be sufficient, especially for neutral organics not subject to rapid metabolism.[56] The elimination component can be made more complex to account for a reduction in depuration with body weight[57] or for fast and slow compartments[49,58,59] (Figure 2). As mentioned, several assumptions inherent with first-order elimination may not hold with metals, and accurate predictions may require incorporation of threshold effects.

These types of models have successfully described bioaccumulation of PCBs, kepone, and methylmercury by fish in both freshwater and estuarine environments.[57,60-62] The only validation of a bioenergetic-based toxicokinetic model for sediment-ingesting invertebrates is a set of experiments with *Macoma nasuta.*[20,40,63,64] In these nonequilibrium exposures, predicted HCB tissue residues were within 92% to 114% of those observed in laboratory exposures. The good fit of the predicted to the observed tissue residues for fish and a sediment-ingesting invertebrate illustrates the utility of the approach. The next steps in validation should include direct testing of the assumption of additivity of uptake routes, and comparisons between predicted and observed tissue residues from field studies and steady-state laboratory exposures.

Model Utility and Limitations

The strength of the bioenergetic-based toxicokinetic approach is its flexibility in predicting tissue residues under any equilibrium or nonequilibrium exposure scenario. As with the first-order kinetic model, the bioenergetic-based models can predict the time course of uptake. The bioenergetic-based models can also predict uptake when the pollutant concentration varies, such as transient exposure events. Because uptake routes are assumed to be independent, this approach lends itself to modeling multiple input routes, such as exposure to bedded and resuspended sediments, and to assessing uptake from ingested sediment vs. interstitial water.[20] Finally, by linking bioaccumulation to bioenergetics, it is possible to predict how environmental factors, such as food and temperature, and intrinsic factors, such as growth rate, will affect tissue residues.

The greatest limitation is the difficulty in estimating the values of the input parameters. The small size and burrowing nature of infaunal invertebrates make it difficult to conduct physiological measurements. Selective ingestion of fine, high-TOC particles[40] complicates estimating the ingested sediment dose. Development of an exposure chamber, the "clambox," has largely overcome these problems for the surface deposit-feeding clam *Macoma nasuta*.[65] This chamber separates the incurrent and excurrent siphons, allowing the collection of ventilated water and defecated sediment. Unfortunately, similar methodology is more difficult to apply to most other species, and further research is required to develop toxicokinetic techniques.

Because of the extensive data needs and the ongoing methods development, the bioenergetic-based toxicokinetic approach presently has limited utility as a routine monitoring or predictive tool. These models are appropriate when detailed analysis of sediment or biological effects on bioaccumulation are required, as a method to generate insights into the factors regulating sediment bioavailability and as a method to test assumptions of the equilibrium partitioning approach to deriving sediment quality criteria.[66] Development of a data base of values for key parameters could allow the toxicokinetic approach to be used as a regulatory tool, especially in cases when nonequilibrium exposures are expected.

CONCLUSIONS

Considerable gains have been made over the last decade in our ability to predict tissue residues in infaunal organisms, especially for neutral organics. The regulator or scientist can choose among six approaches — field collection, bioaccumulation tests, BAFs, the equilibrium partitioning model, kinetic process models, and bioenergetic-based toxicokinetic models — some of which have several variations.

As an aid to making defensible and cost-effective decisions as to which technique to use, the data requirements of the six approaches are summarized in Table 4. Data requirements can be used as a rough approximation of cost. All the approaches require the analysis of sediment pollutant concentrations, so some analytical chemistry capability is required. Though required only for the equilibrium partitioning model, both lipid concentration and TOC should be measured routinely. Literature values from similar species for missing data may suffice as approximations if published values are available. When writing an expert system to predict human cancer risk from contaminated sediment, we found k_s values for only

Table 4. Minimum Data Requirements for the Six Approaches to Predicting Tissue Residues in Sediment-Dwelling Organisms

Parameter/Approach[a]	Field	Lab	BAF	Equil	1st-order[b]	Bioener
Sediment pollutant concentration	yes	yes	yes	yes	yes	yes
Tissue residue	yes	yes	no	no	no	no
Sediment TOC	no	no	no	yes	no	no
Lipid concentration	no	no	no	yes	no	no
Bioaccumulation factor	no	no	yes	no	no	no
Accumulation factor	no	no	no	yes	no	no
Sediment uptake rate coefficient (k_s)	no	no	no	no	yes	no
Elimination rate constant (k2)	no	no	no	no	yes	yes[c]
Efficient uptake sediment	no	no	no	no	no	yes
Flux sediment	no	no	no	no	no	yes
Interstitial water pollutant concentration	no	no	no	no	no	yes
Efficient uptake interstitial water	no	no	no	no	no	yes
Flux interstitial water	no	no	no	no	no	yes

[a] Methods: Field = field measurements at the site or a surrogate site; Lab = bioaccumulation test on sediment from site or surrogate site; BAF = bioaccumulation factor; Equil = equilibrium partitioning model; 1-st order = first-order kinetic model; Bioener = bioenergetic-based toxicokinetic model.

[b] More complex forms of kinetic process models may require additional data.

[c] The bioenergetic model requires an estimate of the rate of elimination, which can be a k2 or a more mechanistic function.

Table 5. Utility and Appropriate Exposure Scenarios of the Six Approaches[a] to Predicting Tissue Residues. Assessments Are for the Approaches as Discussed in the Text

	Field	Lab	BAF	Equil	1st-order[b]	Bioener
Ability to monitor tissue residues in existing sediment	high[c]	high	mod	mod	mod	mod-low
General ability to predict tissue residues	low[d]	low-mod[d]	low-mod	mod-high	mod-high	high
Predict steady-state tissue residue of neutral organics	yes	yes	yes	yes	yes	yes
Predict steady-state tissue residue of heavy metals	yes	yes	yes	no	yes[e]	yes
Predict how tissue residues vary among sediment types	no	no	no	yes	no	yes[f]
Predict how tissue residues vary among species	no	no	no	yes	no	yes[f]
Predict time course of uptake and depuration	no	yes[g]	no	no	yes	yes
Model transient, nonsteady-state pollutant exposures	no	no	no	no	yes	yes

Table 5. Utility and Appropriate Exposure Scenarios of the Six Approaches[a] to Predicting Tissue Residues. Assessments Are for the Approaches as Discussed in the Text (continued)

	Field	Lab	BAF	Equil	1st-order[b]	Bioener
Model uptake with varying pollutant concentrations	no	no	no	no	no	yes
Model uptake from multiple uptake routes (e.g., resuspended and bedded sediment)	no	no	no	no	no	yes
Predict effects of growth	no[h]	no[h]	no	no	yes	yes
Model how uptake varies with organism bioenergetics	no	no	no	no	no	yes
Mechanistic realism of model	na[i]	na	low	low-mod	mod	high
Existing data base of model parameters	na	na	high	mod-high	low	low
Expertise required to implement approach	mod	mod	low	low	mod	high

[a] Methods: Field = field measurements at the site or at surrogate site. Lab = bioaccumulation test on sediment from site or surrogate. BAF = bioaccumulation factor. Equil = thermodynamic partitioning model. 1st order = first-order kinetic model. Bioener = bioenergetically-based toxicokinetic model.
[b] Assessment limited to first-order model. More complex kinetic process models overlap the bioenergetic models in capability.
[c] Ranking: low = low ability to predict tissue residues; mod = moderate ability to predict tissue residues; high = high ability to predict tissue residues.
[d] Prediction of the direct methods is by sampling or testing a site with comparable environmental and pollutant conditions.
[e] Unclear whether k2 adequately represents depuration of metals.
[f] Requires efficiency of uptake and flux values appropriate for the sediment or organism type.
[g] Sampling periods in addition to the standard 28-d sampling are required to generate a time course of uptake.
[h] Incorporates growth, but does not usually allow prediction to other exposure scenarios.
[i] Not applicable for the direct approaches.

4 of the 15 neutral organics.[7] Until a greater data base of key parameters is developed, laboratory studies often will be required.

The decision of which approach to use should be based heavily on the exposure scenario and the required accuracy. A summary of the factors to consider and the relative efficacy of the different approaches are summarized in Table 5. The assessments in Table 5 are a guide, and other researchers might have different rankings.

In addition to Tables 4 and 5, several general guidelines are possible. All the approaches predict steady-state tissue residues. Because of its ease of application, the equilibrium partitioning approach lends itself as a screening tool for neutral organics. When accurate predictions for steady-state tissue residues of neutral organics are required, either site-specific AFs should be generated for all the key compounds or another model should be used. Using BAFs as a screening tool for polar organics and metals is more problematical. Unless a site-specific BAF is available, bioaccumulation tests and/or kinetic models are the better choices for predicting tissue residues of these compounds.

The assumption of steady state is not appropriate if there is insufficient time to obtain steady state or if the goal is to predict temporal changes in tissue residues. In these cases, kinetic process or bioenergetic-based models are the appropriate approaches. These models also should be used to predict the effects of growth on tissue residues. Incorporation of growth is most important when predicting the elimination of slowly depurated compounds. In addition, the assumption of steady state is violated if there are substantial variations in sediment pollutant concentrations or in the physiological state of the organism. Nontrivial changes in pollutant concentration are presumably most common for rapidly degraded compounds and in areas exposed to rapid erosion or deposition. Bioenergetic-based models are the appropriate approach to model tissue residues under changing conditions, although nonlinear kinetic models can also incorporate varying sediment concentrations. The bioenergetic-based models, or more complex kinetic process models, are the correct approach to predicting bioaccumulation from multiple uptake routes.

The progress over the last decade is encouraging, but our understanding of sediment bioavailability and bioaccumulation is far from complete. This knowledge gap is underscored by the 80-fold range in AFs for a single PCB congener[22] and the 600-fold range in k2 for B(a)P.[7] Multiple causes contribute to this uncertainty, including analytical nonstandardization, violations of model assumptions, and unrealistic or incomplete model formulations. To reduce this uncertainty, future research needs to identify the key sources of variation and then refine the models and measurements of the input parameters accordingly.

ACKNOWLEDGMENTS AND DISCLAIMER

I greatly appreciate the efforts of my co-workers — Dr. Bruce Boese, Judy Pelletier, Bob Randall, Karl Rukavina, David Specht, and Martha Winsor — who contributed to generating much of the data this paper draws upon. Dr. Peter Landrum, Dr. Ted DeWitt, Dr. Allen Burton, Dr. Bill Benson, Norm Rubinstein, Judy Pelletier, Martha Winsor, and Tricia Lawson reviewed the manuscript, but they all continue to eschew any responsibility for my mistakes. Martha Winsor helped with all math beyond addition and subtraction. Judy Pelletier drew the figures. The patience of Dr. Allen Burton is gratefully acknowledged. Portions of this paper were modified from "Guidance Manual: Bedded Sediment Bioaccumulation Tests" and "Computerized Risk And Bioaccumulation System." The recommendations made here do not constitute official policy by the U.S. Environmental Protection Agency. Mention of trade names or commercial products does not constitute endorsement by the U.S. Environmental Protection Agency. Environmental Research Laboratory, Narragansett contribution number N161.

REFERENCES

1. Dexter, R., and J. Field. "A Discussion of Sediment PCB Target Levels for the Protection of Aquatic Organisms," in *Oceans 89,* Vol. 2, (Piscataway, NJ: Institute of Electrical and Electronic Engineers, 1989), pp. 452–456.
2. Young, D. "Report on the Assessment and Application of Pollutant Biomagnification Potential in Near Coastal Waters," U.S. EPA Report No. 600/x-88/295, ERL-Narragansett, p. 195 (1988).
3. Weseloh, D. V., S. M. Teeple, and M. Gilberton. "Double-Crested Cormorants of the Great Lakes: Egg-Laying Parameters, Reproductive Failure, and Contaminant Residues in Eggs, Lake Huron 1972–1973," *Can. J. Zool.* 61:427–436 (1983).
4. Spies, R. B., and D. W. Rice. "Effects of Organic Contaminants on Reproduction of the Starry Flounder *Platichtys stellatus* in San Francisco Bay. II. Reproductive Success of Fish Captured in San Francisco Bay and Spawned in the Laboratory," *Mar. Biol.* 98:191–200 (1988).
5. Lee, H., II, and R. C. Randall. "Cancer Risk Associated with Sediment Quality," in *Proc. 1st Ann. Meeting Puget Sound Research,* Vol. 2 (Seattle, WA: Puget Sound Water Quality Authority, 1988), p. 676.
6. U.S. Environmental Protection Agency. "Assessing Human Health Risks from Chemically Contaminated Fish and Shellfish: A Guidance Manual," U.S. EPA Report No. 503/8-89-002, Office of Marine and Estuarine Protection, Washington, D.C., p. 91 and appendices (1989).
7. Lee, H., II, M. Winsor, J. Pelletier, and R. C. Randall (J. Bertland, and B. Coleman, programmers). "Computerized Risk and Bioaccumulation System," U.S. EPA Report, ERL-Narragansett Contribution No. N137, p. 23 and computer program (1990).
8. Lee, H., II, "A Clam's Eye View of the Bioavailability of Sediment-Associated Pollutants," in *Organic Substances in Sediment and Water*, J. F. McCarthy, Ed. (Chelsea, MI: Lewis Publishers, 1991), pp. 73–93.
9. Hedgpeth, J. W. "Models and Muddles: Some Philosophical Observations," *Helgoland. Meeresunt.* 30:92–104 (1977).
10. Lee, H., II, B. L. Boese, J. Pelletier, M. Winsor, D.T. Specht, and R. C. Randall. "Guidance Manual: Bedded Sediment Bioaccumulation Tests," U.S. EPA Report No. 600/x-89/302, ERL-Narragansett, p. 232 (1989).
11. Pearson, T. H., and R. Rosenberg. "Macrobenthic Succession in Relation to Organic Enrichment and Pollution of the Marine Environment," *Oceanogr. Mar. Biol. Annu. Rev.* 16:229–311 (1978).
12. Rhoads, D. C., P. L. McCall, and J. Y. Yingst. "Disturbance and Production on the Estuarine Seafloor," *Am. Sci.* 66:557–586 (1978).
13. Varanasi, U., W. L. Reichert, J. E. Stein, D. B. Brown, and H. R. Sanborn. "Bioavailability and Biotransformation of Aromatic Hydrocarbons in Benthic Organisms Exposed to Sediment from an Urban Estuary," *Environ. Sci. Technol.* 19:836–841 (1985).

14. McElroy, A., J. W. Farrington, and J. M. Teal. "Influence of Mode of Exposure and the Presence of a Tubiculous Polychaete on the Fate of Benz[a]anthracene in the Benthos," *Environ. Sci. Technol.* 24:1648–1655 (1990).
15. Williams, A. B., and K. H. Bynum. "A Ten-Year Study of Meroplankton in North Carolina Estuaries: Amphipods," *Ches. Sci.* 13:175–192 (1972).
16. Williams, A. B., and H. J. Porter. "A Ten-Year Study of Meroplankton in North Carolina Estuaries: Occurrence of Postmetamorphal Bivalves," *Ches. Sci.* 12:26–32 (1971).
17. U.S. EPA/ACE. "Ecological Evaluation of Proposed Discharge of Dredged Material into Ocean Waters; Implementation Manual for Section 103 of PL 92-532," Environmental Effects Laboratory, U.S. Army Engineer Waterways Experiment Station, Vicksburg, MS (1977).
18. U.S. EPA/ACE. "Evaluation of Dredged Material Proposed for Ocean Disposal (Testing Manual)," U.S. EPA Report No. 503/8-91/001, Office of Marine and Estuarine Protection, Washington, D.C. (1991).
19. Landrum, P. F. "Bioavailability and Toxicokinetics of Polycyclic Aromatic Hydrocarbons Sorbed to Sediments for the Amphipod *Pontoporeia hoyi*," *Environ. Sci. Tech.* 23:588–595 (1989).
20. Boese, B. L., H. Lee, II, D. T. Specht, R. C. Randall, and M. Winsor. "Comparison of Aqueous and Solid-Phase Uptake for Hexachlorobenzene in the Tellinid Clam, *Macoma nasuta* (Conrad): A Mass Balance Approach," *Environ. Toxicol. Chem.* 9:221–231 (1990).
21. Foster, G. D., S. M. Baski, and J. C. Means. "Bioaccumulation of Trace Organic Contaminants from Sediment by Baltic Clams *(Macoma balthica)* and Soft-Shell Clams *(Mya arenaria)*," *Environ. Toxicol. Chem.* 6:969–976 (1987).
22. Lake, J. L., N. I. Rubinstein, H. Lee, II, C. A. Lake, J. Heltshe, and S. Pavignano. "Equilibrium Partitioning and Bioaccumulation of Sediment-Associated Contaminants by Infaunal Organisms," *Environ. Toxicol. Chem.* 9:1095–1106 (1990).
23. Tracey, G., B. K. Taplin, D. J. Cobb, W. J. Berry, D. J. Keith, W. S. Boothman, and N. I. Rubinstein, "Dredged Material Assessment Methods: Evaluation of the Revised Implementation Manual for Section 103 of the Marine Protection Research and Sanctuaries Act of 1972 (PL92-532)," U.S. EPA Report, ERL-Narragansett Contribution No. ERLN-1234, p. 161 (1991).
24. Rubinstein, N. I., R. J. Pruell, B. K. Taplan, J. A. LiVolsi, and C. B. Norwood. "Bioavailability of 2,3,7,8-TCDD, 2,3,7,8-TCDF and PCBs to Marine Benthos from Passaic River Sediments," *Chemosphere* 20:1097–1102 (1990).
25. Rubinstein, N. I., R. J. Pruell, B. K. Taplin, J. A. LiVolsi, and C. B. Norwood. "Bioavailability of 2,3,7,8-TCDD, 2,3,7,8-TCDF and PCBs to Marine Benthos from Contaminated Sediment," paper presented at 10th Society of Environmental Toxicology and Chemistry Meeting, Toronto, Canada (1989).
26. American Society for Testing and Materials. "Standard Practice for Conducting Bioconcentration Tests with Fishes and Saltwater Bivalve Molluscs" (Philadelphia: American Society for Testing and Materials, 1984), E 1022–84.
27. Rubinstein, N. I., E. Lores, and N. Gregory. "Accumulation of PCBs, Mercury, and Cadmium by *Nereis virens, Mercenaria mercenaria,* and *Palaemontes pugio* from Contaminated Sediments," *Aquat. Toxicol.* 3:249–260 (1983).
28. McKay, D., and S. Paterson. "Calculating Fugacity," *Environ. Sci. Tech.* 15:1006–1014 (1981).

29. McKay, D., and S. Paterson. "Fugacity Revisited," *Environ. Sci. Tech.* 16:654A–660A (1982).
30. Lake, J. L., N. I. Rubinstein, and S. Pavignano. "Predicting Bioaccumulation: Development of a Partitioning Model for Use as a Screening Tool in Regulating Ocean Disposal of Wastes," in *Fate and Effects of Sediment-Bound Chemicals in Aquatic Systems,* K. L. Dickson, A. W. Maki, and W. A. Brungs, Eds. (New York: Pergamon Press, 1987), pp. 151–166.
31. McElroy, A. E., and J. C. Means. "Factors Affecting the Bioavailability of Hexachlorobiphenyls to Benthic Organisms," in *Aquatic Toxicology and Hazard Assessment: 10th Vol.,* STP 971, W. J. Adams, G. A. Chapman, and W. G. Landis, Eds. (Philadelphia: American Society for Testing and Materials, 1988), pp. 149–158.
32. McFarland, V. A., and J. U. Clarke. "Testing Bioavailability of Polychlorinated Biphenyls Using a Two-Level Approach," in *Water Quality R&D: Successful Bridging Between Theory and Application,* R. G. Willey, Ed. (Davis, CA: Hydrologic Engineering Research Center, 1986), pp. 220–229.
33. Rubinstein, N. I., J. L. Lake, R. J. Pruell, H. Lee, II, B. Taplin, J. Heltshe, R. Bowen, and S. Pavignano. "Predicting Bioaccumulation of Sediment-Associated Organic Contaminants: Development of a Regulatory Tool for Dredged Material Evaluation," U.S. EPA Report No. 600/x-87/368, ERL-Narragansett, p. 54 and appendices (1987).
34. Ferraro, S., H. Lee, II, R. Ozretich, and D. T. Specht. "Predicting Bioaccumulation Potential: A Test of a Fugacity-Based Model," *Arch. Environ. Contamin. Toxicol.* 19:386–394 (1990).
35. Ferraro, S., H. Lee, II, L. Smith, R. Ozretich, and D. T. Specht. "Accumulation Factors for Eleven Polychlorinated Biphenyl Congeners," *Bull. Environ. Contam. Toxicol.* 46:276–283 (1990).
36. Landrum, P. F., and W. R. Faust. "Effect of Variation in Sediment Composition on the Uptake Rate Coefficient for Selected PCB and PAH Congeners by the Amphipod, *Diporeia* sp.," in *Aquatic Toxicology and Risk Assessment, 14th Vol.* (Philadelphia: American Society for Testing and Materials), in press.
37. Lee, H., II et al. Work in progress. U.S. EPA, ERL-Narragansett, Newport, OR.
38. Opperhiuzen, A., and D. T. H. M. Sijm. "Bioaccumulation and Biotransformation of Polychlorinated Dibenzo-p-dioxins and Dibenzofurans in Fish," *Environ. Toxicol. Chem.* 9:175–186 (1990).
39. Landrum, P. Personal communication.
40. Lee, H., II, B. L. Boese, J. Pelletier, and R. C. Randall. "A Method to Estimate Gut Uptake Efficiencies for Hydrophobic Organic Pollutants in a Deposit-Feeding Clam," *Environ. Toxicol. Chem.* 9:215–219 (1990).
41. Gobas, F., C. Derek, and D. MacKay. "Dynamics of Dietary Bioaccumulation and Fecal Elimination of Hydrophobic Organic Chemicals in Fish," *Chemosphere* 17:943–962 (1988).
42. Boese, B. L., H. Lee, II, M. Winsor, D. T. Specht, and R. C. Randall. "Testing Sediment Accumulation Factors: The Effects of Sediment TOC and Selective Deposit-Feeding," paper presented at the 11th Society Environmental Toxicology and Chemistry Meeting, Arlington, VA (1990).
43. Randall, R. C., H. Lee, II, R. Ozretich, J. L. Lake, and R. J. Pruell. "Evaluation of Selected Lipid Methods for Normalizing Pollutant Bioaccumulation," *Environ. Toxicol. Chem.* 10:1431–1436 (1991).

44. Spacie, A., and J. L. Hamelink. "Alternate Models for Describing the Bioconcentration of Organics in Fish," *Environ. Toxicol. Chem.* 1:309–320 (1982).
45. Davies, R. P., and A. J. Dobbs. "The Prediction of Bioconcentration in Fish," *Water Res.* 18:1253–1262 (1984).
46. Stehly, G. R., P. F. Landrum, M. G. Henry, and C. Klemm. "Toxicokinetics of PAHs in *Hexagenia,*" *Environ. Toxicol. Chem.* 9:167–174 (1990).
47. Niimi, A. J., and C. Y. Cho. "Elimination of Hexachlorobenzene (HCB) by Rainbow Trout *(Salmo gairdneri),* and an Examination of Its Kinetics in Lake Ontario Salmonids," *Can. J. Fish. Aquat. Sci.* 38:1350–1356 (1981).
48. Branson, D. R., G. E. Blau, H. C. Alexander, and W. B. Neely. "Bioconcentration of 2,2′,4,4′-Tetrachlorobiphenyl in Rainbow Trout as Measured by an Accelerated Test," *Trans. Am. Fish. Soc.* 4:785–792 (1975).
49. Landrum, P. F., and D. Scavia. "Influence of Sediment on Anthracene Uptake, Depuration, and Biotransformation by the Amphipod *Hyalella azteca,*" *Can. J. Fish. Aquat. Sci.* 40:298–305 (1983).
50. Landrum, P. F., and R. Poore. "Toxicokinetics of Selected Xenobiotics in *Hexagenia limbata,*" *J. Great Lakes Res.* 14:427–437 (1988).
51. Luoma, S. N., and G. W. Byran. "A Statistical Study of Environmental Factors Controlling Concentrations of Heavy Metals in the Burrowing Bivalve *Scrobicularia plana* and the Polychaete *Nereis diversicolor,*" *Estuar. Coast. Shelf Sci.* 15:95–108 (1982).
52. Di Toro, D. M., J. D. Mahony, D. J. Hansen, K. J. Scott, M. B. Hicks, S. M. Mayr, and M. S. Redmond. "Toxicity of Cadmium in Sediments: The Role of Acid Volatile Sulfide," *Environ. Toxicol. Chem.* 9:1489–1504 (1990).
53. Byran, G. W. "Some Aspects of Heavy Metal Tolerance in Aquatic Organisms," in *Effects of Pollution on Aquatic Organisms,* A. P. M. Lockwood, Ed. (New York: Cambridge University Press, 1976), pp. 7–34.
54. Lee, H., II, B. L. Boese, M. Winsor, R. C. Randall, and D. T. Specht. "Passive Sorption of HCB from Interstitial Water by the Clam *Macoma nasuta,*" paper presented at 9th Society of Environmental Toxicology and Chemistry Meeting, Arlington, VA (1988).
55. Landrum, P. F., and C. R. Stubblefield. "Role of Respiration in the Accumulation of Organic Xenobiotics by the Amphipod, *Diporeia* sp.," *Environ. Toxicol. Chem.* 10: 1019–1028 (1991).
56. Boese, B. L., M. Winsor, H. Lee, II., D. T. Specht, and K. R. Rukavina. "Depuration Kinetics of Hexachlorobenzene in the Clam, *Macoma nasuta,*" *Comp. Biochem. Physiol.* 96C:327–331 (1990).
57. Norstrom, R. J., A. E. McKinnon, and A. S. deFreitas. "A Bioenergetic Based Model for Pollutant Accumulation by Fish. Simulation of PCB and Methylmercury Residue Levels in Ottawa River," *J. Fish. Res. Board Can.* 33:248–267 (1976).
58. Goldstein, R. A., and J. W. Elwood. "A Two-Compartment, Three-Parameter Model for the Absorption and Retention of Ingested Elements by Animals," *Ecology* 52:935–939 (1971).
59. Agren, G. I. "Simple Elimination Models: Theoretical Estimates of the Effect of Variations in Elimination Rates," *Ecol. Model.* 12:281–295 (1981).
60. Jensen, A. L., S. A. Spigarelli, and M. M. Thommes. "PCB Uptake by Species of Fish in Lake Michigan, Green Bay of Lake Michigan, and Cayuga Lake, New York," *Can. J. Fish. Aquat. Sci.* 39:700–709 (1982).

61. Thomann, R. V., and J. P. Connolly. "Model of PCB in the Lake Michigan Lake Trout Food Chain," *Environ. Sci. Toxicol.* 18:65–71 (1984).
62. Connolly, J. P., and R. Tonelli. "Modeling Kepone in the Striped Bass Food Chain of the James River Estuary," *Estuar. Coast. Shelf Sci.* 20:349–366 (1985).
63. Boese, B. L., H. Lee, II, and D. T. Specht. "The Efficiency of Uptake of Hexachlorobenzene from Water by the Tellinid Clam *Macoma nasuta,*" *Aquat. Toxicol.* 12:345–356 (1988).
64. Winsor, M., B. L. Boese, H. Lee, II, R. C. Randall, and D. T. Specht. "Determination of the Ventilation Rate of Interstitial and Overlying Water by the Clam *Macoma nasuta,*" *Environ. Toxicol. Chem.* 9:209–213 (1990).
65. Specht, D. T., and H. Lee, II. "Direct Measurement Technique for Determining Ventilation Rate in the Deposit-Feeding Clam *Macoma nasuta* (Bivalvia, Tellinacea)," *Mar. Biol.* 101:211–218 (1989).
66. Di Toro, D. M., C. S. Zarba, D. J. Hansen, W. J. Berry, R. C. Swartz, C. E. Cowen, S. P. Pavlou, H. E. Allen, N. A. Thomas and P. R. Paquin. "Technical Basis for Establishing Sediment Quality Criteria for Non-ionic Organic Chemicals by Using Equilibrium Partitioning." *Environ. Toxicol. Chem.* 10: 1541–1583 (1991).

CHAPTER 13

Sediment Bioaccumulation Testing with Fish

Michael J. Mac and Christopher J. Schmitt

INTRODUCTION

Bottom sediments serve as reservoirs for many persistent chemicals in aquatic systems, and consequently, sediments in polluted areas often contain high concentrations of these chemicals. Because sediments may also serve as sources of contaminants to aquatic organisms, their management is critical to the successful and cost-effective remediation of polluted harbors, rivers, and nearshore areas. Sediment-associated contaminants can be toxic to bottom-dwelling organisms; such toxicity can be ascertained by the bioassays discussed in previous chapters. These contaminants can also be bioaccumulated by aquatic organisms to the extent that elevated levels in tissues result in chronic effects.

To make intelligent decisions on the removal or disposal of contaminated sediments from a particular site, two principal questions about bioaccumulation must be answered: (1) are contaminants in the sediments available for bioaccumulation by aquatic organisms; and (2) what concentration of these contaminants may be accumulated by organisms from different levels of the food chain? These questions are most important where persistent chemicals such as polychlorinated biphenyls (PCBs) and mercury enter food chains and biomagnify to concentrations that may induce chronic toxicity, particularly in predators at higher trophic levels, or accumulate to concentrations that render fish and wildlife unsafe for human consumption.

The status of sediment assessment is unsettled. Only during the last few years has research been adequate on several sediment bioassays so that bioassessment strategies can be confidently recommended to replace, or at least augment, bulk sediment criteria. As part of any bioassessment, the problem of bioaccumulation must be addressed. How best to accomplish this is still under study, and in this chapter we discuss methods for conducting bioaccumulation bioassays with fish; the advantages and disadvantages of using fish rather than invertebrates; and problems associated with bioaccumulation testing, with a special emphasis on statistical treatment.

BIOACCUMULATION TESTING WITH FISH RATHER THAN INVERTEBRATES

Test organisms most often recommended in sediment bioassays, including bioaccumulation tests, are benthic or epibenthic invertebrates.[1-5] This is justifiable because these organisms, which live in closest proximity to the contaminant source (i.e., the sediment), are expected to be most affected. Different organisms, however, may have unique features that determine their utility for testing the accumulation of contaminants from sediments. The most comprehensive assessment of sediments would therefore include bioaccumulation testing with both benthic invertebrates and fish, particularly fish species that have some association with sediment.

Fish offer some unique advantages over many benthic invertebrates in bioaccumulation testing. First, fish often resuspend sediment, which realistically increases contaminant availability. Second, fish provide adequate tissue mass for extensive chemical analyses, and because of their high lipid content, tissue mass requirements for organic contaminant analyses are further reduced. Whereas tissue mass needed for analysis may not be limiting in many marine invertebrates, the availability of sufficiently large, indigenous benthic invertebrates for year-round testing of freshwater sediments is greatly limited. Third, fish have several routes of exposure to sediment-associated contaminants: ingesting sediment or infaunal organisms, accumulating dissolved chemicals from the water column through the gills and skin, and, possibly, by providing a substrate (gill) for the dissociation of hydrophobic chemicals from sediment particles. The contribution of contaminants from diet and respiration to the overall bioaccumulation process, and factors that influence their contribution, such as molecular size and octanol-water partition coefficient, are understood adequately enough to allow predictive modeling of bioaccumulation in fish.[6-8] Fourth, fish can easily be caged in the field for in situ assessments where accumulation rates similar to the laboratory have been reported.[9] Although some invertebrates, particularly marine species, may possess some of these attributes, no species really combines them with other useful features of fish, such as available laboratory culture methods, established use in sediment bioassays, and extensive acute toxicity data bases.

In bioaccumulation tests, fish may not provide the same results as invertebrates. Fish exposure may be different because of their mobility. This mobility enables fish to ameliorate the variable distribution of contaminants in nonhomogeneous sediments, and their dose would be represented by the concentration of sediments over a large area. Conversely, the exposure received by sessile benthic organisms is more representative of their immediate surroundings, making the dose received easier to measure. Fish also have different metabolic capabilities than invertebrates and may readily transform parent compounds that were available from sediment. Unless one analyzes for specific metabolites, some compounds that were accumulated by less metabolically active invertebrates may seem unavailable to fish.

BIOACCUMULATION IN REGULATORY TESTING

Freshwater

In the U.S., current regulations for evaluating the suitability of dredged material from freshwater for disposal are basically the U.S. Environmental Protection Agency (EPA) 1977 bulk sediment criteria.[10] These criteria rank sediments as nonpolluted, moderately polluted, or heavily polluted, with ranges of values for heavy metals, volatile solids, and oil and grease. These are the only available criteria for regulatory procedures, and they do not employ the bioassessment-based strategy recommended by many experts.[1-5] These bulk criteria could best be described as obsolete because of the present knowledge of sediment toxicity testing. Over the past decade, several guidance documents for evaluating freshwater sediments have been published as products of workshops or various committees.[1,4,5,11] These documents were compiled by scientists familiar with sediment evaluation techniques, and they universally recommended a bioassessment approach, often in a tiered testing protocol, which included bioaccumulation testing with both infaunal invertebrates and fish. The federal government does not require bioaccumulation tests for evaluating the disposal of freshwater sediments. However, in the EPA Assessment and Remediation of Contaminated Sediments Program,[5] the most recent evaluation of contaminated sediments in the Great Lakes includes bioaccumulation testing in a multiple-assay scheme.

In Canada the development of regulatory guidelines for freshwater sediment evaluation is also evolving. The Ontario Ministry of the Environment recommended bulk criteria values[12] that would not require bioaccumulation testing. Environment Canada, however, is examining a more biological-effects-based strategy that emphasizes benthic community structure and includes testing for bioaccumulation.

Marine

The "Implementation Manual"[13] published by the EPA and the U.S. Army Corps of Engineers in 1977 was used to evaluate sediments for ocean disposal under the

Ocean Dumping Regulations of the Marine Sanctuaries, Research, and Protection Act (MSRPA;40 CFR 220-228). However, this manual was recently revised.[2] Both the 1977 manual and its revision recommend bioaccumulation testing. The new manual, which includes multiple testing tiers, recommends several bioaccumulation measures. In the second tier of the testing protocol, an evaluation of the theoretical bioaccumulation potential of nonpolar organic contaminants is recommended. This technique estimates the equilibrium concentration of contaminants in organisms, based on the concentration of contaminants in the sediment, the total organic carbon content of the sediment, and the lipid content of the organisms. In Tier III, a laboratory bioaccumulation test is recommended, with either polychaetes or bivalve molluscs as test organisms. The methods of Lee et al.[14] (see Chapter 12 of this volume) may be followed where they are consistent with the manual.[2] The manual states that the purpose of the Tier III test is to determine bioaccumulation potential only and does not imply that steady-state levels are attained. In Tier IV, the assessment of steady-state bioaccumulation is recommended, using either field reconnaissance of contaminant concentrations in indigenous organisms or the American Society for Testing and Materials (ASTM) standard bioconcentration test with fish and mollusks,[15] which uses time-series sampling. In summary, marine sediment assessment will continue to have bioaccumulation testing in its protocol, and benthic invertebrates are the preferred test organism.

AVAILABLE METHODS FOR TESTING FISH

Fish have not been used extensively in testing of sediment toxicity (see Chapters 7 and 8 of this volume), but several species have been used successfully in bioaccumulation tests in the laboratory. Only one fish bioaccumulation test, which uses the fathead minnow *(Pimephales promelas)* to measure availability of contaminants in sediments from freshwater,[9] has been published as a test method. The recommended protocol for this test is a 10-d exposure to whole sediment under flow-through conditions (0.1 L/min) at 20°C. Organisms are not fed during exposure, and no purging is required. Concurrent exposure to reference or control sediments is necessary for statistical comparisons. Although a specific 10- or 28-d test for marine fish has not been developed, several species have been used successfully in laboratory research on factors affecting bioaccumulation. These include the spot, *Leiostomus xanthurus,*[16] the sheepshead minnow, *Cyprinodon variegatus,* and the Atlantic silverside, *Menidia menidia.*[17] Rubinstein et al.[17] found that the sheepshead minnow accumulated PCBs to the greatest degree and with the lowest variability; they also recommended the silverside if sensitivity to toxicity was a critical factor. None of these marine species is a bottom dweller, however, which limits their utility for the testing of bioaccumulation from sediments.

In addition to laboratory assays, field assessment of bioaccumulation is another viable evaluation technique. This can be accomplished either by sampling resident organisms from an area or by exposing caged organisms in situ. The Ocean Disposal Guidelines recommend field assessment as an option in certain situations;[2,13] however, these guidelines also recommend that mobile organisms, such as fish, not be used. Fish species that have limited ranges or are confined by physical boundaries could be used however. Such local fish populations would be representative of conditions in a harbor, estuary, embayment, or river reach. For the evaluation of sediment sources of contamination, bottom-dwelling fish such as common carp *(Cyprinus carpio),* bullhead (*Ictalurus* spp.), or suckers (*Catostomidae* spp.) in freshwater, or the flatfishes *(Bothidae, Pleuronectidae, Soleidae)* in marine environments are preferred. For evaluation of a specific location, caged organisms can be used. Caged fathead minnows have been used to evaluate contaminant availability from sediments in short-term (8- and 10-day) and long-term (62-day) exposures.[9,18,19] Short-term exposures of caged fathead minnows have also been used to discriminate between sediment and water column sources of contaminants[9] and to identify potential point sources.[19] Selection of the organism, whether fish or infaunal invertebrate, and the type of exposure (laboratory, in situ caged, or feral) should be dictated by the information required. Feral fish give the best qualitative information about the numbers and types of chemicals in an area. Laboratory exposures can identify sediments as a source of contaminants, but this has also been accomplished with in situ cage exposures.[9]

Another technique that is in the developmental stages for estimating bioaccumulation is the Theoretical Bioaccumulation Potential (TBP) Method.[2,20-22] With TBP one can approximate the equilibrium concentration of nonpolar organic chemical residues in organisms exposed to contaminated sediments. The contaminant concentration in the sediment (SED) is normalized for the amount of organic carbon (%TOC) in the sediment and the percent lipid content (%L) in the organism with the formula

$$\mathrm{TBP} = 4\left(\mathrm{SED}/\%\mathrm{TOC}\right)\%\mathrm{L}$$

The value of the constant in this formula has varied slightly among researchers; the value of 4 is from the "Revision of the Marine Implementation Manual"[2] and is supported by the findings of Lake et al.[23] Most support for this method, however, is based on invertebrate data, and its applicability to fish is uncertain. The manual states that TBP is used where the sediment in question is the only source of contaminants to the organism. This would rarely be the case with fish, and the example in the manual that uses rainbow trout *(Oncorhynchus mykiss)* to demonstrate TBP seems inappropriate. TBP or an alternative model is advantageous in sediment assessment because it estimates the maximum contaminant concentration attainable in biota and allows sediment assessment strategies to just test for bioavailability, which is much easier. Supportive information for applying TBP to fish will be difficult to gather from the field because the mobility of fish complicates sediment exposure assessment, and whether fish can attain an equilibrium concentration in a laboratory exposure to sediments when dietary exposure is included is uncertain.[8,24]

Another technique for the measurement of bioavailability of polar and nonpolar organic chemicals is solvent-filled artificial membranes.[25,26] These membranes, which rely on either dialysis or partitioning across an artificial membrane, mimic biological membranes to some extent by restricting the size and chemical characteristics of the molecules that can pass. The membranes contain an organic solvent or lipid into which chemicals passively diffuse. This technique may accurately measure the dissolved fraction of organic chemicals, which is readily available, but does not take into account other routes of accumulation, such as ingestion.

METHODOLOGICAL PROBLEMS

Length of Test

The length of bioassays for sediment bioaccumulation has been controversial because the short tests (10 day) and the longer tests (28 day) have benefits and drawbacks. In the "1977 Implementation Manual,"[13] the 10-day solid-phase exposure was recommended for measuring bioaccumulation potential. In the revision the recommended duration for bioaccumulation tests with benthic invertebrates is 10 days if metals are the only contaminants of concern, but 28 days for organic contaminants.[2] The revised manual states that the purpose of these tests is to measure bioavailability, and reaching a steady-state concentration is not important. This clarification is critical because data from 28-day exposures should not be mistaken as a steady-state measure. Organic contaminants in sediments vary widely in octanol/water partition coefficients (K_{ow}), and thus display a wide range of chemical behaviors. The attainment of even a pseudo steady state could not be guaranteed for all chemicals. This is even more critical with fish bioaccumulation testing because in the case of chemicals with log K_{ow} ≥ 4, accumulation from food becomes significant[8] and steady state may never be reached.[24]

Evidence of the relation between exposure time and K_{ow} can be demonstrated in the results of a study by Rathbun et al.[19] that examined the accumulation of PCB congeners. Recent information indicates that the toxicological significance of PCB contamination can only be addressed on a congener-specific basis because some of the congeners, particularly those that are either nonortho, or mono-ortho substituted, are much more toxic than others.[27,28] Thus, addressing PCBs as individual congeners is critical in accumulation testing. Rathbun et al.[19] used caged fathead minnows to assess PCB contamination in and around a confined disposal facility in Saginaw Bay, Lake Huron. As part of this study,[19] resident fathead minnows from within the confined disposal facility were also analyzed for PCB congeners. If one assumes that the resident fish represented steady state, one can calculate the percentage of steady

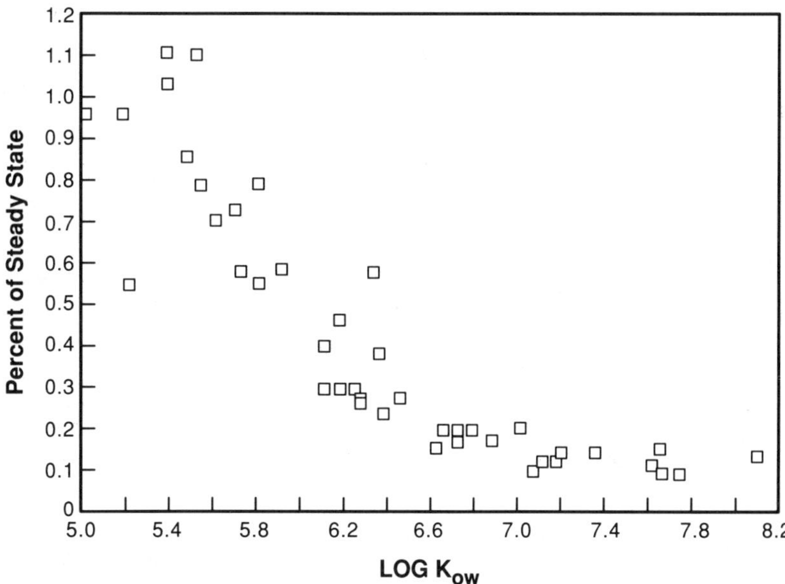

Figure 1. Percent of PCB concentration attained by fathead minnows exposed for 8 d inside Saginaw Bay confined disposal facility, compared with resident fish taken from within the facility. PCB congeners represented by log of the octanol/water partition coefficient (K_{ow}).

state achieved by caged fish in the 8-day exposure. Figure 1 shows the percent of steady state attained for 42 congeners that were highest in concentration and could be quantified as single congeners in relation to calculated K_{ow} values (from Hawker and Connell[29]). After 8 days, concentrations of the low K_{ow}, di- and trichlorobiphenyls in caged fathead minnows had reached nearly 100% of the concentrations in resident fish. This percentage decreased rapidly to about 50% for the higher K_{ow} tetrachlorobiphenyls and 30% for the pentachlorobiphenyls (Figure 1). Concentrations of all biphenyls with six or more chlorines were less than 20% of those in resident fish. Increasing the length of exposure would likely increase concentrations in caged fish of all congeners that were at less than steady state. However, congeners of different log K_{ow} would still vary in their percentage of steady state, and it is doubtful that congeners with higher log K_{ow} values would be anywhere near steady state.

Without either exposure to contaminated food or sufficient time to do the long-term exposure needed to reach steady state, regulatory testing of sediments for bioaccumulation by fish can only adequately address the question of bioavailability. The "Implementation Manual"[2] recommends in Tier IV the "determination of steady-state bioaccumulation," using the ASTM bioconcentration test.[15] But this method uses a kinetic approach with time-series sampling of exposed organisms, including a depuration phase, to calculate "...apparent steady-state or projected steady-state bioconcentration factors...."[15] There is no consideration of dietary uptake of contaminants, and therefore, values obtained with this test should not be considered comparable to field-collected fish.

If bioaccumulation testing with fish only addresses the question of bioavailability, then the length of the test should be geared toward answering this question most accurately and efficiently. The amount of published information on short-term accumulation by fish from sediment is limited. In freshwater sediments, Mac et al.[9] showed statistically significant accumulation of total polychlorinated biphenyls (PCBs) by fathead minnows from sediments containing concentrations ≥0.1 μg/g (dry wt). In the 8-day exposures of Rathbun et al.,[19] fish caged in the confined disposal facility accumulated 94 of 98 congeners (or coeluting congeners). Although changes in the dredging schedule cut this exposure short of the intended 10 days, accumulation and bioavailability were amply demonstrated in 8 days.

With this understanding, the implications of extending the length of the test past 10 days for fish should be examined. If we expect other compounds besides the PCBs to behave similarly, based on their partition coefficients, then increasing the time of exposure will provide a higher percentage of steady state attained for some chemicals and increase the probability of detecting others that may be present at low, but significant, concentrations. Limits of detection for chemical analysis will certainly influence this probability. Conversely, the lower log K_{ow} compounds might be lost from the limited sediment reservoir in a test tank, especially under flow-through conditions. These compounds may also be depurated by the organism. Consequently, there is a tradeoff between the information gained and lost in a longer exposure; the cost effectiveness of the added information should be evaluated.

To Purge or Not

One of the generally accepted parts of a sediment bioaccumulation test is purging, a period following the exposure during which the test animals are removed from contaminated sediment and placed in clean tanks, with or without uncontaminated sediment. The purpose of purging is to allow exposed animals to clear the gut of ingested test sediments, which may add contaminants to the organism that are not actually incorporated into tissue. Ingestion of sediment by deposit-feeding invertebrates can influence the contaminant concentration.[30] When the alternative of feeding for a period on clean sediment is considered, however, the potential exists for the dilution of the actual body burden. Purging was originally part of the 10-day test[9] with fathead minnows,[13] but testing showed that purging had less of an effect on fish than on invertebrates,[9] mostly because the weight of ingested sediment relative to body weight is far greater in invertebrates. Regardless of the test organism, purging allows for the potential depuration of chemicals. Thus, in the accumulation test[9] with fathead minnows, it was recommended to not purge the fish because the test is short and bioavailability is the question. Not purging maximizes the ability to measure statistically significant accumulation of chemicals. The gut contents of a prey organism, such as the fathead minnow, is ingested by a predator and thus should be measured as part of the body burden of the prey. This may be more important for models that use results from bioaccumulation tests to predict food chain accumulation. Omission of the purging period also reduces the cost of the assay.

STATISTICAL CONSIDERATIONS

The statistical procedures for evaluating the results of bioaccumulation tests apply equally to protocols with either fish or invertebrates.

Assumptions of Parametric Statistical Procedures

Most parametric statistical procedures assume that the dependent variables have a normal distribution and uncorrelated errors and that they are measurable to the same degree of accuracy and precision over the entire measurement scale. In bioaccumulation testing and measurement, we generally find that few tissue residues are normally distributed; the variances of most residue variables tend, at best, to increase predictably with concentration (i.e., constant CV); censored data (the occurrence of values below the limit of quantitation) are the rule rather than the exception; and animals die during exposures, sometimes unexplainably, which can lead to missing cells. All of these problems are surmountable, but with varying degrees of difficulty. With currently available computer software, missing data are no longer the problem they once were, and non-normality, heterogeneity of variance, and censored data can be handled with appropriate transformations or nonparametric statistical methods.

The Log Transformation and Some Alternatives

Tissue concentrations of contaminants are often best described by lognormal or other skewed distributions.[31,32] Consequently, a log transformation of the tissue concentrations generally improves the degree to which the data meet the assumption of parametric statistical methods derived from normal theory. The log transformation often offers the added advantage of variance stabilization. As noted previously, tissue concentrations are typically characterized by variances that increase with increasing concentration, which violates the assumption of variance homogeneity implicit in the analysis of variance (ANOVA), the analysis of covariance (ANCOVA), regression, and other common parametric procedures. In such instances, the log transformation is among those recommended to stabilize the variance across the range of values present,[33,34] thereby permitting its partitioning for hypothesis testing purposes. Log transformation may also increase robustness; it reduces the effects that a few extremely large or small values may have on the sample mean and variance and, consequently, tends to increase the power of statistical tests to reject a null hypothesis. This also tends to reduce the temptation of the investigator to view stray observations as outliers. The frequency distributions of metal concentrations in fish have been described as "outlier prone,"[31] a description that certainly describes nonpolar organic chemical residues as well.[35] Although there are statistical methods available for the evaluation of outliers,[36] the use of such procedures in a regulatory decision-making environment would surely invite suspicion.[37] The former approach of viewing the distributions as outlier prone and using an appropriate, robust transformation seems intuitively more palatable. Log transformation may actually represent more than "statistical sleight-of-hand" for fixing problem data; theoretically, the concentrations of substances in the environment are determined by geometric processes (such as expanding populations of cells), which lead to lognormal distributions.[38-41] Tests of this theory have proven it to be equivocal, however.[42]

Several alternatives to the log transformation are also attractive, although some may be overly complex for routine application in a regulatory context. Power transformations[43,44] can be "tuned" to a particular data set, an advantage when a suite of bioacummulable contaminants, each with its own distribution, may be present. A variation of the log transformation, proposed by Berry,[37] similarly adapts individual data sets by adding a constant to each value before transformation. Not to be confused with *ad hoc* selection of a constant (say, 1.0) often used to avoid taking the logarithm of zero,[35,45,46] this transformation is robust and relatively straightforward while having the added benefit of being especially well suited to dealing with censored data, a commonplace occurrence in bioaccumulation testing where comparisons with a reference sediment are required. The rank transformation also effectively normalizes skewed distributions (the ranks are normally distributed), deals with censored values (they are assigned the same rank), and is robust, albeit at a cost of some sensitivity and one error degree of freedom.[37,47] Relatively powerful and robust nonparametric analogues[48,49] to parametric procedures also constitute an alternative approach.

In general, we agree with Berry;[37] in addition to being robust and powerful, an "attractive" statistical procedure should be available in software packages with little or no extra programming and should be easy to understand. Most of the transformations and procedures discussed here meet these criteria.

Hypothesis Testing

The first necessity of a sediment evaluation program, be it for toxicity, bioaccumulation, or some suite of criteria, is an answer from the regulators and managers to the following question: is the quality of the environment to be protected (i.e., a dredged-material disposal area and its living resources) sufficiently high to warrant the imposition of strict regulations stipulating "no further degradation" of existing conditions? This is obviously not a question to be addressed solely by ecologists and toxicologists; rather, it is one best answered by informed policy makers through the regulatory process. The answer does, however, profoundly influence the nature of the tests to be conducted and their interpretation. If existing conditions are to be maintained, then an approach comparing test sediments with reference sediments from the disposal area (as specified in the "EPA/COE Implementation Manual"[2]) is appropriate. If, however, conditions are degraded to the extent that the disposal area supports a depauperate fauna, or fishery products from the area are deemed unsafe for human consumption, then the regulatory policy may be more stringent. In this situation, the reference sediments either need to come from another source, or the comparisons must instead be made with the control responses (i.e., such as "clean" sediment).

Once the nature of the comparison to be made has been established, the bioaccumulation test is conducted. According to the "Implementation Manual,"[2] replicate groups of animals (commonly, $n = 5$) are exposed to test, reference, and control sediments for a specified time. In the simplest version of the test, animals are enumerated and, if survival was acceptable (i.e., the sediment was not acutely toxic), the survivors are removed and purged (or not, per previous discussion), composited, and analyzed for the contaminant(s) of concern. Mean concentrations in the survivors from the test sediments are compared statistically to those from the reference sediments using t tests or ANOVA (after assuring through transformation and statistical testing that parametric procedures are appropriate), and if more than one sediment is being tested, a multiple-means comparison procedure (usually Dunnett's, which is specifically designed for comparisons with a single control or reference mean). A test sediment is rejected for unrestricted open-water disposal if the mean concentration of contaminant(s) in exposed organisms at the end of the test is significantly ($P < 0.05$) greater than that of the animals from the reference sediment. A variation of this approach employs comparisons of results from long-term (28-day) exposures, where steady state may have been attained, with known values, such as FDA action levels or levels of concern. Again, a sediment is rejected only if the mean concentration in the test organisms exceeds the critical value (at the 0.05 level of significance). Still another version of the test accounts for the slow rate of uptake for some contaminants that might not attain steady state even after 28 days. For such compounds, animals are sampled for analysis at intervals during the test as well as at the end, and nonlinear regression is used to estimate steady-state concentrations and 95% confidence limits. The test sediment is rejected for open-water disposal if the predicted steady-state contaminant concentration in organisms exposed to the sediment is significantly greater than either the estimated steady-state concentration in reference-exposed animals or some predetermined critical value.

For marine sediments, such tests are conducted in accordance with 40 CFR 220-228, issued by EPA under the Marine Protection, Research, and Sanctuaries Act (MPRSA) of 1972 (PL 92-532), which specifically prohibits the disposal of materials containing toxics (including mutagens, carcinogens, and teratogens) in the ocean "...as other than trace contaminants." Section 227.6 stipulates that "...the potential for significant undesirable effect of these constituents shall be determined by application of results of bioassays on liquid, suspended particulate, and solid phases of wastes... ." For suspended and solid-phase testing, Section 227.6 declares that material is suitable for ocean disposal if, "...at the 95 percent level, that, when the materials are dumped, no significant undesirable effects will occur due either to chronic toxicity or to bioaccumulation...."

Throughout the history of regulatory dredged-material testing (i.e., in the implementation manuals[2,13]), the interpretation of the "95% level of confidence" passage of Section 227.6 has been that an α level of 0.05 is appropriate for statistical comparisons and decisionmaking. Such an approach is founded in the statistics of experimentation and hypothetico-deductive reasoning where experiments are designed and conducted such that differences from control constitute positive outcomes, and the statistical hypotheses are

$$H_0 : \overline{x}_{tmt} = \overline{x}_{control}$$

$$H_1 : \overline{x}_{tmt} \neq \overline{x}_{control}$$

For dredged material evaluation, the analogous one-sided hypotheses are

$$H_0 : \overline{x}_{test} \leq \overline{x}_{ref}$$

$$H_1 : \overline{x}_{test} > \overline{x}_{ref}$$

With ANOVA (and Dunnett's test if there is more than one treatment), H_0 is tested at a specified level of α (say, 0.05), where α is the probability of falsely rejecting a true H_0 (i.e., of claiming differences when there are in fact none or claiming a sediment to be unacceptably contaminated when it is not) — a Type I error. In the reductionist methods of research, this is often more serious than a Type II error, which causes acceptance of H_0 when it is false (i.e., failing to recognize a significant treatment effect or a contaminated sediment as such). Because the focus of the statistics of experimentation is frequently on Type I errors, ANOVA and Dunnett's procedure are recommended because they protect at their nominal α levels against such errors. Unfortunately, these methods may or may not protect very effectively against Type II errors (β), the probability of which varies, but may greatly exceed α (as noted in the revised "Implementation Manual"[2]). In dredged-material evaluation (and in other forms of environmental decisionmaking), the potential consequences of Type II errors are often far more serious than those of Type I errors.[50] If a Type I error is committed, H_0 is falsely rejected, i.e., a sediment proposed for unrestricted disposal is falsely declared to be significantly different from the reference value, and an alternative disposal practice (such as expeditious capping with clean material, or upland disposal) would be required. A Type II error, however, incorrectly accepts H_0 and permits the unrestricted disposal of sediments that contain harmful constituents that may threaten fish and wildlife or human health. The cost of a Type I error is a short-term monetary loss; a Type II error, however, may cause a long-term or irretrievable loss of a valuable resource or may endanger human health.[50]

In general, MPRSA (and 40 CFR 220-228) have been interpreted such that the well-defined null hypothesis is that all material is suitable for unrestricted disposal unless the hypothesis is discredited (by means of a statistical test with a nominal α of 0.05). This approach, which effectively removes from the disposal permit applicant the burden of proving the safety of the material, has been reinforced through numerous administrative and regulatory policies.[51] There is a growing awareness in a number of fields (pharmaceutical and cosmetic testing, environmental policy, etc.), however, that proof of safety and proof of hazard, although related, are not equivalent and that regulatory agencies charged with ensuring public safety and environmental quality have failed to do so by removing from the applicant the burden of proof.[50,52-54] Consequently and because the evaluation of sediment may culminate in an administrative decision, a more appropriate approach would be to minimize a combination of Type I (rejecting unrestricted disposal when such is in fact appropriate) and Type II (allowing unrestricted disposal when such is inappropriate) errors. Statistical treatment of a problem through consideration of not only sample information, but also the possible errors that could be made and the potential losses or costs incurred by making those errors, is known as decision theory, which was formulated for precisely such problems.[55]

Sample Size

We deliberately left the discussion of sample size until the end of this chapter because the preceding questions needed to be addressed first. Policy issues, such as the magnitude of the difference between test and reference sediments that will be considered "meaningful" (as opposed to statistically significant), which will in turn depend on the relative costs associated with Type I and Type II errors, and the degree of uncertainty that decision makers will accept, should underlie discussions of statistical power. Alldredge[56] logically presents the statistics of such an approach.

And finally, a sample for bioaccumulation testing purposes is normally construed as a composite of like individuals exposed as a group to a test or reference sediment.[2,13] The representativeness and "replication" (vs. pseudoreplication) of the sediment sample are issues that lie beyond the scope of this discussion. However, the number of animals per replicate, which is normally stated as a "given" in bioaccumulation testing ($n = 5$ in the implementation manuals), is not. A recent study of *p,p*-DDT in blue mussels *(Mytilus edulis)* showed that the number of individuals in a composite sample may profoundly influence variability among replicate samples and that the number of samples needed for minimum CV ($n = 20$ in mussels) may be larger than called for in most testing protocols.[57] In fact, it may be another variable that needs to be optimized within the overall testing strategy.

ACKNOWLEDGMENTS

This paper benefitted greatly from discussion by CJS with M.S. Kaiser, mathematical statistician. Contribution 792 of the National Fisheries Research Center–Great Lakes.

REFERENCES

1. "Procedures for the Assessment of Contaminated Sediment Problems in the Great Lakes," Report of the Sediment Subcommittee to the Great Lakes Water Quality Board, International Joint Commission, Windsor, Ontario (1988).
2. "Evaluation of Dredged Material Proposed for Ocean Disposal," U.S. EPA Report EPA-503-8-91/001 (1991).
3. Long, E. R., and P. M. Chapman. "A Sediment Quality Triad: Measures of Sediment Contamination, Toxicity, and Infaunal Community Composition in Puget Sound," *Mar. Pollut. Bull.* 16(10):405–415 (1985).
4. Dillon, T. M., and A. B. Gibson. "Bioassessment Methodologies for the Regulatory Testing of Freshwater Dredged Material," U.S. Army Engineers, Waterways Experiment Station, Misc. Paper EL-86-6 (1986).
5. Ross, P. E., G. A. Burton, Jr., J. C. Filkins, J. P. Giesy, Jr., C. G. Ingersoll, P. F. Landrum, M. J. Mac, T. J. Murphy, J. Rathbun, V. E. Smith, H. Tatem, and R. W. Taylor. "Assessment and Remediation of Contaminated Sediments I: Background and Approach," *J. Aquat. Ecosyst. Health* 1:(in press) (1992).
6. Muir, D. C. G., and A. L. Yarechewski. "Dietary Accumulation of Four Chlorinated Dioxin Congeners by Rainbow Trout and Fathead Minnows," *Environ. Toxicol. Chem.* 7(3):227–236 (1988).
7. Clark, K. E., F. A. P. C. Gobas, and D. Mackay. "Model of Organic Chemical Uptake and Clearance by Fish from Food and Water," *Environ. Sci. Technol.* 24(8):1203–1213 (1990).
8. Connolly, J. P., and C. J. Pedersen. "A Thermodynamic-Based Evaluation of Organic Chemical Accumulation in Aquatic Organisms," *Environ. Sci. Technol.* 22(1):99–103 (1988).
9. Mac, M. J., G. E. Noguchi, R. J. Hesselberg, C. C. Edsall, J. A. Shoesmith, and J. D. Bowker. "A Bioaccumulation Bioassay for Freshwater Sediments," *Environ. Toxicol. Chem.* 9(11):1407–1416 (1990).
10. "Guidelines and Register for Evaluation of Great Lakes Dredging Projects," Report of the Dredging Subcommittee to the Great Lakes Water Quality Board, International Joint Commission, Windsor, Ontario (1982).
11. "Standard Guide for Conducting Sediment Toxicity Tests with Freshwater Invertebrates," American Society for Testing and Materials, Standard Guide No. E-1383-90 (1990).
12. Persaud, D., R. Jaagumagi, and A. Hayton. "The Provincial Sediment Quality Guidelines. Ontario Ministry of the Environment," unpublished (1990).
13. "Ecological Evaluation of Proposed Discharge of Dredged Material into Ocean Waters; Implementation Manual for Section 103 of Public Law 92-532 (Marine Protection, Research, and Sanctuaries Act of 1972)," U.S. Army Engineers Waterways Experiment Station (1977).

14. Lee, H., II, B. L. Boese, J. Pelletier, M. Winsor, D. T. Specht, and R. C. Randall. "Guidance Manual: Bedded Sediment Bioaccumulation Tests," U.S. EPA Report, EPA/600/x-89/302 (1989).
15. "Standard Practice for Conducting Bioconcentration Tests with Fishes and Saltwater Bivalve Molluscs," American Society for Testing and Materials, Standard Practice No. E-1022-84 (1984).
16. Rubinstein, N. I., W. T. Gilliam, and N. R. Gregory. "Dietary Accumulation of PCBs from a Contaminated Sediment Source by a Demersal Fish *(Leiostomus xanthuras),*" *Aquat. Toxicol.* 5:331–342 (1984).
17. Rubinstein, N. I., W. T. Gilliam, and N. R. Gregory. "Evaluation of Three Fish Species as Bioassay Organisms for Dredged Material Testing," U.S. Army Engineers Waterways Experiment Station, Misc. Paper D-84-1 (1984).
18. Rice, C. P., and D. S. White. "PCB Availability Assessment of River Dredging Using Caged Clams and Fish," *Environ. Toxicol. Chem.* 6(4):259–274 (1987).
19. Rathbun, J. E., R. G. Kreis, Jr., E. L. Lancaster, M. J. Mac, and M. J. Zabik. "Pilot Confined Disposal Facility Biomonitoring Study: Channel/Shelter Island Diked Facility, Saginaw Bay, 1987," U.S. EPA Large Lakes Research Station, ERL-Duluth, unpublished (1988).
20. McFarland, V. A. "Activity-Based Evaluation of Potential Bioaccumulation from Sediments," in *Dredging 84,* (New York: American Society of Civil Engineers, 1984), 1:461–467.
21. McFarland, V. A., and J. U. Clarke. "Testing Bioavailability of Polychlorinated Biphenyls from Sediments Using a Two-Level Approach," in *Proceedings, Sixth Seminar,* U.S. Army Engineers Committee on Water Quality, New Orleans, pp. 220–229 (1986).
22. Lake, J. A., N. I. Rubinstein, and S. Pavignano. "Predicting Bioaccumulation: Development of a Simple Partitioning Model for Use as a Screening Tool for Regulating Ocean Disposal Wastes," U.S. EPA Report, EPA/600/D-89/253 (1987).
23. Lake, J. A., N. I. Rubinstein, H. Lee, II, C. A. Lake, J. Heltshe, and S. Pavignano. "Equilibrium Partitioning and Bioaccumulation of Sediment-Associated Contaminants by Infaunal Organisms," *Environ. Toxicol. Chem.* 9(8):1095–1106 (1990).
24. van der Oost, R., H. Heida, and A. Opperhuizen. "Polychlorinated Biphenyl Congeners in Sediments, Plankton, Molluscs, Crustaceans, and Eel in a Freshwater Lake: Implications of Using Reference Chemicals and Indicator Organisms in Bioaccumulation Studies," *Arch. Environ. Contam. Toxicol.* 17:721–729 (1988).
25. Huckins, J. N., M. W. Tubergen, and G. K. Manuweera. "Semipermeable Membrane Devices Containing Model Lipid: A New Approach to Monitoring the Bioavailability of Lipophilic Contaminants and Estimating Their Bioconcentration Potential," *Chemosphere* 20(5):533–552 (1990).
26. Sodergren, A. "Solvent-Filled Dialysis Membranes Simulate Uptake of Pollutants by Aquatic Organisms," *Environ. Sci. Technol.* 21(9):855–859 (1987).
27. Safe, S. "Polychlorinated Biphenyls (PCBs), Dibenzo-p-dioxins (PCDDs), Dibenzofurans (PCDFs), and Related Compounds: Environmental and Mechanistic Considerations Which Support the Development of Toxic Equivalency Factors (TEFs)," *Crit. Rev. Toxicol.* 21(1):51–88 (1990).

28. Smith, L. M., T. R. Schwartz, K. Feltz, and T. J. Kubiak. "Determination and Occurrence of AHH-Active Polychlorinated Biphenyls, 2,3,7,8-Tetrachloro-p-dioxin, and 2,3,7,8-Tetrachlorodibenzofuran in Lake Michigan Sediment and Biota. The Question of Their Relative Toxicological Significance," *Chemosphere* 21(9):1063–1085 (1990).
29. Hawker, D. W., and D. W. Connell. "Octanol-Water Partition Coefficients of Polychlorinated Biphenyls," *Environ. Sci. Technol.* 22(4):382–387 (1988).
30. Boese, B. L., H. Lee, II, D.T. Specht, and R.C. Randall. "Comparison of Aqueous and Solid-Phase Uptake for Hexachlorobenzene in the Tellinid Clam *Macoma nasuta* (Conrad): A Mass Balance Approach," *Environ. Toxicol. Chem.* 9(8):221–231 (1990).
31. Giesy, J. P., and J. G. Wiener. "Frequency Distributions of Trace Metal Concentrations in Five Freshwater Fishes," *Trans. Am. Fish. Soc.* 106(4):393–403 (1977).
32. Eberhardt, L. L., R. O. Gilbert, H. L. Hollister, and J. M. Thomas. "Sampling for Contaminants in Ecological Systems," *Environ. Sci. Technol.* 10(9):917–925 (1976).
33. Kempthorne, O. *The Design and Analysis of Experiments* (New York: Wiley, 1952).
34. U.S. Environmental Protection Agency. "Guidelines on Sampling and Statistical Methodologies for Ambient Pesticide Monitoring," Monitoring Panel, Federal Working Group on Pest Management, Washington, D.C. (1974).
35. Schmitt, C. J., J. L. Zajicek, and P. Peterman. "National Contaminant Biomonitoring Program: Residues of Organochlorine Chemicals in Freshwater Fishes of the United States, 1976–1984," *Arch. Environ. Contam. Toxicol.* 19:748–782 (1990).
36. Grubbs, F. E. "Procedures for Detecting Outlying Observation in Samples," *Technometrics* 11(1):1–21 (1969).
37. Berry, D. A. "Logarithmic Transformations in ANOVA," *Biometrics* 43(2):439–456 (1987).
38. Duvall, J. S., Jr., T. Schwartzer, and J. A. S. Adams. "Lognormal Distribution of Trace Elements in the Environment," in *Trace Substances in Environmental Health,* Vol. 4, D. D. Hemphill, Ed. (Columbia, MO: University of Missouri, 1971), pp. 120–131.
39. Liebscher, K., and H. Smith. "Essential and Non-Essential Trace Elements," *Arch. Environ. Health* 17:881–890 (1968).
40. Koch, A. L. "The Logarithm in Biology. I. Mechanisms Generating the Lognormal Distribution Exactly," *J. Theor. Biol.* 12:276–290 (1966).
41. Koch, A. L. "The Logarithm in Biology. II. Distributions Simulating the Lognormal." *J. Theor. Biol.* 17:881–890 (1969).
42. Pinder, J. E., III, and J. P. Giesy. "Frequency Distributions of the Concentrations of Essential and Nonessential Elements in Largemouth Bass, *Micropterus salmoides,*" *Ecology* 62(2):456–468 (1981).
43. Taylor, J. M. G. "Power Transformations to Symmetry," *Biometrika* 72:145–152 (1985).
44. Green, R. H. "Power Analysis and Practical Strategies for Environmental Monitoring," *Environ. Res.* 50:195–205 (1989).
45. Kushner, E. J. "On Determining the Statistical Parameters for Pollution Concentration from a Truncated Data Set," *Atmos. Environ.* 10:975–979 (1976).
46. Ohlendorf, H. M., E. E. Klaas, and T. E. Kaiser. "Environmental Pollutants and Eggshell Thinning in the Black-Crowned Night Heron," in *Wading Birds,* Research Report No. 7 (New York: National Audubon Society, 1978), pp. 63–82.

47. Conover, W.J., and R. L. Iman. "Rank Transformations as a Bridge Between Parametric and Nonparametric Statistics," *Am. Stat.* 35:124–133 (1981).
48. Prentice, R. L., and P. Marek. "A Qualitative Discrepancy Between Censored Data Rank Tests," *Biometrics* 35(4):861–867 (1979).
49. Conover, W. J. *Practical Nonparametric Statistics,* 2nd ed. (New York: Wiley, 1980), p. 493.
50. Peterman, R. M. "Statistical Power Analysis Can Improve Fisheries Research and Management," *Can. J. Fish. Aquat. Sci.* 47(1):2–15 (1990).
51. Belsky, M. H. "Environmental Policy Law in the 1980's: Shifting Back the Burden of Proof," *Ecol. Law Q.* 12(1):1–88 (1984).
52. Bross, I. D. "Why Proof of Safety is Much More Fifficult Than Proof of Hazard," *Biometrics* 41(3):785–793 (1985).
53. Millard, S. P. "Proof of Safety vs. Proof of Hazard," *Biometrics* 43(3):719–725 (1987).
54. Millard, S. P. "Environmental Statistics and the Law: Room for Improvement," *Am. Stat.* 41(4):249–253 (1985).
55. Berger, J. O. *Statistical Decision Theory and Bayesian Analysis* (New York: Springer-Verlag, 1980), p. 617.
56. Alldredge, J. R. "Sample Size for Monitoring of Toxic Chemical Sites," *Environ. Monit. Assess.* 9:143–154 (1987).
57. Baez, B. P. F., and M. S. G. Bect. "DDT in *Mytilus edulis:* Statistical Considerations and Inherent Variability," *Mar. Pollut. Bull.* 20(10):469–499 (1989).

CHAPTER 14

Integrative Assessments in Aquatic Ecosystems

Peter M. Chapman, Elizabeth A. Power, and G. Allen Burton, Jr.

INTRODUCTION AND OVERVIEW

What Are Integrative Assessments?

Integrative assessments are defined as investigations involving attempts to integrate measures of environmental quality to make an overall assessment of the status of the system. The measures can include two or more of the following components: sediment toxicity tests, sediment chemical analyses, tissue chemical analyses, pathological studies, and community structure studies. The purpose of such investigations is to determine environmental quality; such may be defined in terms of relative and/or current status, but particularly relates to ecosystem health. Specifically, integrative assessments involve more than one generic measure of environmental quality. For instance, a study that involved only sediment toxicity tests, even though several different tests were involved, would not be considered an integrated assessment. Such a study must minimally involve two, and realistically three or more, of the above five components.

Integrative assessments are more than the sum of their parts; the total amount of information about a system extracted by an integrative assessment will be of greater utility than the sum of information from the individual components.[1,2] Integrative assessments use a preponderance/burden of evidence approach where the conclusions drawn from individual components are considered relative to one another; i.e., is there concordance and do the results of each component support those obtained by the others? In addition, by looking at the same problem from different viewpoints, one is more likely to determine what the mechanism is. This is the argument made for using a holistic approach to ecosystem health assessment, as discussed later in this chapter.

This chapter is *not* a comprehensive review of all possible integrative assessments. Rather, this chapter describes the conceptual basis for integrative assessments and provides selected specific, recent examples of an integrative concept and an application of that concept. The following chapters concerning case studies show additional examples of successful integrative assessments.

Why Conduct Integrative Assessments?

Integrative assessments are conducted because we generally do not have adequate knowledge of cause and effect to determine environmental quality. This is true for the complex mixtures of chemicals that commonly characterize contaminated sediments and, in particular, for long-term effects of low levels of contaminants (e.g., nutrients, organochlorines, TBT, hydrocarbons).[3] Although, individually, each of the five major possible components of integrative assessments provide useful information, none of them provide complete information (Table 1). The reason for multicomponent integrative assessments is the reality that there is no "magic bullet";[4] the environment is highly complex and dynamic, so burden of evidence remains the only reasonable method presently available to define sediment quality relative to ecosystem health.

Although Giesy and Hoke[5] note "At this time there is no consensus on the correct approach for the assessment of the toxicity of sediment-associated contaminants," they also note the need to answer three basic questions: (1) what are the contaminant(s); (2) what is the extent of contamination; and (3) what are the potential effects on biota? The latter question is arguably the most important. Although differentiating areas that are severely degraded from those that are pristine does not require complex integrative assessments, the majority of areas with contaminated sediments fall into a middle ground (probably greater than 70% of all contaminated sediments) . Chapman[6] characterizes this "gray zone" as one where effects may or may not be detected, depending on the measure used and the variability of the results. It is difficult to determine environmental significance in such areas. Outside this "gray zone" a high concordance can be expected, even between separate components of an integrative assessment (e.g., between laboratory toxicity tests and in situ biological responses).[7] Within this "gray zone", answers that lack agreement are provided even by different measures of the same component (e.g., laboratory toxicity tests).[8,9]

Ideally, all five components would be used to provide as complete a picture as possible of the environmental significance of contaminated sediments. This is especially desirable because contaminated sediments may affect both organisms

Table 1. Components of Integrative Assessments

Component	Information	
	Provided	**Lacking**
Sediment toxicity (SEDTOX)	· Laboratory responses by organisms exposed to test conditions	· Field responses · Responses to tests not conducted and organisms not exposed
Sediment chemistry (SEDCHEM)	· Presence and levels of measured chemicals	· Chemical (bio)availability · Presence and levels of chemicals not measured
Tissue chemistry (TISCHEM)	· Presence and levels of measured chemicals in organisms and tissues · Bioavailability	· Effects of chemicals in tissues · Presence of transformed chemicals · Presence and levels of unmeasured chemicals in organisms and tissues
Pathology (PATHOL)	· Presence and levels of measured responses in organisms and tissues	· Effects of pathological conditions · Presence and levels of unmeasured responses in organisms and tissues
Community structure (COMMSTRU)	· Presence and numbers of taxa and individuals	· Causality (e.g., natural versus anthropogenic) · Ecosystem-level relevance

Acronyms are as used in subsequent tables and figures.

dwelling within the sediments[10] and, by exchange with overlying waters,[11] organisms living in the water column. In reality, such extensive (and expensive) studies are rarely conducted and may not be necessary, depending on the questions being addressed (Table 2).

Three-component assessments involving measures of contamination (sediment chemistry), toxicity (sediment bioassays), and in situ alteration (usually benthic infaunal community structure; rarely pathology) appear to be particularly useful. For instance, Warren-Hicks et al.[12] found that links between adverse ecological effects and hazardous waste sites are best established "...by demonstrating a pattern of effects between ecological, toxicological, and chemical data." This particular combination of measures has been named the Sediment Quality Triad (Triad) (discussed later in this chapter), and its usefulness has been demonstrated in both marine and freshwater systems.

Ecosystem Health

As discussed in several chapters of this book, concern with contaminated sediments reflects an overall concern for the biosphere and for the ecosystems and communities that comprise it. Integrative assessments are conducted at the

Table 2. Major Questions Addressed by Several-Component Integrative Assessments

Assessment	Major Questions
3-Component:	
* SEDTOX · SEDCHEM · PATHOL	Are sediments contaminated, toxic, and do pathological effects occur related to these sediments?
* SEDTOX · SEDCHEM · COMMSTRU	Are sediments contaminated, toxic, and do associated communities show adverse effects?
SEDCHEM · PATHOL · COMMSTRU	Are sediments contaminated, and do pathological effects and community alterations occur related to these sediments?
SEDCHEM · PATHOL · TISCHEM	Are sediments contaminated, are organisms contaminated, and do pathological effects occur related to this contamination?
TISCHEM · COMMSTRU · PATHOL	Are organisms contaminated, and do pathological effects and community alterations occur related to this contamination?
4-Component:	
** SEDTOX · SEDCHEM · PATHOL · TISCHEM	Are sediments contaminated and toxic, are organisms contaminated, and do pathological effects occur related to this contamination and toxicity?
** SEDTOX · SEDCHEM · COMMSTRU · PATHOL	Are sediments contaminated and toxic, and do pathological effects and community alteration occur related to this contamination and toxicity?
TISCHEM · COMMSTRU · PATHOL · SEDCHEM	Are sediments and organisms contaminated, and do pathological effects and community alterations occur related to this contamination?
5-Component:	
** SEDTOX · SEDCHEM · TISCHEM · PATHOL · COMMSTRU	Are sediments contaminated and toxic, are organisms contaminated and do pathological effects and community alterations occur related to this contamination and toxicity?

Assumes: (1) PATHOL, TISCHEM and COMMSTRU are on bottom-dwelling or bottom-associated organisms; (2) SEDCHEM measurements include chemicals relevant to biological measurements; (3) SEDTOX end-points are relevant to PATHOL and/or COMMSTRU measurements.

* = Sediment Quality Triad (sediment chemistry, sediment toxicity and some measure of in situ bioeffects).

** = All elements of Sediment Quality Triad included.

community or ecosystem levels and are generally intended to determine environmental quality in terms of ecosystem health. However, ecosystem health is not readily definable.

Rapport[13] describes ecosystems as being "...something more than a community of species but less than the biosphere." He describes the primary requirements for a healthy ecosystem as system integrity and sustainability,[13] but describes ecosystem health as "an arcane concept," given the long time span ecosystems operate on.[14] However, he suggests that the notion of "integrity" appears paramount and refers to the capability of a system to remain intact, to self regulate in the face of stress, and to evolve towards increasing complexity and integration.[14]

Kelly and Harwell[15] contend that an ecosystem "...can be perceived and defined only in an operational context..." and suggest that one measure of an ecosystem's health is a change in selected indicators relative to a baseline state. This suggestion implies that a baseline state representing acceptable ecosystem health has been and can be chosen. The problem with such choices, i.e., the reference area concept, is discussed later. Chapman[16] defines an ecosystem as "...any grouping of different, interacting species which is recognized as such by qualified and competent ecologists." Clearly, this definition precludes reliance only on "hard" science and requires societal involvement, which will also be discussed later under the reference area concept.

In addition, ecosystem health implies some level of stability. Preston[17] notes that "stability lies in the ability to bounce back." In other words, stability is persistence in the face of adversity. But change is also a normal part of ecosystem development; ecosystem health does not imply stasis, and thus integrative assessments must allow for normal ecosystem succession. Our ability to determine "stability" or degradation, therefore, is complicated by the natural, unknown, and chaotic dynamics of ecosystem processes.

Rapport[14] describes several considerations in the detection and diagnosis of ecosystem pathology, including observed abnormalities, indicator and integrator organisms, size spectra, ecosystem distress syndrome, and index of biotic integrity. However, he points out that one drawback to all of the above measures is that they are based on retrospective monitoring of symptoms rather than anticipatory responses. The natural resilience of ecosystems and the long time spans of measurable change results in the limitations to our "early warning" capabilities.[14] Ryder[18] points out that there are instances where biotic components appear to be in the peak of "health", when in actuality they are on the threshold of precipitous decline.

IUCN/UNEP/WWF[19] describe ecosystems as healthy when they have a high level of biodiversity, productivity, and habitability. Biodiversity (biological diversity) refers to the variety of all life, wild and domesticated. An alternative way to define ecosystem health is to determine what the characteristics are of an unhealthy ecosystem. Rapport[13] noted the following symptoms of ecosystem breakdown:

"...reduced primary productivity, loss of nutrients, loss of sensitive species, increased instability in component populations, increased disease prevalence, changes in the biotic size spectrum to favor smaller life-forms, and increased circulation of contaminants." Measurement of components in Table 1 will allow identification of those symptoms that are applicable to sediment communities (i.e., excluding primary productivity and nutrient availability). Various combinations of these five components will provide information on environmental health, with the exact combinations dependent on the questions addressed and logistical constraints, such as cost.

Ecological stress triggered in an ecosystem is an elusive and poorly defined concept, and early detection of reduced health is more difficult than commonly thought.[18] Historically, ecologists have used reduced diversity as a measure of stress, but it has been shown that the relationships between diversity, stability, and stress are far more complex than previously imagined.[14,20,21] Warwick[22] made several assumptions leading to a conceptual model for predicting changes in a macrobenthic community with gradually increasing pollution. For example, communities tend towards equilibrium, and the biomass becomes dominated by a few species that are large in biomass but few in number. There is a more even distribution of numbers of individuals among species than biomass among species. Warwick[22] suggests that, under the effects of pollution, the larger species are eliminated because they contribute more biomass to the community than the more numerically dominant smaller species. These assumptions have been challenged by McManus and Pauly,[23] who developed what they describe as a more objective procedure to graphically interpret the community data. However, most authors specifically state that there are no "rules" to predict the effects of stress on ecosystems and that predictive models must be field validated for a particular system.

Holistic vs. Reductionist

Ideally, integrative assessments should emphasize a holistic (top-down) approach, using reductionist (bottom-up) approaches to assist in elucidating reasons (where such are actually necessary).[24] Top-down studies are based on field data and evaluate impacts at the community or population level. Their strength lies in describing the real world, but they are weak in diagnostic and predictive information. In contrast, bottom-up studies are based on simplistic laboratory data, relying on extrapolation to explain complex real-world effects. However, if properly used and designed, they provide predictive and diagnostic information to complement holistic studies. In the context of integrative assessments, in situ bioeffects measurements provide holistic information, while laboratory sediment toxicity tests provide reductionist information.

Assessment of effects on ecosystems requires the development of approaches that incorporate, but are not restricted to, laboratory tests. For instance, Levin et al.[25] call for approaches that "...go beyond laboratory tests." The most useful integrative assessments (1) combine holistic and reductionist approaches, (2) merge biological (effects) and chemical (causes) approaches, and (3) use both field and laboratory approaches and methods.

Such combinations provide the best means to both address known problems and anticipate future problems (i.e., early warning). They can be both reactive and proactive. But as discussed in Chapter 1, proactivity must be based on appropriately sensitive measures that are not overly protective and unrealistic.

ASSESSMENT COMPONENTS

As previously discussed, integrative assessments consist of five possible components, each of which provide useful, but incomplete, information (cf., Table 1). A total of ten two-component combinations are possible, of which seven have the strongest linkages between components (Figure 1). Excluded are sediment toxicity and tissue chemistry, which are not directly linked; neither are sediment toxicity and pathology unless sediment tests are developed that reflect potential pathological effects; and pathology and community structure are also excluded, as the latter is generally concerned with fish, while the former is generally concerned with benthic infauna (cf., Table 3). Multicomponent (three or more) integrative assessments are most useful if they involve the nine particular combinations derived from these seven two-component combinations (Figure 1).

Two-component integrative assessments are effectively addressed in previous chapters (cf., Chapters 4–12) and are limited in the information they provide. For instance, sediment chemistry and laboratory-derived toxicity measures provide no indication of real-world relevance; tissue chemistry and pathology provide no link to sediment contamination. Becker et al.[26] compared sediment bioassays and alterations of benthic macroinvertebrate assemblages, but could only arrive at firm conclusions when sediment chemistry data were also included. Swartz et al.[27] similarly needed all three components. Chapman and Long[28] noted that bioassays are most useful as part of a comprehensive approach to marine pollution assessment. The relationship (and validation) of laboratory-derived effect measures (i.e., surrogate and indigenous organism/community responses) has been documented in several three-component assessments of freshwater systems.[29-33] Two-component studies are best used in screening and/or tiered testing, as described later; this chapter is primarily concerned with more complex assessments.

Sediment chemistry measures are important in four of the seven recommended two-component combinations. This is not to denigrate the importance of biological effects measurements, but without chemical data it is difficult, if not impossible, to approach the problem of causality. Whereas biological measures (with the exception of tissue chemistry, which provides information on a phenomenon, not an effect) provide information on effects, chemical measures provide information on potential causes.

2-Component[a]:

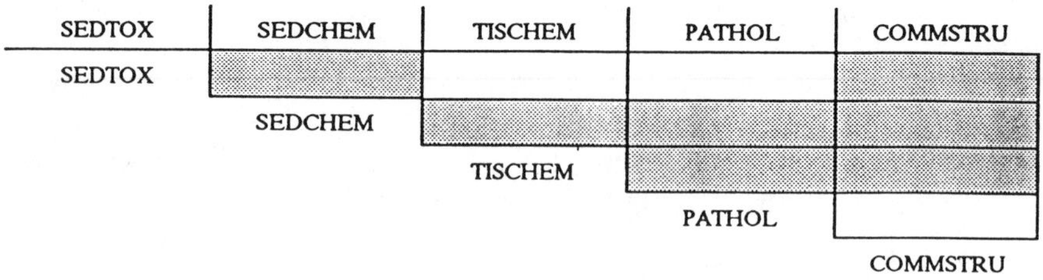

SEDTOX	SEDCHEM	TISCHEM	PATHOL	COMMSTRU
SEDTOX				
	SEDCHEM			
		TISCHEM		
			PATHOL	
				COMMSTRU

a. Useful = Components directly and strongly related; all possible combinations (2-way) shown above - most useful combinations are shaded.

3-Component (Based on seven shaded combinations above; components directly and strongly related.)

SEDTOX • SEDCHEM • PATHOL

SEDTOX • SEDCHEM • COMMSTRU

SEDCHEM • PATHOL • COMMSTRU

SEDCHEM • PATHOL • TISCHEM

TISCHEM • COMMSTRU • PATHOL

4-Component (Based on 3-way combinations.)

SEDTOX • SEDCHEM • PATHOL • TISCHEM

SEDTOX • SEDCHEM • COMMSTRU • PATHOL

TISCHEM • COMMSTRU • PATHOL • SEDCHEM

5-Component (The only possible combination.)

SEDTOX • SEDCHEM • TISCHEM • PATHOL • COMMSTRU

Figure 1. Most useful integrative assessments.

Different information is provided by each of the nine multiple-component assessments. Five of these assessments involve the Sediment Quality Triad (SQT), which is defined as any three-component integrative assessment that includes sediment toxicity, sediment chemistry, and some measure of in situ bioeffects (Figure 2).

Table 3. Examples of Component-Specific Targets and Measures

Component	Most Probable Target(s)	Example Measures
SEDTOX[a]	· Benthic infauna · Sensitive indicator organisms · Commercially important organisms · Ecologically important organisms	· Survival · Sublethal effects · Chronic (life-cycle) effects · Mutagenic, cytotoxic, genotoxic responses
SEDCHEM	· Sediments	· Individual contaminants (organic and inorganic) · Sedimentary features (e.g., grain-size) · Ancillary analyses (e.g., TOC, AVS)
TISCHEM[b]	· Bottom-fish · Benthic infauna · Benthic epifauna	· Individual contaminants (organic and inorganic) · Ancillary analyses (e.g., size, weight, age, lipid content)
PATHOL	· Bottom-fish	· Individual pathological conditions · Ancillary analyses (e.g., size, weight, age)
COMMSTRU	· Benthic infauna	· Taxa presence/abundance · Dominance · Diversity

[a] May involve direct exposure to sediments, elutriates or pore waters (e.g., Chapters 9 and 10), or prior treatment of the sediment (e.g., fractionation — Samoiloff et al., 1983; extraction — Larson, 1989).
[b] Analyses may involve whole organisms and/or specific tissues, notably muscle and liver or hepatopancreas.

An ideal integrative assessment would provide, through its individual components, relevant information on the following levels of biological organization:

- Subcellular
- Cellular
- Organism
- Population
- Community
- Ecosystem

However, such remains an ideal rather than a reality. Such an assessment would be extremely costly. It is debatable whether the additional information gained would be worth the cost, and there presently is no effective means to link all the information provided by these different (and probably nonlinear) levels of organization.

Although there are major differences in tolerance and response between individuals, life stages, species, taxa, populations, communities, and ecosystems,[34-36] it is clearly impractical to determine the responses of specific bioassays to all possible

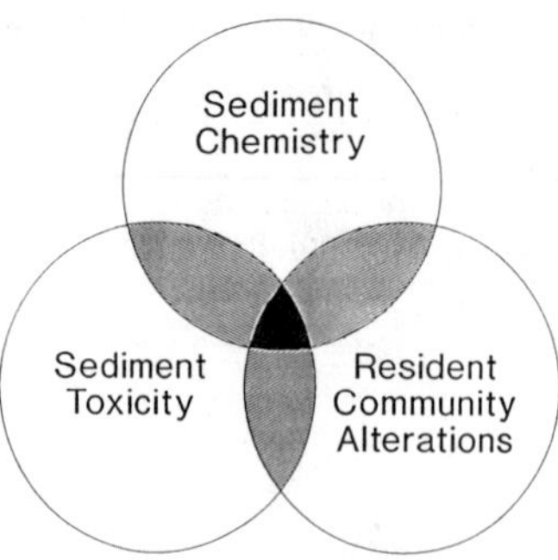

Figure 2. Conceptual model of the Sediment Quality Triad, which combines data from chemistry, toxicity, and in situ studies. Chemistry and bioassay estimates are based on laboratory measurements with field-collected sediments. In situ studies preferentially include measures of benthic community structure; they may also include measures of resident organism pathology and bioaccumulation/metabolism.

perturbations, individually and in mixtures. Thus, realistic (nonideal) integrative assessments must allow for a high level of uncertainty[37] that can only be reduced by identifying and concentrating on key processes and species and, wherever possible, making direct measurements and avoiding surrogates. At the species level, three key processes should be measured: survival, growth, and reproduction.[38] Key processes at the population, community, and ecosystem level tend to be specific rather than generic.[16] For example, the availability of substrate is important for sessile benthic communities, but not for pelagic communities. The best integrative assessments are based on a test battery involving structural and functional endpoints and crossing trophic levels (cf., Chapters 9 and 10).

REFERENCE AREAS AND THE QUESTION OF SIGNIFICANCE

In science, questions of significance generally revolve around statistics. A statistical interpretation of significance represents a relatively value-free approach based on isolating impacts from natural variation. However, environmental significance is not the unique prerogative of science; societal considerations are implicit, important, and often paramount.

> The use of knowledge coming out of the scientific approach is not scientific, it is political. The failure of ecological impact assessment is not being able to generate information which can be used at the political level. We are forced to consider not only social impact assessment itself, but the social values attached to ecological aspects and the importance of ecological concerns from a sociological perspective.[39]

The problem of setting the "correct" significance levels is discussed by Barnthouse et al.[40] in terms of the "...relationship between statistical, biological and societal significance...." They emphasize the "...need to consider societal values in determining the significance of toxicity test data..." because "...given extremely precise data, very small and biologically inconsequential changes might be found statistically significant." Clearly, statistics is important but not paramount,[2] yet there is an increasing emphasis and reliance on statistics for integrative assessments.[41]

Societal considerations cannot be ignored because they

1. Drive the formation of the end result: acceptable environmental quality
2. Provide for funding and support from stakeholders and major decision makers
3. Allow setting of societal goals on which significance levels and statistical design must be based. Whether scientific studies are based on a greater probability of false positives (Type I errors) or of false negatives (Type II errors) must depend on societal decisions as to the level of conservatism desired.

It is probably most difficult to develop consensus on the ecological significance of an impact. Definitions range from loss of habitat to decreased productivity, local extinction, and decreased ecosystem stability. As discussed earlier in this chapter, all impact definitions relate back to the definition of ecosystem health. Recognizing that most definitions contain inherent value judgements, societal judgements affect the question of significance even from an ecological point of view.

Integrative assessments generally involve comparisons to one or more reference areas. Ideally, reference areas are, by definition, as similar as possible to the areas under investigation, but not subject to anthropogenic influences and not contaminated or toxic. The choice of a reference area implicitly determines the acceptable condition for the area under investigation. If there is no significant difference from the reference area, then no remedial or other action is necessary; the degree of remediation or other action recommended and/or undertaken depends on this comparison. Thus, the choice of a reference area is critical and far too important to be left solely to science. Societal determinations of ecosystem health need to be considered in determining acceptable environmental conditions, and they need to be given significant weight in the final study design.

In addition, there are many possible ways to characterize responses to contaminated sediments, and a wide variety of parameters that can be measured (ranging from the subcellular to the community and population level). Because of the complexity of the environment, the issues of causality and relevance between a disturbance and a measured response assume major importance. The possibility that many of the measured responses may not be anthropogenic in nature must always be a consideration.[42,43] Further, adaptation, recovery, and normal succession (i.e., change) are also potential modifying factors. Laboratory toxicity tests often use the most sensitive species under worst-case conditions,[44] which may result in overprotection and overregulation.

Ecosystems comprise a mosaic of patches recovering from disturbance[45] and are dynamic; they are in a constant state of change. Klerks and Levinton[46] provide evidence for the evolution of resistance to heavy metals and provide arguments for the fact that evolution of resistance to toxic substances strongly alters the perception of the type of answer obtained from toxicity tests. At worst, such adaptive responses can change the predictions completely. For instance, Chapman et al.[47] found that benthic communities in the vicinity of an oil platform showed no evidence of adverse effects (e.g., no lower diversity or detrimental changes in dominance and abundance) despite the fact that the sediments were chemically contaminated and toxic in laboratory tests. They hypothesized adaptation as one possible explanation. Similarly, Keilty and Landrum[48] found that the freshwater oligochaete, *Stylodrilus heringianus*, can adapt to low-level toxicity such that resident populations are unaffected by the sediments that they live in, but which kill the same species collected from an uncontaminated area. Organisms adapted to toxic contamination can survive and function, but are probably stressed such that they have a lower resilience to additional stress. In other words, they are closer to the threshold at which adverse effects become manifest than are populations from uncontaminated areas. Integrative assessments are clearly necessary to make such determinations.[47]

ATTEMPTS AT INTEGRATIVE ASSESSMENTS

Information on the responses of ecosystems to stress has increased greatly in the last few decades, although this has not necessarily resulted in greater understanding of causality. There are many ways to characterize a response to stress, ranging from unequivocally adverse responses (e.g., extensive species mortalities) to responses whose significance (or cause) is highly questionable (e.g., minor enzymatic changes). Integrative assessments generally involve structural (e.g., species richness) rather than functional characteristics (e.g., decomposition and biogeochemical cycling). Clearly, there is a need for functional assessment methodologies for contaminated sediments.

To date, only two comprehensive five-component integrative assessments of sediment quality have been completed, namely the Superfund investigations in Puget Sound, Washington and the ARCS program studies in the Great Lakes (discussed in Chapters 17 and 18).[1,2,49,50] An experimental study involving all five components and two transects in the North Sea is presently nearing completion and will be the subject of future publications.[51] Because of the expense and magnitude of five-component studies, they are expected to remain relatively rare for the near term.

Four-component studies are more common, but most extensively used are three-component configurations that have been described as the Sediment Quality Triad[6,29-31,47,52-61] (also see Chapter 19). The Triad, a concept, and the Apparent Effects Threshold (AET),[6,56,57,62-64] an application of this concept, are discussed below as examples of integrative assessments.

Sediment Quality Triad

The triad concept is described in detail by Chapman.[59] It comprises an effects-based approach to describing sediment quality, which typically incorporates measures of sediment chemistry, sediment toxicity, and benthic infaunal community structure. This combination of components can be both descriptive and numeric and serves to

- Identify and differentiate pollution-degraded areas from reference conditions
- Determine the extent of pollution-induced degradation of sediments in a non-numerical, multiple-chemical mode, particularly in areas intermediate between degraded and nondegraded conditions (the "gray zone")
- Determine contaminant concentrations always associated with effects (= numerical sediment quality criteria; cf., following section on Apparent Effects Threshold, AET) and those never associated with effects
- Prioritize and rank areas and their environmental significance (e.g., is remediation necessary?)
- Describe ecological relationships between sediment properties and biota at risk
- Predict where degradation will occur, based on levels of contamination and toxicity.

This approach has been used in marine and estuarine waters on the west coast of North America (e.g., Vancouver Harbor, B.C.;[61] Puget Sound, WA;[28] San Francisco Bay, CA[55]), in the Gulf of Mexico,[47] in the North Sea,[51] and in the Chesapeake Bay.[65] Freshwater studies have been conducted in Oklahoma, South Dakota, Ohio, and the Great Lakes.[29-31] Possible conclusions provided by using this approach are detailed in Table 4. Similar tables can be readily constructed for any other integrative assessment (including four and five components) and should be prepared prior to study initiation. Such tables readily form the basis of hypothesis formulation and testing, which must be the basis of any credible scientific assessment.

The following assumptions apply to the triad and to similar integrative assessments and, with minor modifications, to all other integrative assessments:

- The approach incorporates (1) interactions between contaminants in complex sediment mixtures, including additivity, antagonism and synergism; (2) actions of unidentified toxic chemicals; and, (3) effects of environmental factors that influence biological responses (including toxicant concentrations);
- The chemical contaminant concentrations selected for any particular study are appropriate indicators of overall chemical contamination;
- The toxicity tests selected for any particular study are appropriate indicators of all possible, relevant toxic modes of action;
- The selected in situ bioeffects measures (e.g., selected benthic community structure parameters) are appropriate indicators of biological effects.

Table 4. Information Provided by Differential Triad Responses

Contamination	Toxicity	Alteration	Possible Conclusions
+	+	+	Strong evidence for pollution-induced degradation
–	–	–	Strong evidence that there is no pollution-induced degradation
+	–	–	Contaminants are not bioavailable
–	+	–	Unmeasured chemicals or conditions exist with the potential to cause degradation
–	–	+	Alteration is not due to toxic chemicals
+	+	–	Toxic chemicals are stressing the system
–	+	+	Unmeasured toxic chemicals are causing degradation
+	–	+	Chemicals are not bioavailable or alteration is not due to toxic chemicals

Responses are shown as either positive (+) or negative (–), indicating whether or not measurable (e.g., statistically significant) differences from reference conditions/measures are determined. (Adapted from Chapman, 1990.)

Clearly, choices made by investigators prior to study initiation can severely influence the effectiveness of integrative assessments (as they can of any study). Other important *a priori* choices include the use (or not) of synoptic data and the compositing (or not) of sediments. Synoptic data (e.g., sediments collected at the same time, testing done on all samples at the same time) are preferable, as they minimize uncertainty. Sediment sample compositing of sediments, which has been generally used to date,[56,57] should be avoided if the intent is to determine environmental variability rather than overall response.[61] Laboratory toxicity test procedures and sediment handling effects on assay responses should be defined to allow for valid extrapolations to in situ conditions (see Chapters 3, 8, and 10). Similarly, based on study aims, assessments analyzing tissue chemical concentrations and pathology need to determine whether these measures are to be done synoptically on the same organisms and tissues or whether chemical measures are to comprise a composite value.

Collection and homogenization of sediments affects not just the vertical distribution of contaminants, but also other sediment components, such as grain size, organic carbon, and sulfides, all of which influence bioavailability. For example, sulfides may be oxidized during testing, releasing bound metals into the interstitial water and resulting in toxicity that is not normally expressed in the real environment. Swartz et al.[27] demonstrated the heterogeneity of toxicity in marine sediments. Recent, more detailed work in freshwater by Stemmer et al.[66] emphasized the reality of heterogeneity, on micro- as well as macroscales. Sasson-Brickson and Burton[44] examined sediment chemistry, in situ toxicity using *Ceriodaphnia dubia*, community measures of macroinvertebrates and fish, and laboratory toxicity in a freshwater system. Results of this triad approach showed that traditional sample collection and laboratory toxicity test methods may alter the water and sediment toxicity that occurs in situ. Test method variables have also been shown to influence the results of toxicity testing. Stemner et al.[67] found that the spiking method, sediment storage conditions, and sediment holding time affected the toxicity response of *Daphnia magna* to selenium.

Ideally, replicate measures of sediment chemistry should be taken; realistically, this is often not possible due to cost. However, replication of selected representative chemicals (e.g., a metal and an organic) would provide useful interpretative information and should be encouraged. Similar comments apply to tissue chemical analyses.

A posteriori choices center on methods of data analysis and presentation. These can be simplistic, as illustrated in Figure 3. Components are shown as composite ratios relative to reference values; these ratios are determined by dividing the values of specific variables by the corresponding reference area values, assuming some measure of additivity, plotting the resultant dimensionless numbers on a scale with a common origin, and placing them at 120° from each other to form a triangle. More complex methods involving multivariate statistical/descriptive approaches are being developed, based on noncentered principal components analysis (PCA).[61]

Although the triad (and other integrative assessment methods) is both labor intensive and expensive, the following strengths render it (and such assessments in general) extremely cost effective for the level of information provided:

- It provides empirical (based on observation, not theory) evidence of sediment quality;
- It allows ecological interpretation of physical, chemical, and biological properties relative to the real environment;
- It does not require *a priori* assumptions concerning specific methods of interaction between organisms and chemicals;
- It uses a preponderance-of-evidence approach rather than relying on single measurements;
- Because of its comprehensive nature, additional follow-up studies are either unnecessary or very limited;
- The data generated can be used to develop effects-based classification indices; and
- It can be used for any sediment type.

The triad approach is suitable for in-place pollutant control, for source control, and for disposal applications. Major limitations are those of its component parts. For instance, sediment toxicity tests provide some degree of variability even in interlaboratory calibration experiments. Mearns et al.,[68] Chapman et al.,[55] and Long et al.[9] found that the *Rhepoxynius abronius* sediment toxicity test had coefficients of variation (CV) ranging from 21.4 to 22.4%. Although this range compares favorably with water column tests,[68] it still indicates variability. However, in comparison,

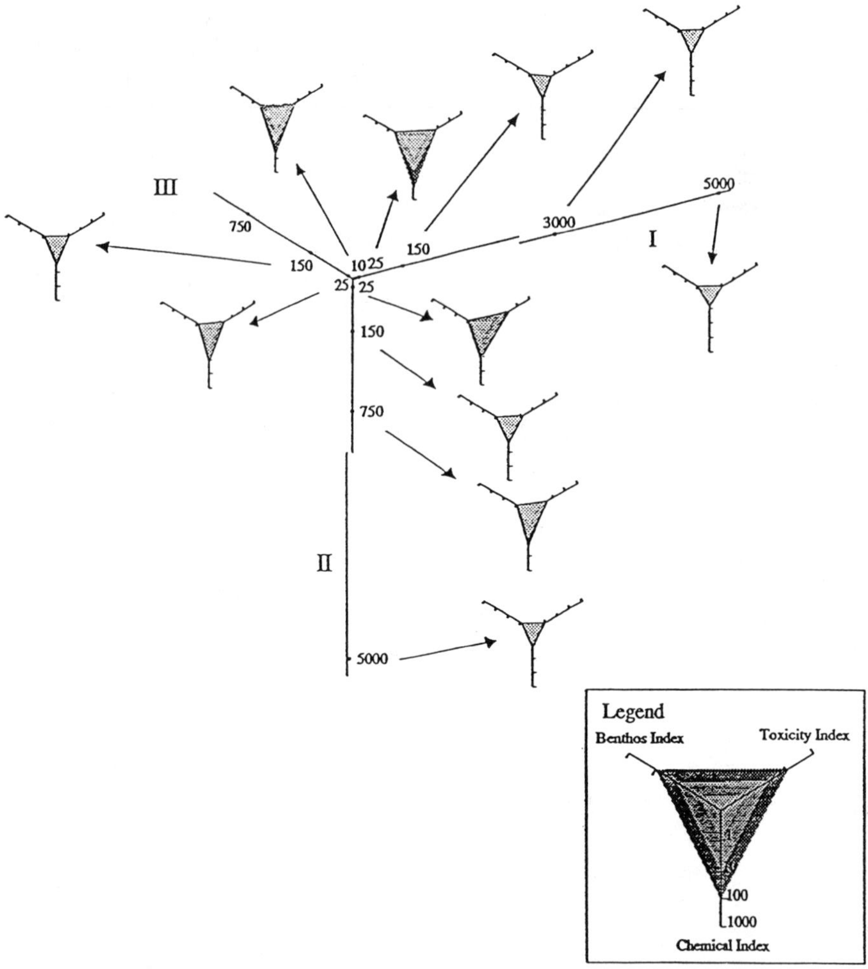

Figure 3. Simple representation of Sediment Quality Triad data for a central oil platform and a remote platform (gradient II, station 750).[47] Station numbering represents meters distant from the center; all data as ratios to combined reference stations I 5000 and II 5000.

Kovacs and Ferguson[69] found that, in freshwater, CVs ranged from 17 to 69% in an interlaboratory calibration with *Daphnia magna*. Variability in toxicity tests means that the results cannot be considered as a single number, but rather a range of values, which results in uncertainty, particularly in the "gray zone."[67] Further, some sediment toxicity tests may be influenced by grain-size effects.[70]

Major limitations to the triad, which are representative of similar integrative assessments, are

- Statistical (and other) criteria are in the developmental stage;
- Rigorous criteria for integrating different measures for each component (e.g., different toxicity tests, chemical measures, and benthic community parameters) have not been developed;
- Results may be strongly influenced by the presence of unmeasured chemicals that may or may not covary with measured chemicals;
- It is highly labor intensive and expensive; and
- Expert judgement is required to interpret the results.

However, if properly designed, strong linkages can be made between different components, which greatly assists the interpretation of results. Chapman and Long[28] make the point that the absence of species that show effects in laboratory toxicity tests from communities in (or associated with) contaminated sediments provides the strongest possible evidence for pollution-induced degradation. Their presence and an absence of laboratory effects would, conversely, provide equally strong evidence for a lack of pollution-induced degradation. Because marine amphipods are particularly sensitive to pollution,[71] they have been used and are being used, extensively in sediment toxicity tests undertaken as part of integrative assessments where their presence (or absence) in the benthic community is measured.[27,72]

Apparent Effects Threshold (AET)

The AET represents a specific application of the Sediment Quality Triad. Similar or other applications can be derived, and are expected to be derived, from this and other integrative assessments. Although the usefulness of this application and of its component parts is the subject of a current debate,[57,73] in the absence of any viable alternatives capable of producing numerical data, AET data are presently being used in a regulatory context in Puget Sound (cf., Chapter 18). Further details of the AET, which is only briefly discussed here, are available from a variety of sources.[6,50,62,64,74,75]

The AET is derived using empirical data from each of the three triad components to identify individual chemical concentrations above which the measured biological effects (e.g., sediment toxicity endpoints, depressions in benthic infaunal community abundances) would always be expected to occur. Specifically, they are derived by determining, for a given chemical within a data set, the chemical sediment concentration above which a particular adverse biological effect is always statistically significant ($p \leq 0.05$) relative to designated reference conditions. An example of the AET approach with two chemicals, lead and 4-methylphenol, is shown in Figure 4. Data are from samples collected at various locations and analyzed for a single specified laboratory toxicity response in relation to measured chemical concentrations. For lead, the toxicity response would always be expected to occur above 660 ppm, so that concentration becomes the AET value for that toxicity test. For 4-methylphenol, the AET is 3600 ppb.

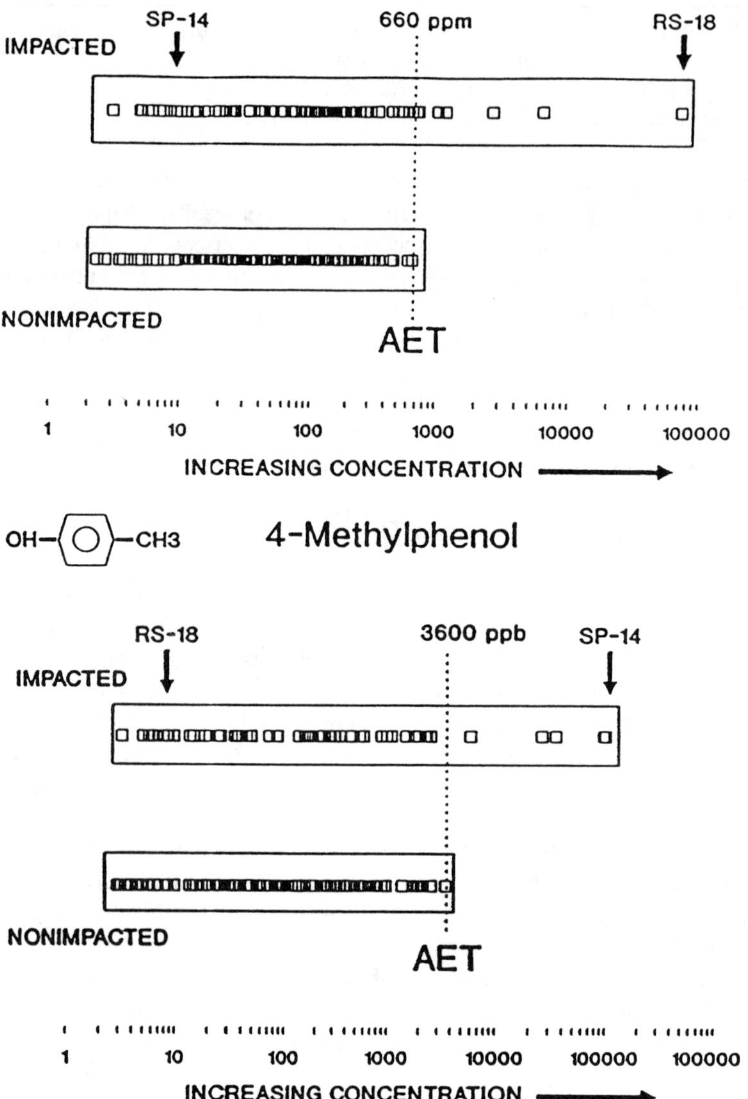

Figure 4. The AET approach[64] applied to sediments tested for lead and 4-methylphenol concentrations and toxicity response during bioassays.

Sediments are clearly separable into two classes: those that exhibit statistically significant responses relative to reference conditions ("impacted" conditions), and those exhibiting no statistical differences ("nonimpacted" conditions). Within the "impacted" group of stations, there is a threshold above which significant sediment toxicity occurs in all samples. Below this threshold, significant sediment toxicity occurs in some samples, but not in others. The AET is defined as the highest concentration of that chemical in the sediments that did *not* exhibit sediment toxicity.

AET levels are developed independently for each of the bioeffects measures used. For instance, if three toxicity tests are used, then three separate AETs will be developed. Separate AETs are also determined for each measure of benthic infaunal community structure. Use of the AET thus requires decisions as to whether the lowest (LAET) or highest (HAET) will be applied or whether some middle ground will be chosen. Ultimately, the best professional judgement, based on burden of evidence, must be applied to determine how the results of this method are applied in regulatory and other contexts. At this point, the question of significance (previously discussed) becomes paramount.

As with all methodologies, the AET has its supporters and its detractors. However, it is the only example, to date, of how a sediment integrative assessment can be used to derive numerical data for assessing sediment quality. Whether numerical data are desirable is another question.[6]

TIERED TESTING

Current approaches to environmental assessment encourage the term "cost effectiveness". In other words, what are the least monies that can be spent, or spent in a phased fashion, to achieve program goals? Accordingly, there is increasing emphasis on tiered testing. Tiered testing simply means that all testing is not done synoptically or even concurrently; initial testing is done to determine areas for in-detail study, which may (or may not) involve a complete integrative assessment. For instance, Reynoldson and Zarull[41] recommend using the results of tiered testing to reduce sampling needs and, hence, costs. Alternatively, the initial survey may provide coincident data related to one component of such an assessment.

Proceeding further with the example of the triad, tiered testing may involve either initial sediment chemistry or sediment toxicity measures. Community structure data, though extremely useful in cases of conventional pollution (e.g., low dissolved oxygen),[76] in the case of toxic chemical mixtures are most useful for definitive, postscreening studies.[10] Possible examples of tiered testing are provided in Figure 5. The problem with any initial screening test is the reality that it is impossible to prove conclusively that there are no chemicals or no effects. However, careful design can provide reasonable surety that major chemicals and effects will be detected.

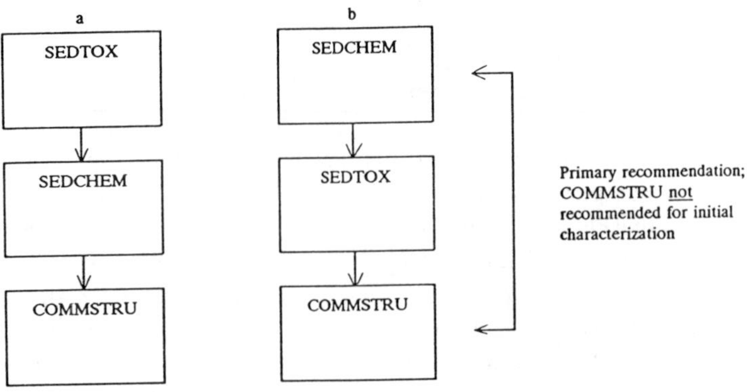

a. Example of possible sub-components, toxicity:

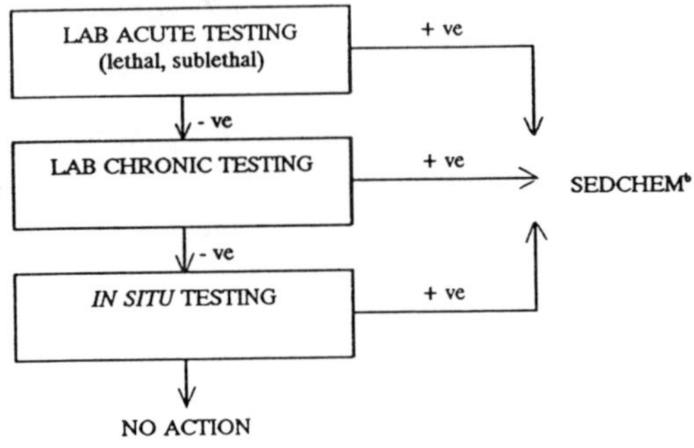

- ve = no effect; + ve = effects

b. Example of possible sub-components, chemistry:

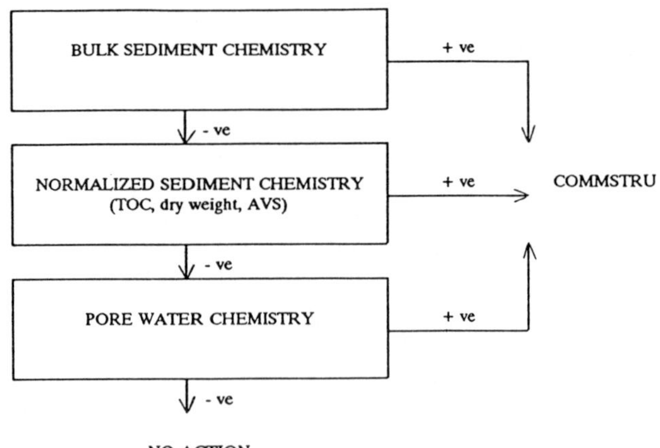

- ve = concentrations not elevated; + ve = elevated concentrations

Figure 5. Example of possible tiered testing approaches (based on SEDTOX-SEDCHEM-COMMSTRU, 3-component assessment).

Examples of useful tiered testing approaches related to several-component integrative assessments are provided by NRC[50,77] and in Chapter 18.

Although tiered testing meets the stated primary requirements of remedial cleanup, namely for "simple and inexpensive" sediment testing and classification procedures,[77] it must be emphasized that simple initial screening procedures are far from definitive. For instance, concordance between toxicity tests and actual in situ conditions may not be achieved for a number of reasons, including contaminant bioavailability related to in-place, as opposed to disturbed, sediments (in other words, changes to the sediments during collection and testing);[44,67] heterogeneity in sediment contamination and toxicity (i.e., grab samples may not be representative or comparable);[66] sediment effects (e.g., grain size) on test organisms; the behavior of infaunal organisms (e.g., avoidance); the different mechanisms of toxic action (i.e., the need for suites of species in testing); and the evolution or change towards tolerance. Also, infauna may avoid contaminated sediments that may only comprise a layer in the sediments. For instance, Johns et al.[78] found that the polychaete *Nephtys incisa* exhibited avoidance when exposed to layered contaminated sediments. As previously discussed, current procedures generally involve collecting sediments to depth and exposing animals to a homogeneous mixture that destroys redox gradients and alters toxicant(s) bioavailability. Further, the depth of sediment burrowing varies with the size of animals and species, thus exposure under natural conditions will vary depending on the type of contamination.

The use of sediment chemistry for screening contains additional possibilities for error because sediment analyses are not effects based. Bulk sediment chemistry is more often (and more easily and cheaply) measured than pore water chemistry, yet the latter comprises the major route of uptake of contaminants to benthic organisms, which can result in bioeffects (cf., Chapter 1). The example of possible chemistry subcomponents shown in Figure 5 is unlikely to be fully used in tiered testing, although a real possibility is normalized sediment chemistry data.

Tissue chemical contamination is not predictable in terms of either concentrations at which bioeffects occur[79] or in terms of concentration from contaminated sediments.[80] This component should not be used for initial screening.

Pathology has been used for screening and as the basis for further, more detailed testing.[81-83] But not all pathological effects are anthropogenic.[42,84] Further, pathology should be used in case studies, but not for site-specific screening, because bottom fish, almost universally used for pathological studies involving contaminated sediments, are motile, and their condition is not representative of conditions at a specific, localized site.

SUMMARY AND CONCLUSIONS

Ultimately, judgements on environmental quality, assuming the persistence of habitat,[16] can only be determined by the responses or condition of multiple (but never single or a large number of) measures conducted as part of integrative assessments.

Reliance on single measures of any component is not recommended. The uncertainty and variability inherent in both ecosystems and our methods of measurement require a range of values and a burden-of-evidence approach.

The triad and AET, discussed in this chapter as examples of integrative assessments, can and must be modified to determine particular parameters to measure and monitor for in specific situations and ecosystems. Most importantly, specific (e.g., chemicals of concern, sensitive organisms at risk) and generic knowledge (e.g., method of chemical action, organism responses) must be applied.

The variability of ecosystems must be taken into account. This information is necessary to aid in separating responses to anthropogenic disturbance from natural variability (i.e., distinguishing signal from noise). Two methods can be used: observational approaches and experimental studies. To date, sediment assessments at the whole-ecosystem level have generally involved observational approaches; experimental studies have been primarily restricted to subcellular, cellular, and whole-organism levels of organization.

At present there are no simple means to determining environmental health related to contaminated sediments. There will always be a need for integrative assessments. These are, and will continue to be, most useful where they are flexible and responsive to study goals, ecological realities, and social concerns.

RECOMMENDATIONS

The following major recommendations describe critical future research and development needs:

- Development of better understanding of spatial and temporal variance in contamination and toxicity, of resilience, and of recovery
- Standardization of techniques for each integrative assessment component, and the development of better (faster, more reliable) techniques including functional, not just structural, measures
- Development and improvement of techniques for determining concordance between assessment components (including, but not limited to, statistical analyses)
- Research to assist in determining key structures and processes related to measures and whether weightings should be assigned in particular instances (e.g., are chronic effects of more importance than acute lethality or vice versa?)
- Determination of the appropriateness of different endpoints of different toxicity tests, selected chemical contaminants, selected measures of benthic community structure, and other potential measures of in situ biological alteration
- Determination of the limits to predictions
- Development of theoretical insights and frameworks for the application of integrative assessments.

This latter "need" is arguably the most critical. Extensive effort is presently being expended on reductionist research related to new techniques and methodologies.

Such research is useful and necessary. However, relatively little research effort concerned with contaminated sediments is being expended on holistic approaches to integrative assessments. This is a critical failing; without major effort in this area, there will be a multiplicity of methods and a paucity of frameworks for their application and interpretation. This research imbalance must be remedied before the problems inherent in contaminated sediments can be adequately determined and ultimately remediated.

REFERENCES

1. Harris, H. J., P. E. Sager, S. Richman, V. A. Harris, and C. J. Yarbrough. "Coupling Ecosystem Science with Management: A Great Lakes Perspective from Green Bay, Lake Michigan, USA," *Environ. Manag.* 11:619–625 (1987).
2. Harris, H. J., P. E. Sager, H. A. Regier, and G. R. Francis. "Ecotoxicology and Ecosystem Integrity: The Great Lakes Examined," *Environ. Sci. Technol.* 24:598–603 (1990).
3. Howells, G., D. Calamari, J. Gray, and P. G. Wells. "An Analytical Approach to Assessment of Long-Term Effects of Low Levels of Contaminants in the Marine Environment," *Mar. Pollut. Bull.* 21:371–375 (1990).
4. Cairns, J., Jr. "The Myth of the Most Sensitive Species," *BioScience* 36:670–672 (1986).
5. Giesy, J. P., and R. A. Hoke. "Freshwater Sediment Toxicity Bioassessment: Rationale for Species Selection and Test Design," *J. Great Lakes Res.* 15:539–569 (1989).
6. Chapman, P. M. "Current Approaches to Developing Sediment Quality Criteria," *Environ. Toxicol. Chem.* 8:589–599 (1989).
7. Eagleson, K. W., D. L. Lenat, L. W. Ausley, and F. B. Winborne. "Comparison of Measured Instream Biological Responses with Responses Predicted Using the *Ceriodaphnia dubia* Chronic Toxicity Test," *Environ. Toxicol. Chem.* 9:1019–1028 (1990).
8. Williams, L. G., P. M. Chapman, and T. C. Ginn. "A Comparative Evaluation of Sediment Toxicity Using Bacterial Luminescence, Oyster embryo, and Amphipod Sediment Bioassays," *Mar. Environ. Res.* 19:225–249 (1986).
9. Long, E. R., M. F. Buchman, S. M. Bay, D. J. Breteler, P. M. Chapman, J. E. Hose, A. L. Lissner, J. Scott, and D. A. Wolfe. "Comparative Evaluation of Five Toxicity Tests with Sediments from San Francisco Bay and Tomales Bay, California," *Environ. Toxicol. Chem.* 9:1193–1214 (1990).
10. Bilyard, G. R. "The Value of Benthic Infauna in Marine Pollution Monitoring Studies," *Mar. Pollut. Bull.* 18:581–585 (1987).
11. Larsson, P. "Contaminated Sediments of Lakes and Oceans Act as Sources of Chlorinated Hydrocarbons for Release to Water and Atmosphere," *Nature* 317:347–349 (1985).
12. Warren-Hicks, W., B. R. Parkhurst and S. S. Baker, Jr., Eds. *Ecological Assessment of Hazardous Waste Sites: A Field and Laboratory Reference Document,* U.S. Environmental Protection Agency, Corvallis, OR EPA 600/3-89/013 (1989).
13. Rapport, D. J. "What Constitutes Ecosystem Health?" *Pers. Biol. Med.* 33:120–132 (1989).

14. Rapport, D. J. "Challenges in the Detection and Diagnosis of Pathological Changes in Aquatic Ecosystems," *J. Great Lakes Res.* 16:609–618 (1990).
15. Kelly, J. R., and M. A. Harwell. "Indicators of Ecosystem Response and Recovery," in *Ecotoxicology: Problems and Approaches*, S. A. Levin, M. A. Harwell, J. R., Kelly and K. D. Kimball, Eds. (New York: Springer-Verlag, 1988), pp. 9–35.
16. Chapman, P. M. "Ecosystem Health Synthesis: Can We Get There from Here?" *J. Ecosyst. Health Manage.* (in press) (1991).
17. Preston, F. "Diversity and Stability in the Biological World," in *Diversity and Stability in Ecological Systems*, G. M. Woodwell, and H. H. Smith, Eds. Brookhaven National Laboratories Symposium #22 (1969).
18. Ryder, R. A. "Ecosystem Health, a Human Perception: Definition, Detection, and the Dichotomous Key," *J. Great Lakes Res.* 16:619–624 (1990).
19. IUCN/UNEP/WWF. "Caring for the World: A Strategy for Sustainability," United Nations Conservation Union, United Nations Environment Programme, and World Wide Fund for Nature, 2nd draft (June 1990).
20. Connell, J. H. "Diversity in Tropical Rain Forests and Coral Reefs," *Science* 199:1302–1310 (1978).
21. Huston, M. "A General Hypothesis of Species Diversity," *Am. Nat.* 113:81–101 (1979).
22. Warwick, R.M., M. Pearson, and T.H. Ruswahyuni. "Detection of Pollution Effects on Marine Macrobenthos: Further Evaluation of the Species Abundance/Biomass Method," *Mar. Biol.* 95:193–200 (1987).
23. McManus, J. W., and D. Pauly. "Measuring Ecological Stress: Variations on a Theme by R. M. Warwick," *Mar. Biol.* 106:305–308 (1990).
24. Power, E. A., K. R. Munkittrick, and P. M. Chapman. "An Ecological Impact Assessment Framework for Decision-Making Relative to Sediment Quality," in *Aquatic Toxicology and Risk Assessment: Vol. 14,* STP 1124, M. A. Mayes, and M. G. Barron, Eds. (Philadelphia: American Society for Testing and Materials, 1991), pp. 48–64.
25. Levin, S. A., M. A. Harwell, J. R. Kelly, and K. D. Kimball. "Ecotoxicology: Problems and Approaches," in *Ecotoxicology: Problems and Approaches*, S. A. Levin, M. A. Harwell, J. R. Kelly, and K. D. Kimball, Eds. (New York: Springer-Verlag, 1988), pp. 3–7.
26. Becker, D. S., G. R. Bilyard, and T. C. Ginn. "Comparisons Between Sediment Bioassays and Alterations in Benthic Macroinvertebrate Assemblages at a Marine Superfund Site: Commencement Bay, Washington," *Environ. Toxicol. Chem.* 9:669–685 (1990).
27. Swartz, R. C., W. A. DeBen, K. A. Sercu, and J. O. Lamberson. "Sediment Toxicity and the Distribution of Amphipods in Commencement Bay, Washington, USA," *Mar. Pollut. Bull.* 13:359–364 (1982).
28. Chapman, P. M., and E. R. Long. "The Use of Bioassays as Part of a Comprehensive Approach to Marine Pollution Assessment," *Mar. Pollut. Bull.* 4:81–84 (1983).
29. Burton, G. A., Jr., and B. L. Stemner. "Evaluation of Surrogate Tests in Toxicant Impact Assessments," *Tox. Assess.* 3:255–269 (1988).
30. Burton, G. A., Jr. "Evaluation of Seven Sediment Toxicity Tests and Their Relationships to Stream Parameters," *Tox. Assess.* 4:149–159 (1989).
31. Burton, G. A., Jr., L. Burnett, M. Henry, S. Klaine, P. Landrum, and M. Swift. "A Multiassay Comparison of Sediment Toxicity of Three 'Areas of Concern'," Abstr. Annu. Meet. Soc. Environ. Toxicol. and Chem., Arlington, VA (1990).

32. Norberg-King, T. J., and D. I. Mount. "Validity of Effluent and Ambient Toxicity Tests for Predicting Biological Impact: Skeleton Creek, Enid, Oklahoma." U.S. Environmental Protection Agency, Duluth, MN (1986).
33. Mount, D. I., N. A. Thomas, T. J. Norberg, M. T. Barbour, T. H. Roush, and W. F. Brandes. "Effluent and Ambient Toxicity Testing and Instream Community Response on the Ottawa River, Lima, Ohio," U.S. Environmental Protection Agency, Duluth, MN (1984).
34. Chapman, P. M., R. N. Dexter, R. M. Kocan, and E. R. Long. "An Overview of Biological Effects Testing in Puget Sound, Washington: Methods, Results and Implications," in *Aquatic Toxicology and Hazard Assessment: Seventh Symposium*, STP 854, R. D. Cardwell, R. Purdy, and R. C. Bahner, Eds. (Philadelphia: American Society for Testing and Materials, 1985), pp. 344–363.
35. Chapman, P. M. "Summary of Biological Effects in Puget Sound — Past and Present," in *Oceanic Processes in Marine Pollution: Vol 5, Urban Wastes in Coastal Marine Environments,* D. A. Wolfe, and T. P. O'Connor, Eds. (Malabar, FL: Robert E. Krieker, 1988), pp. 169–183.
36. Long, E. R., and M. R. Buchman. "An Evaluation of Candidate Measures of Biological Effects for the National Status and Trends Program," NOAA Tech. Memo, NOS OMA 45, U.S. Department of Commerce, Rockville, MD (1989).
37. Schaeffer, D. J., D. K. Cox, and R. A. Deem. "Variability of Test Systems Used to Assess Ecological Effects of Chemicals," *Water Sci. Technol.* 19:39–45 (1987).
38. Scott, K. J. "Effects of Contaminated Sediments on Marine Benthic Biota and Communities," in *Marine Board, National Research Council Symposium/Workshop on Contaminated Marine Sediments* (Washington, D.C.: National Research Council, 1989), pp. 132–154.
39. Beanlands, G. E., and P. N. Duinker. "An Ecological Framework for Environmental Assessment in Canada. Institute for Resource and Environmental Studies and Federal Environmental Assessment Review Office, Canada," ISNB 7703-0460-5, p. 132 (1983).
40. Barnthouse, L. W., G. W. Suter, II, and A. E. Rosen. "Inferring Population-Level Significance from Individual-Level Effects: An Extrapolation from Fisheries Science to Ecotoxicology," in *Aquatic Toxicology and Environmental Fate: Eleventh Volume*, STP 1007, G. W. Suter, II, and M. A. Lewis, Eds. (Philadelphia: American Society for Testing and Materials, 1989), pp. 289–300.
41. Reynoldson, T. B., and M. A. Zarull. "The Biological Assessment of Contaminated Sediments — The Detroit River Example," *Hydrobiologia* 188/189:463–476 (1989).
42. Kurelec, B., A. Garg, S. Krca, M. Chacko, and R. C. Gupta. "Natural Environment Surpasses Polluted Environment in Inducing DNA Damage in Fish," *Carcinogenesis* 10:1337–1339 (1989).
43. Ames, B. N., and L. S. Gold. "Too Many Rodent Carcinogens: Mitogenesis Increases Mutagenesis," *Science* 249:970–971 (1990).
44. Sasson-Brickson, G., and G.A. Burton, Jr. "In Situ and Laboratory Sediment Toxicity Testing with *Ceriodaphnia dubia*," *Environ. Toxicol. Chem.* 10:201–207 (1991).
45. Pickett, S. T. A., and P. S. White, Eds. *The Ecology of Natural Disturbance and Patch Dynamics* (New York: Academic Press, 1985).

46. Klerks, P. L., and J. S. Levinton. "Effects of Heavy Metals in a Polluted Aquatic Ecosystem," in *Ecotoxicology: Problems and Approaches*, S. A. Levin, M. A. Harwell, J. R. Kelly, and K. D. Kimball, Eds. (New York: Springer-Verlag, 1988), pp. 41–67.
47. Chapman, P. M., R. N. Dexter, H. B. Anderson, and E. A. Power. "Evaluation of Effects Associated with an Oil Platform, Using the Sediment Quality Triad," *Environ. Toxicol. Chem.* 10: 407–424 (1991).
48. Keilty, T. J., and P. F. Landrum. "Population-Specific Toxicity Responses by the Freshwater Oligochaete, *Stylodrilus heringianus*, in Natural Lake Michigan Sediments," *Environ. Toxicol. Chem.* 9:1147–1154 (1990).
49. Barrick, R. C., D. S. Becker, D. P. Weston, and T. C. Ginn. "Commencement Bay Nearshore/Tideflats Remedial Investigation," U.S. Environmental Protection Agency, Seattle, WA, EPA 910/9-85-134a (1985).
50. Ginn, T. "Assessment of Contaminated Sediments in Commencement Bay (Puget Sound, Washington)," in *Marine Board, National Research Council Symposium/ Workshop on Contaminated Marine Sediments*, (Washington, D.C.: National Research Council, 1989), pp. 425–439.
51. Chapman, P. M. Unpublished data.
52. Swartz, R. C., D. W. Shults, G. R. Ditsworth, W. A. DeBen, and F. A. Cole. "Sediment Toxicity, Contamination, and Macrobenthic Communities Near a Large Sewage Outfall," in *Validation and Predictability of Laboratory Methods for Assessing the Fate and Effects of Contaminated Aquatic Ecosystems*, STP 865, T. P. Boyle, Ed. (Philadelphia: American Society for Testing and Materials, 1985), pp. 152–175.
53. Long, E. R., and P. M. Chapman. "A Sediment Quality Triad: Measures of Sediment Contamination, Toxicity and Infaunal Community Composition in Puget Sound," *Mar. Pollut. Bull.* 16:405–415 (1985).
54. Chapman, P. M., R. N. Dexter, and S. F. Cross. "A Field Trial of the Sediment Quality Triad in San Francisco Bay," NOAA Tech. Memo, NOA OMA 25. U.S. Department of Commerce, Rockville, MD (1986).
55. Chapman, P. M., R. N. Dexter, and E. R. Long. "Synoptic Measures of Sediment Contamination, Toxicity and Infaunal Community Structure (The Sediment Quality Triad) in San Francisco Bay," *Mar. Ecol. Prog. Ser.* 37:75–96 (1987a).
56. Chapman, P. M., R. C. Barrick, J. M. Neff, and R. C. Swartz. "Four Independent Approaches to Developing Sediment Quality Criteria Yield Similar Values for Model Contaminants," *Environ. Toxicol. Chem.* 6:723–725 (1987b).
57. Chapman, P. M., E. R. Long, R. C. Swartz, T. H. DeWitt, and R. Pastorok. "Sediment Toxicity Tests, Sediment Chemistry and Benthic Ecology Do Provide New Insights into the Significance and Management of Contaminated Sediments — A Reply to Robert Spies," *Environ. Toxicol. Chem.* 10: 1–4.
58. Chapman, P. M. "Sediment Quality Criteria from the Sediment Quality Triad: An Example," *Environ. Toxicol. Chem.* 5:957–964 (1986).
59. Chapman, P. M. "The Sediment Quality Triad Approach to Determining Pollution-Induced Degradation," *Sci. Tox. Environ.* 97-8: 815–825 (1990).
60. Long, E. R. "The Use of the Sediment Quality Triad in Classification of Sediment Contamination," in *Marine Board, National Research Council Symposium/Workshop on Contaminated Marine Sediments* (Washington, D.C.: National Research Council, 1989), pp. 78–93.

61. Cross, S. F., J. M. Boyd, P. M. Chapman, and R. O. Brinkhurst. "A Multivariate Approach to Assessing the Spatial Extent of Benthic Impacts Established Using the Sediment Quality Triad," manuscript in preparation.
62. PTI Environmental Services. "Briefing Report to the EPA Science Advisory Board: The Apparent Effects Threshold Approach," report prepared for the U.S. Environmental Protection Agency, Region 10, Seattle, WA, PTI Environmental Services, Bellevue, WA (1988).
63. Shea, D. "Developing National Sediment Quality Criteria," *Environ. Sci. Technol.* 22:1256–1261 (1988).
64. Barrick, R. C., H. Beller, S. Becker, and T. Ginn. "Use of the Apparent Effects Threshold Approach (AET) in Classifying Contaminated Sediments," in *Marine Board, National Research Council Symposium/Workshop on Contaminated Marine Sediments* (Washington, D.C.: National Research Council, 1989), pp. 64–77.
65. Chapman, P. M., and E.A. Power. Unpublished data.
66. Stemner, B. L., G. A. Burton, Jr., and G. Sasson-Brickson. "Effect of Sediment Spatial Variance and Collection Method on Cladoceran Toxicity and Indigenous Microbial Activity Determinations," *Environ. Toxicol. Chem.* 9:1035–1044 (1990a).
67. Stemner, B. L., G. A. Burton, Jr., and S. Leibfritz-Frederick. "Effect of Sediment Test Variables on Selenium Toxicity to *Daphnia magna*," *Environ. Toxicol. Chem.* 9:381–389 (1990b).
68. Mearns, A. J., R. C. Swartz, J. M. Cummins, P. A. Dinnel, P. Plesha, and P. M. Chapman. "Interlaboratory Comparison of a Sediment Toxicity Test Using the Marine Amphipod *Rhepoxynius abronius*," *Mar. Environ. Res.* 19:13–37 (1986).
69. Kovacs, T. G., and S. M. Ferguson. "An Assessment of the Ontario Ministry of the Environment Protocols for Conducting *Daphnia magna* Acute Lethal Toxicity Tests with Pulp and Paper Mill Effluents," *Environ. Toxicol. Chem.* 9:1081–1093 (1990).
70. DeWitt, T. H., G. R. Ditsworth, and R. C. Swartz. "Effects of Natural Sediment Features on Survival of the Phoxocephalid Amphipod, *Rhepoxynius abronius*," *Mar. Environ. Res.* 25:99–124 (1988).
71. Bellan-Santini, D. "Relationship Between Populations of Amphipods and Pollution," *Mar. Pollut. Bull.* 11:224–227 (1980).
72. Swartz, R. C., F. A. Cole, D. W. Schults, and W. A. DeBen. "Ecological Changes in the Southern California Bight Near a Large Sewage Outfall: Benthic Conditions in 1980 and 1983," *Mar. Ecol. Prog. Ser.* 31:1–13 (1986).
73. Spies, R. B. "Sediment Bioassays, Chemical Contaminants and Benthic Ecology: New Insights or Just Muddy Water?" *Mar. Environ. Res.* 27:73–75 (1989).
74. Battelle. "Overview of Methods for Assessing and Managing Sediment Quality," report prepared for the U.S. Environmental Protection Agency Office of Marine and Estuarine Protection, Washington, D.C. Battelle New England, Duxbury, MA (1988).
75. Science Advisory Board. "Report of the Sediment Criteria Subcommittee: Evaluation of the Apparent Effects Threshold (AET) Approach for Assessing Sediment Quality," U.S. EPA, Science Advisory Board, SAB-EETFC-89-027, p. 18 (1989).
76. Pearson, T. H., and R. Rosenberg. "Macrobenthic Succession in Relation to Organic Enrichment and Pollution of the Marine Environment," *Oceanogr. Mar. Biol. Ann. Rev.* 16:229–311 (1978).
77. NRC. "Contaminated Marine Sediments — Assessment and Remediation," National Research Council, National Academy Press, Washington, D.C., p. 493 (1989).

78. Johns, D. M., R. Gutjahr-Gobell, and P. Schauer. "Use of Bioenergetics to Investigate the Impact of Dredged Material on Benthic Species: A Laboratory Study with Polychaetes and Black Rock Material," Tech. Rept. D-85-7, U.S. Army Engineers Waterways Experiment Station, Vicksburg, MS (1985).
79. Payne, J. F., J. Kiceniuk, L. L. Fancey, V. Williams, G. L. Fletcher, A. Rahimtula, and B. Fowler. "What is a Safe Level of Polycyclic Aromatic Hydrocarbons for Fish: Subchronic Toxicity Study on Winter Flounder (*Pseudopleuronectes americanus*)," *Can. J. Fish. Aquat. Sci.* 45:1983–1993 (1988).
80. Bierman, V. J., Jr. "Equilibrium Partitioning and Biomagnification of Organic Chemicals in Benthic Animals," *Environ. Sci. Technol.* 24: 1407–1412 (1990).
81. Malins, D. C., B. B. McCain, D. W. Brown, S.-L. Chan, M. S. Myers, J. T. Landahl, P. G. Prohaska, A. J. Friedman, L. D. Rhodes, D. G. Burrows, W. D. Gronlund, and H. O. Hodgins. "Chemical Pollutants in Sediments and Diseases of Bottom-Dwelling Fish in Puget Sound," *Environ. Sci. Technol.* 18:705–713 (1984).
82. Baumann, P. C., and J. C. Harshbarger. "Frequencies of Liver Neoplasia in a Feral Fish Population and Associated Carcinogens," *Mar. Environ. Res.* 17:324–327 (1985).
83. Becker, D. S., T. C. Ginn, M. L. Landolt, and D. B. Powell. "Hepatic Lesions in English Sole (*Parophrys vetulus*) from Commencement Bay, Washington (USA)," *Mar. Environ. Res.* 23:153–173 (1987).
84. Mix, M. C. "Cancerous Diseases in Aquatic Animals and Their Association with Environmental Pollutants: a Critical Literature Review," *Mar. Environ. Res.* 20: 1–141 (1986).

CHAPTER 15

Management Framework for Contaminated Sediments (The U.S. EPA Sediment Management Strategy)

Elizabeth Southerland, Michael Kravitz, and Thomas Wall

OVERVIEW

The U.S. Environmental Protection Agency (EPA) has authority, under a variety of statutes, to manage contaminated sediments (see Table 1). Until recently, however, EPA had not focused on sediment quality issues except in relation to the disposal of dredged material removed during navigational dredging.

In 1985 and 1987, EPA's Office of Water conducted inventories of contaminated sediment sites.[1,2] Most of the information found for these inventories was from areas in the Northeast, along the Atlantic and Gulf Coasts, and in the Great Lakes region. No estimates could be made of the national total of river miles impaired by contaminated sediments or the cubic yards of problem sediments because sediment monitoring data were not available for much of the country. The 1987 inventory listed the following sources of known sediment contamination: chemical, steel, metal working, electroplating, engines and automotive, and nuclear energy production plants; paper mills; tanneries; refineries; electrical component and capacitor manufacturers; wood preservers; municipal sewage treatment plants; combined sewer overflows; agricultural and urban stormwater runoff; atmospheric deposition; seepage from waste disposal facilities; mining; shipping; and ocean dumping. Based on this extensive list of sources, most waterbodies serving major urban and industrial areas in the U.S. probably contain contaminated sediments. Waterbodies with known or suspected problems include ocean waters, estuaries, rivers/streams, lakes, and reservoirs, in all regions of the country.

The EPA inventories found that heavy metals and metalloids (e.g., arsenic), PCBs, pesticides, and PAHs are the most frequently reported contaminants in sediment. In the pesticide category, DDT and its derivatives, dieldrin and chlordane, appear to be the most widespread contaminants. Significant ecological impacts were often reported at the contaminated sediment sites, including impairment of the reproduction, structure, and health of aquatic life communities. Potential human health problems were also noted at a number of sites because contaminants in sediment accumulated in edible fish to levels that led to fish consumption advisories or bans. In 1987 the amendments to the Clean Water Act reflected Congressional concern about contaminated sediment problems. Congress passed Section 118(c), which authorized $25 million to conduct a demonstration program in the Great Lakes on how to remediate contaminated sediments at high priority sites.

Faced with this evidence of the extent and severity of contaminated sediment problems, EPA's Office of Water organized a Sediment Steering Committee chaired by the Assistant Administrator of Water and composed of senior managers in all the offices involved in contaminated sediments and EPA's 10 regional offices. In January 1990 this committee decided to prepare an agencywide sediment management strategy to coordinate and focus the agency's resources on contaminated sediment problems. Option papers on how to improve the agency's efforts to assess, prevent, remediate, and manage the disposal of contaminated sediment were developed. EPA discussed these option papers with other federal agencies and representatives of state governments with experience in sediment contamination. The views of these government officials were presented to the steering committee when they made a preliminary selection of the options in May 1991. The agency plans to propose a sediment management strategy in the *Federal Register* and request public comment on it by the winter of 1993.

EPA's sediment management strategy is designed around the following four principles:

1. In-place sediment should be protected from contamination to ensure that the beneficial uses of the nation's surface waters are maintained for future generations;
2. Protection of in-place sediment should be achieved through pollution prevention and source controls;
3. In-place sediment remediation will be limited to high-risk sites where natural recovery will not occur in an acceptable time period and where the cleanup process will not cause greater problems than leaving the site alone;
4. In implementing the strategy, EPA will use consistent methods to assess sediment contamination, but will vary trigger levels for regulatory decisions on a geographic basis, taking into account surface water uses and technical/economic feasibility issues.

The following sections of this chapter describe how current EPA programs handle the assessment, prevention, remediation, and dredged material management of contaminated sediment. The descriptions of current program activities are largely based on internal papers prepared by the staff of these programs. After each current program description, the changes that will be made in that program under the sediment management strategy will be discussed.

ASSESSMENT STRATEGY

This section summarizes EPA's authorities and programs for general assessments of sediment quality and generally applicable assessment procedures (e.g., bulk sediment chemical measurements, bioassays, and benthic community surveys) that may be used in a variety of programs.

Existing Assessment Programs

The Clean Water Act (CWA) is the single most important law dealing with the environmental quality of all U.S. surface waters, both marine and fresh. This act sets a national goal to restore and maintain the physical, chemical, and biological integrity of the nation's waters. Under the CWA, EPA works with the states to monitor the quality of surface water. States receive federal grants to conduct water monitoring activities and report the results of these assessments to EPA. EPA, in turn, provides monitoring guidance and technical support to the states and sponsors special studies addressing issues of national concern.[3] EPA maintains its own capabilities for surface water monitoring, primarily in its ten Regional Environmental Service Divisions and its Office of Research and Development laboratories.

EPA and state monitoring programs have traditionally focused on detecting water column problems caused by point-source dischargers. Nonpoint-source loadings, which are often episodic and unpredictable, are not well addressed. Similarly, monitoring for toxic substances in the water column, fish tissue, and sediment may not be extensive in some states.[3]

In general, EPA does not routinely sample sediment quality at fixed stations. More typically, data on sediment quality comes from intensive surveys or special studies. These studies and surveys are undertaken on an *ad hoc* basis, when sufficient resources can be mustered.

The U.S. Geological Survey (USGS) and National Oceanic and Atmospheric Administration (NOAA) routinely gather bulk sediment chemistry data. However, because of questions about the bioavailability of sediment contaminants (discussed below under criteria), these data may be useful for general overviews of sediment quality, but of limited use in a regulatory context. As a measure of how much sediment quality data EPA, other federal agencies, and the states collect, EPA's Storage and Retrieval Data system (STORET) contains data on 231,443 sediment samples from 43,221 sample locations. In contrast, STORET contains data on 10,189,466 water column samples from 481,814 sample stations.[4]

No national directive requires the states to monitor sediments for contamination, and EPA provides the states with little guidance on sediment monitoring. Nevertheless, EPA's 1988 report to Congress on the status of the nation's waters indicates that 34 states identified 533 separate instances of sediment contamination.[3] As methods of analyzing sediments and interpreting analytical results are still in their infancy, and as states may not have reported all available sediment data, these figures probably understate the extent of the problem.[3]

To make decisions as to whether a chemical or biological measurement of sediment quality indicates impairment, states and EPA must compare their data against some yardstick that indicates whether sediment quality is acceptable. Under Section 304 of the Clean Water Act, EPA has broad authority to develop chemical-specific criteria to protect water quality and to develop and use other assessment techniques. In general, EPA and the states are using three approaches to assess sediment quality: *field assessments* (which may involve benthic community structure analysis, fish tissue analyses, or caged animal studies to measure bioaccumulation and toxic effects), *laboratory bioassays* (to measure toxicity and bioaccumulation), and *chemical-specific measurements* (as a less expensive surrogate for bioassays and field work).

Primarily because of a desire to use, or require the use of, less expensive assessment techniques to the greatest extent possible, EPA has focused its research on sediment bioassays and chemical-specific criteria. Bioassays that measure short-term lethality and bioaccumulation are already available and have been used in the CWA and Marine Protection, Research, and Sanctuaries Act (MPRSA) dredging programs. The first bioassays to measure long-term, sublethal effects are under development by EPA's Office of Research and Development (ORD) and should be available in the near future.

Bioassays are advantageous because they provide a measure of the combined effect of all the contaminants in a sediment sample. They do not, however, indicate which compounds are contributing to the observed effect, which means that regulators in prevention, control, and remediation programs may need to do follow-up testing to answer this question. As one approach to identifying causative agents of toxicity, EPA is building on its work with effluents to develop sediment "toxicity identification evaluations" in which sediment samples are physically and/or chemically manipulated (e.g., by addition of EDTA to bind free metals) and tested again for any changes in toxicity.

EPA has been very interested in chemical-specific criteria for sediments as a relatively inexpensive and quick way to identify which contaminants are causing chronic effects in aquatic life. EPA has developed an equilibrium partitioning methodology for applying the extensive data base that supports water column criteria to nonionic organic sediment contaminants and is proposing five criteria in 1992. An equilibrium partitioning method for developing metals criteria is currently under development. The state of Washington has used an apparent effects threshold approach to develop its sediment management standards. This approach relies on an extensive, location-specific data base (e.g., Puget Sound) that correlates observed toxicity to bulk sediment concentrations of contaminants. California is in the process of evaluating various methodologies for developing chemical-specific criteria for sediments.

State water quality standards, adopted under the authority of Section 303 of the Clean Water Act, must be considered when EPA, other federal agencies, and states make decisions on actions that could affect water quality. Although they have rarely been used outside of dredging programs, biological assessments of sediment quality could be used to measure compliance with the narrative water quality standard that all states have adopted, which prohibits the discharge of "toxic materials in toxic amounts" (or similar language). Except for Washington, states have not yet adopted chemical-specific standards for sediments, in large measure because EPA has not published final sediment criteria. In March 1991 the state of Washington adopted sediment management standards for Puget Sound. In the Washington standards, the chemical-specific criteria developed using the apparent effects threshold method are overridden if laboratory bioassays do not confirm toxicity effects.

In addition to the generally applicable authorities for assessing sediments under the Clean Water Act, and the tools that have been developed, or are under development, to implement these authorities, a number of other programs can assess sediment. These include dredged material disposal programs under CWA Section 404, MPRSA, the Resource Conservation and Recovery Act (RCRA) and the Toxic Substances Control Act (TSCA), as well as remediation programs under RCRA and the Comprehensive Environmental Response, Compensation, and Liability Act (CERCLA). Details on sediment assessment procedures for these programs are provided in the following sections of this chapter.

Changes to Assessment Programs

In a recent review of the state of knowledge regarding the extent and severity of sediment contamination, the National Academy of Sciences (NAS) Committee on Contaminated Marine Sediments concluded that assessments based on available data provide only a partial picture of the total problem.[5] The NAS committee recommended that the federal government initiate a program to delineate areas with contaminated sediment, at a sufficient level of detail to set national cleanup priorities. In 1990 several members of Congress also introduced bills that would require EPA to conduct an inventory of contaminated-sediment sites. In the agencywide sediment management strategy, EPA is committing to develop a national inventory of sites and a pilot inventory of potential sources of sediment contamination, based on existing data. The two types of inventories are not mutually exclusive and should be complementary because the source data base can be used to predict where sediments are contaminated in unsampled areas.

The inventories will be designed so that EPA's prevention and remediation programs can use them to focus their resources on cleaning up the top priority sites and sources. The dredged material management programs can use this information to identify sites where dredged material must receive more comprehensive testing before it is considered safe for unrestricted disposal. Details on how each of these programs could use a site/source inventory are included in the sections on prevention, remediation, and dredged-material programs.

In the agencywide sediment management strategy, EPA is committing to develop a consistent, tiered testing strategy that will include a minimum set of sediment chemical criteria, bioassays, and bioaccumulation tests that all programs will agree to use in determining if sediments are contaminated. The tests will be arranged in a hierarchy of successive tiers of more extensive and costly data generation. Testing is to continue through the tiers as long as the information in the preceding tier is not adequate for deciding whether the sediment is contaminated. Tiered testing promotes the optimal use of resources by requiring the least effort where the potential for unacceptable adverse effects is clear, and focuses effort where more specific or comprehensive investigation is needed to determine the potential for those effects.

EPA recently documented ten sediment quality assessment methods that could be used in the agencywide tiered testing system.[6] Some of the methods focus on determining relationships between the numerical concentrations of sediment contaminants and adverse biological effects; others directly measure the effect of bulk sediments on biota, through toxicity tests and bioaccumulation tests. EPA's Sediment Steering Committee is evaluating several models of a tiered testing system that will be used agencywide. The steering committee is also struggling (Summer, 1991) with the more difficult issue of how each program will use the tiered test results. The committee wants stringent source controls and pollution prevention measures that will protect aquatic life at the chronic, no-effect level and prevent bioaccumulation in fish to levels requiring fish consumption advisories. It may not be economically feasible, however, to use these same stringent levels as a mandatory cleanup level for all the contaminated sediment at a site. It may only be feasible to remediate "hot spots" within an area to an acute protection level and allow natural recovery to clean the area over time to the chronic protection level. Now that EPA has committed to using a consistent testing system, each program will have to decide how to use the results of the tests as a trigger for regulatory actions.

PREVENTION STRATEGY

When a site is discovered where sediments are contaminated (e.g., where harmful aquatic life or human health effects are known or predicted), one of the first questions that must be answered is whether ongoing sources are contributing to the problem. If so, prevention or control measures should be considered.

In situations where a single source influences the section of a water body where sediments are contaminated, upstream and downstream toxicity tests may be sufficient to tie the source to the problem. In general, however, it is very useful to know the specific contaminants that are contributing to the harm or the risk. When multiple sources for a given contaminant exist (e.g., multiple contributors of a particular contaminant in a single harbor), a number of approaches may help tie a contaminant to its source. For example, concentration gradients of the contaminant may be observed across the sediment surface, with the highest levels observed below the responsible outfall. Or contaminants may increase at certain depths in the sediment column, indicating the history of the production or use of a contaminant by a particular facility. It may be possible to "fingerprint" a facility by matching the occurrence or relative concentrations of other contaminants that were emitted concurrently with the contaminant of interest.[7]

The scale of contamination may guide the choice of a particular set of prevention and control measures. If a sediment contaminant is causing harm or risk at numerous sites nationwide, it may be relatively inefficient to deal with the problem on a site-by-site basis. Instead, nationally applicable responses should be considered, such as prohibitions or use restrictions under TSCA or FIFRA, technology-based effluent limitations for industrial dischargers, or a national initiative to revise water-quality-based limits in NPDES permits. If atmospheric deposition appears to be a primary source of contamination, responses under the Clean Air Act should be considered.

Where sediment contamination is a concern at particular sites, but not on a national scale, case-by-case assessments and response actions may be appropriate. Based on narrative and chemical-specific criteria and standards, EPA or a state can develop NPDES permit limits for discharges from industrial sources, municipal sewage treatment plants, stormwater outfalls, and combined sewer overflows. States that have nonpoint-source control programs can take actions to reduce nonpoint-source contributions to sediment contamination.

Federal Insecticide, Fungicide, and Rodenticide Act (FIFRA) — Existing Program

Under FIFRA, EPA can ban, cancel, suspend, not register, or restrict the use of a pesticide or other biocide if its use poses risks of unreasonable adverse effects on human health and/or the environment, taking into account the economic, social, and environmental costs and benefits of use. There is clear legal authority to ban or restrict pesticides or other biocides (hereafter referred to as pesticides) with a potential to contaminate sediments.

There are three regulatory processes under FIFRA: (1) registration, (2) reregistration, and (3) section 6 cancellation/suspension (special review). Registration covers new pesticides, new uses, experimental uses, emergency exemptions, state and local uses, etc. Reregistration addresses pesticides registered prior to 1984. The special review process is used for registered pesticides that have been alleged to pose unreasonable risks. Special review can be triggered by EPA's review of registrants' submitted data or reports of adverse effects from the public, the states, other federal agencies, academic research, etc.

The testing requirements and risk assessment procedures are similar for all three regulatory processes. The tests that applicants are required to perform by regulation (40 CFR 158) depend upon the pesticide's intended pattern of use. For example, the tests required for an orchard spray use are generally different from those required for an aquatic-use pesticide. Additional tests or studies may be required if the agency believes they are necessary to adequately assess risks in support of registration, reregistration, or special review.

The toxicity of sediments is not usually addressed in existing test procedures and risk assessments for registration, reregistration, and special review. Existing procedures within the Office of Pesticide Programs (OPP) focus on the water column to assess impacts on bottom-dwelling organisms.

The following three environmental fate data requirements, imposed to support registration and reregistration of aquatic-use pesticides, are the only ones required that involve the analysis of sediment for the test pesticides and its degradates: anaerobic aquatic metabolism, aerobic aquatic metabolism, and aquatic field dissipation. Another study, the field accumulation in aquatic nontarget organisms, routinely involves the analysis of the water column and several types of fish, including bottom feeders, but generally not the analysis of benthic organisms or sediment.

Currently, these aquatic environmental fate data requirements are not normally imposed to support terrestrial uses of pesticides. In cases where the potential for dissolved runoff and/or erosion of contaminated soil particles to surface waters from terrestrial-use sites appears to be significant, these data requirements can be imposed. EPA has imposed these requirements for approximately 10% of the registration requests for pesticides with terrestrial uses.

If EPA decides to estimate the impact of stormwater runoff of terrestrial pesticides, models are used to predict environmental concentrations (EECs) of pesticides in the water column and sediment of various aquatic environments resulting from various pesticide application scenarios. Environmental concentrations are estimated from the use of a dissolved runoff/soil erosion model (currently either the Pesticide Root Zone Model [PRZM] or the Simulation for Water Resources in Rural Basins [SWRRB]) in combination with an aquatic pollutant transport model, either the Exposure Analysis Modeling System (EXAMS) or the Water Analysis Simulation Program (WASP). Two other models, Chemicals, Runoff, Erosion from Agricultural Management Systems (CREAMS), and Groundwater Loading Effects of Agricultural Management Systems (GLEAMS), an upgrade of CREAMS that includes pesticide degradation, leaching, and edge-of-field loading, are also under consideration.

The risks to aquatic organisms associated with contaminated sediment are qualitatively assessed from sediment data and the sediment model EECs. Quantitative risk assessments associated with contaminated sediment are normally not possible because aquatic toxic-effect levels such as EC_{50}s and LC_{50}s are generally reported in terms of water column concentrations, not sediment concentrations.

Furthermore, most of the freshwater aquatic toxicity data, including those required for pesticide registration, are for nonbenthic organisms.

Sediment data and EECs can also be qualitatively considered in estimating the effect of pesticide use on the aquatic food chain and on terrestrial wildlife (particularly birds) that depend on the aquatic food chain. Microcosm studies can be required as special testing to estimate the effects to community structure and interspecies interactions, while food-chain studies can be required to evaluate the distribution and accumulation of pesticides through a food web. Ultimately, if the available environmental fate and aquatic/terrestrial wildlife toxicity data indicate a potential for substantial risks, field ecological data requirements can be imposed. This frequently involves a comparison of aquatic organism and/or terrestrial wildlife populations in areas of pesticide use to those in control areas with no pesticide use, and may also include tissue analysis.

At the present time, sediment data or EECs are not used in assessing the potential exposure and risks to humans that may result from recreational fishing and consumption of fish and shellfish. EPA is required to establish maximum allowable pesticide concentrations or action levels in fish and shellfish only when an aquatic pesticide is used where commercial fishing occurs. Action levels are established on commodities that are transported in interstate commerce; recreational fishing is considered to involve only local consumption.

The Office of Pesticide Programs does not currently sponsor sediment monitoring programs, though, in some cases, it has required such data from registrants. As an example, the registrant will be monitoring sediment, fish tissue, and water as a result of the special review of tributyltin. OPP also compiles any sediment, fish, or water data collected by states and foreign countries in its pesticide monitoring inventory data base. These data are then used to determine whether a special review is needed for a pesticide.

Changes to FIFRA Program

EPA's Sediment Steering Committee has concluded that FIFRA could be used as a pollution prevention tool for sediment contamination if changes are made in the guidance and procedures used for pesticide registration, reregistration, and special review. In general, the committee does not feel that changes are needed in the statutory authority of FIFRA, to accomplish this goal.

If EPA's national inventory of contaminated-sediment sites lists sites with aquatic life or human health problems caused by pesticides currently on the market, EPA will use this information to select pesticides for reregistration. In order to focus FIFRA on preventing sediment contamination, 40 CFR 158 must be revised to require sediment toxicity testing. The pesticide assessment guidelines and standard evaluation procedures that describe the tests and field studies that must be run on pesticides will also be revised to include sediment methods. Once EPA develops the agencywide tiered testing system, these chemical criteria and biological tests will be among the methods specified in the FIFRA regulation and guidance. The states will also need to use these sediment testing methods because they have the responsibility, under FIFRA, of investigating pesticide runoff and determining how much pesticide is in water, sediment, and fish. The states compare these levels to EPA levels of concern and decide if enforcement action is necessary.

Revisions in 40 CFR 158 could also be made to routinely require sediment fate data on terrestrial-use pesticides. Another procedural change EPA is considering is to require manufacturers to perform the aquatic field dissipation and field accumulation studies at the same locations so that correlations between pesticide concentrations in the water column, sediment, and aquatic organisms can be made. Analyses of benthic organism tissues could also be added to the requirements to complete the link from sediments to fish.

EPA could also decide to revise 40 CFR 158 to require human health evaluations for terrestrial-use pesticides and to extend these evaluations to recreational fishing situations. A statutory change in FIFRA would probably be required to evaluate recreational fishing in addition to commercial fishing.

Toxic Substances Control Act (TSCA) — Existing Program

Under TSCA, EPA must reduce "unreasonable risk of injury to human health or the environment presented by chemical substances and mixtures." In reducing this risk, EPA can employ a range of risk management tools, ranging from simple information gathering, to restrictive use requirements, to an outright ban of the chemical.

Today, the U.S. chemical industry manufactures or imports tens of thousands of commercial chemicals in amounts greater than 10,000 lb/year. While it is probably a good assumption that many of these chemicals have low toxicity, there is a large universe of chemicals between those we already know are hazardous and those we have good reason to believe pose little risk. EPA assesses risk under both new chemical and existing chemical programs. Each year the New Chemical Program reviews about 3000 premanufacturing notices in order to identify and control risks. Since its inception, this program has controlled or banned the distribution and use of many hazardous chemicals. While this premanufacturing review and control represents a very successful *direct* application of "pollution prevention" in its truest sense, the program's biggest impact may be in *indirectly* influencing the chemical industry's R&D programs toward the development of safer chemicals.

Where EPA requires additional data to support a finding of unreasonable risk, it can obtain that information by issuing a test rule. The test rule outlines the protocols that must be followed to evaluate whether the chemical will cause environmental harm, or provides guidelines to industry for developing those protocols. For the first time, in 1989, EPA's Office of Toxic Substances (OTS) documented two sediment toxicity test protocols for use in evaluating the environmental impact of new or old chemicals. These bioassay methods are currently under EPA internal review and have not yet been published in the *Federal Register.* OTS also has two microcosm protocols that include the effects of sediment on organisms: a generic freshwater microcosm test (CFR 797.3050) and a site-specific aquatic microcosm test (CFR 797.3100). Sediment quality testing is not, however, included in the Significant New Use Rule that was recently proposed in the *Federal Register.*

Even after sediment protocols are published in the *Federal Register,* these methods cannot be included in the protocols for a specific chemical's test rule unless EPA provides a convincing written justification as to why such testing is required. Absent any monitoring data, the justification would have to rely on modeling studies that predicted the sediment concentrations resulting from the discharge of the chemical. OTS has done little modeling over the years to predict bed sediment concentrations because of the following reasons: (1) methods have not been available to relate sediment concentrations to human health or ecological impacts, (2) the available models require site-specific data that are rarely available for new chemicals, and (3) the models are too labor intensive to be used in the 90-day review period for new chemicals.

In assessing risk, EPA does use models to estimate water column concentrations of chemicals released to surface waters. These models provide reasonable worst-case scenarios that can be used to estimate human exposure through ingestion of water and fish and aquatic-species exposure. Any contributions of contaminants from sediments into the water column are not addressed; neither are effects on benthic organisms caused by contaminated sediment.

Changes to TSCA Program

EPA's Sediment Steering Committee has concluded that TSCA could be used as a pollution prevention tool for sediment contamination if changes are made in the guidance and procedures used to determine if new or existing chemicals pose an "unreasonable risk of injury to human health or the environment." The committee does not feel that changes are needed in the statutory authority of TSCA to accomplish this goal.

In order to focus TSCA on preventing sediment contamination, EPA will include sediment toxicity tests in the test rules that specify the protocols that chemical manufacturers must follow. These tests should also be added to the Significant New Use Rule that has recently been proposed in the *Federal Register.* Once EPA develops an agencywide tiered testing system, these chemical and biological methods will be among the methods required in the TSCA test rules and the Significant New-Use Rule. When EPA's national inventory of contaminated sediment sites lists locations with aquatic life or human health problems caused by known chemicals, EPA will use this information to select chemicals for review.

EPA also wants to improve its environmental fate models that predict sediment contamination. Research on sediment criteria is making progress in relating sediment concentrations to ecological impacts. EPA could use structure-activity relationships and the criteria methodology for known chemicals, to estimate the ecological impacts of new chemicals. This approach may eliminate one reason why TSCA reviews have failed to consider sediment contamination in the environmental fate models. If EPA decides to develop generic environmental data bases to run the sediment models and predict effects, this should eliminate the other two obstacles to sediment modeling (the lack of site-specific data and the need for extensive model set-up time).

Clean Air Act (CAA) — Existing Program

Atmospheric deposition may be an important source of sediment contaminants. A 1990 report by EPA's Science Advisory Board identified the atmospheric loading of toxic pollutants to aquatic systems as a high-risk problem on a regional scale and, in particular, cited the role of atmospheric deposition as a source of toxics in the Great Lakes.[8] Some of these atmospheric-borne pollutants bind to sediments and become sediment contaminants.

The 1990 amendments to the CAA include new, more stringent requirements for controlling toxic air pollutants. These new requirements will address stationary-source emissions that may be sources of sediment contamination. EPA will have to review the effect of these substantial new controls before making a decision to take further steps to reduce sediment contamination via atmospheric deposition.

Existing Technology-Based Point-Source Control Program

National Pollutant Discharge Elimination System (NPDES) permits are the primary means for preventing the discharge of pollutants from point sources. Under Sections 301, 304, 306, and 307 of the CWA, EPA has set minimum, technology-based requirements for municipal dischargers (e.g., primary and secondary treatment standards) and sets similar requirements for industrial dischargers (e.g., best available technology economically achievable; pretreatment standards for existing sources). EPA does not directly consider sediment contamination when developing technology-based requirements. However, the treatment that is required may reduce loadings of pollutants that partition into sediments, persist, and cause effects.

Changes to Technology-Based Point-Source Control Program

EPA's Sediment Steering Committee has concluded that the CWA effluent guidelines program could be used to control sediment contamination. EPA will use information from the national inventory of sites and sources of sediment contamination to decide which industries will be studied and considered for regulation with a national rule-making. EPA will evaluate the discharge of pollutants by all facilities in each industrial category selected for review and will, as appropriate, set discharge limitations and/or pollution prevention requirements.

Existing Water Quality-Based Point-Source Control Programs

Under Section 402 of the CWA, NPDES permits must also require compliance with limitations more stringent than technology-based requirements if they are necessary to ensure the maintenance of balanced populations of indigenous fish, shellfish, and wildlife, and to protect human health. Water quality criteria (as implemented through state water quality standards) and aquatic toxicity tests are the benchmarks for setting water-quality-based effluent limitations.

Municipal sewage treatment systems and industrial facilities are the two primary categories of active point sources regulated by the NPDES program. Many municipal and industrial facilities already have stringent water quality-based limits for toxics and toxicity in their permits, or will get such limits when their permits expire and are reissued. The current procedures for writing these permits do not consider sediment contamination, but the limitations they impose may have indirect benefits for sediments.

Municipal separate storm sewer systems and combined sewer overflows are widely recognized as major sources of degraded water quality and may significantly contribute to sediment contamination. In some parts of the country (e.g., in Region 10), EPA and states have tried to carefully assess storm sewer and combined sewer overflow (CSO) contributions to sediment problems and have found that these sources appear to be directly tied to ongoing sediment contamination.

At this time, municipal separate storm sewer systems and combined sewer overflows are not well controlled by the NPDES permit program. EPA has developed a CSO strategy that, while not directly addressing sediments, may substantially reduce pollutant discharges that contaminate sediments.

The Water Quality Act of 1987 provides a new mandate for controlling pollutants that began as nonpoint sources, but that are subsequently channeled into point sources. The Act requires that EPA and the states develop permits for discharges from all municipal separate storm sewer systems that serve populations of more than 100,000, and that they permit stormwater discharges associated with industrial activity as well. The stormwater rule promulgated in October 1990 provides wide-reaching authority for the prevention of contaminated sediments in urban areas and also specifies authority to control nonpoint sources from silvicultural and mining sources.

Any off-site discharges of pollutants into waters of the U.S. from RCRA and CERCLA facilities must be in compliance with a NPDES permit. As with other NPDES permits, these permits do not contain limitations that directly address sediment contamination, but may control pollutants that cause sediment problems.

Changes in Water Quality-Based Point-Source Programs

EPA's Sediment Steering Committee has concluded that the NPDES permitting program could be used to require sediment quality-based effluent limitations for industries, municipal sewage treatment plants, municipal separate storm sewers, and combined sewer overflows. The committee does not feel that changes are needed in the statutory authority of the CWA to accomplish this, since EPA's Office of General Counsel rendered an opinion in 1984 that sediment quality criteria are components of water quality criteria.

When EPA develops the national inventory of sources of contaminated sediment, the Office of Water can target NPDES permitting to those dischargers and pollutants that are most responsible for causing aquatic life or human health problems. The national inventory of contaminated sediment sites can also be used to target the geographic areas that are at greatest risk from sediment contamination caused by ongoing point sources. The resulting NPDES permit limits to prevent sediment contamination may be more stringent than those to protect water quality. Without Congressional authorization to reopen permits for sediment limits, NPDES permit authorities would probably have to wait for existing permits to expire (once every 5 years) to write new limits.

In order to focus the NPDES permitting program on preventing sediment contamination, sediment quality criteria and assessment procedures must be available. Now that EPA has decided to develop an agencywide tiered testing system, the biological methods in this system can be used in NPDES permitting to identify point-source discharges causing, or contributing to, whole-sediment toxicity. Sediment toxicity identification evaluations can then be performed to determine the chemicals causing the toxicity. If chemical-specific sediment quality criteria are available for the problem chemicals, wasteload allocation modeling can be used to calculate effluent limits that meet the criteria. If sediment criteria are not available, whole-sediment toxicity limits can be placed on the discharger.

Both combined sewers and separate stormwater sewers can be targeted for controls to reduce ongoing sediment contamination. Guidance will be needed on how to identify and control sources of sediment contamination that discharge to these sewers. Since these sewer systems receive a large component of nonpoint-source runoff, compliance with end-of-pipe limitations may require local implementation of nonpoint-source controls and practices.

EPA is also considering actions to ensure that discharges from CERCLA sites and RCRA facilities comply with NPDES sediment quality-based requirements. In the case of CERCLA, only "off-site" discharges from Superfund sites to surface waters are directly controlled under NPDES. "On-site" discharges are required to meet applicable or relevant and appropriate requirements (ARARS), which include conditions equivalent to those under NPDES. In the case of RCRA, hazardous waste facilities that have point-source discharges are either permitted under NPDES or no discharge is allowed. Run-on and run-off controls are also required at active facilities to control nonpoint-source contributions to surface waters. EPA is evaluating the need to extend nonpoint-source controls to "interim status" facilities that have been shut down, but are still contaminating sediments through stormwater runoff or leaching.

Existing Nonpoint-Source Control Program

Section 319 of the Clean Water Act, established by the Water Quality Act of 1987, provides a framework for preventing and managing all nonpoint sources of water pollution. Under Section 319 of the Clean Water Act, all states are required to complete a comprehensive assessment of their navigable waters and evaluate the effects of all categories and sources of nonpoint pollutants. In the December 1987 "Nonpoint Source Guidance," EPA encouraged states to provide information regarding those waters not meeting beneficial uses, including those not meeting beneficial uses due to contaminated sediments. The guidance classified contaminated sediments as a nonpoint-source pollution category. In so doing, EPA made contaminated sediment prevention and remediation efforts eligible for funding under Section 319.

State nonpoint-source (NPS) management programs are to include plans for preventing and managing nonpoint sources of pollution by encouraging, assisting, or requiring the implementation of the best management practices (BMPs). At their own discretion, states can enact legislation or regulations for control of nonpoint sources. Section 319 does not provide any federal authority to promulgate or enforce regulations to control nonpoint sources; however, it gives EPA authority to award grant funds to states as an incentive for nonpoint-source control, including control of sources that cause sediment contamination.

Most state management programs did not specifically address the prevention or remediation of contaminated sediment in their program milestones, nor did the states apply for 319 grant monies to fund such activities. The states, however, have proposed projects that will ultimately afford prevention benefits. These projects are relatively restricted in scale because of the limited funding available from 319 grants.

EPA currently has authority, under Section 303(d) of the CWA, to require states to establish total maximum daily load (TMDL) limits for waters not meeting beneficial uses. If states fail to develop TMDLs, EPA must develop them for the state. Under a TMDL, pollutant loads are allocated to both point and nonpoint sources. Currently, TMDLs are designed to meet water quality standards, but they could also be developed to meet sediment quality standards when these have been adopted by the states.

Changes in Nonpoint-Source Program

EPA's Sediment Steering Committee has concluded that the CWA nonpoint-source program could be used to control sediment contamination. Once EPA develops a national inventory of contaminated sediment sites, the nonpoint-source program can use this inventory to target locations for which grants and technical assistance could be provided to the states to prevent ongoing sediment contamination.

EPA is looking at several ways CWA Section 319 can be used to focus nonpoint-source grants and technical assistance on high-priority contaminated sediment sites. EPA will use the 5% national incentive set aside for Section 319 grants to fund sediment contamination prevention efforts. In addition, EPA is considering using Section 319 demonstration project funds to explore new technologies and approaches for controlling contamination of sediment.

In high-priority contaminated sediment sites, EPA could ensure that states develop TMDLs based on sediment quality criteria that include load allocations for nonpoint sources of this contamination. In those cases where states do not have the authority to impose nonpoint-source controls, or do not invoke the authorities they possess, EPA can develop the TMDL under Executive Order 12088 and CWA Section 313. Any agency, federal or not, which failed to comply with the TMDL would be subject to legal challenges under an administrative procedures act.

EPA is working with other federal agencies to promote the prevention of sediment contamination. EPA will encourage the U.S. Department of Agriculture (USDA) to include high-priority contaminated-sediment sites in that agency's water quality initiative for demonstrating the effectiveness of BMPs. In addition, EPA will work with NOAA to include prevention of contaminated sediments as an objective in coastal zone management activities.

REMEDIATION STRATEGY

EPA may remediate sediments under CERCLA, RCRA, CWA, the Rivers and Harbors Act, and TSCA. The agencywide sediment management strategy emphasizes that sources of contamination should be controlled prior to remediation efforts, but that it may be necessary to proceed with remediation ahead of effective source controls if the contaminated sediments pose a sufficiently great environmental hazard. In making remediation decisions, the strategy also points out that it is important to consider whether contaminated sediments at a site can be transported to downstream or offshore areas if left in place, thereby increasing the size of the contaminated area and making future remediation efforts much more difficult. Other factors to consider include the time frame for natural recovery, the potential for containment mobilization during remediation, and the feasibility and cost of various treatment and removal options.

Clean Water Act — Existing Program

The Clean Water Act (CWA), which authorizes EPA to restore and maintain the chemical, physical, and biological integrity of the nation's waters, offers a number of possible remedies to the problems of contaminated sediment. Sections 309, 311, and 504 are discussed under "Enforcement-Based Remediation" below. Section 115 of the act directs the administrator of EPA to identify the location of in-place pollutants and, through the Secretary of the Army, to make contracts for the removal of contaminated sediments. The $15 million authorized by this section has only been requested once, and all the funds were spent in the 1970s.

Changes in CWA Program

EPA's Sediment Steering Committee has concluded that the agency should request appropriations for CWA Section 115. Since this section has only been implemented once, EPA has developed no procedures to identify cleanup sites or to establish cleanup levels under this authority. Once EPA develops a national inventory of contaminated-sediment sites, the top priority harbors and navigable waters on this list could be used as the CWA Section 115 list of sites. When EPA develops an agencywide tiered testing system, this testing can be used in the Section 115 program to help select cleanup levels and monitor the effectiveness of remedial actions.

In addition to implementing Section 115, EPA has been considering piggybacking remediation projects onto the U.S. Army Corps of Engineer's (COE) navigation maintenance projects. Funds would more effectively be used by adding remediation to the standard COE's dredging activities. Piggybacking projects could save the costs associated with dredge mobilization and demobilization and possibly with some sediment testing. The national inventory of contaminated sediment sites could be compared with the list of areas where the COE is scheduled to dredge for navigation, to identify candidate sites for remediation. Under the Water Resources Development Act of 1990, sites could be remediated in association with navigation maintenance when the nonfederal sponsor agrees to pay 50% of the removal costs and 100% of the disposal costs. In addition, EPA is considering negotiation with the COE to undertake sediment remediation outside the limits of its Federal Standard Project when such remediation would clearly result in a long-term federal cost savings. With this situation, additional dredging for remediation purposes could be performed as long as the project has a high benefit/cost ratio. If regulatory changes were made revising the federal standard that currently requires the least cost, environmentally acceptable engineering alternative, piggyback cleanups could be performed at a much higher percentage of sites.

The major obstacle to the Section 115 and piggybacking projects is the lack of funding. Only a very few relatively small areas could be remediated using the $15 million that was authorized under this section. Piggybacking would also require the additional funding if the nonfederal sponsor was unable to pay. Both Section 115 and piggybacking projects would also require development of a formalized system of coordination between EPA and the COE.

CERCLA — Existing Program

Under CERCLA, the EPA Office of Emergency and Remedial Response (OERR) has established a comprehensive program for identifying, investigating, and remediating hazardous-waste sites. Preliminary site assessments are conducted on sites where a problem is suspected. Results from the assessment are used to determine a "Hazard Ranking System" (HRS) score. The HRS is a detailed, systematic scoring system for estimating the level of danger to human health or the environment, resulting from a given contaminated site. HRS scores are used by EPA to determine which sites should be placed on Superfund's national priority list (NPL) as priorities for cleanup funds and for more detailed evaluations. As part of the 1990 revision of HRS, sediment contamination was assigned a numerical weight that will now be used to determine site scores and listing on the NPL. For NPL sites, EPA carries out a detailed analysis of risks posed by the site to human health and the environment, and the feasibility of various remedial action alternatives to reduce the risk. The analysis is carried out through the preparation of a remedial investigation/feasibility study (RI/FS). The *Risk Assessment Guidance for Superfund, Volume II—Environmental Evaluation Manual* presents a broad framework for the assessment of human health and environmental impacts.[9] This strategy is not designed specifically for sediments, but rather for the purpose of assessing all exposure routes from contamination at Superfund sites.[7] Currently, the only sediment testing required for this stage of investigation is a measurement of bulk-sediment chemical concentrations at the impacted area and at a "background" area. Other guidance documents pertaining to Superfund's risk assessment process are "Guidance for Conducting Remedial Investigations and Feasibility Studies under CERCLA" and *Superfund Exposure Assessment Manual,* but neither of these provide details on how to assess sediment contamination.[10,11]

According to Section 121(d)(2)(A) of CERCLA, Superfund remedial actions must meet any federal standards, requirements, criteria, or limitations that are determined to be legally applicable, or relevant and appropriate, requirements (ARARs). This section also provides that water quality criteria established under Section 304 or 303 of the CWA should be attained by cleanup where the criteria are relevant and appropriate. ARARs are established for the site as part of the initial scoping stage of the RI/FS. If no ARARs exist for a particular contaminant, site-specific levels are developed from risk assessments or guidance documents.

The analysis of remedial alternatives, the selection of a remedy, and the chosen cleanup level for a site are documented in a "record of decision" (ROD). A recent evaluation of 486 RODs identified 69 sites with contaminated sediments, even though the HRS used at that time for listing sites did not give weight to sediment problems. Of these sites, sediment remediation was selected for 49 sites. The types of remedial action selected were categorized into two subsets consisting of excavation followed by treatment (30 sites), or excavation and containment (includes capping). The treatment technologies most commonly employed were incineration and solidification/stabilization.[12] It should be noted that EPA is currently evaluating a variety of remedial options, both for localized "hot spots" as well as for large volumes of moderately contaminated sediments.

Changes in CERCLA Program

EPA's Sediment Steering Committee has concluded that Superfund is an effective sediment remediation program in the agency. The committee feels that changes in procedures and guidance will focus more attention on contaminated sediments.

Once EPA develops national inventories of sources and sites of sediment contamination, the Superfund NPL program managers could use these data bases to identify areas that need to be evaluated for scoring on the Hazard Ranking System. Superfund's removal action program could also use both a source and a site data base to identify areas requiring cleanup.

When EPA develops an agencywide tiered testing system, this testing will be used in conducting remedial investigation/feasibility studies and monitoring after remediation to evaluate cleanup effectiveness. The Superfund program will determine, on a site-specific basis, whether it is economically or technically feasible to require all contaminated sediments to be cleaned up to the same level. At certain sites it may be more cost-effective to clean up "hot spots" and allow the rest of the area to cover over with clean sediments after stringent point- and nonpoint-source controls are enacted.

RCRA — Existing Program

RCRA provides EPA with the authority to assess whether releases from a hazardous waste treatment, storage, or disposal facility have contaminated sediments and to require "corrective action," including possible remediation, if contamination is discovered. RCRA "corrective action" authorities (summarized under "Enforcement-Based Remediation" below) apply to releases regardless of when the waste was placed in a unit. EPA inspects hazardous waste facilities that have applied for a RCRA permit, as well as "interim status" facilities (those that received interim permits for handling hazardous wastes, but ceased operations before the deadline for submitting applications for a final RCRA permit). These inspections are called "RCRA facility assessments" (RFAs). If an RFA suggests that a release to surface waters has occurred, hazardous waste permit writers can prepare permit conditions or enforcement orders requiring facility operators or owners to conduct extensive RCRA facility investigations (RFIs) to determine the extent of any contamination. If the RFI indicates that the facility caused contamination of sediments, the permit could be modified to require sediment remediation. At present, the RFI guidance warns about potential ecological and human health problems from contaminated sediments, but it does not specify how to determine if these problems exist.

Changes in RCRA Program

EPA's Sediment Steering Committee has concluded that the RCRA corrective action program could be a major player in remediating contaminated sediments. The committee does not feel that changes are needed in the statutory authority of RCRA to accomplish this, but that changes in guidance and procedures will enhance its role.

Once EPA has developed a national inventory of sources of contaminated sediment, and hazardous waste facilities appear on the list, the Office of Solid Waste (OSW) will use this information to schedule RCRA facility assessments for active facilities. If the inventory of sources indicates that interim status facilities (which are closed) pose substantial risk to human health or the environment, regional hazardous waste program managers will direct resources to clean up these problems. The hazardous waste program will also use the national inventory of contaminated sites if hazardous waste facilities are identified as contributing to the problem at specific sites.

When EPA develops an agencywide tiered testing system, this testing will be specified in the "RFI Guidance." This guidance will also be revised to include sediment criteria in the list of applicable health and environmental criteria. If sediment criteria are exceeded, interim corrective measures and/or a corrective measures study would generally be required, unless the owner or operator of the facility presents convincing site-specific data that demonstrates no further action is necessary. If states adopt sediment standards, these standards would probably be considered in the RFI automatically because of the requirement to consider state-established criteria. The RCRA program will determine, on a site-specific basis, whether it is economically or technically feasible to require all contaminated sediments to be cleaned up to the same level. At certain sites it may be more cost effective to clean up "hot spots" within a site and allow the rest of the area to cover over with clean sediments after point- and nonpoint-source controls are enacted.

Enforcement-Based Remediation — Existing Program

EPA has authority to seek enforcement-based sediment cleanup under CERCLA, RCRA, CWA, the Rivers and Harbors Act, and TSCA. EPA can use its statutory authority to (1) compel parties to clean up the sites they have contaminated, (2) recover costs from responsible parties for EPA-performed cleanups, and (3) coordinate with natural resource trustees to seek restitution from responsible parties for natural resources damages. Agency actions to obtain sediment remediation may be enhanced through the coordinated use of contractor listing (40 CFR Part 15), debarment and suspension (40 CFR Part 32), state or local laws and regulations, and the agency's criminal enforcement authority.

CWA

Section 309 of the CWA authorizes EPA to commence civil action for appropriate relief, including permanent or temporary injunction, for enumerated violations, including any discharges in violation of permit limits. Negotiated settlements in such cases have incorporated sediment cleanup as part of the relief. Settlements negotiated for CWA Section 301(a) unauthorized-discharge violations have also incorporated cleanup as part of the injunctive relief.

Section 311 authorizes the President to act to remove, or arrange for the removal of, an actual or threatened discharge of oil or hazardous substances into "navigable waters," adjoining shorelines or waters of the contiguous zone, that may affect natural resources of the U.S. or present an imminent and substantial danger to the public health or welfare.

Section 504 provides a possibility for injunctive relief under the act if it can be shown that polluted sediments present an "imminent and substantial endangerment" to the health of persons, or the livelihoods of persons, whose employment might be affected by contaminated sediments.

The Rivers and Harbors Act

The Rivers and Harbors Act (Refuse Act of 1899) provides that criminal and injunctive relief may be sought against dischargers of "refuse" into navigable and tributary waters of the U.S., except when they are improving navigation or constructing public works that are considered necessary and proper by officers of the U.S. or that are authorized by the Secretary of the Army. Courts have broadly interpreted this act to absolutely prohibit discharges other than discharges in compliance with a permit under the Clean Water Act. This authority is administered by the COE, and the Department of Justice is mandated to take enforcement action under the act.

CERCLA

CERCLA provides one of the most comprehensive authorities available to EPA to obtain sediment cleanup, reimbursement of EPA cleanup costs, and compensation to natural resource trustees for damages to natural resources affected by contaminated sediments. Once EPA determines that a release, or substantial threat of release, of hazardous substances to the environment has occurred, or may occur, EPA may undertake response action necessary to protect public health and the environment or compel the potentially responsible parties (PRPs) to undertake the cleanup. PRPs are owners or operators of facilities where hazardous substances are located, or the generators or transporters of hazardous substances disposed of at a facility. PRPs are liable without regard to whether they were negligent (strict liability), and if the hazardous substances have intermingled indivisibly in the sediments, all PRPs are individually liable for all cleanup costs (joint and several liability). CERCLA broadly defines "hazardous substances", and the statute is applicable no matter when the hazardous substances were disposed.

Section 104 authorizes EPA to initiate short-term removal actions ($2,000,000 or less, or less than 12 months) where hazardous substances have been released or where contaminants present an imminent and substantial danger to public health and welfare. These actions are financed by CERCLA's Superfund, and cost recovery may be sought from potentially responsible parties. No cost recovery is available for a federally permitted release. Section 104 has been used to address sediment contamination as part of the overall remedy for several sites.

Section 106 authorizes EPA to require the attorney general to secure such relief as is necessary to abate an imminent and substantial threat to the public health or welfare, or the environment, because of an actual or threatened release of a hazardous substance. If sediment contamination is the cause of the threat, and the contamination is not solely the result of a federally permitted release or there is evidence that the discharged contaminants were released in excess of permitted levels, then an action under Section 106 to compel responsible parties to perform cleanups may be appropriate. Failure or refusal to comply with the Section 106 order, without sufficient cause, subjects the responsible parties to treble damages and penalties of up to $25,000 a day. Under this authority, responsible parties have been ordered to address contaminated sediments.

Section 107 provides that the U.S. may recover all costs of CERCLA response actions, when not inconsistent with the national contingency plan, as well as damages for injury to natural resources and costs of health assessments. Generally, liable parties include persons who owned, operated, treated, transplanted, and/or disposed of hazardous substances. EPA cannot recover response costs or damages resulting from a federally permitted release, under Section 107, unless there is evidence that the contamination was not solely from pollutants permitted under Section 402 of the CWA or that discharges exceeded the permitted levels.

Section 122 directs EPA to facilitate agreements with PRPs that are in the public interest and consistent with the national contingency plan, in order to expedite effective response actions and minimize litigation. Settlements under this section may include compensation for, or remediation of, natural resources damages if the Department of Interior, the state, or another designated natural resources trustee is a party to the settlement.

CERCLA's Superfund can be used to finance long-term remedial actions only at facilities that have been placed on the National Priorities List (NPL). Cost recovery may be sought from PRPs, but may not be available for a federally permitted release.

RCRA

Under RCRA, EPA has broad authority to address contaminated sediments if the contamination is linked to an RCRA-regulated facility.

Section 3004(u) directs EPA to require, as a condition of a facility's permit, corrective action to address all releases of hazardous waste or constituents from any solid waste management unit, regardless of when the waste was placed in the unit. This authority includes requiring the permittee to address contaminated sediment as part of its corrective action plan.

Section 3004(v) authorizes EPA to establish standards requiring corrective action for releases from a facility that have migrated beyond the boundaries of the facility (e.g., offsite sediments), where necessary, to protect human health or the environment, unless the facility's owner or operator demonstrates that it was unable to obtain access to the contaminated areas.

Section 3008(h) authorizes EPA to issue orders requiring interim status facilities to take corrective action, or such other response measures that are necessary, to protect human health or the environment from a release of hazardous waste.

Several facilities have been required to investigate contaminated sediments in streambeds and lakes, pursuant to consent orders entered into under 3008(h) and permit conditions issued under 3004(u) and 3004(v).

Section 7003 authorizes EPA to bring suit against persons who contribute(d) to past or present handling, storage, treatment, transportation, or disposal of any solid or hazardous waste that may present an imminent and substantial threat to health or the environment. EPA may further order such persons to take other actions as may be necessary to mitigate the health or environmental threat. This authority has been used to enter into consent orders whereby the facility has agreed to investigate contaminated sediments.

TSCA

TSCA requires the removal of sediments that were contaminated by TSCA-regulated waste after the applicable effective date for the particular substance found in the sediment. Any party that undertakes to remove or handle sediments containing TSCA-regulated materials must follow the regulations promulgated under TSCA for the handling of these substances, regardless of the date of contamination.

Changes in Enforcement Programs

To date, EPA has only used Section 309 of the CWA and Section 106 of CERCLA, in conjunction with its violating facility listing authority, to require the cleanup of contaminated sediment. In addition, settlements negotiated under Section 301 of the CWA provisions on unauthorized discharge violations have included cleanup as part of the injunctive relief. EPA's Sediment Steering Committee has committed to using the other laws discussed above to mount an enforcement-based remediation strategy.

Once EPA develops national inventories of contaminated sediment sites and sources, this information will be used to identify the parties responsible for sediment contamination and to evaluate whether an enforcement action is appropriate. When EPA develops an agencywide tiered testing system, this testing will be used to identify areas requiring remediation and to help provide the cleanup goal for the enforcement-based remediation.

EPA's major concern over implementing this enforcement-based remediation strategy is the possibility of evidentiary problems, particularly if the contaminants in the sediment come from multiple sources, if the responsible parties are nonviable, if the discharge occurred before reliable records existed, or if a portion of the contamination occurred due to a federally permitted release. One way of dealing with these situations is to forego enforcement-based cleanup and turn the agency's efforts to assisting trustees in seeking compensation or mitigation for natural resources damages. Another way of handling these situations is to seek statutory changes that will (1) define liability when multiple responsible parties are involved and harm is divisible, (2) resolve uncertainties with respect to liability when contamination resulted from NPDES or other permitted discharges, (3) clarify recovery of funds under CWA Section 309, and (4) define cleanup objectives. EPA will consider using natural resource damages and statutory changes in the next few years.

DREDGED MATERIAL MANAGEMENT STRATEGY

The maintenance of our nation's waterways for navigation requires the dredging and disposal of large volumes of material. The COE estimates that the annual volume of dredged material ranges from 250 to 450 million yd^3.[13,14] Based on current assessment techniques, the COE considers about 3% of the annual volume of dredged material highly contaminated and an additional 30% moderately contaminated.[15] It is expected that the more sensitive tests included in the current revision of the ocean dumping testing manual will result in a higher percentage of dredged material being classified as contaminated.

Disposal of dredged material into waters of the U.S. and ocean waters are regulated mainly under Section 404 of the CWA and Section 103 of the MPRSA, respectively. Materials that are to be disposed of within the baseline of the territorial sea (including estuaries, rivers, the Great Lakes, and other rivers and lakes) are regulated entirely under the CWA. Outside of the territorial sea (beyond 3 nautical miles from the baseline), only MPRSA applies. In general, within the territorial sea (0 to 3 nautical miles seaward of the baseline), the CWA regulates fill material (which may include dredged material discharged for the purpose of fill, such as beach nourishment), while MPRSA regulates dredged material disposal.[7] Other regulations/agreements that contain provisions relating to the disposal of dredged material are TSCA, RCRA, the Great Lakes Water Quality Agreement (GLWQA), and the London Dumping Convention.

MPRSA — Existing Program

MPRSA gives EPA the authority to designate acceptable dump sites for the disposal of dredged material in the territorial waters of the U.S. The COE is given the authority to issue ocean dumping permits where it is determined that "such dumping will not unreasonably degrade or endanger human health, welfare, or amenities; the marine environment, ecological systems; or their economic potential."

EPA requires that ocean dumping of dredged material be done at EPA-designated dump sites. The evaluation process for designating sites is formalized in an environmental impact statement (EIS) for dredged material dump-site designation.

Permits for the ocean dumping of dredged material, issued by the COE, are reviewed by EPA according to several environmental criteria and either approved or denied by EPA. These criteria have been codified in the Ocean Dumping Regulations.[16] EPA and the COE have jointly developed a tiered testing protocol to determine whether dredged material from a particular project meets the criteria. The protocol is presented in the Evaluation of *Dredged Material Proposed for Ocean Disposal — Testing Manual,* known as the "Green Book."[17] The protocol consists of four tiers, including evaluation of existing data on potential sources of contamination, sediment chemical analyses, acute bioassays and bioaccumulation tests, and biological community field studies. At each successive tier the level of effort involved is more intensive in both time and effort. Testing continues through the tiers until sufficient evidence is available on whether dredged sediments are contaminated or not.

An international agreement, the London Dumping Convention, contains provisions that affect all dumping seaward of the baseline of the territorial sea and includes specific limitations and bans on the dumping of certain substances. Under Annex 1, the dumping of dredged material containing extremely hazardous substances (e.g., organohalogen compounds, cadmium, mercury, oil, high-level radioactive wastes) in "other than trace amounts" is forbidden. In U.S. waters the London Dumping Convention is implemented by the ocean dumping regulations and testing protocols contained in the "Green Book."

CWA — Existing Program

EPA oversees the program that regulates the discharge of dredged or fill material in Section 404 waters. The current national guidance for Section 404 testing procedures is contained in the 404(b)(1) guidelines,[18] which provide only a general framework under which testing is to be performed; there is no broadly applied analog to the "Green Book." However, EPA and the COE are currently developing a comprehensive national testing manual for evaluating dredged material to be discharged into Section 404 waters.

Under the Section 404 program, unlike the ocean-dumping program, discharge sites for dredged material are specified on a project- or permit-specific basis. As with the assessment of dredged material, the evaluation of impacts on the discharge sites is done in accordance with requirements of the Section 404(b)(1) guidelines. One of the primary requirements of the guidelines is that no discharge can be permitted if there is a practicable alternative with less adverse impact on the aquatic environment (unless the identified alternative poses other significant environmental problems). In addition, no discharge can be permitted under the guidelines if it violates state water quality standards, violates toxic effluent standards, jeopardizes the continued existence of an endangered species, or violates any requirements enacted to protect a federally designated marine sanctuary. Discharges are also not in compliance with the guidelines if they will cause or contribute to significant degradation of the waters of the U.S. Finally, the guidelines require that all appropriate and practicable measures to minimize potential harm to the aquatic ecosystem be taken.[7]

The existing COE approach to conducting cost/benefit analyses for dredged-material disposal is limited to evaluating project costs and benefits for recreation or commercial navigation. Environmental costs and benefits are not monetized. It is also COE policy to assure that dredged material disposal occurs in the least costly, environmentally acceptable manner, consistent with the engineering requirements of the project. This is known as the "federal standard." For coastal areas, unconfined disposal in bays, river mouths, estuaries, and near-coastal environments is considered less costly than disposal off the continental shelf, due to the closeness of the dredging site and to the subsequent fuel/transportation savings. For inland waters, disposal in lakes and rivers is considered less costly than disposal in upland or confined disposal facilities (CDF), due to the closeness of the dredging site and the subsequent transportation/land acquisition savings.

EPA Regions — Existing Programs

Several EPA regions have developed local guidance for dredged-material testing that are in various stages of implementation. Regional guidance documents generally follow the pattern of the "Green Book," with modifications for freshwater systems, local biota, and chemicals of concern. The most comprehensive dredged material programs have been developed by EPA in the Great Lakes and Puget Sound because of special concerns about contaminated sediment in those areas. These programs are described here in more detail.

The Great Lakes Water Quality Agreement (GLWQA) between the U.S. and Canada contains provisions relating to the disposal of dredged material. Under this agreement, the Dredging Subcommittee of the U.S.-Canada International Joint Commission's (IJC) Water Quality Board published an evaluation procedure for selecting disposal options for Great Lakes dredging projects.[19] The basic scheme is similar to the tiered protocol developed under MPRSA for ocean disposal. Though not formally adopted by EPA, this evaluation procedure is often employed by EPA Region V, along with interim guidance for sampling and testing related to navigational maintenance dredging under CWA Section 404.[20]

EPA's Region X uses a dredged material testing sequence developed under an interagency effort known as the Puget Sound Dredged Disposal Analysis (PSDDA) program. PSDDA's focus is on identifying and managing disposal sites and developing evaluation procedures to determine the suitability for unconfined, open-water disposal of sediments dredged from navigation projects throughout the sound. Procedures for regulating/managing sediments that are unsuitable for open-water disposal are currently under development. The PSDDA approach for determining the acceptability of dredged material for open-water disposal is a tiered testing sequence that incorporates the requirements of the CWA Section 404(b)(1) guidelines and MPRSA. As in the case of the MPRSA "Green Book," the PSDDA tiered testing includes evaluation of existing data on potential sources of contamination; sediment chemical analyses, acute bioassays, and bioaccumulation tests; and field studies. The unique aspect of the PSDDA system is that the Tier II chemical analyses compare bulk sediment concentrations to sediment criteria developed specifically for the Puget Sound, using the apparent effects threshold method.

TSCA — Existing Program

Under TSCA, dredged material containing PCB concentrations of 50 ppm or greater must be incinerated or placed in a TSCA-approved landfill or disposed of by an alternate method prescribed by the regional administrator.

RCRA — Existing Program

Under RCRA, any characteristic hazardous waste (a waste that has one or more hazardous characteristics per RCRA-prescribed testing methods) must be disposed of by RCRA-approved methods. Contaminated sediments may be toxic under the toxicity characteristic leaching procedure (TCLP) test, a test that compares the concentrations of various chemicals in the leachate from the dredged material with levels established to protect human health and the environment. When a leachate concentration exceeds the established level, then the dredged material exhibits the "toxicity characteristic" and generally must be managed in a RCRA-permitted or interim-status facility (a Subtitle C hazardous-waste landfill, or equivalent) and must conform to treatment standards consisting of a specified level of hazardous constituent or specific treatment technology. The RCRA hazardous-waste program has a statutory directive to prohibit land disposal of untreated hazardous wastes beyond specified dates for particular wastes. This prohibition, authorized under Section 3004 of RCRA, is known as the "land ban". As the specified dates for most hazardous wastes have passed, EPA has prohibited or restricted these wastes from land disposal without treatment.[7]

Changes in Dredged-Material Management Programs

EPA's Sediment Steering Committee has recommended that EPA encourage the COE to give more weight to environmental protection in their analyses of alternative dredged-material disposal methods. Changes in the COE statutory authority may be needed to accomplish this goal.

Once EPA develops national inventories of contaminated sediment sites or sources, the ocean dumping program and the Section 404 dredge and fill program could use this as Tier 1 data that would trigger additional Tier 2 and 3 testing for applicants who wish to dredge material from areas near the listed sources or in the listed sites. Once EPA develops an agencywide tiered testing system, the testing required for the ocean dumping and Section 404 dredge and fill program might have to be revised if the agencywide system includes tests different from those used in these programs.

EPA will also ask the COE to consider revising their cost/benefit analyses and redefining the federal standard. The environmental costs of various disposal alternatives and the benefits to aquatic life could be added to these cost/benefit analyses. As an example, dredged material could more frequently be used for beach nourishment, marsh creation/restoration, and barrier island formation, if the environmental benefits of these alternatives were calculated and the environmental costs of other disposal alternatives were taken into account.

EPA will seek to have the COE reevaluate the federal standard and allow the use of a disposal option with a small incremental expenditure yielding greater environmental benefit. EPA will also pursue discussions with the COE concerning the application of the sequencing aspect of the Section 404(b)(1) guidelines to the selection of disposal sites for dredged material. Dredged material disposal sites could be selected first with the intent of avoiding adverse impacts to aquatic resources, so long as the alternative disposal site would not result in other significant adverse environmental consequences. The remaining unavoidable impacts would be mitigated to the extent appropriate and practicable.

EPA realizes that all costs associated with dredging and disposal will decrease only if the contaminant input to waters decreases significantly. The sediments could then, over time, become less polluted, and the dredging, transport, disposal, and monitoring of dredged material would be less complicated and, consequently, less expensive. EPA is considering support for "anticipatory dredging" at sites where sources of sediment contamination have been controlled. Anticipatory dredging involves the removal of hot spots upstream, at depth, and inside channels, before the contaminants in these sediments can migrate into navigation channels.

The major factor preventing the COE from enhancing the environmental protectiveness of their disposal alternatives is the economic impact. The economy of the U.S. relies heavily on a functioning commercial fleet, and the COE has focused on deepening the ports, since the overall trend in shipbuilding is towards increased size and draft for more cost efficiency. If environmentally protective disposal options prove to be too expensive or lead to a lack of disposal sites, dredging operations and the use of navigational waterways could be severely limited.

CONCLUSION

In the next few years EPA plans to make significant progress in implementing the sediment management strategy. The agency will develop a national inventory of contaminated sediment sites that will describe the extent and severity of the problem. The inventory will be used to focus EPA programs on high-risk areas and to provide the public with an estimate of the national extent of the contamination. EPA programs will also agree on consistent sediment testing methods for use in preventing, remediating, and managing the dredged material disposal of contaminated sediments. These methods will enable the pollution prevention programs under FIFRA and TSCA to eliminate problem chemicals and will allow the abatement and control programs to regulate ongoing discharges. The same methods will be used to identify where remediation is needed and will allow EPA, other federal agencies, and state and local governments to focus cleanup efforts on the high-risk sites. Finally, these same methods will ensure that dredged sediments are clean so they can be used for beneficial purposes, such as beach nourishment and wetlands restoration. Throughout the entire process, EPA will ensure an active sediment research program to continually improve our assessment methods and remediation techniques.

REFERENCES

1. "National Perspective on Sediment Quality," Office of Water Regulations and Standards, U.S. EPA Report (1985).
2. "An Overview of Sediment Quality in the United States," Office of Water Regulations and Standards, U.S. EPA Report (1987).
3. "National Water Quality Inventory — 1988 Report to Congress," Office of Water, U.S. EPA Report-440-4-90-003 (1990).
4. Taylor, P. U.S. EPA, personal communication (1990).
5. *Contaminated Marine Sediments — Assessment and Remediation*, National Academy Press, Washington D.C. (1989), p. 5.
6. "Sediment Classification Methods Compendium," Watershed Protection Division, U.S. EPA Draft Final Report (1989).
7. "Managing Contaminated Sediments: EPA Decision-Making Processes," Sediment Oversight Technical Committee, U.S. EPA Report 506/6-90/002 (1990).
8. "The Report of the Ecology and Welfare Subcommittee," in "Reducing Risk: Setting Priorities and Strategies for Environmental Protection," U.S. EPA Report SAP-EC-90-021 (1990).
9. "Risk Assessment Guidance for Superfund, Volume II — Environmental Evaluation Manual, Interim Final," Office of Emergency and Remedial Response, U.S. EPA Report 540/1-89/001 (1989).
10. "Guidance for Conducting Remedial Investigations and Feasibility Studies under CERCLA, Interim Final," Office of Emergency and Remedial Response, U.S. EPA Report 540/G-89/004 (1988).
11. "Superfund Exposure Assessment Manual," Office of Emergency and Remedial Response, U.S. EPA Report 540/1-88/001 (1988).
12. "Summary Report of FY82 Through FY89 Records of Decision Documenting Sediment Contamination," U.S. EPA Internal Working Document (1990).

13. Engler, R. Complete statement before the Subcommittee on Water Resources, Committee on Public Works and Transportation, U.S. House of Representatives. Department of the Army, Office of the Assistant Secretary of the Army (Civil Works), p. 7 (1989).
14. "The Dredging Research Program," Department of the Army, Corps of Engineers, Dredging Research DRP-88-1, p. 1–7 (1988).
15. "Wastes in Marine Environment," U.S. Congress, Office of Technology Assessment Report OTA-0-334 (1987).
16. "Ocean Dumping," U.S. Code of Federal Regulations, Vol. 40, Parts 220–228 (1977).
17. "Evaluation of Dredged Material Proposed for Ocean Disposal — Testing Manual," U.S. Army Corps of Engineers and U.S. EPA Report 503/8-91/001 (1991).
18. "Guidelines for Specification of Disposal Sites for Dredged or Fill Material," U.S. Code of Federal Regulations, Vol. 40, Part 230 (1980).
19. "Guidelines and Register for Evaluation of Great Lakes Dredging Projects," Report of the Dredging Subcommittee to the Water Quality Programs Committee of the Great Lakes Water Quality Board, International Joint Commission, Ontario (1982).
20. "Interim Guidance for the Design and Execution of Sediment Sampling and Testing Efforts Relative to Navigational Maintenance Dredging in Region V," Region V Planning and Management Division, U.S. EPA (1988).

CHAPTER 16

Assessment and Management of Contaminated Sediments in Puget Sound

Thomas C. Ginn and Robert A. Pastorok

INTRODUCTION

Puget Sound is a fjord-like estuary in Washington State that is connected to the Pacific Ocean by the Strait of Juan de Fuca (Figure 1). This complex estuarine system includes many individual sub-basins, straits, embayments, and man-made waterways. The main basin of Puget Sound reaches a maximum depth of 280 m and is separated from several other basins by shallower sills. The entire series of inlets and embayments results in a shoreline of over 2,000 km, with a total area of 220 km^2. Salinity typically ranges from 27 to 30 ppt.

Puget Sound is bordered by several major urban, industrial areas that depend on deepwater port facilities, such as Elliott Bay (Seattle), Sinclair Inlet (Bremerton), and Commencement Bay (Tacoma). In addition to port activities, these urban areas have pulp and paper mills, shipbuilding facilities, metal refineries, chemical manufacturing firms, and many other industrial activities. Most of the industrial sites are concentrated along relatively shallow waterways ranging in depth from 10 to 15 m.

Sediments are transported into Puget Sound by 12 major river systems, including the Skagit, Snohomish, Duwamish, Puyallup, and Nisqually rivers on the eastern shore. Most of the estimated 3 million metric tons of riverine sediments entering the sound annually is fine-grained, glacial material, much of which is discharged into the major urban embayments, such as Everett Harbor, Elliott Bay, and Commencement

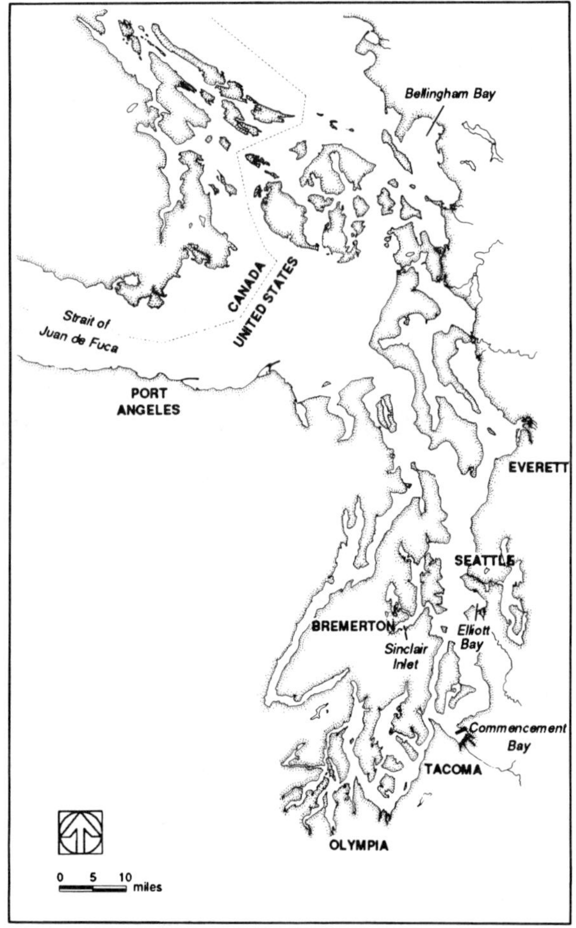

Figure 1. Puget Sound and adjacent waters.

Bay. The erosion of glacial till material from shoreline bluffs represents another major source of sediments to the sound.

Sediment contamination has received extensive study and evaluation in Puget Sound since about 1980. The earliest comprehensive assessments[1-4] demonstrated that Puget Sound sediments were contaminated by many organic and inorganic chemicals, including polychlorinated biphenyls (PCBs), polycyclic aromatic hydrocarbons (PAHs), and metals, such as copper, lead, mercury, and silver. Assessments of the historical data base at that time indicated that elevated levels of toxic chemicals were primarily associated with the major industrial embayments. Within the nearshore areas (e.g., waterways), localized sediment "hot spots" were identified that had greatly elevated contaminant concentrations (e.g., >1,000 ×

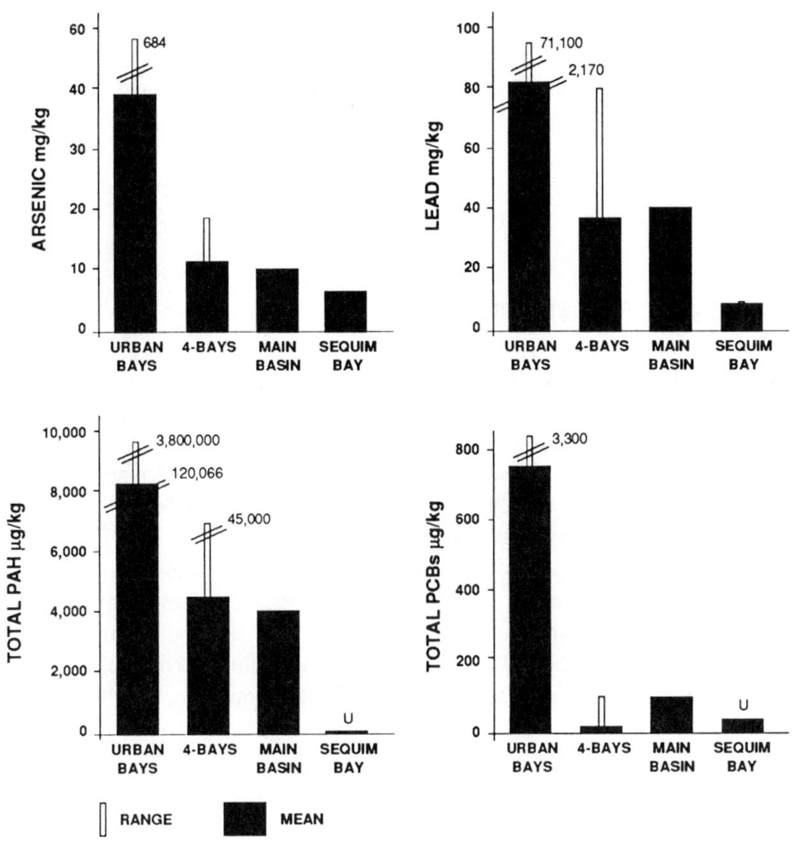

NOTE:
Urban bays are Commencement Bay (Tacoma), Elliott Bay (Seattle), Everett Harbor (Everett), and Sinclair Inlet (Bremerton). 4 bays are Dyes Inlet, Gig Harbor, Oak Harbor, and Port Angeles. Sequim Bay is a reference area.

Figure 2. Puget Sound sediment contamination.

background levels). Although contaminant concentrations in sediments tended to decrease rapidly with distance from the nearshore sources, widespread low-level contamination ($\approx 10 \times$ background) was also documented in the deepwater sediments of the main basin of Puget Sound (Figure 2).

In the early 1980s several kinds of biological effects were documented in areas of sediment contamination in Puget Sound. Studies of the livers of bottom fishes, such as English sole *(Parophrys vetulus)* in Elliott Bay, the Duwamish River, and Commencement Bay, revealed elevated prevalence of several abnormalities, including cancerous tumors and preneoplastic lesions.[5] Sediment bioassays using organisms such as the amphipod *(Rhepoxynius abronius)*[6] and the Pacific oyster

(Crassostrea gigas)[7] also demonstrated the potential for significant toxic effects in the urban embayments.[8,9] Concern for the bioaccumulation of sediment contaminants increased following the documentation of elevated levels of PCBs in marine birds and mammals from the sound.[10,11] In addition, fish, crabs, and bivalves in contaminated urban areas have been found to accumulate pollutants in their muscle tissue and internal organs.[12-14] In several of these areas, local health departments have advised residents to limit their consumption of recreationally harvested seafood.

SEDIMENT MANAGEMENT PROGRAMS

Results of the studies of sediment contamination and biological effects led to increased regulatory and public awareness of the contaminated sediment problems in Puget Sound and to the formation of several programs to address those problems. The assessment and management of contaminated sediments in the Puget Sound area falls under several agency and interagency programs administered at the federal and state levels. The success of these programs in effectively assessing and managing contaminated sediments has resulted from two important factors: (1) a strong public and agency commitment to clean up Puget Sound and (2) a strong commitment at the federal, state, and local levels for interagency cooperation to develop and implement innovative technical approaches. These cooperative activities have included the coordination of assessment approaches and data collection efforts, interagency workgroups, and research conferences.

Puget Sound Water Quality Authority

The present Puget Sound Water Quality Authority (PSWQA) was established in 1985 by the Washington State legislature. The primary responsibility of PSWQA was to develop and oversee the implementation of a comprehensive management plan for Puget Sound. The initial Puget Sound Water Quality Management Plan (the Plan) was adopted by PSWQA in 1986 following an intensive process of issue identification, scoping, and public comment. In developing the Plan, PSWQA was assisted by an advisory committee and a scientific review panel. The Plan has been updated every 2 years, with the most recent version being adopted in 1991.[15] One area targeted for action under the original Plan involves contaminated sediments and dredging. The Plan's goal involves the reduction and ultimate elimination of adverse environmental effects of contaminated sediments by reducing pollutant discharges and by remediating existing contaminated sediments. Important parts of the plan that deal with contaminated sediments include the development of sediment quality standards, the inventory of contaminated sediments, and decisions regarding sediment cleanup. The 1989 Plan also included the establishment of an ambient monitoring program for the sound, which focuses on contaminated sediments.

Sediment Quality Standards

The Washington State Department of Ecology (Ecology) was given the responsibility, under the Plan, for developing regulatory standards for designating sediments having acute or chronic adverse effects on aquatic organisms or posing a significant risk to human health. Final sediment quality standards for marine waters were issued in 1991 for 47 chemicals or groups of chemicals.[16] The sediment quality standards are intended to provide a regulatory goal for the state by identifying surface sediments that have no adverse effects on biological resources. The primary use of the standards is to establish an inventory of sediment sites that pass or fail the applicable standards.

The sediment quality standards are based on a biological effects-based approach that uses the lowest apparent effects threshold (AET) values of four biological indicators for all chemicals except phenanthrene, whose standard is based on the equilibrium partitioning (EP) approach (Table 1). An AET is defined as the concentration of a single chemical (or chemical class) in sediments above which a particular biological effect has always been observed (and thus is predicted to be observed in other areas with similar concentrations of that chemical).

The standard for phenanthrene is based on extrapolation from the U.S. Environmental Protection Agency (EPA) national water quality criterion for phenanthrene, assuming EP of nonionic organic chemicals between pore water and sediment particles. Puget Sound AET values have been developed for amphipod *(Rhepoxynius abronius)* mortality, bivalve *(Crassostrea gigas)* larval abnormality, and Microtox® *(Photobacterium phosphoreum)* bacterial luminescence bioassay endpoints and for abundances of major taxa of indigenous benthic infauna. AET values are based on statistical relationships of chemical and biological variables and do not provide proof of cause-effect relationships for individual chemicals. Therefore, an important part of the overall establishment of sediment cleanup standards is that direct biological testing (e.g., sediment bioassays, assessment of indigenous macroinvertebrate communities) may be used to confirm or override the results of direct comparisons of sediment contaminant concentrations with the numerical standards. The technical basis of AET is discussed further in a later section on assessment approaches.

Sediment standards were also established for the 47 chemicals for three sediment management activities:

- **Source Control Standards**. The maximum level of sediment contamination allowed from ongoing discharges within a sediment impact zone (SIZ_{max})
- **Cleanup Screening Level (CSL)**. The maximum degree of contamination allowed before site cleanup is required
- **Minimum Cleanup Level (MCUL)**. The maximum degree of contamination allowed to remain after site cleanup.

Table 1. Sediment Management Standards for the State of Washington

Chemical Parameter	Sediment Quality Standards	Sediment Cleanup Standards
Metals (mg/kg dry weight)		
Arsenic	57	93
Cadmium	5.1	6.7
Chromium	260	270
Copper	390	390
Lead	450	530
Mercury	0.41	0.59
Silver	6.1	6.1
Zinc	410	960
Nonpolar Organic Compounds (mg/kg organic carbon)[a]		
Aromatic Hydrocarbons		
LPAH[b]	370	780
Naphthalene	99	170
Acenaphthylene	66	66
Acenaphthene	16	57
Fluorene	23	79
Phenanthrene	100	480
Anthracene	220	1200
2-Methylnaphthalene	38	64
HPAH[c]	960	5300
Fluoranthene	160	1200
Pyrene	1000	1400
Benz(a)anthracene	110	270
Chrysene	110	460
Total benzofluoranthenes[d]	230	450
Benzo(a)pyrene	99	210
Indeno(1,2,3-c,d)pyrene	34	88
Dibenz(a,h)anthracene	12	33
Benzo(g,h,i)perylene	31	78
Chlorinated Benzenes		
1,2-Dichlorobenzene	2.3	2.3
1,4-Dichlorobenzene	3.1	9
1,2,4-Trichlorobenzene	0.81	1.8
Hexachlorobenzene	0.38	2.3
Phthalate Esters		
Dimethyl phthalate	53	53
Diethyl phthalate	61	110
Di-*n*-butyl phthalate	220	1700
Butyl benzyl phthalate	4.9	64
Bis(2-ethylhexyl) phthalate	47	78
Di-*n*-octyl phthalate	58	4500
Miscellaneous		
Dibenzofuran	15	58
Hexachlorobutadiene	3.9	6.2
N-nitrosodiphenylamine	11	11
Total PCBs	12	65

Table 1. Sediment Management Standards for the State of Washington (continued)

Chemical Parameter	Sediment Quality Standards	Sediment Cleanup Standards
Ionizable Organic Compounds (µg/kg dry weight; parts per billion)		
Phenol	420	1200
2-Methylphenol	63	63
4-Methylphenol	670	670
2,4-Dimethyl phenol	29	29
Pentachlorophenol	360	690
Benzyl alcohol	57	73
Benzoic acid	650	650

[a] The listed chemical parameter criteria represent concentrations in parts per million, "normalized," or expressed, on a total organic carbon basis. To normalize to total organic carbon, the dry-weight concentration for each parameter is divided by the decimal fraction representing the percent total organic carbon content of the sediment.

[b] The LPAH criterion represents the sum of the following "low-molecular-weight polynuclear aromatic hydrocarbon" compounds: naphthalene, acenaphthylene, acenaphthene, fluorene, phenanthrene, and anthracene. The LPAH criterion is not the sum of the criteria values for the individual LPAH compounds as listed.

[c] The HPAH criterion represents the sum of the following "high-molecular-weight polynuclear aromatic hydrocarbon" compounds: fluoranthene, pyrene, benz(a)anthracene, chrysene, total benzofluoranthenes, benzo(a)pyrene, indeno(1,2,3,-c,d)pyrene, dibenz(a,h)anthracene, and benzo(g,h,i)perylene. The HPAH criterion is not the sum of the criteria values for the individual HPAH compounds as listed.

[d] The total benzofluoranthenes criterion represents the sum of the concentrations of the "B," "J," and "K" isomers.

The sediment standards for these three criteria (SIZ_{max}, CSL, and MCUL) are also based primarily on the AET approach (Table 1). In this case, "minor adverse effects" are allowed, and each individual standard equals the second-lowest AET for that chemical. The SIZ_{max}, CSL, and MCUL are therefore specified at higher chemical concentrations than the sediment quality standards and would allow incrementally higher levels of biological effects than the sediment quality standards.

Sediment cleanup standards are based on a range of allowable levels of contamination that are applied on a site-specific basis. The low end of the range is defined by the sediment quality standards, and the upper end of the range is defined by the minimum cleanup levels. The site-specific cleanup standards are intended to be as close as practical to the sediment quality standards, with consideration of net environmental effect, cost, and technical feasibility of any cleanup action. Evaluation of the natural recovery of contaminated sediments is also an important part of the determination of site-specific cleanup levels.

Inventory of Contaminated Sediments

The Plan calls for Washington State to maintain an inventory of sediments in Puget Sound that do not meet the sediment quality standards. The inventory will be presented in a map format and will identify the chemicals responsible for violations of the standards at each site. The Plan also includes directives for the development of a sediment ranking system to prioritize sites that exceed the standards.

Cleanup of Contaminated Sediments

The Plan requires the development of a uniform decision-making process for determining appropriate remedial actions for sediments exceeding the sediment quality standards. The process will include evaluation of the cost of sediment cleanup options, the no-action alternative, and the natural recovery of sediments following source control. Remedial alternatives will include a variety of in situ treatment options (e.g., capping) as well as removal and treatment or disposal.

Sediment Monitoring

The Plan also required development of the Puget Sound Ambient Monitoring Program (PSAMP) that includes, among many components, a comprehensive assessment of soundwide sediment contamination and biological effects. The goal of PSAMP is to provide long-term monitoring of Puget Sound resources, using consistent methods and a systematic approach. The approach of PSAMP is to use the Sediment Quality Triad[9] at a series of fixed and floating stations located away from toxic hot spots, for an integrated assessment of contamination and effects. Sediment monitoring under PSAMP was initiated in 1989. The first year results revealed the presence of many contaminants in sediments at sites throughout Puget Sound. However, there were few toxic effects measured by bioassays (amphipod and Microtox®) or by alterations of benthic macroinvertebrate assemblages.[17]

Puget Sound Estuary Program

In 1988 Puget Sound was designated by EPA as an estuary of national significance, under Section 320 of the Clean Water Act. The resultant Puget Sound Estuary Program (PSEP) is comanaged by EPA Region 10, Ecology, and PSWQA. Although PSEP activities are directed towards several water quality issues in Puget Sound, a substantial part of the effort has been associated with the assessment and management of contaminated sediments. One of the major activities of PSEP involves the Urban Bay Action Program, which is conducted jointly by EPA with Ecology and other state and local agencies. The Urban Bay Action Program provides a coordinated mechanism for conducting the following activities in a target embayment: (1) data compilation and identification of problem areas; (2) identification of agency activities and management gaps, and development of a source-control action plan; (3) implementation of remedial actions (source control and sediment cleanup); and (4) evaluation of results and revision of the action plan. The activities of the program are implemented primarily by Urban Bay Action

Teams, which are composed of technical staff from various regulatory, resource, and planning agencies. From 1985 to 1990, action plans were either completed or are in the final phases for seven urban embayments in Puget Sound. Individual cleanup and regulatory accomplishments are described in a summary report.[18]

PSEP has played a central role in the development of standardized protocols for conducting field and laboratory assessments of contaminated sediments in Puget Sound. The protocols were developed using a consensus-building approach in a workshop format with participation by regional scientists from agencies, consulting firms, analytical laboratories, and academic institutions. The protocols are produced in a loose-leaf format and are revised periodically to incorporate new information. The following protocols have been developed for the assessment of contaminated sediments: chemical analyses of inorganic and organic chemicals in sediments, sediment bioassays (amphipod mortality, bivalve and echinoderm larvae abnormality, Microtox®, and polychaete growth), and benthic macroinvertebrate studies.[19]

Puget Sound Dredged Disposal Analysis

The Puget Sound Dredged Disposal Analysis (PSDDA) is a multiagency cooperative program involving the U.S. Army Corps of Engineers, the Washington Department of Natural Resources, EPA, and Ecology. The primary objective of PSDDA is to develop environmentally safe and publicly acceptable options for the unconfined, open-water disposal of dredged material. PSDDA has developed a series of procedures and guidelines for evaluating dredged material[20] and has designated disposal sites in three areas of Puget Sound.

An important part of the PSDDA program has been the development of a series of chemical and biological testing procedures to determine the acceptability of dredged material for open-water, unconfined disposal in Puget Sound. The evaluation procedures are based on a tiered approach that incorporates data on sediment chemistry and the results of various acute and chronic bioassays. The approach of PSDDA is to use chemistry data for the basic decision-making guidelines and to rely on biological toxicity testing when the chemistry data indicate the potential for adverse biological effects. Two kinds of chemical guidelines were established for the evaluations:

- A relatively low chemical screening level (SL). The material is acceptable for open-water disposal if the concentrations of all contaminants of concern are below the respective SL values.
- A maximum level (ML). The material is unacceptable for open-water disposal if the concentration of any one contaminant of concern is above the respective ML value.

If all contaminants are below the SL values, no further testing is required. If the concentration of one or more contaminants falls between the SL and ML values, a series of acute bioassays and a bioaccumulation test may be performed to determine if the material meets disposal guidelines. If one or more chemicals exceed the ML values, the dredger has the option to conduct the standard acute bioassays, in addition to sensitive sublethal tests, to potentially override the chemistry results. ML values were set by PSDDA at the highest AET among four biological indicators. SL values were set as 10% of each ML. However, if 10% of the ML is less than the average reference-area concentration for Puget Sound, the reference level is used as the SL. If 10% of the ML exceeds the lowest AET, the latter value is used as the SL.

Superfund

Three of the industrialized embayments with sediment contamination in Puget Sound have been designated as Superfund sites by EPA. The earliest Superfund designation was for the Commencement Bay Nearshore/Tideflats site in Tacoma, WA. In 1983, EPA entered into a cooperative agreement with Ecology (with Ecology as the lead agency) to conduct a remedial investigation/feasibility study (RI/FS) of contaminated sediments in the nearshore areas of Commencement Bay and the associated waterways. Commencement Bay represents an extremely complex site, with numerous sources of contaminants and very complex patterns of sediment contamination in the waterways. The RI/FS included extensive studies of sediment contamination, pollutant sources, and biological effects.[21] The investigation also included the development of a multifaceted decision-making approach to identify and prioritize problem areas and contaminants, which later formed the basis for the EPA Urban Bay Approach.[22] The Commencement Bay RI/FS was completed in 1989, and remedial design activities were initiated thereafter (see subsequent section).

Other Superfund sites in Puget Sound are following Commencement Bay in the overall process. Eagle Harbor, the site of a wood treatment facility with creosote contamination, is presently in the final stages of an RI/FS. The Harbor Island site in Seattle is a complex industrial area located in the lower Duwamish River and Elliott Bay. The Harbor Island RI/FS was begun in 1989 and is now entering a sampling and analysis phase.

ASSESSMENT APPROACHES

Reliable, cost-effective indicators of contamination and associated biological effects are required for the efficient management of contaminated sediments. Basic chemical measurements, combined with toxicological assessment techniques (e.g., sediment bioassays) and benthic macroinvertebrate surveys, have provided the most useful information for the management of Puget Sound sediments.[9,23] This kind of integrative approach supported the development of empirical sediment-quality values. Empirical relationships are the basis for threshold chemical criteria, such as

AET, that are useful in establishing the concentration above which adverse effects can be expected to occur. In addition to empirical threshold approaches, probabilistic models are currently being developed to estimate ecological and human health risks associated with contaminated sediments in Puget Sound.

Chemical and Biological Indicators

Chemical measurements are necessary to investigate the composition and magnitude of sediment contamination and to identify contaminant sources. However, chemical data alone are often insufficient to characterize or predict potential biological effects. Toxicity assessments based only on chemical data are complicated by possible synergistic and antagonistic interactions among chemicals. Moreover, EPA-approved chemical analysis techniques are available only for a small fraction of the thousands of potentially toxic chemicals that may contaminate sediments of highly industrialized waterways.

Sediment Chemistry

Measurements of contaminant concentrations in Puget Sound have often focused on the EPA priority pollutant organic compounds and metals.[21,24-28] Earlier studies documented high concentrations of metals in surface sediments near specific sources, and low to moderate concentrations soundwide.[24] Barrick[25] quantified individual PAH concentrations in sediments of central Puget Sound. Using "fingerprinting" techniques, the study was able to quantify and distinguish the relative contribution of hydrocarbon flux from sewage discharges when compared to natural hydrocarbon sources. Recent studies have assessed concentrations of organic compounds in sediments, including PAHs,[21,26-28] PCBs,[21,26-28] nitrogen-containing aromatic compounds,[29] chlorinated benzenes and butadienes,[21,26-28] various chlorinated and organophosphorus pesticides,[30] tributyltin,[30] and dioxins.[31] Most of these studies have demonstrated that the highest levels of sediment contamination occur in shallow, near-shore areas that are in close proximity to pollutant sources.

Bioaccumulation

Measurements of contaminant concentrations in tissues of fish and shellfish have demonstrated the usefulness of bioaccumulation measures as indicators of sediment contamination.[13] Many studies have focused on the edible tissues of English sole *(Parophrys vetulus)* or Dungeness crab *(Cancer magister)*,[21,27,31] although contamination of other commercially and recreationally harvested fish and shellfish species has received attention recently.[12-14] Ginn and Barrick[13] investigated the relationships between organic contaminant concentrations in muscle and liver tissues of English sole *(Parophrys vetulus)* and contaminant concentrations in sediments collected in the same area as the fish (Figure 3). The analyses demonstrated a positive

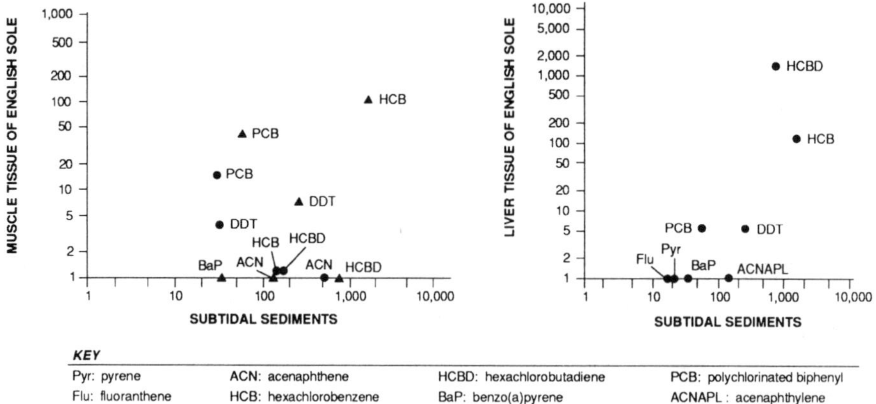

Figure 3. Comparisons of concentrations of contaminants in tissues of English sole and subtidal sediments. (Axes represent ratios of concentrations in waterways to reference areas.)

relationship between sediment and tissue levels for chlorinated compounds, but not for PAHs, which are readily metabolized by fishes. Malins et al.[28,32] demonstrated positive correlations between (1) the prevalence of liver neoplasms in English sole *(Parophrys vetulus)* and the concentration of aromatic hydrocarbons in sediments, and (2) the prevalence of liver neoplasms and the concentration of organic free radicals (a derivative of a PAH compound linked with a protein) in liver microsomes of English sole *(Parophrys vetulus).*

In laboratory exposure studies, Varanasi et al.[33] assessed the bioavailability and biotransformation of aromatic hydrocarbons in benthic organisms exposed to contaminated sediment from the mouth of the West Waterway (Duwamish River). The extent of metabolism of radiolabeled benzo(a)pyrene (i.e., clam [*Macoma nasuta*] < amphipod [*Eohaustorius washingtonianus*] < amphipod [*Rhepoxynius abronius*] ≤ shrimp [*Panadalus platyceros*] ≈ English sole [*Parophrys vetulus*]) was negatively correlated with tissue concentrations of 3- to 6-ring aromatic hydrocarbons, except in amphipods. The amphipod species accumulated higher concentrations of aromatic hydrocarbons than did clams, possibly because other factors, such as feeding strategy and excretion rate, also influenced body burden. A portion of the benzo(a)pyrene (and presumably other aromatic hydrocarbons) extracted chemically from sediments was not bioavailable.

Biological Effects

Biological effects indicators used to assess Puget Sound sediments range from physiological variables (or biomarkers, such as the induction of mixed-function oxidases in organisms exposed to sediment contaminants)[34] to acute mortality bioassays,[35,36] chronic sublethal bioassays,[34] and surveys of benthic communities.[39] Chapman et al.[35] and Long[36] summarized results of various bioassays used in Puget Sound. Most authors[35-37] suggest that a tiered battery of bioassays is optimal for predicting biological effects in the field. Williams et al.[38] demonstrated a significant overall concordance (Kendall's coefficient = 0.64; $p < 0.001$) among the results of the oyster larvae *(Crassostrea gigas)* abnormality, amphipod *(Rhepoxynius abronius)* mortality, and Microtox® *(Photobacterium phosphoreum)* bioassays. However, they noted that substantial variations between paired combinations of these bioassays may be attributable to differential sensitivity to individual contaminants, differences in exposure routes, and the heterogeneous distribution of contaminants in sediments.

Pastorok and Becker[37] compared 13 endpoints from 7 sediment bioassays and ranked their statistical sensitivity in terms of frequency of detecting significantly greater toxicity than in corresponding tests of reference area sediments. The Microtox® *(Photobacterium phosphoreum)* organic extract test and the echinoderm *(Dendraster excentricus)* embryo test were the most sensitive bioassays, based on the number of significant ($p \leq 0.05$) effects detected relative to reference area sediments. The order of sensitivity of the other tests (from highest to lowest) was as follows: Microtox® saline extract > amphipod *(Rhepoxynius abronius)* mortality > polychaete (*Neanthes* sp.) biomass > polychaete mortality > *Rhepoxynius* nonreburial > *Eohaustorius* nonreburial = geoduck *(Panopa generosa)* mortality = echinoderm *(Dendraster excentricus)* chromosomal abnormality. When all bioassay endpoints were compared using the magnitude of response, a relatively low concordance was observed among the various tests. The lack of overall concordance among bioassays with diverse endpoints most likely resulted from differential sensitivity to different kinds of chemicals among bioassays related to the specific endpoint, the test species, the life stage (i.e., embryo, juvenile), or the duration of exposure (i.e., 15 min to 20 days). The lack of overall concordance among a wide range of endpoints supports an approach based on a battery of bioassays to test sediment toxicity.

Surveys of benthic macroinvertebrate assemblages that estimate the magnitude of depressions in major taxa abundances have also proved useful in ranking contaminated-sediment problem areas.[23,39] Compared to laboratory toxicity bioassays, the relevance of surveys of indigenous populations is often more obvious to regulators and to the public. However, in interpreting such results, caution is needed to avoid confounding natural factors that obscure the potential effects of contaminants or lead to false positives. For example, comparisons of benthic taxa abundances between potentially contaminated sites and reference areas should be stratified by season and by habitat features (e.g., sediment grain size and total organic carbon content).

Other useful biological measures for assessing contaminated sediments include pathological indices, such as the prevalence of liver lesions in English sole *(Parophrys vetulus)*.[5,28,40] Although the prevalence of liver lesions may be influenced

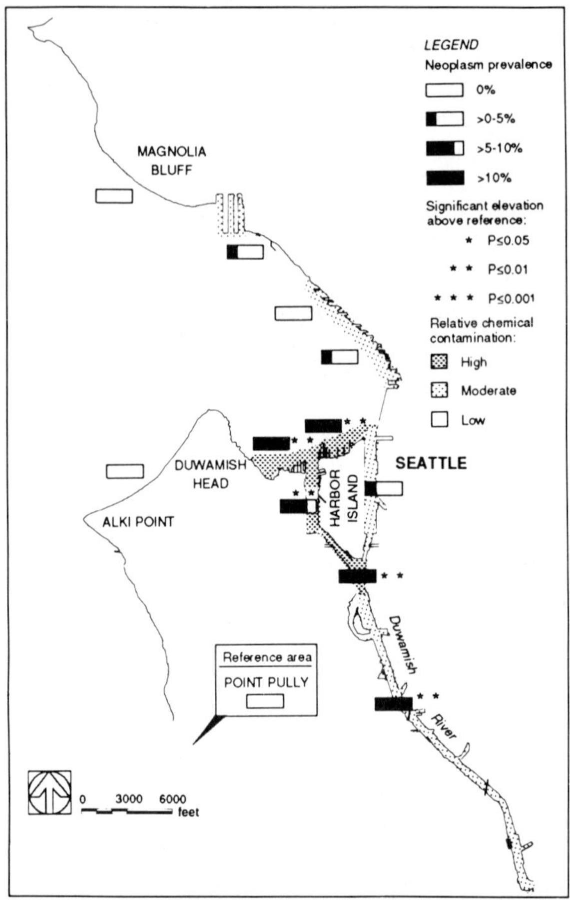

Figure 4. Spatial distribution of liver neoplasm prevalence in English sole and chemical contaminants within Elliott Bay.

by factors other than chemical contamination, the spatial pattern of liver lesion prevalence reflects the geographic distribution of contaminated sediments (Figure 4).

Urban Bay Approach

The strategy for assessing sediment quality in the PSEP Urban Bay Action Programs and at state and federal Superfund sites in Puget Sound relies primarily on the Sediment Quality Triad.[9,22,39] In some cases the basic triad has been enhanced by including measures of bioaccumulation and fish liver lesions as indicators of contaminated sediment problem areas. Because many megainvertebrates (e.g., crab) and fish species used for bioaccumulation or pathology indicators are mobile, these indicators are applied only to areawide assessments on the scale of an individual waterway or portion of a waterway (i.e., a channel approximately 1 to 2 km long). In general, these indicators cannot be used to discriminate sediment conditions among individual stations sampled for the triad. Typically, bioaccumulation and pathology indicators are applied generically to a group of discrete stations distributed widely within an area, or to the area as a whole. Despite fish (or crab) movements, spatial heterogeneities in bioaccumulation, and pathology indicators within an entire bay or among adjacent waterways are clearly related to heterogeneities in the degree of sediment contamination (Figure 4).[5, 21]

Typically, elevation above reference values are calculated for individual variables (e.g., chemical concentrations, bioassay endpoints, or macroinvertebrate taxa abundances in the triad) and integrated in a matrix approach and multiattribute scoring system to identify and rank problem areas. The environmental evaluation is one element of the overall strategy to identify and control sources of contaminants (Figure 5). Evaluation of information in the form of a matrix (e.g., indicator variable by study area) enables the decisionmaker to answer the following questions:

- Is there a significant increase in contamination, toxicity, or biological effects at any study site?
- What combination of indicators is significant?
- What are the relative magnitudes of the elevated indices (i.e., which represent the greatest relative hazards)?

The decision to evaluate potential sources of contamination and the need for possible remedial alternatives applies only to those sites that exceed a minimum action level. An *action level* is a level of contamination or biological effect that defines a problem (e.g., as defined in the "Washington State Sediment Management Standards"). Individual stations that exceed action level guidelines are grouped into problem areas, based on consideration of chemical distributions (including data from recent historical studies), the character and proximity of potential contaminant sources, and geographic and hydrographic boundaries. Ginn[22] described the action levels used to define problem areas in the Commencement Bay Nearshore/Tideflats Superfund project.

Sediment Quality Values

In Puget Sound, sediment quality values based on the AET approach[37,41] have been used to characterize the severity of sediment contamination and to prioritize problem chemicals.[21,22,27,37,41] AET values have been developed for 64 organic and inorganic chemicals from a large historical data base of the observed relationships between biological effects and chemical concentrations.

As described previously, sediment quality standards have been adopted for 47 of the total of 64 chemicals for which AET values have been developed. The selection

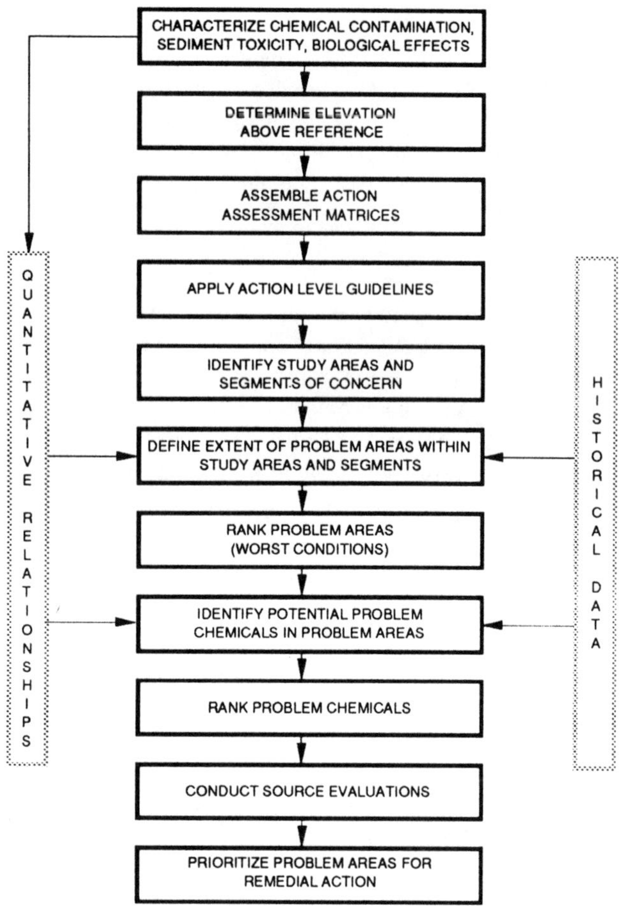

Figure 5. Decision-making framework for assessing contaminated sediments in urban bays.

of the 47 chemicals was based on their relative importance in Puget Sound and the confidence associated with the empirical database for each chemical.

The AET approach does not provide absolute proof of cause and effect for individual chemicals, but uses a preponderance of evidence of the association of chemical contamination and adverse biological effects in assessing the ecological risks of contaminated sediments. The identification of problem chemicals by the AET approach represents a best effort, with available information, to discern between measured chemicals that appear to be associated with adverse biological effects and chemicals that do not show such relationships. In addition, because all potentially toxic chemicals cannot be measured routinely, sediment quality assessments must rely, to some extent, on the regulation or management of surrogate chemicals. If, for example, an unmeasured chemical (or group of chemicals) varies consistently in the environment with a measured chemical, then the AET values established for the measured contaminant will indirectly apply to, or result in the management of, the unmeasured contaminant. In such cases, a measured contaminant would act as a surrogate for an unmeasured contaminant (or group of unmeasured contaminants).

The performance of the Puget Sound AET values has been evaluated by estimating their sensitivity (i.e., proportion of impacted stations that were correctly predicted) and efficiency (i.e., proportion of stations designated as impacted that were correctly predicted).[41] Sensitivity ranged from 58 to 93%, and efficiency ranged from 37 to 72%, depending on the biological indicator. Overall reliability (i.e., percentage of all predictions that are correct) is highest for the Microtox® and oyster larvae AETs. Both the oyster larvae and Microtox® bioassays rely on water-mediated exposure to contaminants and do not represent direct sediment exposures (e.g., contact or ingestion). Nevertheless, these toxicity test responses display a high concordance with alterations in benthic macroinvertebrate assemblages (81% concordance for oyster larvae; 68 percent concordance for Microtox®; $p < 0.01$).[23]

The EPA Office of Water Criteria and Standards is investigating the use of EP theory to develop national sediment quality criteria.[42] A theoretical model is used to describe EP of nonpolar organic chemicals between sediment organic matter and interstitial water. The model is used to calculate the interstitial water concentration of a contaminant associated with a given bulk sediment concentration of that contaminant. The interstitial water concentration of that contaminant is then compared to the corresponding EPA ambient water quality criterion. In an evaluation of the reliability of AET and EP sediment quality values, the sensitivity and efficiency of the EP sediment quality approach to predicting biological effects in Puget Sound (as indicated by sediment toxicity tests and infaunal abundances) were generally less than the sensitivity and efficiency of the corresponding AET values.[41]

Ecological and Human Health Risk Modeling

The AET method represents an empirical effects-based approach to assessing sediment contamination by complex mixtures of chemicals. Although the AET approach is consistent with toxicological theory, the theoretical basis for specific AET values has not been established because of the difficulty of modeling complex chemical interactions and the lack of data for many chemicals. In contrast to the AET approach, probabilistic modeling approaches incorporate dose-response relationships for single chemicals,[43,44] or uncertainty analysis of criteria exceedance,[44] to estimate the risk of adverse effects associated with measured contaminant concentrations. Current efforts to develop risk assessment models relevant to the management of Puget Sound sediments include both uncertainty analysis and risk-benefit tradeoff models.[45,46]

As part of an ecological risk assessment of contaminated sediments in Elliott Bay,[45] a fugacity model (based on EP theory) was initially used to predict steady-state concentrations of chemicals in air, water, sediments, interstitial water, and biota. Toxicity guidelines for sediments included mean acute values and corresponding chronic values (derived from application of an acute-to-chronic effects ratio), which were calculated from EPA water quality criteria guidelines, using the fugacity model. Each predicted contaminant concentration was compared with a series of toxicity guidelines for aquatic species, based on different degrees of species protection, to assess ecological risk. Quantitative comparisons of this approach and other predictive approaches (e.g., AET and EP) have not been conducted.

Limited attempts have been made to evaluate human health risks associated with contaminated sediments in Puget Sound.[47,48] Models to estimate human health risks associated with consumption of chemically contaminated fish and shellfish are well developed,[48,49] but prediction of contaminant concentrations in aquatic organisms, based on sediment chemistry data, is a critical source of uncertainty.[48] For example, some authors assume that contaminant concentrations in seafood organisms are in equilibrium with sediment interstitial water, to estimate contaminant concentrations in edible tissues. Lee and Randall[47] suggest that sediment criteria based on human health risk assessment models are generally more restrictive (i.e., specify lower tissue contaminant concentrations) than corresponding criteria to protect aquatic organisms. Other work[50,51] has focused on assessing human health risks associated with observed concentrations of contaminants in edible tissues of fish and shellfish, but the link to sediment contamination has not been definitely established. Pastorok and Schoof[40] concluded that data on the prevalence of liver lesions in bottom fish of Puget Sound were valuable in identifying and ranking sediments contaminated by carcinogenic chemicals, but that such data could not be used to estimate human health risk from sediment contaminants.

SEDIMENT REMEDIAL ACTIVITY

Many of the sediment regulatory programs in Puget Sound will eventually result in the remediation of contaminated sediments. In general, evaluation of the need for remediation of contaminated sediments has involved two important considerations:

1. Have existing pollutant sources contributing to the contamination been controlled?
2. Will natural recovery processes (e.g., natural sedimentation, bioturbation, contaminant degradation) result in an elimination of the problem with time?

Given the high potential costs of sediment remedial alternatives, these considerations ensure that resources will not be expended on sediments that will just be recontaminated by ongoing sources and that any remedial activity will take full advantage of natural recovery processes.

Because of the many ongoing sources of contaminants in the industrialized embayments of Puget Sound, PSEP has focused primarily upon source-control activities as an initial step toward sediment cleanup. Following the development of action plans, the Urban Bay Action Teams have completed hundreds of inspections and site surveys, which have resulted in the issuance of numerous regulatory orders for control of contaminant discharges.

In Commencement Bay, source control and the development of a comprehensive sediment remediation program have received considerable emphasis by the Urban Bay Action Team under the RI/FS. The overall remedy selected for Commencement Bay combines the following key elements: site-use restrictions, source control, natural recovery, sediment remedial action, and monitoring. For the eight high-priority problem areas in Commencement Bay, the "Record of Decision" specifies performance-based sediment quality objectives rather than a precise cleanup action for each site. The sediment quality objectives are based primarily on AETs for the indicator chemicals for each area. The sediment quality objectives are translated into remedial action levels by incorporating the degree of natural recovery predicted to occur within 10 years following source control. The degree of natural recovery is derived from a mathematical model, SEDCAM,[53] that predicts recovery as a function of source loading, sedimentation rate, and surficial sediment mixing. Areas that have present-day sediment contaminant concentrations in excess of the remedial action levels for that area and are not expected to recover naturally within 10 years are subject to remedial action. Examples of ranges of remedial action levels for some problem chemicals include PCBs, 240 to 300 μg/kg dry weight; high-molecular-weight PAHs, 20 to 32 mg/kg dry weight; copper, 470 to 1,100 mg/kg dry weight; and arsenic, 97 to 160 mg/kg dry weight.

For the entire site, a total volume of 1,181,000 yd^3 of contaminated sediments was estimated to be potentially subject to remedial action. For six of the eight problem areas of Commencement Bay, the selected remedy incorporates multiple sediment remedial options associated with the confined disposal of contaminated sediments. The confined disposal alternative for the six areas could include specific remedial actions, such as in situ capping or dredging followed by confined aquatic disposal, nearshore disposal, or upland disposal. In one other area (St. Paul Waterway), the selected remedial action involved only in situ capping. At the mouth of City Waterway, no sediment action was specified because the sediments were predicted to recover naturally within 10 years following source control.

As of 1990, most of the Commencement Bay problem areas are being evaluated under the remedial design phase of the Superfund process. One problem area at the St. Paul Waterway was remediated by in situ capping in 1988. Extensive in-plant source-control activities and the installation of a deepwater outfall were also implemented to reduce the subsequent contamination of the cap materials. Problem chemicals at the St. Paul Waterway site resulted primarily from pulp mill operations. Approximately 118,000 yd^2 of shallow subtidal sediments had concentrations of 4-methylphenol above the remedial action level of 1,300 μg/kg.

The St. Paul Waterway problem area was covered with a cap of clean fill material 1.5- to 3.7-m thick, with adjacent areas of organic wood debris being covered with a cap 0.6-m thick. A monitoring program was initiated in 1988–89 to evaluate the physical, chemical, and biological characteristics of the cap material.[54] To evaluate cap integrity, action levels for individual chemicals were established at 80% of the respective lowest AET value. Chemical analyses of sediment core samples indicate that the sediment cap has effectively isolated the contaminated sediments, with no indication of upward movement of contaminants through the cap material. However, at one station there is evidence that the original contaminated sediments may have become mixed with the cap material during construction. Studies of benthic infauna demonstrated that within 1 year following cap construction, the assemblages were dominated by opportunistic polychaetes (e.g., *Tharyx multifilis*) and bivalve mollusks (e.g., *Axinopsida serricata* and *Macoma* spp.) that are typical of the Commencement Bay waterways. Total infaunal abundances ranged from 1580 to 4400/m^2, which is within the range found in Puget Sound reference areas.

A recent confined disposal project for contaminated dredged material along the Seattle waterfront provides an additional example of sediment remediation because of its similarity to the proposed nearshore confined disposal alternatives for other sites. The project was referred to as the "Terminal 91 Short Fill" and consisted of filling an abandoned terminal slip with contaminated sediments that were determined to be unsuitable for unconfined, open-water disposal. About 100,000 m^3 of contaminated sediments were placed behind a nearshore berm and capped with clean fill material. A computer model was used in the design phase to evaluate tidal pumping of soluble contaminants in the fill material. Initial monitoring results indicate that contaminants are being confined in the filled area and that a normal epibenthic biological assemblage has become established on the berm face.[55]

FUTURE DIRECTION

In the 1990s, the management of contaminated sediments in Puget Sound is entering a new phase. Many of the contaminated problem areas have been identified. Assessment methods will likely be streamlined and tailored to a process for prioritizing contaminated sites, evaluating the need for sediment impact zones and site-specific cleanup levels, and modeling the natural recovery of sediments under various source-control scenarios. Methods for evaluating the tradeoffs between risk reduction and remediation costs need to be developed further to ensure cost effectiveness of remedial efforts.

Implementation of Washington State Marine Sediment Standards

The sediment management standards that have been developed for Puget Sound involve two main elements: (1) the sediment quality standards and (2) source control and cleanup standards (Table 1). The sediment quality standards establish the long-term sediment quality goals for Puget Sound. The overall goal is to achieve sediment contaminant levels that have no adverse effects on biological resources. To allow for technical and cost limitations on cleanup actions, Ecology also established the source control and cleanup standards, which are set at higher chemical-concentration and biological-effects levels than those defined by the long-term sediment quality goal. Source control and cleanup standards apply to sediments within a sediment impact zone (SIZ), which provides a variance for a specific discharge activity to allow consideration of cost and technical feasibility in meeting sediment quality standards. An SIZ may be authorized by the state when an ongoing or proposed discharge violates, or has the potential to violate, the sediment quality standards. Violation of the SIZ authorization will require the following maintenance activities (in order): evaluating compliance with known, available, and reasonable treatment methods and best management practices; altering the size and/or degree of effects in the SIZ; requiring additional source-control or sediment cleanup actions.

The overall sediment management process is illustrated in Figure 6. The first step is to create an inventory of all locations in Puget Sound where concentrations of contaminants in sediment exceed the sediment quality standards (long-term goal). Site boundaries are defined by the distribution of individual contaminants exceeding the sediment quality standards. Next, groups of stations are defined as "station clusters of potential concern" based on exceedance(s) of the CSL within each station cluster. Following the screening step, a hazard assessment is performed on all identified station clusters of potential concern. During this stage, additional information is assembled to further characterize the clusters. Clusters that continue to exceed the CSL, as identified above, are then to be defined as cleanup sites. Sites are then prioritized to efficiently allocate resources to remediate those contaminated sediments that pose the greatest environmental and public health threat.

The final step in the contaminated sediments management process is to select the appropriate cleanup alternative(s) for the sites determined to warrant such action(s). As part of this step, a site may be divided into discrete units that are distinguishable in terms of their habitat and/or resource values, or other unique characteristics. A unique cleanup standard (i.e., the contaminant concentration to be achieved by a remedial action) will then be developed for the site as a whole or for each identified site unit, as appropriate. The cleanup standard for each unit of the site as a whole will always be as close as practicable to the cleanup goal (i.e., the sediment quality standards), based on a consideration of the potential for natural recovery of the sediments over time and the cost and engineering feasibility of the available remedial action alternatives. In all cases, the MCUL defines the upper constraint on the unit-specific or site-specific cleanup standard. The MCUL is defined as the maximum allowable chemical concentrations and biological effects permissible at the cleanup site, to be achieved by 10 years after the completion of active site cleanup. The incorporation of a time of compliance criterion (i.e., 10 years) into the definition of the MCUL further enhances management flexibility by allowing the potential for natural recovery to be considered when defining the appropriate remedial action response for a particular site.

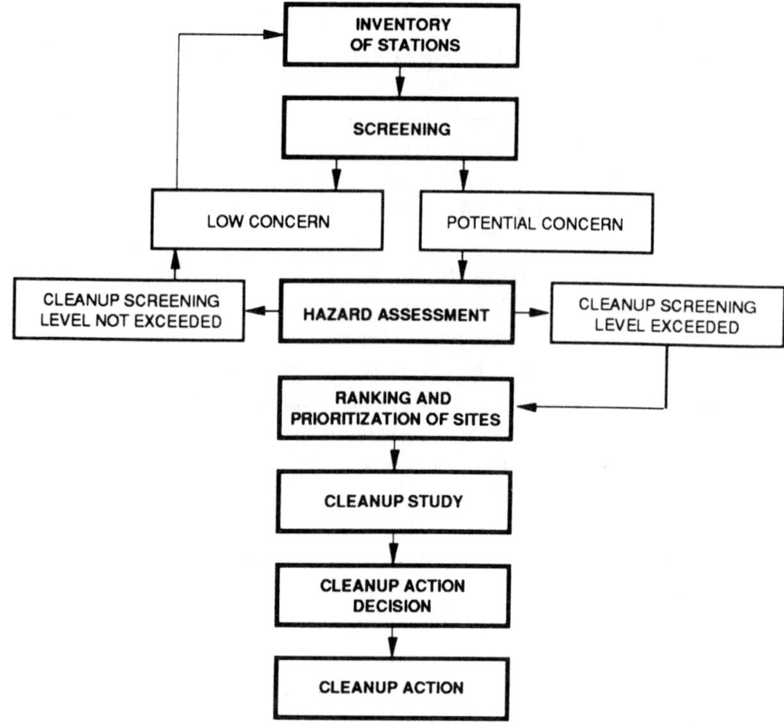

Figure 6. Decision-making framework for contaminated sediment cleanup.

An area of contaminated sediments that meets the site-specific cleanup standards, but still exceeds the sediment quality standards, will be designated as a sediment recovery zone (SRZ) during the sediment cleanup decision process. Requirements for ongoing monitoring of the recovery process in SRZs may be included in site cleanup decisions.

Modeling Impact Zones and Natural Recovery

To establish SIZs and SRZs, models are needed to predict the transport and fate of contaminants discharged into the marine environment. Specifically, these models should be capable of (1) determining the areal extent and degree of contamination attributable to a discharge and (2) simulating natural recovery processes and multiple source loading to predict long-term trends in chemical concentrations in surface sediments and associated changes in the extent and degree of contamination. No single available model can address the entire range of source types and receiving water conditions that exist in Puget Sound. Available models that are being considered for use in managing Puget Sound sediments include the Cornell Mixing Zone Expert System (CORMIX) and the Water Quality Analysis Simulation

Program 4 (WASP4).[56] CORMIX simulates the initial mixing of an effluent in the water column (the nearfield model), and WASP4 simulates longer-term sediment transport processes (the farfield model). Together, these models simulate processes such as hydrodynamic mixing and dilution of effluent discharged into a water body, sediment transport, ionization/precipitation, adsorption/desorption, hydrolysis/hydration, photolysis, oxidation/reduction, volatilization, and microbial degradation.

Biological Criteria

Puget Sound surface sediments with concentrations of contaminants that exceed the sediment quality standards are designated as having adverse effects, pending confirmation via biological testing. Two acute and one chronic effects tests are used as biological effects criteria to confirm or overturn initial designations based on chemical criteria. Compliance with the biological effects criteria is determined in each case by statistical comparisons with reference sediment results and by the magnitude of the response to test sediments. Any sediment that fails one of the biological tests is designated as failing the sediment quality standards, regardless of the initial designation. Conversely, if all of the required biological tests show no adverse effects, the sediment is designated as passing the sediment quality standards. Biological tests used in classifying sediments are listed in Table 2.

The incorporation of quantitative biological effects criteria into the Washington State sediment quality standards represents a progressive stage in the implementation of environmental regulatory standards. Biological effects criteria provide an effective management tool by directing regulatory efforts toward direct resource assessments, by enhancing classification systems for water use impairment, and by providing for integrated assessments of physical, chemical, and biological integrity of ecological systems.[57]

Ecological and Human Health Risk Modeling

Modeling of ecological and human health risks associated with chemical contaminants in the marine environment is in an early stage of development. Most ecological approaches are based on the application of toxicity criteria or dose-response relationships for single species to estimate acute mortality or a chronic effect.[43,45] Human health risk assessment approaches generally follow standardized EPA models that were developed for terrestrial hazardous waste sites, while relying on estimates of contaminant concentrations in tissues of seafood species, based either on actual measurements or on predictions from EP theory.

Future developments in ecological risk modeling may link population dynamics models with exposure estimates and dose-response functions to estimate population responses associated with specific chemical concentrations in the environment.[44] Although the potential value of population dynamics models for assessing contaminant-induced stress on commercial and recreational fisheries has been demonstrated,[58] such approaches have not been applied to Puget Sound.

Table 2. Biological Tests Used to Classify Sediments

Test (exposure)	Species/Life Stage	Endpoint	Test Conditions: Temperature (°C)	Salinity (ppt)	pH	Dissolved Oxygen (mg/L)	Reference[a]
Acute Effects							
Amphipod (10-day)	*Rhepoxynius abronius* (adult)	Mortality	15 ± 1	28 ± 1	8 ± 1	≥5	Swartz et al.[6]
Bivalve (2-day)	*Crassostrea gigas* (larva)	Developmental abnormality Mortality	12 ± 1	28 ± 1	8 ± 1	≥5	Chapman and Morgan[7]
Echinoderm (2-day)	*Mytilus edulis* (larva) *Dendraster excentricus* (larva) *Strongylocentrotus purpuratus*	Developmental abnormality Mortality	15 ± 1	29 ± 1	8 ± 1	≥5	Dinnel and Stober[59]
Sublethal Effects							
Polychaete (20-day)	*Neanthes* sp. (juvenile)	Biomass	20 ± 1	28 ± 1	8 ± 1	≥5	Johns et al.[60]
Microtox® (15-min)	*Photobacterium phosphoreum* (bacterium)	Decrease in luminescence	15 ± 1	NA[b]	NA	NA	Bulich et al.[61] Beckman Instruments[62]
Benthic macro-invertebrates	Species assemblages (all stages)	Density of major taxa	NA: field-collected samples				PSEP[19]

[a] Also see PSEP Protocols.[19]
[b] Not applicable.

The modeling of human health risks in the marine environment will likely focus on improving techniques to estimate contaminant concentrations in tissues of seafood species. Multimedia (e.g., sediment, water, and seafood) exposure models and methods to assess uncertainties will also be needed. Further developments will probably also involve dynamic modeling of temporal changes in contaminant concentrations and spatial variation in exposure parameters (e.g., fishing activity and level of seafood contamination).

Modeling Cost-Benefit Tradeoffs in Sediment Management

Modeling approaches are needed to evaluate tradeoffs between benefits (i.e., risk reduction) achieved by sediment management actions and costs associated with those actions. PSDDA[43] developed a framework for risk analysis of dredged material disposal options, but risk-cost balancing was not evaluated explicitly. Bogardi et al.[46] and Ebasco[45] are examples of ongoing efforts to develop risk-cost management models based on Puget Sound data. Both of these approaches address human health as well as ecological risks of contaminated sediments.

Bogardi et al.[46] present a model to aid managers in choosing a dredged material disposal plan that minimizes human health risks and ecological risks in a cost-effective manner. Data on risks and costs are integrated using a multicriterion decision-making method called composite programming. Tradeoffs are evaluated at several tiers of the system, including evaluation of the risk of toxicity vs. the risk of burial to aquatic organisms, the risk to fish vs. the risk to shellfish, cancer risk vs. noncancer risk for humans, human health risk vs. ecological risk, and overall composite risk vs. cost. Sensitivity analysis shows that the model provided a robust solution for the choice of confined aquatic disposal as a dredged-material disposal option.

Ebasco[45] is developing models to relate the costs of contaminated sediment remediation for alternative remedial designs to the economic impact of a given level of sediment contamination remaining after remediation. Ultimately, sampling and analysis efforts should be designed to provide integrated data sets for risk assessments conducted during remedial investigations and natural resource damage assessments, as well as cost-benefit modeling relevant to remedial design phases of feasibility studies.

SUMMARY AND RECOMMENDATIONS

In the Puget Sound region, several state and federal programs have been successfully implemented to manage contaminated sediments. The success of these programs is exemplified by the development of a comprehensive management plan, adoption of numerical sediment quality standards, and establishment of a soundwide monitoring program. Moreover, source control activities and sediment remediation projects have also been implemented under several programs. The success of the various programs has resulted largely from a commitment to interagency cooperation, strong public support, and a desire to move ahead in technical and program development in the absence of national sediment quality criteria or program guidance. Key elements of interagency cooperation include the use of common bioassessment techniques and standardized sampling and analytical protocols. The basic approach for the development of sediment quality values (i.e., AET) is also shared among the federal and state agencies, leading to a common technical basis for sediment management actions.

The management of contaminated sediments in Puget Sound is now moving from a planning and technical development phase into a plan implementation and cleanup phase. Although a firm groundwork has been established (and many regulatory successes have already occurred), there are several areas that will need further improvement or development. Although strong empirical relationships have been developed for sediment quality values based on biological effects, there has been no corresponding development of human health-based sediment criteria. Technical development is needed in the areas of theoretical modeling or direct empirical assessment of the bioaccumulation of sediment contaminants, which could form the basis for sediment criteria protective of human health. Predictive models that are used to define source-sediment relationships and estimate natural recovery rates are in the early stages of development. These models will require continued technical development and validation as they are used in sediment management programs.

The development of standardized protocols for the sampling and analysis of sediments has resulted in the collection of comparable data among the various regulatory programs. The existing protocols should be thoroughly evaluated as they are used in sediment monitoring programs, and appropriate refinements should be incorporated into future versions. All of the sediment bioassays that have been used in Puget Sound programs have limitations and uncertainties, usually associated with exposure conditions, endpoint relevance, or the effects of sediment factors other than target chemicals. These bioassays should undergo continued evaluation, with the objective of selecting the optimal suite of bioassay responses to be used in sediment toxicity studies.

The evaluation of sediment remedial alternatives is in the early stages of development. A primary consideration in the selection of a specific remedial alternative involves the application of cost-benefit analysis to balance the monetary and environmental costs of remediation against the environmental benefits of the completed action. An organized framework is presently needed for conducting such analyses and selecting optimal remedial actions. As more remedial actions are completed, it will also be important to implement well-designed monitoring programs to evaluate the long-term effectiveness of each selected action.

REFERENCES

1. Malins, D. C., B. B. McCain, D. W. Brown, A. K. Sparks, and H. O. Hodgins. "Chemical Contaminants and Biological Abnormalities in Central and Southern Puget Sound," NOAA Tech. Mem. OMPA-2 (1980).
2. Long, E. R. "An Assessment of Marine Pollution in Puget Sound," *Mar. Pollut. Bull.* 13:380–383 (1982).
3. Riley, R. G., E. A. Crecelius, M. L. O'Malley, K. H. Abel, and D. C. Mann. "Organic Pollutants in Waterways Adjacent to Commencement Bay (Puget Sound)," NOAA Tech. Mem. OMPA-12 (1981).
4. Dexter, R. N., D. E. Anderson, E. A. Quinlan, L. S. Goldstein, R. M. Strickland, R. M. Kocan, M. Landolt, J. P. Pavlou, and V. R. Clayton. "A Summary of Knowledge of Puget Sound," NOAA Tech. Mem. OMPA-13 (1981).
5. Becker, D. S., T. C. Ginn, M. L. Landolt, and D. B. Powell. "Hepatic Lesions in English Sole *(Parophrys vetulus)* from Commencement Bay, Washington (USA)," *Mar. Environ. Res.* 23:153–173 (1987).
6. Swartz, R. C., W. A. DeBen, J. K. Phillips, J. O. Lambertson, and F. A. Cole. "Phoxocephalid Amphipod Bioassay for Marine Sediment Toxicity," in *Aquatic Toxicology and Hazard Assessment: Seventh Symposium,* R. D. Cardwell, R. Purdy, and R. C. Bahner, Eds. (Philadelphia: American Society for Testing and Materials, 1985).
7. Chapman, P. M., and J. D. Morgan. "Sediment Bioassays with Oyster Larvae," *Bull. Environ. Contam. Toxicol.* 31:438–444 (1983).
8. Swartz, R. C., W. A. DeBen, K. A. Sercv, and J. O. Lamberson. "Sediment Toxicity and the Distribution of Amphipods in Commencement Bay, U.S.A.," *Mar. Pollut. Bull.* 13:359–364 (1982).
9. Long, E. R., and P. M. Chapman. "A Sediment Quality Triad: Measures of Sediment Contamination, Toxicity and Infaunal Community Composition in Puget Sound," *Mar. Pollut. Bull.* 16:405–415 (1986).
10. Riley, R. G., E. A. Crecelius, R. E. Fitzner, B. L. Thomas, J. M. Gurtisen, and N. S. Bloom. "Organic and Inorganic Toxicants in Sediments and Marine Birds from Puget Sound," NOAA Tech. Mem. NOS OMS1 (1982).
11. Calambokidis, J., J. Peard, G. H. Steiger, J. C. Cubbage, and R. L. DeLong. "Chemical Contaminants in Marine Mammals from Washington State," NOAA Tech. Mem. NOS OMS6 (1984).
12. Gahler, A. R., J. M. Cummins, J. N. Blazovich, R. H. Riech, R. L. Arp, C. E. Gangmark, S. V. W. Pope, and S. Filip. "Chemical Contaminants in Edible Nonsalmonid Fish and Crabs from Commencement Bay, Washington," U.S. EPA Report 910/9-82-093 (1982).
13. Ginn, T. C., and R. C. Barrick. "Bioaccumulation of Toxic Substances in Puget Sound Organisms," in *Oceanic Processes in Marine Pollution, Volume 5; Urban Wastes in Coastal Marine Environments,* Wolfe, D. A., and T. P. O'Connor, Eds. (Malabar, FL: Robert E. Krieger, 1988), pp. 157–168.
14. Landolt, M. L., F. R. Hafer, A. Nevissi, G. van Belle, K. Van Ness, and C. Rockwell. "Potential Toxicant Exposure Among Consumers of Recreationally Caught Fish from Urban Embayments of Puget Sound," NOAA Technical Memorandum NOS-OMA-23 (Rockville, MD: National Oceanic and Atmospheric Administration, 1985).

15. "1991 Puget Sound Water Quality Management Plan" Puget Sound Water Quality Authority, Seattle, WA (1991).
16. "Sediment Management Standards — Chapter 173–204 WAC," Washington State Department of Ecology, Olympia, WA (1991).
17. Copping, A., A. Frahm, and V. Stern. "Puget Sound Update: First Annual Report of the Puget Sound Ambient Monitoring Program," Puget Sound Water Quality Authority, Seattle, WA (1990).
18. "The Puget Sound Urban Bay Activities Program — Action Team Accomplishments, October 1985–March 1990," J. Faigenblum, Ed., Washington State Department of Ecology, Olympia, WA (1990).
19. PSEP. "Recommended Protocols and Guidelines for Measuring Selected Environmental Variables in Puget Sound," U.S. Environmental Protection Agency, Seattle, WA (1986).
20. PSDDA. "Evaluation Procedures Technical Appendix," Puget Sound Dredged Disposal Analysis, Seattle, WA (1988).
21. Tetra Tech, Inc. "Commencement Bay Nearshore/Tideflats Remedial Investigation," EPA Report 910/9-85-1346, Tetra Tech, Inc., Bellevue, WA (1985).
22. Ginn, T. C. "Assessment of Contaminated Sediments in Commencement Bay (Puget Sound, Washington)," in *Contaminated Marine Sediments — Assessment and Remediation* (Washington D.C.: National Academy Press, 1989), pp. 425–439.
23. Becker, S. D., G. R. Bilyard, and T. C. Ginn. "Comparisons Between Sediment Bioassays and Alterations of Benthic Macroinvertebrate Assemblages at a Marine Superfund Site: Commencement Bay, Washington," *Environ. Toxicol. Chem.* 9:669–685 (1990).
24. Crecelius, E. A., M. H. Bothner, and R. Carpenter. "Geochemistry of Arsenic, Antimony, Mercury, and Related Elements in Sediments of Puget Sound," *Environ. Sci. Technol.* 9:325–33 (1975).
25. Barrick, R. C. "Flux of Aliphatic and Polycyclic Aromatic Hydrocarbons to Central Puget Sound from Seattle (Westpoint) Primary Sewage Effluent," *Environ. Sci. Technol.* 16:682–692 (1982).
26. Crecelius, E. A., and H. C. Curl, Jr. "Temporal Trends of Contamination Recorded in Sediments of Puget Sound," in *Proceedings: First Annual Meeting on Puget Sound Research* (Seattle, WA: Puget Sound Water Quality Authority, 1988).
27. PTI and Tetra Tech. "Elliott Bay Action Program: Analysis of Toxic Problem Areas," PTI Environmental Services, Bellevue, WA (1988).
28. Malins, D. C., B. B. McCain, D. W. Brown, S.-L. Chan, M. S. Myers, J. T. Landahl, P. G. Prohaska, A. J. Friedman, L. D. Rhodes, D. G. Burrows, W. D. Gronlund, and H. O. Hodgins. "Chemical Pollutants in Sediments and Diseases of Bottom-Dwelling Fish in Puget Sound, WA," *Environ. Sci. Technol.* 18:705–713 (1984).
29. Krone, C. A., D. G. Burrows, D. W. Brown, P. A. Robisch, A. J. Friedman, and D. C. Malins. "Nitrogen-Containing Aromatic Compounds in Sediments from a Polluted Harbor in Puget Sound," *Environ. Sci. Technol.* 20:1144–1150 (1986).
30. Crecelius, E. A., D. L. Woodruff, and M. S. Myers. "1988 Reconnaissance Survey of Environmental Conditions in 13 Puget Sound Locations" (Duxbury, MA: Battelle Ocean Sciences, 1989).

31. Romberg, G. P., S. P. Pavlou, R. F. Shokes, W. Hom, E. A. Crecelius, P. Hamilton, J. T. Gunn, R. D. Muaench, and J. Vinelli. "Presence, Distribution, and Fate of Toxicants in Puget Sound and Lake Washington," Metro Toxicant Program Report No. 6A, Toxicant Pretreatment Planning Study Technical Report C1, Municipality of Metropolitan Seattle, WA (1984).
32. Malins, D. C., M. S. Myers, and W. T. Roubal. "Organic Free Radicals Associated with Idiopathic Liver Lesions of English Sole *(Parophrys vetulus)* from Polluted Marine Environments," *Environ. Sci. Technol.* 17:679–685 (1983).
33. Varanasi, U., W. L. Reichert, J. E. Stein, D. W. Brown, and H. R. Sanborn. "Bioavailability and Biotransformation of Aromatic Hydrocarbons in Benthic Organisms Exposed to Sediment from an Urban Estuary," National Marine Fisheries Service, Seattle, WA (1985).
34. Collier, T. K., J. E. Stein, W. L. Reichert, B.-T. L. Eberhart, and U. Varanasi. "Using Bioindicators to Assess Contaminant Exposure in Flatfish from Puget Sound, WA," in *Proceedings: First Annual Meeting on Puget Sound Research* (Seattle, WA: Puget Sound Water Quality Authority, 1988).
35. Chapman, P. M., R. N. Dexter, R. M. Kocan, and E. R. Long. "An Overview of Biological Effects Testing in Puget Sound, Washington: Methods, Results, and Implications," in *Aquatic Toxicology, Proceedings of the Seventh Annual Symposium,* Spec. Tech. Rept. 854 (Philadelphia: American Society for Testing and Materials, 1985), pp. 344–362.
36. Long, E. R. "Sediment Bioassays: A Summary of Their Use in Puget Sound," National Oceanic and Atmospheric Administration, Seattle, WA, unpublished.
37. Pastorok, R. A., and D. S. Becker. "Comparative Sensitivity of Sediment Toxicity Bioassays at Three Superfund Sites in Puget Sound," in *Aquatic Toxicology and Risk Assessment: Thirteenth Volume,* STP 1096, W. G. Landis, and W. H. van der Schalie, Eds. (Philadelphia: American Society for Testing and Materials, 1990), pp. 123–139.
38. Williams, L. G., P. M. Chapman, and T. C. Ginn. "A Comparative Evaluation of Sediment Toxicity Using Bacterial Luminescence, Oyster Embryo, and Amphipod Sediment Bioassays," *Mar. Environ. Res.* 19:225–249 (1986).
39. Barrick, R. C., H. Beller, D. S. Becker, and T. C. Ginn. "Use of the Apparent Effects Threshold Approach (AET) in Classifying Contaminated Sediments," in *Contaminated Marine Sediments — Assessment and Remediation* (Washington, D.C.: National Academy Press, 1989).
40. Pastorok, R. A., and R. A. Schoof. "Limitations on the Use of Fish Liver Neoplasms as Indicators of Human Health Risk," in *Puget Sound Research '91* (Seattle, WA: Puget Sound Water Quality Authority, 1991), pp. 571–581.
41. Barrick, R. C., and H. R. Beller. "Reliability of Sediment Quality Assessment in Puget Sound," in *Proceedings of Oceans '89* (New York: Institute of Electrical and Electronics Engineers, 1989), pp. 421–426.
42. U.S. Environmental Protection Agency. "Equilibrium Partitioning Approach to Generating Sediment Quality Criteria," Draft Briefing Report, U.S. Environmental Protection Agency, Washington, D.C. (1988).
43. PSDDA. "Framework for Comparative Risk Analysis of Dredged Material Disposal Options," Puget Sound Dredged Disposal Analysis, Seattle, WA (1986).

44. Barnthouse, L. W., and G. W. Suter, Eds. *User's Manual for Ecological Risk Assessment*, Publication No. 2679, U.S. Environmental Protection Agency, Washington D.C. (1986).
45. Ebasco. "Metro Toxic Sediment Remediation Project," Review Draft Report, Parametrix, Ebasco Environmental, and Hartman Associates, Bellevue, WA (1991).
46. Bogardi, I. "Risk-Cost Analysis of Dredged Material Management: Test Application," University of Nebraska-Lincoln, Lincoln, NE, unpublished, (1990).
47. Lee, H., II, and R. Randall. "Cancer Risk Associated with Sediment Quality," in *Proceedings: First Annual Meeting on Puget Sound Research* (Seattle, WA: Puget Sound Water Quality Authority, 1988).
48. Becker, D. S., R. A. Pastorok, R. C. Barrick, P. N. Booth, and L. A. Jacobs. "Contaminated Sediments Criteria Report," PTI Environmental Services, Bellevue, WA (1989).
49. U.S. Environmental Protection Agency. "Guidance Manual for Assessing Human Health Risks from Chemically Contaminated Fish and Shellfish," U. S. EPA Report 503/8-89-002, Washington, D.C. (1988).
50. Pastorok, R. A., P. N. Booth, and L. Williams. "Estimating Potential Health Risks of Chemically Contaminated Seafood," *Puget Sound Notes* (May 1986).
51. Williams, L., and C. Krueger. "Assessment of Risks Associated with Eating Recreationally Harvested Puget Sound Seafood," *Puget Sound Notes* (Winter 1989).
52. U.S. Environmental Protection Agency. "Commencement Bay Nearshore/Tideflats Record of Decision," U.S. Environmental Protection Agency (1989).
53. Jacobs, L. A., R. C. Barrick, and T. C. Ginn. "Application of a Mathematical Model (SEDCAM) to Evaluate the Effects of Source Control on Sediment Contamination in Commencement Bay," in *Proceedings: First Annual Meeting on Puget Sound Research* (Seattle, WA: Puget Sound Water Quality Authority, 1988).
54. Parametrix, Inc. "St. Paul Waterway Remedial Action and Habitat Restoration Monitoring Report 1988–1989," Parametrix, Bellevue, WA (1990).
55. Hotchkiss, D. A. "Terminal 91 Short Fill Monitoring Results: Insights for the Future of Contaminated Dredge Material Disposal," Port of Seattle, Seattle, WA, unpublished (1989).
56. Chiou, J.-D., L. A. Jacobs, R. G. Fox, and W. Clark. "Recommended Sediment Impact and Recovery Zone Models," PTI Environmental Services, Bellevue, WA (1990).
57. U. S. Environmental Protection Agency. "Biological Criteria: National Program Guidance for Surface Waters," U.S. EPA Report 440/5-90-004 (April 1991).
58. Barnthouse, L. W., G. W. Suter, II, and A. Rosen. "Risks of Toxic Contaminants to Exploited Fish Populations: Influence of Life History, Data Uncertainty and Exploitation Intensity," *Environ. Toxicol. Chem.* 9:297–311 (1990).
59. Dinnel, P. A., and Q. J. Stober. "Methodology and Analysis of Sea Urchin Embryo Bioassays," Circular No. 85-3, University of Washington, Seattle, WA (1985).
60. Johns, D. M., T. C. Ginn, and D. J. Reish. "Interim Protocol for Juvenile *Neanthes* Bioassay," PTI Environmental Services, Bellevue, WA (1989).
61. Bulich, A. A., M. W. Greene, and D. L. Isenberg. "Reliability of the Bacterial Luminescence Assay for Determination of the Toxicity of Pure Compounds and Complex Effluents," in *Aquatic Toxicology and Hazard Assessment: Proceedings of the Fourth Annual Symposium* (Philadelphia: American Society for Testing and Materials, 1981).

62. Beckman Instruments. “Microtox® System Operating Manual,” Beckman Publication No. 015-555879, Carlsbad, CA (1982).

CHAPTER 17

The Effects of Contaminated Sediments in the Elizabeth River

R. J. Huggett, P. A. Van Veld, C. L. Smith, W. J. Hargis, Jr., W. K. Vogelbein, and B. A. Weeks

INTRODUCTION

The Elizabeth River is a highly industrialized estuary bordered by the cities of Norfolk and Portsmouth, VA. It is located on the southern shore of the James River near its confluence with the Chesapeake Bay (Figure 1). Bottom sediments of the Elizabeth contain a record of past and present chemical discharges and spills. Particularly significant in this respect is the severe contamination of the system by creosote. Since the turn of this century there have been four or five wood treatment facilities located on the rivers' shores, treating telephone poles, railroad ties, and pilings with creosote to control fungal wet and dry rot.[1] During this period accidental spills, point source, and chronic nonpoint inputs of the pesticide mixture have entered the river and accumulated in bottom sediments. Little attention was paid to environmental concerns before the 1970s, so few detailed records of events are available. Study of Elizabeth River sediments in midchannel shows a nearly exponential increase in sedimentary PAH content from the river mouth extending up the Southern Branch to about km 20, above which, concentrations fall off somewhat rapidly. Records show that at least two separate creosote spills occurred between 1963 and 1967.[2-4]

There are numerous other pollutants in the Elizabeth, ranging from trace metals emanating from heavy industries to advanced primary treated sewage effluent from

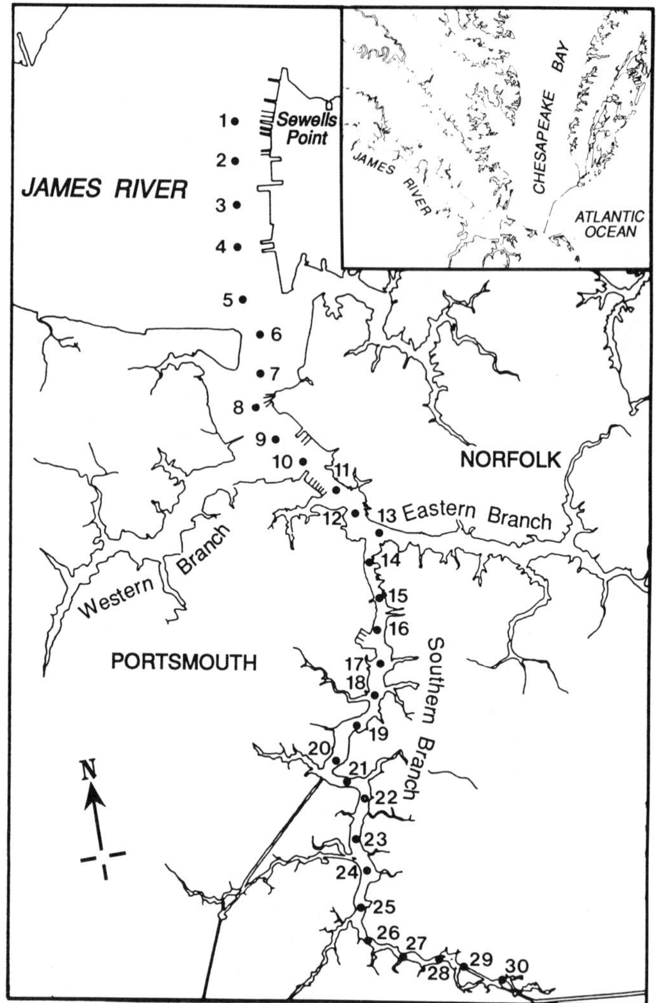

Figure 1. Map of Elizabeth River and Bay. Station numbers represent kilometers from Sewells Point.

the city of Norfolk. Hazardous substances normally associated with municipalities and harbors enter the system from nonpoint sources. The areal distributions of these contaminants, however, are different than that of the creosote compounds, and the biological responses reported in this manuscript are best correlated with the concentration of the creosote compounds.[5]

PHYSICAL DESCRIPTION OF THE SYSTEM

The Elizabeth River is a rather simple system consisting of a main stem and three major branches. It has been changed from the once typical marsh-lined estuary by several centuries of channel dredging, bulkheading, and filling. The most recent stage, the diking and filling of Craney Island (a dredged materials disposal area), has lengthened the mainstem of the river by several kilometers and isolated much of the Port Norfolk area and other port facilities as far as the Lafayette River from the James River. The present river, particularly the mainstem and the Southern and Eastern Branch portions, is characterized by a single deep central channel, fringed by shallows, tidal flats, and developed shorelines. The Eastern Branch divides the industrial center of the city of Norfolk and is lined by industry. The Southern Branch is the longest section of the river and is lined by industries and shipyards. It routes the major small boat traffic of the Intracoastal Waterway around the Dismal Swamp into the Albemarle Sound, North Carolina. The Western Branch is somewhat different. It joins the mainstem near the river's mouth and has multiple relatively shallow channels; its shoreline tends to have few industries, and natural marsh areas are abundant. The physical nature of the Elizabeth River is such that little flushing of contaminants occurs.[6] The tidal currents are relatively slow, and the freshwater influx is low due to canal locks on the upper river, which regulate flow. Dredging is responsible for most of the removal of contaminants in sediments, but this is operative only in and at the edges of the channel. The shipping channels are heavily used and are maintained by the Corps of Engineers. This maintenance consists primarily of removing shallow spots caused by slumping of channel edges, but this activity helps redistribute sediment fines and their associated organics.

CHEMICAL CONTAMINATION IN SEDIMENTS

Sampling and Analysis

Sediment samples collected over a 10-year period from 1980 to 1989 were obtained by gravity-core and by surface-grab devices. Aliquots of wet sediments were desiccated and then Soxhlet-extracted with dichloremethane. The resultant extracts were reduced in volume, and the PAH fractions were separated from biolipids and macromolecules by gel permeation chromatography followed by high-performance liquid chromatography. A normal phase column using gradient elution with hexane and dichloromethane was used to separate the PAH fractions from aliphatic and polar materials. The PAH-containing fraction was further separated by gas chromatography, and compound identifications were made by an Aromatic Retention Index marker system making reference to index tables of authentic standards and previously identified compounds. Quantitations were made by internal standard methods. Mass spectrometry was used to confirm assigned identifications and to make tentative identifications of unknowns and questionable peaks. A detailed description of the analytical methodology is given in our "Analytical Protocol."[7]

The samples were collected for a variety of purposes, including exploratory, intermediate, and long-term monitoring, as well as to directly support other related studies.[1-3,8-14] Because the methodologies of sampling and analysis remained comparable and because no major temporal trends were detected during the decade, it is reasonable to consider all sample sets throughout the river basin in a synoptic overview.

It is known from effluent studies[11,13] of industries having wastewater outfalls on the Elizabeth River that a wide variety of chemicals are entering the system. Many are volatile organic compounds (i.e., alkyl- and halobenzenes, haloalkanes, and haloalkenes), which rapidly evaporate from the water and seldom contact the sediment. Many other petroleum-derived organics stay at the air-water interface and contact sediments mainly at the intertidal zone. Analyses of submerged sediments show that the most abundant and frequent contaminants are polycyclic aromatic hydrocarbons (PAH) and polycyclic aromatic heterocyclic compounds. It is likely that the aliphatic hydrocarbons (i.e., alkanes and isoprenoid hydrocarbons) are also present in substantial abundance due to, among other sources, frequent petroleum spillage in the river. These have been examined previously with the conclusion that the variation of these concentrations is independent of PAH concentration.[4] Aliphatics are relatively nontoxic to biological systems, with the exception of direct physical contact of organisms with crude or refined oils. Therefore, the determination of its abundance and distribution was intentionally excluded in this discussion.

Results

The abundance of total PAH in sediments varies broadly with location. A maximum abundance of 15 g/kg (1.5%) was measured in the Southern Branch, with very high concentrations present at numerous other sites. Figure 2 displays the variation of benzo[a]pyrene, a representative PAH, with river location. The regions between stations 18–21 exhibit contamination many orders of magnitude greater than other locations. These are associated with past and present wood treatment industries and creosote use. A general trend of decreasing concentrations of PAH from this section of the Southern Branch toward the river mouth exists, particularly for samples taken in midchannel, though exceptions can be found.

The individual compounds that comprise the PAH fraction also vary in relative abundance between sites throughout the river. A list of 20 characteristic PAH compounds found in Elizabeth River sediments is given in Table 1, along with their maximum concentrations in the river and their mean concentration at various sites in the Southern Branch. Although hundreds more than these 20 PAH are found in the river sediments, most others (i.e., alkyl-substituted PAH) are less well characterized.

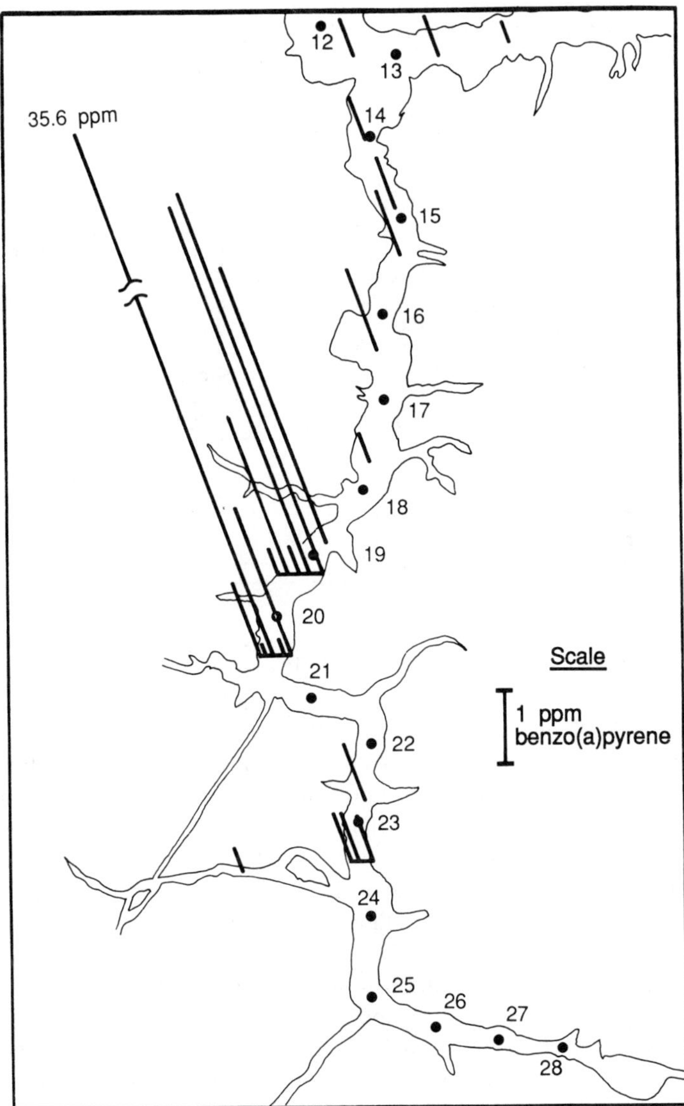

Figure 2. Sediment concentrations of benzo[a]pyrene along the Southern Branch of the Elizabeth River.

Aromatic nitrogen-, oxygen-, and sulfur-containing heterocycles, such as carbazole, dibenzofuran, dibenzothiophene, and naphthobenzothiophenes, are common. The 20 listed compounds represent about two thirds of the total PAH concentration in most sediment samples. Polychlorinated biphenyls, roughly characterized as having

Table 1. Twenty Elizabeth River Polycyclic Aromatic Hydrocarbons Ranked in Order of Maximum Observed Concentrations[a]

Compound	Maximum Conc.	Mean Conc.	% Samples (of 225)
Phenanthrene	2,430,000	40,000	96.88
Naphthalene	1,300,000	15,000	76.79
Anthracene	1,300,000	10,000	91.96
Fluoranthene	1,000,000	24,000	99.55
Fluorene	1,000,000	15,000	83.04
Acenaphthene	810,000	14,000	79.02
Naphthalene, 2-methyl-	720,000	8,400	58.93
pyrene	620,000	17,000	100.00
Naphthalene, 1-methyl-	390,000	5,800	64.29
chrysene	180,000	6,000	99.55
Benz(a)anthracene	140,000	4,300	93.75
Benzo(a)fluorene	120,000	4,500	98.21
Benzo(b)fluorene	120,000	4,600	97.32
Benzofluoranthene	110,000	5,700	98.66
Benzo(e)pyrene	56,000	2,500	98.21
Benzo(a)pyrene	50,000	2,800	99.55
Ideno(1,2,3-cd)pyrene	18,000	1,000	94.20
Perylene	14,000	710	97.77
Benzo(ghi)perylene	14,000	870	98.21
Dibenz(a,h)anthracene	2,500	200	69.64

[a] Concentration in μg/kg sediment.

congener distributions similar to Aroclor 1254- or Aroclor 1260-derived mixtures, were routinely measured in sediments, though in minor amounts (50–500 ppb normal range, and 2000 ppb maximum) relative to the PAH.

The compounds found in the highest absolute concentration are not always those with the highest mean concentrations throughout the river. Inspection of individual samples shows that there are major differences between samples when compounds are ranked in order of relative abundance. The first grouping of compounds (pyrene, fluoranthene, benzofluoranthenes, and chrysene) is broadly represented in all samples in the highest rank cluster (although they are not always the most abundant compounds at certain stations). The compounds in the second grouping (benzopyrenes, benzofluorenes, and benzanthracene) tend to occur at a frequency similar to compounds in the first grouping and in the same samples, but fall in the second-highest rank cluster. The major distinction between the first and second compound groups is in average overall abundance, with the first group being higher (Table 1). By contrast, the third grouping (phenanthrene, anthracene, acenaphthene, fluorene, naphthalene, and C-2 naphthalenes) is seldom represented in the highest rank cluster, and at many stations most of this group are in the lowest rank cluster. The compounds in this grouping are unexpectedly high, both in terms of maximum abundance and mean abundance (Table 1).

Regardless of the total PAH concentration, two archetypes can be seen to emerge as dominant poles: the "recent/unweathered or creosote type", recognizable by the predominance in rank abundance of low-molecular-weight species (i.e., naphthalene, fluorene, acenaphthene, phenanthrene, and anthracene, and many of their alkylated homologs). These compounds are very high in abundance at a few sites: so high in fact that these compounds show high mean abundance throughout the river, in spite of the fact that their abundance ranking is low at many sites. The presence of these few "creosote" sites tends to skew the perception of compound distribution through the river. The other "pyrogenic or weathered creosote" type is characterized by the virtual absence of components of the first type, and the dominance of high-molecular-weight PAH (i.e., pyrene, fluoranthene, benzofluoranthenes, chrysene, benzopyrenes, benzofluorenes, benzanthracene, and others). These latter components are also found in the first sediment type and in wood and coal tars and creosote, but in lower total abundance (though not necessarily in lower total actual concentrations). They are also universally found as products of incomplete combustion of carbonaceous fuels and can be spread by aeolean transport processes and by nonpoint-source runoff to estuarine sediments.

Estimation of the amounts of creosote, or even creosote-derived PAH, in the sediments cannot be done with great accuracy, as many unmeasurables and assumptions are required. The components measured in the dichloromethane-soluble fraction of a fresh-looking creosote inclusion isolated from sediment collected near station 20 only accounted for about 20% of the weight of material, using the described analytical methodology, though PAH contents of creosotes reported in the literature range as high as 80%.[15] Current knowledge does not allow distinction of PAH-derived from highly weathered creosote (i.e., creosote that has lost its low-molecular-weight PAH) from other general pyrogenic sources. Indeed creosote and other allied wood-preservative formulations are manufactured by dilution of certain destructive distillation products of coal (coal tars) with special proprietary mixtures of lower-molecular-weight aromatic distillates.[15] The exact composition of such materials may vary from batch to batch and with end-use formulation. These materials have many components in common with PAH formed in incomplete combustion of wood, petroleum, coal, and other fuels. It does, however, appear that most of the sediment samples examined from the Elizabeth River, which are not grossly and obviously contaminated by the lower-molecular-weight components of creosote, have far more PAH across the higher-molecular-weight spectrum from pyrene to benzo(ghi)perylene than would derive from a creosote source alone. This may be an artifact of differential weathering and biodegradation.[16] However, it may be evidence that the wood treatment facilities and their spills and effluents are not the only sources of higher-molecular-weight PAH in the sediments.

Conclusion

The contamination of Elizabeth River sediments can be understood by a combination of several contributing features. The first feature is the massive nearshore creosote contamination of a few localized sites in the Southern Branch. It appears that the creosote PAH may spread by several mechanisms. The first is by simple flow, a short-lived mechanism that must have occurred shortly after the spills, in which liquid creosote smothered major portions of the bottom and filled holes and crevices. Certain large sections of the Southern Branch of the river, in particular the east side of the river between river km 19.5 and 20.5, are extraordinarily high ($>1 \times 10^6$ ppb) in PAH. This area is adjacent to the sites of two defunct wood treatment plants. A second localized region with even higher sediment PAH concentrations ($>1.5 \times 10^7$ ppb) lies on the west side of the river near station 17 and is adjacent to a wood treatment facility still in operation. These sediments, heavily loaded with fresh or unweathered creosote, exert major local effects on the biological systems with which they come into direct contact, either through exposure or diet.

A pattern of diminished PAH concentration as a function of distance from these sites is present, though exceptions to this generalization are commonplace. Total PAH concentrations may be constant with sediment depth from surface to basement, though some core samples show great variation. In some instances where unusual values were measured, repeated sampling was done at nearby sites. Often quite different results were obtained, showing inhomogeneity on a very small scale. A major feature, particularly evident in certain cross-river transect sample sets, is the alternating concentration pattern: high near shore, low at channel edge, and high again in midchannel. This is apparently the result of two interacting mechanisms: the near-shore sediments are directly contaminated by creosote or PAH from industry or land-runoff, and channel maintenance dredging, usually employed at the channel edge, where contaminated sediments are physically removed.

The second significant feature controlling sediment contamination is the spreading throughout the study area by the turbulent transport (1) of smaller, weathered creosote globules that have broken away from the parent mass and undergone more extensive weathering to remove the more soluble and volatile components, and (2) of the weathered adsorbed creosote on fine-particulate matter. This fraction consists of high-molecular-weight residues containing the environmentally persistent portion of the creosote. The PAH distribution in sediments thus changes character increasingly, from the "creosote" archetype towards the "pyrogenic" archetype, with distance from creosote sources. The silt- and clay-size sediment, with its burden of PAH accumulated by contact with areas of massive creosote intrusion, is easily suspended in water where it behaves according to classic laws of turbulent dispersion. The total concentrations appear to show an exponential decrease with distance from the source(s).

BIOLOGICAL EFFECTS

External Lesions

Description and Relevancy of Assay

External lesions have been described in numerous reports on diseases of finfishes.[17,18] Many are known to be caused by viral, bacterial, fungal, protozoan and metazoan obligate, and opportunistic parasites. Increasingly, they have been associated with exposure to various anthropogenic environmental contaminants. Those most commonly reported in this connection are hyperemia of skin, fins, and eyes; fin erosion; integumental ulcerations; exophthalmia; and tumors of various locations and types. Recently others, such as corneal cloudiness and lens cataracts, have been reported with increasing frequency.[19] The most common external lesions, i.e. fin erosion and ulcerations, are often found in fish collected from severely contaminated environments and have been used as indicators of exposure to toxicant-laden waters and sediments. Other external lesions can serve the same purpose. Where abnormal numbers of possible biological agents (i.e., bacteria, fungi, copepods, and helminths) are observed in such lesions, their relative importance in lesion production must be established.

Sampling

To determine the effects of exposure to PAH and other sediment-contained contaminants in the Elizabeth River, a series of trawl collections was made between 1983 and 1988. Stations were selected to cover the entire 28-km length of the subestuary and to span the area of greatest contamination of sediments by PAH. Reference collections were made in the "cleaner" Nansemond River nearby and in the York River.

During one intensive sampling period (October 1983–November 1984), 11 stations between river Km 7–28 were occupied almost monthly. Single collections were made in October 1985, 1986, and 1988, occupying representative downriver, midriver, and upriver stations chosen from the same stations, to catch the migratory species before they left the subestuary and after exposure to the contaminants. Samples were taken along the edges of the channel, using a lined (1/2-in. stretch-mesh liner), 30 ft, semiballoon trawl in standard tows against and with the tide. All fish were sorted to species, counted, and visually examined for lesions.

Over 95,000 individuals of 49 finfish species were collected from the Elizabeth and some 16,900 of 20 species from the reference Nansemond. Eight dominant species representing migratory and endemic life cycles, as well as oceanic and estuarine distributions, were carefully examined for cataracts, fin erosion, integumental ulcerations, and other externally visible lesions. Lesion prevalences among 71,262 individuals of these species are presented in Table 2. Two of the eight species with one or more lesions (i.e., red hake, *Urophycis chuss*, and gizzard shad, *Dorosoma cepedianum*) occurred in relatively small numbers and were excluded from further consideration.

In addition to the trawl survey, mummichogs *(Fundulus heteroclitus)* were collected in minnow traps in the shallows and adjacent wetlands shoreward on both sides of Station 20, where PAH concentrations were known to be extremely high. Baited minnow traps were deployed during six collecting periods. The easternmost station (20-East) was nearest an abandoned creosote plant and active petroleum

Table 2. External Lesions Observed Vs. Total Individuals (Elizabeth River Collections, Oct.–Dec. 1983; Apr.–Nov. 1984)

Species	Total Examined	Number with Lesions	% with Lesions
Lens Cataracts			
Spot	42,434	1,254	2.96
Atlantic Croaker	8,475	403	4.76
Weakfish	5,980	184	3.08
Spotted Hake	2,925	47	1.61
Gizzard Shad	46	1	2.17
Subtotals	59,860	1,889	
Fin Erosion (Fin Rot)			
Spot	42,434	7	0.02
Atlantic Croaker	8,475	25	0.29
Weakfish	5,980	15	0.25
Hogchoker	10,744	178	1.66
Oyster Toadfish	631	40	6.34
Subtotals	68,264	265	
Ulcerations			
Spotted Hake	2,925	23	0.78
Red Hake	27	1	3.70
Subtotals	2,952	24	
Grand totals (individuals, all species)	71,262	2,178[a]	

[a] Some individuals had more than one type of lesion; hence, they occur more than once in total.

transfer and storage sites; 20-West was across the river. No reference mummichog collections were made. A total of 398 mummichogs were examined for external lesions. External tissues from 37 individuals were mounted in paraffin, (ca., 6 μm sections), stained (Harris' hematoxylin, Mason trichome, and periodic acid-Schiff), and examined microscopically.

Results

The prevalence of three lesions by species and station for fish taken during 1983 and 1984 when most of the stations were occupied are given in Tables 2 and 3. Five species exhibited lens cataracts. The four, collected in sufficient numbers to be considered significant, were Atlantic croaker, *Micropogonias undulatus*, 4.76% affected; weakfish, *Cynoscion regalis*, and spot, *L. xanthurus*, affected nearly equally with 3.08 and 2.96%, respectively; and spotted hake, *U. regius*, with 1.61%. The prevalences of cataracts was generally highest at the midriver stations (20 to 23) where sediment PAH contamination is highest. Cataract prevalence decreased in fish collected upriver and downriver from these sites, in a manner that reflects the sediment PAH gradient. Most hake were collected from the downriver stations. This probably reflects the uneven distribution of this marine species, which prefers high-salinity waters downriver.

Five species bore signs of fin erosion. Most affected were oyster toadfish, *Opsanus tau* (6.33%), and hogchoker, *Trinectes maculatus* (1.66%). The three pelagic sciaenids were far-less affected. Prevalence of fin erosion for all species followed the

Table 3. Prevalence, External Lesions by Stations[a]

Species	Station OBS.	7 No. %	9 No. %	10 No. %	11 No. %	14.5 No. %	17 No. %	20 No. %	21.5 No. %	23 No. %	24 No. %	28 No. %	Totals No. %
Weakfish	T	667	447	862	1036	986	576	244	373	358	150	281	5980
	C	2 0.3	0 0	2 0.2	15 1.4	6 0.6	14 2.4	19 7.8	54 14.5	63 17.6	2 1.3	7 2.5	184 3.1
	FE	0 0	0 0	2 0.2	0 0	0 0	2 0.4	2 0.8	8 2.1	1 0.3	0 0	0 0	15 0.2
Croaker	T	652	3346	614	650	463	398	338	368	512	316	818	8475
	C	15 2.3	5 0.1	12 2.0	13 2.0	28 6.0	61 15.3	39 11.5	82 22.3	118 23.0	11 3.5	19 2.3	403 4.8
	FE	1 0.2	0 0	0 0	1 0.2	4 0.9	2 0.5	13 3.8	3 0.8	1 0.2	0 0	0 0	25 0.3
Spot	T	3322	8712	4030	2176	3132	3115	2361	2803	3302	4153	5326	42434
	C	3 0.1	4 0.0	13 0.3	36 1.6	99 3.2	114 3.7	188 8.0	217 7.7	264 8.0	169 4.1	147 2.8	1254 3.0
	FE	1 0.0	0 0	0 0	0 0	3 0.1	1 0.0	1 0.0	1 0.0	0 0	0 0	0 0	7 0.0
Toadfish	T	112	122	72	38	59	107	41	41	10	23	6	631
	FE	0 0	0 0	5 6.9	1 2.6	3 5.1	14 13.1	10 24.4	6 14.6	1 10.0	0 0	0 0	40 6.3
Hogchoker	T	638	1120	1824	1485	618	875	786	990	1159	645	604	10744
	FE	1 0.2	2 0.2	4 0.2	5 0.3	6 1.0	26 3.0	44 5.6	66 6.7	22 1.9	1 0.2	1 0.2	178 1.7
Spotted Hake	T	771	561	788	266	211	160	99	38	22 0.8	9	0	2925
	C	6 0.8	3 0.5	7 0.9	4 1.5	6 2.8	11 6.9	7 7.1	2 5.3	0 0	1 11.1		47 1.6
	U	0 0	1 0.2	7 0.9	8 3.0	1 0.5	4 2.5	2 2.0	0	0 0	0 0		23 0.8

[a] Symbols: T = total individuals; C = cataracts; FE = fin erosion; U = ulceration.

same trend observed with cataracts; the highest prevalence occurred at midriver stations where sediment PAH are present at the highest concentrations.

Among the collections included in Table 2, integumental ulcerations were limited to the hakes. Although hake exhibited an uneven distribution, the trend towards highest prevalence in midriver fish was also observed for these lesions.

Of the 398 mummichogs (*F. heteroclitus*) examined for external lesions, 42 (10.6%) bore lesions of several types. Twenty-six, or 6.5%, had external parasites and associated lesions. Of those without parasites or parasite-related lesions, eight (2.0%) had non-neoplastic, inflammatory lesions attributed primarily to mechanical injury. Eight bore neoplasms. One individual had separate inflammatory and neoplastic lesions. All of the inflammatory and neoplastic lesions occurred in the anterior halves of the animals, around the head. Six of the neoplasms were diagnosed as either frank or incipient papillomas of the oral area, usually of the mandible. The second type of neoplasm found was a capillary type hemangioendothelioma in the left gill chamber, involving the operculum, pharynx, gills, and, probably, the oral mucosa. A single neoplasm of the distal end of the right pectoral fin was diagnosed as a malignant schwannoma. Prevalence of all externally visible neoplasms was 2.0%.[19]

Conclusions

The highest prevalence of all lesions in all species at the most heavily contaminated sites suggests a chemical etiology of these lesions. These conclusions on the toxic nature of contaminants in the sediments from the midriver sections of the Southern Branch of the Elizabeth are reinforced by previous studies on histopathological lesions observed in the gills of the same fish.[20] Also, laboratory exposure of spot, *L. xanthurus,* to sediments from Station 20 in a flow-through system caused deaths within 7 days.[21,22] Survivors displayed severe fin erosion and integumental ulcerations and cataracts. None of these three externally visible lesions was produced among the control fishes whose tanks contained clean sediments or the overflow from clean sediments.

All external neoplasms in feral animals were found on mummichogs from the shallow waters and tidal wetlands of that reach of the river where the sediments are heavily contaminated by PAH.[23] Earlier reports, such as those of Peters[17] and Overstreet,[18] related the occurrence of tumors to environmental contamination. Their findings of positive connections support our suspicion that the neoplasms noted herein on mummichogs were associated with exposure to pollutants.

Internal Lesions

Description and Relevance

Numerous laboratory and field studies indicate that exposure to toxic chemical agents induces the development of characteristic pathologic disorders in aquatic organisms. In fish these lesions can develop in almost every organ system; however, the liver appears to be one of the most common target organs. Being the predominant detoxifying organ, the liver is a prime target for the damaging effects of many xenobiotic chemical contaminants. For example, a spectrum of hepatocellular lesions, including neoplastic changes, develops in laboratory fish that have been exposed to potent chemical carcinogens.[24-28] Similarly, field surveys conducted in chemically contaminated coastal environments indicate that certain species of fish have very high prevalences of hepatic neoplasms and related tissue lesions.[29-32] These liver pathologies, as a group, are considered to be indicative of exposure to chemical carcinogens; and in feral fish from such heavily contaminated sites as those in Puget Sound, Washington, Boston Harbor, Massachusetts, and the lower Hudson River, New York, they are perhaps the most convincing indicators of exposure to xenobiotic chemical contaminants. In the English Sole, *Parophrys vetulus*, from urbanized areas of Puget Sound, the spectrum of liver lesions has been correlated with high sediment concentrations of PAH, the induction of xenobiotic metabolizing enzyme systems (i.e., cytochrome P-450 monooxygenases), and the elevation of PAH metabolites in the bile of these fish.[33]

Because the effects of exposure to toxic environmental agents are frequently expressed as characteristic tissue lesions, histopathologic investigations are gaining acceptance in pollution monitoring programs. These methodologies offer several advantages: (1) tissue-processing procedures are routine, quick, and relatively inexpensive; (2) effects in multiple-target organs are easily and rapidly evaluated; (3) acute as well as chronic effects can be distinguished; (4) mechanisms of action can sometimes be inferred, especially when concurrent biochemical and physiological measurements are taken; and (5) correlations with potential effects at higher levels of biological organization (i.e., organismal and population levels) are sometimes possible.

Sampling

Initial attempts to apply histopathological methods to fish disease surveys in the Elizabeth River focused on the identification of "sensitive" species, and comparisons of lesion prevalences in specimens from contaminated sites of interest with those from uncontaminated reference localities. A recent study focused on the small nonmigratory teleost, *Fundulus heteroclitus*.[37] Mummichogs were collected at three localities: Wilson Creek, an uncontaminated reference site off Mobjack Bay; 17-East, a moderately contaminated site in the Elizabeth River; and 17-West, an Elizabeth River site heavily contaminated with creosote. Station 17-West is located in a small creek adjacent to an operating wood treatment facility. Station 17-East is directly across the river, less than 600 m away. Sediment PAH concentrations at Wilson Creek, 17-East, and 17-West are 3, 61, and 2200 mg/kg dry sediment, respectively. Specimens were collected using standard minnow traps. Individuals were anesthetized with Tricaine methanesulphonate (MS-222), measured, weighed, and necropsied. To facilitate the morphological evaluation of entire fish, the head, a section of the body wall containing the pronephros, and all visceral organs were fixed for 48 h in Bouin's fluid. Tissues were subsequently processed by routine methods for paraffin histology; sections were cut at 5 μm and stained with Harris' hematoxylin and eosin.[34]

Results

Past histopathological surveys in the Elizabeth River have indicated that certain fish populations are adversely affected by chronic exposure to xenobiotic chemical contaminants. Hargis and Zwerner[19] associated the occurrence of gill lesions in several species of ER fish with exposure to PAH-contaminated sediments. Also, eye lens cataracts in ER sciaenid fishes were attributed to exposure to high sediment PAH concentrations.[20] Similar eye lesions were induced by laboratory exposures of spot to heavily PAH-contaminated ER sediments.[21] Degenerative lesions in the kidney of five estuarine teleosts inhabiting the ER were recently reported.[36] Although renal lesions also occurred in fish from the uncontaminated reference sites, they were significantly more prevalent in ER fish. These higher lesion prevalences were attributed to the effects of petroleum hydrocarbons, although other etiological factors could not be entirely ruled out. Mummichogs, inhabiting 17-West adjacent to an active wood treatment facility, exhibit an extremely high prevalence of toxicopathic liver lesions.[37] Over 70% of the adult fish (TL >75 mm) from this site exhibited putative precancerous hepatocellular lesions, whereas 35% had hepatic neoplasms. Additionally, these fish exhibited a variety of extrahepatic neoplasms, including renal, pancreatic, lymphoid, and vascular tumors. In contrast, adult mummichogs from the two less-contaminated reference sites did not exhibit this spectrum of toxicopathic tissue lesions. Figure 3 compares liver morphology of reference fish with that of fish from 17-West. The livers of reference fish (Figure 4a) were pale tan to reddish brown and exhibited a homogeneous consistency. Histologically, reference fish demonstrated a normal hepatocellular architecture (Figure 4b). In contrast, the livers of fish from the 17-West site exhibited a spectrum of gross pathologic changes (Figure 4c). Histologically, alterations in this population of fish included hepatocellular carcinomas (Figure 4d), altered hepatocellular foci (Figure 4e), and proliferative bile duct lesions (Figure 4f).

Conclusions

These findings suggest a positive association between exposure to xenobiotic chemical contaminants and the wide range of adverse effects manifested at the histological level in Elizabeth River fish. This association is especially strong between high sediment concentrations of wood preservatives (creosote) at 17-West and the greatly elevated prevalence of hepatocellular lesions in mummichogs inhabiting that site. Several studies support these observations. Creosote, an environmentally persistent chemical mixture, is mutagenic[38,39] and induces cancer in rodents.[40] Although studies designed to test the carcinogenicity of creosote in fish have not been conducted, high prevalences of hepatic neoplasms in English Sole have been reported from a creosote-contaminated harbor in Puget Sound.[41] Additionally, laboratory studies have recently demonstrated that certain PAH are carcinogenic to fish.[24,26,27] These studies support the hypothesis that the pathologies observed in ER mummichogs are caused by chronic exposure to xenobiotic chemical contaminants; probably certain constituents of creosote. Establishment of a direct cause-and-effect relationship between creosote exposure and liver cancer in the mummichog will require experimental exposure studies, some of which are presently being conducted at our facility.

Cytochrome P-450

Description and Relevance

Aquatic organisms can accumulate lipophilic organic compounds during exposure to contaminated water, food, and sediment. Following absorption, the fate and effects of these compounds is largely governed by their susceptibility to transformation by various xenobiotic metabolizing enzyme systems. The cytochrome P-450-dependent monooxygenase system is the primary enzyme system involved in foreign compound metabolism.[42] Metabolites produced by monooxygenation are generally more water soluble, less toxic, and more readily eliminated than the parent, although activation of compounds to chemical carcinogens can also occur.[43]

The P-450 monooxygenases are a family of enzymes involved not only in the metabolism of foreign compounds, but in the metabolism of many endogenous compounds (e.g., steroid hormones) as well. A common feature of the system is that exposure to foreign compounds can result in induction (elevation) in the levels and activity of specific forms of P-450.[44,45] Exposure to PAH or to specific coplanar PCBs, for example, can result in the induction of a specific form(s) of P-450 active in PAH metabolism. In fish, the major PAH-inducible form has been designated P-450E,[46] P-450c,[47] or P-450LM$_{4b}$,[48] depending upon the laboratory where purification and characterization of the enzyme were performed.

P-450E catalyzes aryl hydrocarbon hydroxylase (AHH) and ethoxyresorufin 0-deethylase (EROD) activities. Measurements of these activities are used routinely to evaluate the response of P-450E to PAH exposure. More recently, antibodies to purified P-450E and related forms have been produced,[47-49] which allow detection of the enzyme by Western blot analysis.[50] Because P-450E catalyzes EROD activity, high coefficients of correlation between "E" and EROD activity, such as that shown in Figure 4, are often observed. Therefore, measurement of either EROD or "E" alone is often sufficient to provide evidence for induction.

P-450E is sensitive to levels of contaminants encountered in the environment and has been used as an indicator of exposure to PAHs and PCBs in populations of feral

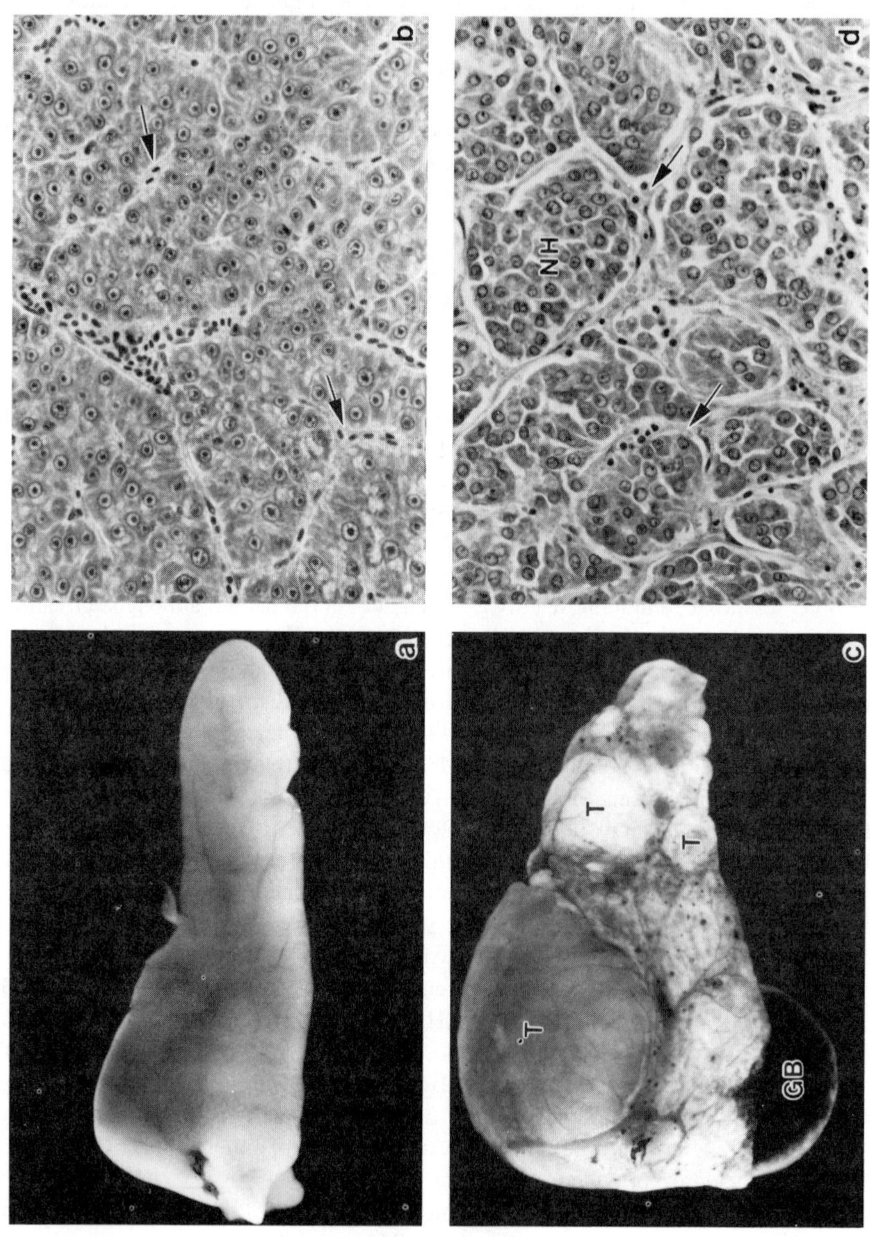
a
b
c
d
NH
T
T
T
GB

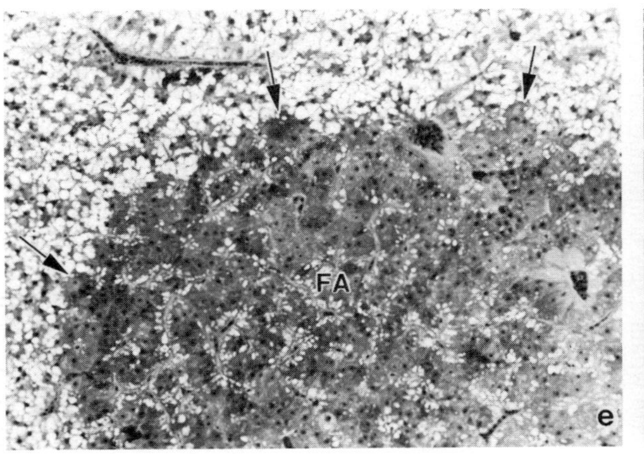

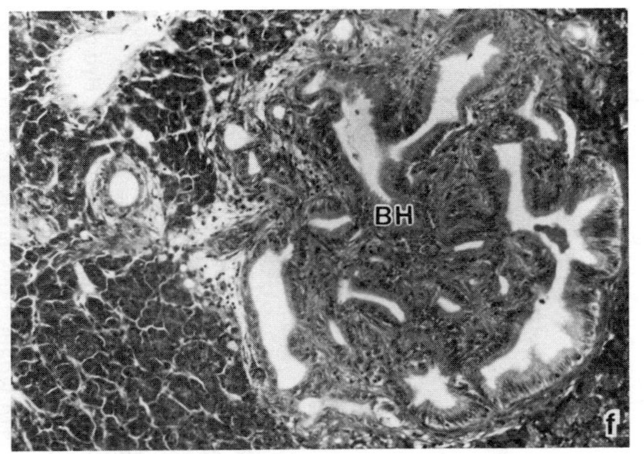

Figure 3. Gross and histological changes in liver of mummichog *(Fundulus heteroclitus)*. (a) Liver of reference fish; (b) histological section of reference liver (arrows = hepatic sinusoids); (c) liver of 17-West mummichog, (T = tumors, GB = gall bladder); (d) histological section of hepatocellular carcinoma (NH = neoplastic hepatocytes, arrows = connective tissue stroma); (e) altered hepatocellular focus (FA) in 17-West mummichog (arrows = border of lesion); (f) proliferative bile duct lesion (BH).

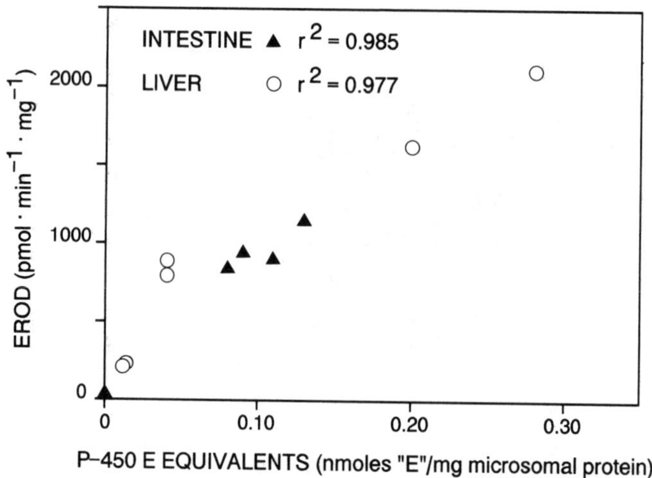

Figure 4. Relationship between EROD activity and immunodetectable P-450E in fish intestine and liver. (From Van Veld, P. A., D. J. Westbrook, B. B. Woodin, R. C. Hale, C. L. Smith, R. J. Huggett, and J. J. Stegeman. "Induced Cytochrome P-450 in Intestine and Liver of Spot *(Leiostomus xanthurus)* from a Polycyclic Aromatic Hydrocarbon Contaminated Environment," *Aquat. Toxicol.* 17:119–132, (1990). With permission.)

fish.[51] Induction of P-450E can be detected within a few hours following exposure and occurs in a dose-response manner.[52] P-450E induction proceeds effects that occur at higher levels of biological organization (e.g., at the tissue level) and can therefore be used as an early warning signal of exposure. A quantitative relationship between monooxygenase induction and higher level effects (e.g., cancer, reproductive success) is not well understood. However, considering the role that this enzyme plays in the activation of chemical carcinogens and steroid hormone metabolism, the potential for these and other such relationships must be considered.

Sampling and Methods

To evaluate the response of P-450E to the PAH gradient in the Elizabeth River, spot, *Leiostomus xanthurus*, a marine teleost, were collected by otter trawl from four stations in the Southern Branch of the Elizabeth River and from two less-contaminated sites in the lower bay. Sediment PAH concentrations at the sites spanned five orders of magnitude from 9 to 96,000 μg PAH per kilogram dry sediment. Fish were immediately submerged in a seawater/ice bath; liver and intestines were removed from the fish on board ship and frozen in liquid nitrogen. In the laboratory, microsomes were obtained by differential centrifugation and were frozen in liquid nitrogen until used. Immunodetectable P-450E was determined by

Western blot analysis, as described in the original paper,[53] using purified P-450E as standards. EROD activity was determined spectrophotometrically, as previously described.[54] Total microsomal protein was measured by the method of Bradford,[55] using bovine serum albumin as standards.

Results

The relationship between sediment PAH concentrations and P-450E is shown in Figure 5. For liver samples there was a strong positive correlation between P-450E and sediment PAH. Liver EROD activity exhibited a similar trend.[53] In intestine, P-450E and associated EROD activity were at, or near, detection limits at the two sites where sediment PAH concentrations were lowest, but were elevated 80 to 100-fold at the four most heavily contaminated sites. Thus, although the data for intestines did not track the PAH gradient as effectively as liver, the fold-increase in P-450E and EROD activity exceeded that of liver. At the moderately contaminated sites, gut P-450E and EROD were approximately equal to that of liver.

Conclusions

The results from the P-450E study indicate the presence of bioavailable PAH in the Elizabeth River fish. Although there may be other environmental contaminants that may have influenced the observed response, PAH are the dominant contaminant in this system. The strong correlation between sediment PAH concentrations and P-450E levels/activity suggest that PAH are the major contaminants contributing to the observed response. Similar results have been observed in a number of field studies.[56]

Most studies of the response of monooxygenase activity to environmental toxicants involve the liver. However, P-450E and related enzymes are present in other organs, including those proximal to the environment, such as the gill[57] and gut.[58] Measurement of activities in extrahepatic organs may yield information on the routes of uptake of environmental agents such as PAH. For example, the strong induction of P-450E in the intestine of Elizabeth River spot suggest that diet may be an important route of uptake in this system. Further studies are needed in the relationship between dose, route of exposure, enzyme induction, and the occurrence of PAH-related disease in Elizabeth River fish.

Immunology

Description and Relevance

The immune system of fish, as in mammals, consists of two major elements, the innate immune system and the adaptive immune system, both of which have considerable interaction. The cells of the innate system include phagocytes that

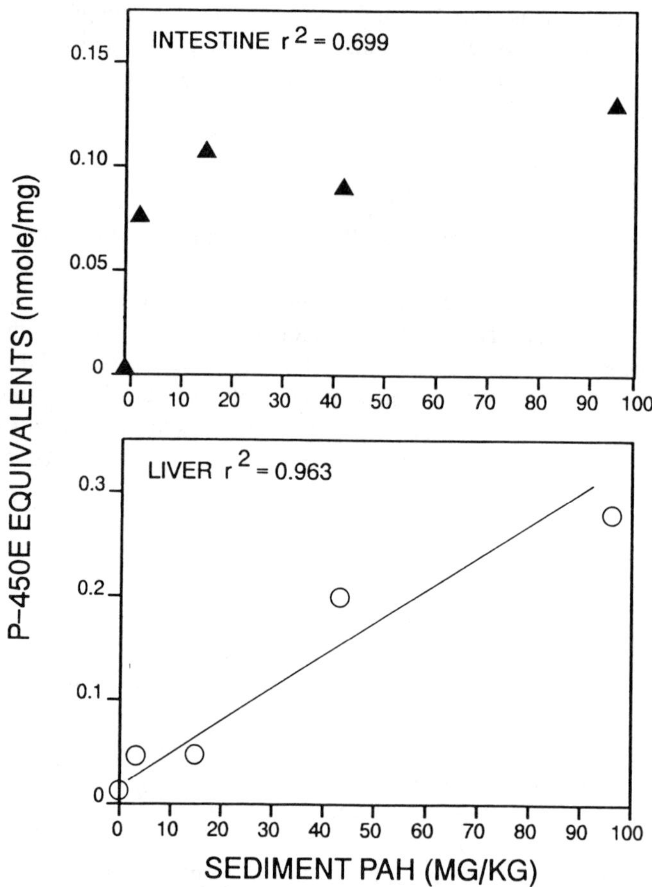

Figure 5. Relationship between Elizabeth River sediment PAH and P-450E equivalents in fish intestine and liver. (From Van Veld, P. A., D. J. Westbrook, B. B. Woodin, R. C. Hale, C. L. Smith, R. J. Huggett, and J. J. Stegeman. *Aquat. Toxicol.* 17:119–132 (1990). With permission.)

consist of polymorphonuclear leukocytes and macrophages in mammals, and the corresponding cells in fish. The latter constitute a first line of defense against invading organisms. There is evidence that the immune system is compromised when fish are exposed to toxic chemicals.[60-62] Such a response could cause fish to be more susceptible to infectious agents and neoplastic cells.

In order for macrophages to perform their function, they must be capable of reaching the target cell (chemotaxis), engulfing the foreign agent (phagocytosis), and killing or immobilizing the agent either by enzymes or by active oxygen radicals. Several of these activities are amenable to measurement in laboratory tests.

Sampling and Methods

Collections of fish were made in the Southern Branch of the Elizabeth River, in the area near site 20 where PAH contamination is most pronounced. Reference fish were collected from the relatively nonpolluted York River. Assays were performed on the macrophages isolated from these fish. The species of fish studied, spot, *Leiostomus xanthurus* (10 to 100 g), and hogchoker, *Trinectes maculatus* (10 to 60 g), were captured by trawl net on several occasions during October 1983 through October 1986. Fish were maintained in tanks of running York River water at ambient temperature until use (no longer than 7 days) and were fed trout chow daily.

Macrophages were obtained from fish by removal of the kidneys, especially the pronephros. Kidneys from spot and hogchoker were pooled (5 to 10 fish per experiment) and macerated in Minimum Essential Medium (MEM). Macrophages were then concentrated by Percoll density gradient centrifugation. Macrophages from the oyster toadfish *Opsanus tau* were also obtained in virtually pure culture by peritoneal lavage.

Chemotaxis was measured by quantitating the percent of macrophages that moved through the pores of a membrane filter toward a stimulus, using a modification of the Boyden double-chamber filter technique.[61] Macrophages were placed in the upper chamber on the surface of a Nuclepore membrane filter with either an 8-μ or 5-μ pore size. The lower chamber contained heat-killed *Escherichia coli* as the chemotactic stimulus. Chemotactic activity was measured by counting the number of macrophages on the upper and lower surfaces of the membrane filter in each field, using a light microscope. At least 100 macrophages were counted on each filter.

Two techniques for assaying the engulfment of foreign cells (phagocytosis) were used. In the first method, a macrophage suspension in MEM (1×10^6 cells) was placed on cover slips in Leighton tubes and allowed to settle and adhere for 60 minutes at 10°C. A suspension of formalin-killed *E. coli* (1×10^8 cells per 0.5 mL) was added to the monolayer of macrophages and incubated at 10°C, with gentle orbital shaking. Individual cover slips were removed at 30-min intervals for 150 min, gently rinsed, and stained with Wright's stain. Phagocytosis was measured by the microscopic enumeration of macrophages with and without endocytosed bacteria and was expressed as percent phagocytosis. At least 100 phagocytes were counted in random fields.[62]

Another method for quantifying engulfment measures the quantity of cells engulfed rather than the fraction of phagocytes that engulf the cells. Yeast cells were stained with Congo red dye and rinsed clear of excess dye by repeated dilution and centrifugation; then the stained cells were exposed to the macrophages to be tested. After a period of incubation, the macrophages were centrifuged to separate them from free yeast cells, rinsed, and then lysed to release the dye contained in the engulfed cells. The dye concentration was read spectrophotometrically.[63]

Phagocytic activity was also measured by quantifying the chemiluminescence produced by increased synthesis of reactive oxygen intermediates (ROI) in activated

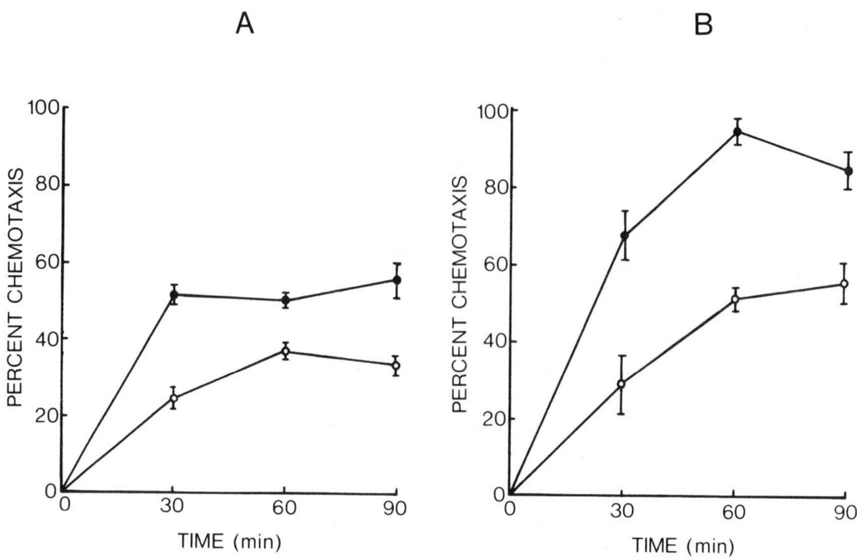

Figure 6. Macrophage chemotaxis in (A) spot and (B) hogchoker from the Elizabeth River and the York River. ● = control fish; ○ = Elizabeth River fish. Each value represents the mean + SEM of four experiments using 6–10 fish per experiment.[16]

macrophages. Chemiluminescence was measured in a liquid scintillation counter in minivials containing macrophages, zymosan as a stimulus, and luminal to amplify the chemiluminescent response. Vials were prepared in triplicate with macrophages from York River spot (controls) and Elizabeth River spot. Chemiluminescence in each vial was measured at 1-min intervals beginning with the addition of the zymosan stimulant. An additional experiment was conducted using York River spot exposed in laboratory aquaria to Elizabeth River sediments. These fish were maintained in 60-gal tanks of filtered, temperature-controlled, flowing seawater and exposed to 10 mg/L Elizabeth River or York River sediment for 80 days. Sediments from the Elizabeth River contained 8 to 10 ppm total resolved PAH, while sediments from the York River contained negligible levels of PAH. Activity in counts per minute (cpm) was plotted against time after stimulation.[64]

Results

Chemotactic activity of macrophages from spot, *Leiostomus xanthurus,* and hogchoker, *Trinectes maculatus,* is shown in Figure 6. Maximum values were reached by 90 min, but activity was significantly reduced in Elizabeth River fish.[61]

Figure 7 shows the effects of the Elizabeth River environment on phagocytosis by macrophages from spot and hogchoker. Phagocytosis was significantly reduced in both species[62] from the Elizabeth River. Macrophages from the oyster toadfish taken

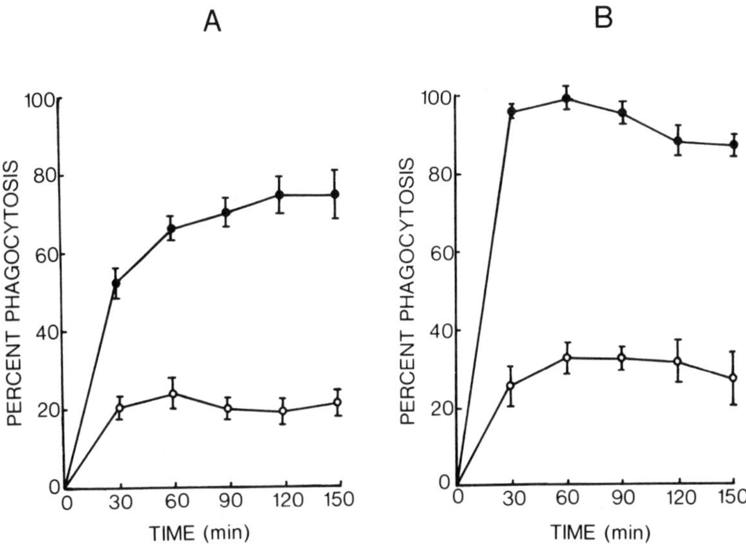

Figure 7. Macrophage phagocytosis in (A) spot and (B) hogchoker from the Elizabeth River and control waters. ●= control fish; ○= Elizabeth River fish. Each value represents the mean + SEM of four experiments using 5–10 fish per experiment.[17]

from stations 4, 12, and 20 in the Elizabeth River showed that the greatest reduction in phagocytosis occurred in fish from stations 12 and 20, i.e., those with highest PAH.[63]

Figure 8A shows the zymosan-stimulated chemiluminescent (CL) response of macrophage suspensions from spot captured in the Elizabeth River and in the York River. Macrophages from the Elizabeth River fish showed no significant response. Figure 8B shows a similar decrease in the zymosan-stimulated CL response of macrophages from the York River spot exposed to Elizabeth River sediments in laboratory aquaria.[64]

Conclusions

Several in vitro macrophage function tests (chemotaxis, phagocytosis, and chemiluminescence) provide information about the integrity of the nonspecific immune response in fish. There is considerable indirect evidence that chronic subclinical levels of toxicants increase the susceptibility of fish to diseases caused by microbial pathogens and neoplastic cells.[64,65]

The normal host defense mechanism against disease in fish involves humoral and cell-mediated immunity, and macrophages are an important component of the cellular immune system, protecting the host by eliminating foreign material. Three functional aspects of macrophage activity (chemotaxis, phagocytosis, and

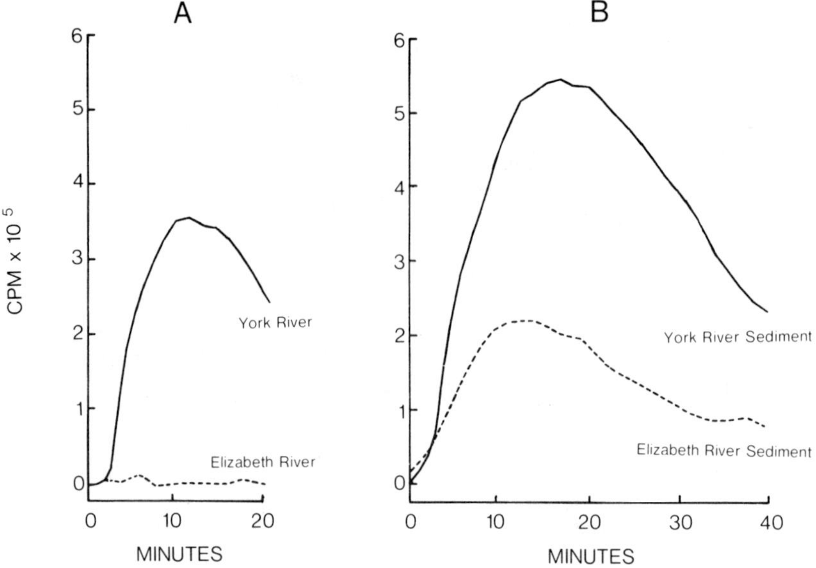

Figure 8. (A) Zymosan-stimulated CL responses of macrophages from spot captured in the Elizabeth River and in the York River; (B) zymosan-stimulated CL response of macrophages from York River spot exposed in the laboratory to Elizabeth River sediment.[18]

chemiluminescence) were markedly reduced in Elizabeth River fish as compared to clean water controls. Furthermore, experiments using control fish exposed to Elizabeth River sediments in the laboratory exhibited the same reduced macrophage activity. Since Elizabeth River sediments contain high levels of PAH in the reaches of the Elizabeth River in which the fish were caught, it is probable that these substances are responsible for the reduced immunological activities. The immunotoxic effects may be reversible, as indicated by the return to control levels after the fish were held in clean water for 3 to 4 weeks. The results reported here suggest that these assays can monitor exposure of fish prior to the occurrence of disease outbreaks.

REFERENCES

1. Lu, M. Z. "Organic Compound Levels in a Sediment Core from the Elizabeth River of Virginia," M.S. Thesis, College of William and Mary, Williamsburg, VA (1982).
2. Smith, C. L., M. Unger, and R. J. Huggett, "Chemistry Section of Biological and Chemical Impacts Related to New Energy Activities in the Elizabeth River, Virginia: Recommendations for Management and Mitigation," Coastal Energy Impact Program — National Oceanic and Atmospheric Administration Report (1984).

3. Bieri, R. H., C. Hein, R. J. Huggett, P. Shou, H. Slone, C. L. Smith, and C. W. Su. "Polycyclic Aromatic Hydrocarbons in Surface Sediments from the Elizabeth River Subestuary," *Int. J. Environ. Anal. Chem.* 26:97–113 (1986).
4. Merrill, E. G., and T. L. Wade. "Carbonized Coal Products as a Source of Aromatic Hydrocarbons to Sediments from a Highly Industrialized Estuary," *Environ. Sci. Technol.* 19:597–603 (1985).
5. Huggett, R. J., M. E. Bender, and M. A. Unger. "Polynuclear Aromatic Hydrocarbons in the Elizabeth River, Virginia," in *Fate and Effects of Sediment-Bound Chemicals in Aquatic Systems*, K. L. Dickson, A. W. Maki, and W. Brungs, Eds., (New York: Pergamon Press, 1987), pp. 327–341.
6. Fang, C. S., C. S. Welch, and H. H. Gordon. "A Surface Circulation Study in Middle Elizabeth River," report prepared by the Virginia Institute of Marine Science for NUS (1975).
7. Virginia Institute of Marine Science. "Analytical Protocol for Hazardous Organic Chemicals in Environmental Samples" (1991).
8. Bieri, R. H., C. Hein, R. J. Huggett, P. Shou, H. Slone, C. L. Smith, and C. W. Su. "Toxic Organic Compounds from the Elizabeth and Patapsco Rivers and Estuaries," prepared by Virginia Institute of Marine Science, Gloucester Point, VA, under contract with U.S. Environmental Protection Agency (1982).
9. deFur, P. O. "Sediment Organic Chemistry — Tributary Monitoring," Virginia State Water Control Board Report (1985).
10. Smith, C. L., P. O. deFur, and R. J. Huggett. "Analysis of Polycyclic Aromatic Hydrocarbons in Elizabeth River Sediments," U.S. Army Corps of Engineers Data Report (1985).
11. deFur, P. O., and C. L. Smith. "Analysis of Effluents and Associated Sediments and Tissue for Toxic Organic Compounds," Virginia State Water Control Board Final Report (1986).
12. Smith, C. L., P. O. deFur, and R. J. Huggett. "Analysis of Polycyclic Aromatic Hydrocarbons in Elizabeth River Sediments," Virginia State Water Control Board Data Report (1987).
13. Hale, R. C., and C. L. Smith. "Analysis of Effluents and Associated Sediments and Tissue for Toxic Organic Compounds," Virginia State Water Control Board Final Report (1988).
14. Greaves, J."Elizabeth River Long-Term Monitoring Program — Phase I — Analysis of Pollutants in Sediments and Blue Crab (*Callinectes sapidus*) Tissues," Virginia State Water Control Board Final Report (1989).
15. Lorenz, L. F., and L. R. Gjovic. "Analyzing Creosote by Gas Chromatography: Relationship to Creosote Specifications," *Proc. Amer. Wood Preserv. Assoc.* 68:32–44 (1972).
16. Enzminger, J. D., and R. C. Ahlert. "Environmental Fate of Polynuclear Aromatic Hydrocarbons in Coal Tar," *Environ. Technol. Lett.* 8:269–278 (1987).
17. Peters, N. "Fischkrankhesten und Gewasserbe Lastung im Kustenberich (Fish Disease and Pollution in Coastal Areas)," *Verhandl Deutschen Zoolog. Geseeschaft* 1981:16–80 (1981).
18. Overstreet, R. M. "Aquatic Pollution Problems, Southeastern U.S. Coasts: Histopathological Indicators," *Aquat. Toxicol. (Amsterdam)* 11:213–239 (1988).

19. Hargis, W. J., Jr., and D. E. Zwerner. "Some Histologic Gill Lesions of Several Estuarine Finfishes Related to Exposure to Contaminated Sediments," in *Proceedings: Understanding the Estuary,* Chesapeake Research Consortium Inc., Gloucester Pt., VA, pp. 474–487 (1988).
20. Hargis, W. J., Jr., and D. E. Zwerner. "Effects of Certain Contaminants on Eyes of Several Estuarine Fishes," *Mar. Environ. Res.* 24: 265–270 (1988).
21. Hargis, W. J., Jr., M. H. Roberts, Jr., and D. E. Zwerner. "Effects of Contaminated Sediments and Sediment-Exposed Effluent Water in an Estuarine Fish: Acute Toxicity," *Mar. Environ. Res.* 14:337–354 (1984).
22. Roberts, M. H., Jr., W. J. Hargis, Jr., C. J. Strobel, and P. L. DeLisle. "Acute Toxicity of Sediments Contaminated with Polycyclic Aromatic Hydrocarbons to the Fish *Leiostomus xanthurus*," *Bull. Environ. Contam. Toxicol.* pp. 233–244 (1989).
23. Hargis, W. J., Jr., D. E. Zwerner, D. A. Thoney, K. L. Kelly, and J. E. Warinner, III. "Neoplasms in Mummichogs from the Elizabeth River, Virginia," *J. Aquat. Anim. Health* 1:165–172 (1989).
24. Hendricks, J. D., T. R. Meyers, D. W. Shelton, J. L. Casteel, and G. S. Bailey. "Hepatocarcinogenicity of Benzo[a]pyrene to Rainbow Trout by Dietary Exposure and Intraperitoneal Injection," *J. Natl. Cancer Inst.* 74:839–851 (1985).
25. Couch, J. A., and L. A. Courtney. "N-nitrosodiethylamine-Induced Hepatocarcinogenesis in Estuarine Sheepshead Minnow *Cyprinodon variegatus*: Neoplasms and Related Lesions Compared with Mammalian Lesions," *J. Natl. Cancer Inst.* 79:299–321 (1987).
26. Hawkins, W. E., W. W. Walker, R. M. Overstreet, T. F. Lytle, and J. S. Lytle. "Dose-Related Carcinogenic Effects of Water-Borne Benzo[a]pyrene on the Livers of Two Small Fish Species," *Ecotox. Environ. Saf.* 16:219–231 (1989).
27. Hawkins, W. E., W. W. Walker, J. S. Lytle, T. F. Lytle, and R. M. Overstreet. "Carcinogenic Effects of 7, 12 Dimethylbenz[a]anthracene in the Guppy *Poecilia reticulata*," *Aquat. Toxicol.* 15:63–82 (1989).
28. Lauren, D. J., S. J. Teh, and D. E. Hinton. "Cytotoxicity Phase of Diethylnitrosamine-Induced Hepatic Neoplasia in Medaka," *Cancer Res.* 50:5504–5514 (1990).
29. Smith, C. E., T. H. Peck, R. J. Klauda, and J. B. McLaren. "Hepatomas in Atlantic Tomcod *Microgadus tomcod* (Walbaum) in the Hudson River Estuary in New York," *J. Fish Dis.* 2:313–319 (1979).
30. Murchelano, R. A., and R. E. Wolke. "Epizootic Carcinoma in the Winter Flounder *Pseudopleuronectes americanus*," *Science* 228:587–589 (1985).
31. Baumann, P. C., W. D. Smith, and W. K. Parland. "Tumor Frequencies and Contaminant Concentrations in Brown Bullheads from an Industrialized River and a Recreational Lake," *Am. Fish. Soc.* 116:79–86 (1987).
32. Myers, M. S., L. D. Rhodes, and B. B. McCain. "Pathologic Anatomy and Patterns of Occurrence of Hepatic Neoplasms, Putative Preneoplastic Lesions, and Other Idiopathic Hepatic Conditions in English Sole *Parophrys vetulus* from Puget Sound, Washington," *J. Natl. Cancer Inst.* 78:333–363.
33. Malins, D. C., B. B. McCain, D. W. Brown, S. Chan, M. S. Myers, et al. "Chemical Pollutants in Sediments and Diseases of Bottom-Dwelling fish in Puget Sound, Washington," *Environ. Sci. Technol.* 18(9):705–713 (1984).
34. Luna, L. G. *Manual of Histologic Staining Methods of the Armed Forces Institute of Pathology*, 3rd ed. (New York: McGraw-Hill, 1968).

35. Hargis, W. J., Jr., and D. E. Zwerner. "Effects of Certain Contaminants on the Eyes of Several Estuarine Fishes," *Mar. Environ. Res.* 24:265–270 (1988b).
36. Thiyagarajah, A., D. E. Zwerner, W. J. Hargis, Jr. "Renal Lesions in Estuarine Fishes Collected from the Elizabeth River, Virginia," *J. Environ. Pathol. Toxicol. Oncol.* 9(3):261–268 (1989).
37. Vogelbein, W. K., J. W. Fournie, P. A. Van Veld, and R. J. Huggett. "Hepatic Neoplasms in the Mummichog *Fundulus heteroclitus* from a Creosote-Contaminated site," *Cancer Res.* 50:5978–5986 (1990).
38. Mueller, J. G., P. J. Chapman, and P. H. Pritchard. "Creosote-Contaminated Sites, Their Potential for Bioremediation," *Environ. Sci. Technol.* 23:1197–1201 (1989).
39. Bos, R. P., C. T. J. Hulshof, J. L. G. Theuws, and P. T. Henerson. "Mutagenicity of Creosote in the *Salmonella/microsome* Assay," *Mutat. Res.* 119:21–25 (1983).
40. Boutwell, R. K., and D. K. Bosch. "The Carcinogenicity of Creosote Oil: Its Role in the Induction of Skin Tumors in Mice," *Cancer Res.* 18:1171–1175 (1958).
41. Malins, D. C., M. M. Krahn, M. S. Myers, L. D. Rhodes, D. W. Brown, C. A. Krone, B. B. McCain, and S. Chan. "Toxic Chemicals in Sediments and Biota from a Creosote-Polluted Harbor: Relationships with Hepatic Neoplasms and Other Hepatic Lesions in English Sole *Parophrys vetulus*," *Carcinogenesis (Lond.)* 6:1463–1469 (1985).
42. Buhler, D. R., and D. E. Williams. "The Role of Biotransformation in the Toxicity of Chemicals," *Aquat. Toxicol.* 11:19–28 (1988).
43. Varanasi, U., J. E. Stein, and M. Nishimoto. "Biotransformation and Disposition of Polycyclic Aromatic Hydrocarbons (PAH) by Fish," in *Metabolism of Polycyclic Aromatic Hydrocarbons in the Aquatic Environment*, U. Varanasi, Ed. (Boca Raton, FL: CRC Press, 1989), pp. 93–149.
44. Black, S. D., and M. J. Coon. "P-450 Cytochromes: Structure and Function," *Adv. Enzymol. Relat. Areas Mol. Biol.* 60:35–87 (1987).
45. Buhler, D. R., and D. E. Williams. "Enzymes Involved in Metabolism of PAH by Fishes and Other Aquatic Animals: Oxidative Enzymes (or Phase I Enzymes)," in *Metabolism of Polycyclic Aromatic Hydrocarbons in the Aquatic Environment*, U. Varanasi, Ed. (Boca Raton, FL: CRC Press, 1989), pp. 151–184.
46. Klotz, A. V., J. J. Stegeman, and C. Walsh. "An Aryl Hydrocarbon Hydroxylating Hepatic Cytochrome P-450 from the Marine Fish (*Stenotomus chrysops*)," *Arch. Biochem. Biophys.* 226:578–592 (1983).
47. Goksoyr, A. "Purification of Hepatic Microsomal Cytochrome P-450 from B-Naphthoflavone-Treated Atlantic Cod (*Gadus morhua*), a Marine Teleost Fish," *Biochem. Biophys. Acta* 840:409–417 (1985).
48. Williams, D. E., and D. R. Buhler. "Benzo[a]pyrene-Hydroxylase Catalyzed by Purified Isozymes of Cytochrome P-450 from B-Naphthoflavone-Fed Rainbow Trout," *Biochem. Pharmacol.* 33:3743–3753 (1984).
49. Park, S. S., H. Miller, A. V. Klotz, P. J. Kloepper-Sams, J. J. Stegeman, and H. V. Gelboin. "Monoclonal Antibodies to Liver Microsomal Cytochrome P-450E of the Marine Fish *Stenotomus versicolor* (scup). Cross Reactivity with 3-Methylcholanthrene Induced Rat Cytochrome P-450," *Arch. Biochem. Biophys.* 249:339–349 (1986).
50. Burnette, W. N. "'Western Blotting.' Electrophoretic Transfer of Proteins from Sodium Dodecyl Sulfate-Polyacrylamide Gels to Unmodified Nitrocellulose and Radiometric Detection with Antibody and Radioiodinated Protein A," *Anal. Biochem.* 112:195–203 (1981).

51. Stegeman, J. J., M. Brouwer, R. T. DiGiulio, L. Forlin, B. Fowler, B. M. Sander, and P. A. Van Veld. "Molecular Responses to Environmental Contamination: Proteins and Enzymes as Indicators of Contaminant Exposure and Effects," in *The Existing and Potential Value of Biomarkers in Evaluating Exposure and Environmental Effect of Toxic Chemicals*, R. J. Huggett, Ed. (Boca Raton, FL: Lewis Publishers, 1991), in press.
52. Van Veld, P. A., J. J. Stegeman, B. R. Woodin, J. S. Patton, and R. F. Lee. "Induction of Monooxygenase Activity in the Intestine of Spot (*Leiostomus xanthurus*), a marine teleost, by dietary PAH," *Drug Metab. Dispos.* 16:659–665 (1988).
53. Van Veld, P. A., D. J. Westbrook, B. R. Woodin, R. C. Hale, R. J. Huggett, C. L. Smith, and J. J. Stegeman. "Induced Cytochrome P-450 in Intestine and Liver of Spot (*Leiostomus xanthurus*) from a Polycyclic Aromatic Hydrocarbon Contaminated Environment," *Aquat. Toxicol.* (1990).
54. Klotz, A. V., J. J. Stegeman, and C. Walsh. "An Alternative 7-Ethoxyresorufin O-Deethylase Assay. A Continuous Spectrophotometric Measurement of Cytochrome P-450 Monooxygenase Activity," *Anal. Biochem.* 140:138–145 (1984).
55. Bradford, M. M. "A Rapid and Sensitive Method for Quantitation of Microgram Quantities of Protein Using the Principle of Protein-Dye Binding," *Anal. Biochem.* 72:248–254 (1976).
56. Payne, J. F., L. L. Fancey, A. D. Rahimtula, and E. L. Porter. "Review and Perspective on the Use of Mixed-Function Oxygenase Enzymes in Biological Monitoring," *Comp. Biochem. Physiol.* 86C:233–245 (1987).
57. Miller, M. R., D. E. Hinton, and J. J. Stegeman. "Cytochrome P-450 Induction and Localization in Gill Pillar (*endothelia*) Cells of Scup and Rainbow Trout," *Aquat. Toxicol.* 14:307–322 (1989).
58. Van Veld, P. A. "Absorption and Metabolism of Dietary Xenobiotics by the Intestine of Fish," *Rev. Aquat. Sci.* 2:185–203 (1990).
59. Ziskowski, J., and R. Murchelano. "Fin Erosion in Winter Flounder," *Mar. Pollut. Bull.* 6:26–29 (1975).
60. Mearns, A. J., and M. J. Sherwood. "Distribution of Neoplasms and Other Diseases in Marine Fishes Relative to the Discharge of Waste Water," in *Aquatic Pollutants and Biological Effects with Emphasis on Neoplasia*, H. F. Kraybill, C. J. Dawe, J. C. Harshbarger, and R. G. Tardiff, Eds., *Ann. N.Y. Acad. Sci.* 298:210–224.
61. Weeks, B. A., J. E. Warinner, P. L. Mason, and D. S. McGinnis. "Influence of Toxic Chemicals on the Chemotactic Response of Fish Macrophages," *J. Fish Biol.* 28:653–658 (1986).
62. Weeks, B. A., and J. Warinner. "Effects of Toxic Chemicals on Macrophage Phagocytosis in two Estuarine Fishes," *Mar. Environ. Res.* 14:327–335 (1984).
63. Seeley, K. R., and B. A. Weeks. "Reduced Phagocytic Capacity in Fish Taken from a Highly Polluted River," *J. Aquat. Anim. Health* (in press).
64. Warinner, J. E., E. S. Mathews, and B. A. Weeks. "Preliminary Investigations of the Chemiluminescent Response in Normal and Pollutant-Exposed Fish," *Mar. Environ. Res.* 24:281–284 (1988).
65. Weeks, B. A., and J. E. Warinner. "Functional Evaluation of Macrophages in Fish from a Polluted Estuary," *Vet. Immunol. Immunopathol.* 12:313–320 (1986).

VIMS Contribution No. 1731.

List of Authors

William H. Benson, Department of Pharmacology and Research Institute of Pharmaceutical Science, The University of Mississippi, University, Mississippi 38677

Robert M. Burgess, Science Applications International Corporation, c/o U. S. Environmental Protection Agency, 27 Tarzwell Drive, Narragansett, Rhode Island 02882

G. Allen Burton, Jr., Department of Biological Sciences, Wright State University, Dayton, Ohio 45435

John Cairns, Jr., Department of Biology and University Center for Environmental and Hazardous Materials Studies, Virginia Polytechnic Institute and State University, Blacksburg, Virginia 24061-0415

Peter M. Chapman, EVS Consultants, 195 Pemberton Avenue, North Vancouver, British Columbia, Canada V7P 2R4

Theodore H. DeWitt, Oregon State University, 2111 S. E. Marine Science Drive, Hatfield Marine Science Center, Newport, Oregon 97365

Robert J. Diaz, Virginia Institute of Marine Science, School of Marine Science, The College of William and Mary, Gloucester Point, Virginia 23062

Richard T. Di Giulio, School of the Environment, Duke University, Durham, North Carolina 27706

James F. Fairchild, U. S. Fish and Wildlife Service, National Fisheries Contaminant Research Center, Columbia, Missouri 65201

Thomas C. Ginn, PTI Environmental Services, Bellevue, Washington 98007

Lars Håkanson, Department of Hydrology, University of Uppsala, V. Ågatan 24, 752 20 Uppsala, Sweden

William J. Hargis, Jr., Virginia Institute of Marine Science, School of Marine Science, The College of William and Mary, Gloucester Point, Virginia, 23062

Robert J. Huggett, Virginia Institute of Marine Science, School of Marine Science, The College of William and Mary, Gloucester Point, Virginia 23062

C. G. Ingersoll, National Fisheries Contaminant Research Center, U. S. Fish and Wildlife Service, Columbia, Missouri 65201

Michael Kravitz, Water Quality Branch, Office of Science and Technology, U. S. Environmental Protection Agency, 401 M Street, S.W., Washington, DC 20460

Janet O. Lamberson, U. S. Environmental Protection Agency, 2111 S. E. Marine Science Drive, Hatfield Marine Science Center, Newport, Oregon 97365-5260

Thomas W. La Point, Department of Environmental Toxicology, Clemson University, P. O. Box 709, Pendleton, South Carolina 29670

Henry Lee II, Pacific Ecosystems Branch, Environmental Research Laboratory – Narragansett, U. S. Environmental Protection Agency, 2111 S. E. Marine Science Drive, Newport, Oregon 97365

Michael J. Mac, U. S. Fish and Wildlife Service, National Fisheries Research Center – Great Lakes, Ann Arbor, Michigan 48105

M. K. Nelson, National Fisheries Contaminant Research Center, U. S. Fish and Wildlife Service, Columbia, Missouri 65201

B. R. Niederlehner, University Center for Environmental and Hazardous Materials Studies, Virginia Polytechnic Institute and State University, Blacksburg, Virginia 24061-0415

Robert A. Pastorok, PTI Environmental Services, Bellevue, Washington 98007

Elizabeth A. Power, EVS Consultants, 195 Pemberton Avenue, North Vancouver, British Columbia, Canada V7P 2R4

Christopher J. Schmitt, National Fisheries Contaminant Research Center, U. S. Fish and Wildlife Service, Columbia, Missouri 65201

K. John Scott, Science Applications International Corporation, 165 Dean Knauss Drive, Narragansett, Rhode Island 02882

C. L. Smith, Virginia Institute of Marine Science, School of Marine Science, The College of William and Mary, Gloucester Point, Virginia 23062

E. P. Smith, University Center for Environmental and Hazardous Materials Studies, and Department of Statistics, Virginia Polytechnic Institute and State University, Blacksburg, Virginia 24061-0415

Elizabeth Southerland, Water Quality Branch, Office of Science and Technology, U. S. Environmental Protection Agency, 401 M Street, S.W., Washington, DC 20460

Richard C. Swartz, U.S. Environmental Protection Agency, 2111 S. E. Marine Science Drive, Hatfield Marine Science Center, Newport, Oregon 97365-5260

P. A. Van Veld, Virginia Institute of Marine Science, School of Marine Science, The College of William and Mary, Gloucester Point, Virginia 23062

W. K. Vogelbein, Virginia Institute of Marine Science, School of Marine Science, The College of William and Mary, Gloucester Point, Virginia 23062

Thomas Wall, Office of Science and Technology, U.S. Environmental Protection Agency, 401 M Street, S.W., Washington, DC 20460

B. A. Weeks, Virginia Institute of Marine Science, School of Marine Science, The College of William and Mary, Gloucester Point, Virginia 23062

Index